Student's Solutions Manual to Accompany

General Chemistry

Second Edition

Jean B. Umland
University of Houston—Downtown

Jon M. Bellama
University of Maryland

Prepared by
Jean B. Umland
University of Houston—Downtown

Juliette A. Bryson
Las Positas College

Byron Christmas
University of Houston—Downtown

West Publishing Company
Minneapolis/St. Paul New York Los Angeles San Francisco

WEST'S COMMITMENT TO THE ENVIRONMENT

In 1906, West Publishing Company began recycling materials left over from the production of books. This began a tradition of efficient and responsible use of resources. Today, 100% of our legal bound volumes are printed on acid-free, recycled paper consisting of 50% new paper pulp and 50% paper that has undergone a de-inking process. We also use vegetable-based inks to print all of our books. West recycles nearly 27,700,000 pounds of scrap paper annually—the equivalent of 229,300 trees. Since the 1960s, West has devised ways to capture and recycle waste inks, solvents, oils, and vapors created in the printing process. We also recycle plastics of all kinds, wood, glass, corrugated cardboard, and batteries, and have eliminated the use of polystyrene book packaging. We at West are proud of the longevity and the scope of our commitment to the environment.

West pocket parts and advance sheets are printed on recyclable paper and can be collected and recycled with newspapers. Staples do not have to be removed. Bound volumes can be recycled after removing the cover.

Production, Prepress, Printing and Binding by West Publishing Company.

TEXT IS PRINTED ON 10% POST CONSUMER RECYCLED PAPER

Printed with Printwise
Environmentally Advanced Water Washable Ink

Printed in the United States of America
03 02 01 00 99 98 97 96 8 7 6 5 4 3 2 1 0

ISBN 0–314–09677–9

Table of Contents

Student's Solutions Manual

CHAPTER 1:
Introduction

All answers in this solutions manual are "best" answers calculated by combining steps to avoid rounding errors. Rounding in intermediate steps may result in your answers differing from these answers in the last digit.

There are different styles for calculating numerical answers. In this Solutions Manual, one more digit than there is in the number that limits the number of significant figures in the answer is generally used for formula masses and constants, to obtain the best possible answers. For example, suppose the problem is to convert 275 g to pounds. According to the table inside the back cover, 1 lb = 453.6 g. In this Solutions Manual, the solution given would be

3 digits in data ↘

$$(275\ \text{g})\left(\frac{1\ \text{lb}}{453.6\ \text{g}}\right) = 0.606\ \text{lb}$$

↖ 4 digits used in solution manual

In the Sample Problems in the text, the same number of digits as there is in the number that limits the number of significant figures in the answer is commonly used for formula masses and constants. This is done to make the calculations in the Sample Problems as simple and easy to understand as possible. In the text, the solution given would be

3 digits in data ↘

$$(275\ \text{g})\left(\frac{1\ \text{lb}}{454\ \text{g}}\right) = 0.606\ \text{lb}$$

↖ 3 digits used in text.

As this very simple example suggests, in most cases the answers obtained using either style will be the same; at worst the answers will differ in the last digit.

Practice Problems

1.1 Because a puddle of liquid does not gather itself into a cube and then freeze, the pictures in Figures 1.1(b) and (c) must have been taken after the picture in Figure 1.1(a).

1.2 In a glass of cola with ice cubes, the ice cubes and the glass are solids. The cola is a liquid. If water vapor has condensed on the outside of the glass, the condensed water is also a liquid. The bubbles rising to the top of the cola and the air above the cola are gases.

1.3 **(a)** Evaporation of rubbing alcohol is a physical change. The liquid alcohol changes to gaseous alcohol and disappears into the air.

(b) When a seed grows into a flower, many complicated chemical reactions take place. Carbon dioxide from the air and water from the soil combine to form carbohydrates using sunlight as a source of energy. Oxygen gas is formed.

(c) Rolling aluminum metal into foil is a physical change; the aluminum is still aluminum, a shiny, silvery–colored metal.

(d) Similarly, when copper is pulled into a wire, the copper is still copper, a shiny, reddish–colored metal.

(e) The burning of a candle is a chemical change. The wax reacts with oxygen from air to form water vapor, carbon (soot), and carbon dioxide gas, all of which disappear into the air. The candle becomes shorter as wax reacts. (The wax around the wick melts and may run down the sides of the candle. Melting of wax is a physical change.)

1.4 Aspirin is a substance. All samples of a substance have the same properties. Therfore, expensive brand–name aspirin is no more effective than cheaper aspirin, no matter what the advertisements say.

1.5 **(a)** Chlorine is an element. It is listed in the table of elements inside the front cover.

(b) Gelatin with fruit in it is a heterogeneous mixture. You can see different phases (fruit and gelatin) and you can separate the components physically, that is, you can pick the fruit out with your fingers.

(c) Sodium is an element.

(d) You can figure out that sugar is a compound. All samples of pure sugar that you've ever met are white solids, taste sweet, dissolve in water, and melt and turn brown when heated, giving off a burnt–sugar odor.

(e) White wine is clear. Therefore, if it is a mixture, it is a solution. It must be a mixture because different samples differ in color, odor, and taste.

1.6 In the table of elements inside the front cover, the elements are listed in alphabetical order by name. The atomic numbers are given in the columns labeled "Atomic Numbers."

(a) The atomic number of aluminum is 13.

(b) The atomic number of carbon is 6.

(c) The atomic number of neon is 10.

(d) The atomic number of silver is 47.

(e) The atomic number of uranium is 92.

1.7 The number of protons in the nucleus is the atomic number of the element. The number of electrons outside the nucleus is equal to the number of protons in the nucleus.

(a) From Practice Problem 1.6, the atomic number of aluminum is 13. Therefore, in an atom of aluminum, there are 13 protons in the nucleus. Because each proton has a unit positive charge, the positive charge on the nucleus is 13+ and there are 13 electrons outside the nucleus.

(b) In an atom of carbon there are 6 protons in the nucleus, the positive charge on the nucleus is 6+, and there are 6 electrons outside the nucleus.

(c) In an atom of neon there are 10 protons in the nucleus, the positive charge on the nucleus is 10+, and there are 10 electrons outside the nucleus.

(d) In an atom of silver there are 47 protons in the nucleus, the positive charge on the nucleus is 47+, and there are 47 electrons outside the nucleus.

(e) In an atom of uranium there are 92 protons in the nucleus, the positive charge on the nucleus is 92+, and there are 92 electrons outside the nucleus.

1.8 In the table of elements inside the front cover, the symbols for the elements are listed in the third column.

(a) The symbol for iron is Fe.

(b) The symbol for argon is Ar.

(c) The symbol for phosphorus is P.

(d) The symbol for lead is Pb.

(e) The symbol for uranium is U.

1.9 **(a)** The symbol Zn stands for zinc.

(b) The symbol Se stands for selenium.

(c) The symbol Pu stands for plutonium.

(d) The symbol Xe stands for xenon.

(e) The symbol Ba stands for barium.

1.10 According to Table 1.1, the reactivity of the Group IA elements toward water increases down the group in the periodic table. Because cesium (Cs) reacts explosively with water and francium (Fr) is the next lower element, francium will probably react very explosively with water. (This prediction is correct.)

1.11 **(a)** The noble gases are the elements in the column of the periodic table labeled 0; therefore, the noble gas is Ne (neon).

(b) The alkali metals are the elements in the column of the periodic table labeled IA; the alkali metal is K (potassium).

(c) The alkaline earth metals are the elements in the column of the periodic table labeled IIA; the alkaline earth metal is Ba (barium).

(d) The halogens are the elements in the column of the periodic table labeled VIIA; the halogen is Cl (chlorine).

(e) Nitrogen (N) is in Group VA.

(f) In the periodic table, elements in vertical columns (groups) have similar properties. Like Br, Cl is in Group VIIA. Therefore, the properties of Cl should be similar to those of Br.

1.12 Referring to the periodic table in Figure 1.10 or to the periodic table inside the front cover, the elements in the third period in order of increasing atomic number are: Na (sodium), Mg (magnesium), Al(aluminum), Si (silicon), P (phosphorus), S (sulfur), Cl (chlorine), and Ar (argon).

1.13 The elements in the actinide series at the bottom of the periodic table belong between elements 88 and 103 in the seventh period. Therefore, the element with atomic number 94 is in the seventh period.

1.14 **(a)** In Figure 1.10 and in the periodic table inside the front cover, metals are in gray spaces. Calcium, molybdenum, and uranium are metals. Metals are toward the left side and toward the bottom of the periodic table.

(b) Semimetals are in blue spaces; antimony is a semimetal. Semimetals are between metals and nonmetals in the periodic table.

(c) Nonmetals are in purple spaces; argon and bromine are nonmetals. Nonmetals are toward the right side and toward the top of the periodic table.

1.15 **(a)** Main group elements are elements in A groups and in Group 0. Of the elements in Practice Problem 1.14, antimony, which is in Group VA, argon, which is in Group 0, bromine, which is in Group VIIA, and calcium, which is in Group IIA, are main group elements.

(b) Transition elements are elements in Groups IIIB through IIB. Elements in the lanthanide and actinide series are inner transition elements. Of the elements in Practice Problem 1.14, molybdenum, which is in Group VIB, is a transition element and uranium, which is in the actinide series, is an inner transition element.

1.16 **(a)** In one molecule of CO_2 there are one carbon atom and two oxygen atoms.

(b) In one molecule of NH_3, there are one nitrogen atom and three hydrogen atoms.

(c) In one molecule of H_2SO_4, there are two hydrogen atoms, one sulfur atom, and four oxygen atoms.

(d) In one molecule of HCl, there are one hydrogen atom and one chlorine atom. In three molecules of HCl there will be three times as many of each kind of atom, or three hydrogen atoms and three chlorine atoms.

1.17 **(a)** In a molecule of nitrous oxide, there are two nitrogen atoms and one oxygen atom. The formula for nitrous oxide is N_2O.

(b) To make two molcules of nitrous oxide, two times two nitrogen atoms, or four nitrogen atoms, are needed.

(c) In a molecule of sulfur dioxide, there are one sulfur atom and two oxygen atoms. The formula for sulfur dioxide is SO_2.

1.18 The elements nitrogen and chlorine have diatomic molecules. Their formulas are N_2 and Cl_2, respectively.

1.19 In molecules of elements, all atoms are the same kind. Molecules of compounds have more than one kind of element in them. Therefore, the formulas Br_2 and S_8 represent elements and the formulas HF, CS_2, and KI represent compounds.

1.20 Ions are charged particles; molecules have no charge. Therefore, NO_2 and NH_3 are molecules and NO_2^- and NH_4^+ are ions.

1.21 Anions are negatively charged ions and cations are positively charged ions. Therefore, F^- is an anion and K^+ and H^+ are cations.

1.22 **(a)** The atomic number of sulfur is 16 and a sulfur atom has 16 electrons. From the periodic table, the noble gas with the closest atomic number is argon, atomic number 18. A sulfur atom with 16 electrons must gain two electrons to have 18 electrons. Therefore, sulfide ion has a 2– charge and the formula for a sulfide ion is S^{2-}.

(b) The atomic number of aluminum is 13 and an aluminum atom has 13 electrons. From the periodic table, the noble gas with the closest atomic number is neon, atomic number 10. An aluminum atom with 13 electrons must lose three electrons to have only 10 electrons. Therefore, an aluminum ion has a 3+ charge and the formula for an aluminum ion is Al^{3+}.

1.23 Compounds must have zero net charge. Therefore, in the compound lithium oxide, for each oxide ion with a 2– charge, there must be two lithium ions with 1+ charges.

$$2(1+) + 1(2-) = 0$$

The formula for lithium oxide is Li_2O.

1.24 Compounds must have zero net charge. Therefore, in the compound aluminum fluoride, for each aluminum ion with a 3+ charge, there must be three fluoride ions with 1– charges.

$$1(3+) + 3(1-) = 0$$

The formula for aluminum fluoride is AlF_3.

1.25 **(a)** The units of calcium oxide are calcium ions and oxide ions. First we must figure out the formulas of these ions. Then we must combine the ions so that the compound has no net charge. The formula for calcium ion is Ca^{2+} because a calcium atom, atomic number 20, must lose two electrons to have 18 electrons like argon. The formula for oxide ion is O^{2-} because an oxygen atom, atomic number 8, must gain two electrons to have ten electrons like neon. For each calcium ion with a 2+ charge, there must be one oxide ion with a 2– charge, and the formula for calcium oxide is CaO.

(b) The formula for aluminum ion is Al^{3+} because an aluminum atom, atomic number 13, must lose three electrons to have ten electrons like neon. The formula for sulfide ion is S^{2-} because a sulfur atom, atomic number 16, must gain two electrons to have 18 electrons like argon. For two aluminum ions with a 3+ charge, there must be three sulfide ions with a 2– charge,

$$2(3+) + 3(2-) = 0,$$

and the formula for aluminum sulfide is Al_2S_3.

1.26 When atoms of nonmetals combine to form compounds, molecules result. Compounds of reactive metals with nonmetals are ionic. Therefore, the units of CS_2 and HF are molecules. The units of MgO, KI, and Mg_3N_2 are ions.

1.27 By analogy to the carbonate and hydrogen carbonate ions (Table 1.2), the hydrogen oxalate ion must have a hydrogen and one less unit of negative charge than the oxalate ion, $C_2O_4^{2-}$. Therefore, the formula for the hydrogen oxalate ion is $HC_2O_4^-$.

1.28 By analogy to the sulfate and sulfite ions (Table 1.2), the chlorite ion has one less oxygen and the same charge as the chlorate ion. ClO_3^-. Therefore, the formula for the chlorite ion is ClO_2^-.

1.29 **(a)** From Table 1.2, the formula for the ammonium ion is NH_4^+. The formula for sulfide ion is S^{2-} because a sulfur atom, atomic number 16, must gain two electrons to have 18 electrons like argon. The compound ammonium sulfide must have zero charge. Therefore, for each sulfide ion, there must be two ammonium ions, and the formula for ammonium sulfide is $(NH_4)_2S$.

(b) The parentheses around the ammonium ion in the formula for ammonium sulfide mean that there are two ammonium ions in each unit of ammonium sulfide. In each ammonium ion there are one N atom and four H atoms. Therefore, in one unit of $(NH_4)_2S$, threre are 2(1 N) = 2 N, 2(4 H) = 8 H, and 1 S.

1.30 Inorganic compounds are compounds of all elements except carbon [and a few simple carbon compounds such as $CaCO_3$, $Ca(HCO_3)_2$, CO, and CO_2]. Organic compounds contain carbon. Thus HCl, $CaCl_2$ and H_2O are inorganic, and C_2H_2 and C_3H_6O are organic.

1.31 If one of two elements in a compound is a metal and the other a nonmetal, the name of the metal is given first. The name of the nonmetal is given second and the ending of its name is changed to –ide (see Table 1.3).

(a) The compound of potassium and iodine is called potassium iodide.

(b) The compound of bromine and magnesium is called magnesium bromide.

1.32 **(a)** The Roman numeral II in the name vanadium(II) chloride tells us that in this compound the formula for the vanadium ion is V^{2+}. From Table 1.3, the formula for chloride ion is Cl^-. The formula for vanadium(II) chloride is VCl_2.

(b) From Table 1.3, the formula for oxide ion is O^{2-}. The formula for vanadium(II) oxide is VO.

(c) The name vanadium(V) oxide tells us that in this compound the formula for the vanadium ion is V^{5+}. The formula for vanadium(V) oxide is V_2O_5.

1.33 **(a)** From Table 1.3, the charge on a sulfide ion is 2–. The charge on the chromium ion in CrS must be 2+ for the net charge on a unit of CrS to be zero. The name of the compound is chromium(II) sulfide.

(b) The charge on the chormium ion in Cr_2S_3 must be 3+ for the net charge on a unit of Cr_2S_3 to be zero. The name of the compound is chromium(III) sulfide.

(c) From Table 1.3, the charge on an oxide ion is 2–. The charge on the titanium ion in TiO must be 2+ for the net charge on a unit of TiO to be zero. The name of the compound is titanium(II) oxide.

(d) The charge on the titanium ion in TiO_2 must be 4+ for the net charge on a unit of TiO_2 to be zero. The name of the compound is titanium(IV) oxide.

1.34 **(a)** From Table 1.4, the prefix di– means two. The formula for disulfur dichloride is S_2Cl_2.

(b) From Table 1.4, the prefix tri– means three. A prefix of mono– (one) is understood for phosphorus in phosphorus tribromide, and thus the formula is PBr_3.

1.35 **(a)** The name of N_2O is dinitrogen monoxide.

(b) The name of P_2S_5 is diphosphorus pentasulfide.

1.36 **(a)** In a chemical equation, the reactants are the substance or substances on the left side of the arrow. In the equation

$$2AgNO_3(aq) + H_2S(g) \rightarrow Ag_2S(s) + 2HNO_3(aq)$$

the formulas of the reactants are $AgNO_3$ and H_2S.

(b) The products are the substance or substances on the right side of the arrow. The formulas of the products are Ag_2S and HNO_3.

(c) Two atoms of Ag, two atoms of N, six atoms of O, two atoms of H, and 1 atom of S appear on each side of the equation.

(d) The symbol (aq) following the formula shows that a substance is in aqueous solution. The substances $AgNO_3$ and HNO_3 are in aqueous solution.

(e) The symbol (s) following the formula shows that a substance is a solid. The substance Ag_2S is a solid.

1.37 A chemical equation must show the same number of each kind of atom on both sides. The only one of the three expressions that does so is (a), which shows one C, two O, and two H on each side. The expression (b) shows one O on the left and two O on the right. The expression (c) shows one H and one Cl on the left and two H and two Cl on the right.

1.38 **(a)** The expression $Mg(s) + O_2(g) \rightarrow MgO(s)$ shows two O on the left and only one O on the right. The first step in balancing it is to place a coefficient of two in front of the formula MgO:

$$Mg(s) + O_2(g) \rightarrow 2MgO(s)$$

Now the expression shows the same number of O on both sides but one Mg on the left and two Mg on the right. Placing a coefficient of two in front of the symbol Mg balances the equation, which is

$$2Mg(s) + O_2(g) \rightarrow 2MgO(s)$$

with two Mg and two O on each side.

(b) To balance the expression $N_2(g) + H_2(g) \rightarrow NH_3(g)$, place a coefficient of two in front of NH_3:

$$N_2(g) + H_2(g) \rightarrow 2NH_3(g)$$

Now there are two N on each side but the numbers of H are different. Place a coefficient of 3 in front of H_2 to get six H on each side:

$$N_2(g) + 3H_2(g) \rightarrow 2NH_3(g)$$

Check your work by counting the number of atoms of each kind to the left and right of the arrow. There are two N and six H on each side. The expression is an equation.

(c) Recognizing that SO_4 (sulfate – Table 1.2) appears in both H_2SO_4 and Na_2SO_4 saves some counting of atoms. There are one Na on the left and two Na on the right. Place a coefficient of two in front of NaOH to balance Na:

$$H_2SO_4(aq) + 2NaOH(aq) \rightarrow Na_2SO_4(aq) + H_2O(l)$$

Now there are four H on the left and only two H on the right. Place a coefficient of 2 in front of H_2O:

$$H_2SO_4(aq) + 2NaOH(aq) \rightarrow Na_2SO_4(aq) + 2H_2O(l)$$

The expression is an equation. There are four H, one S, six O, and two Na on each side. Note that there may be more than one series of steps that leads to the simplest equation. For each reaction, only one simplest equation can be written. Any series of steps that leads to this equation is satisfactory.

1.39 The equation says that on electrolysis an aqueous solution of sodium chloride reacts to yield hydrogen gas, chlorine gas, and an aqueous solution of sodium hydroxide.

1.40 **Step 1.** The reactants are copper metal and oxygen gas. Copper(II) oxide is the product. Heat is required for the reaction to take place. The word equation is

$$\text{copper metal} + \text{oxygen gas} \xrightarrow{\text{heat}} \text{solid copper(II) oxide}$$

Step 2. The formulas are

$$Cu + O_2 \xrightarrow{\text{heat}} CuO$$

Step 3. Balanced, the expression is

$$2Cu + O_2 \xrightarrow{heat} 2CuO$$

Step 4. Check. There are two Cu and two O on each side.

Step 5. Add symbols. The equation for the reaction is

$$2Cu(s) + O_2(g) \xrightarrow{heat} 2CuO(s)$$

1.41 The elements Li and Na are in the same group in the periodic table (Group IA) as K. Therefore, the reactions that take place when $LiNO_3$ and $NaNO_3$ are heated are probably similar to the reactions that take place when KNO_3 is heated. The equations for the reactions are:

$$2LiNO_3(s) \xrightarrow{heat} 2LiNO_2(s) + O_2(g)$$

$$2NaNO_3(s) \xrightarrow{heat} 2NaNO_2(s) + O_2(g)$$

Additional Practice Problems

1.83 (a) The formula for the manganese(III) ion is Mn^{3+} and the formula for the oxide ion is O^{2-}. Combining these ions so that charge cancels gives Mn_2O_3 for the formula of manganese(III) oxide.

(b) The prefix di– means two and the prefix mono– means one. The formula for dibromine monoxide is Br_2O.

(c) The formula for strontium ion is Sr^{2+} because a strontium atom, atomic number 38, must lose two electrons to have 36 electrons like krypton. The formula for the nitrate ion is NO_3^- (Table 1.2). The formula for strontium nitrate is $Sr(NO_3)_2$.

(d) The formula for sodium ion is Na^+ because a sodium atom, atomic number 11, must lose one electron to have 10 electrons like neon. The formula for the perchlorate ion is ClO_4^- (Table 1.2). The formula for sodium perchlorate is $NaClO_4$.

(e) The formula for the ammonium ion is NH_4^+ and the formula for the carbonate ion is CO_3^{2-} (Table 1.2). The formula for ammonium carbonate is $(NH_4)_2CO_3$.

1.85 (a) In the left part of the photograph, iron filings are clinging to the magnet. The original mixture of iron filings and powdered sulfur can be separated with a magnet.

(b) In the right part of the photograph, which was taken after the mixture had been heated, no iron filings are clinging to the magnet. Iron is no longer present. A chemical reaction took place when the mixture was heated.

1.87 Magnesium, calcium, and strontium are in Group IIA in the periodic table. The reactivity toward water increases down the group, as magnesium does not react with water but calcium reacts slowly. The next element, strontium, would be expected to react more rapidly than calcium.

1.89 Compounds must have a net charge of zero. A sodium ion, Na^+, has a 1+ charge and a sulfide ion, S^{2-}, has a 2– charge. Therefore, two sodium ions must combine with each sulfide ion.

Stop and Test Yourself

1. **(b)** iii and v

Processes iii and v involve changes of one or more substances into one or more other substances. Such changes are chemical changes. Processes i, ii, and iv do not involve the changing of substances to other substances. They are physical changes.

2. **(a)** i, iii, and v

A solution is a homogeneous mixture, that is, it contains only one phase. A given sample of a homogeneous mixture has the same properties throughout, but different samples may have different properties. Homogeneous mixtures can be separated into their components by physical means.

3. **(d)** ii, iii, and iv

The formulas CO and CaC_2 each show two different elements (C and O and Ca and C, respectively). Therefore, they are the formulas of compounds. The formula Cl_2 shows only one element (Cl). Therefore, it is the formula of an element. The other two choices, Co and C, are symbols for elements.

4. **(e)** i, iv, and v

From Figure 1.10 or the periodic table inside the front cover, we see that Ne and P are both nonmetals (purple spaces), while K is an alkali metal (gray space), Th is an actinide inner transition metal, and Zr is a transition metal (gray spaces).

5. **(a)** As (33)

As (33) has properties most similar to P (15) because they are in Group VA of the periodic table. A group in the periodic table is a vertical listing of elements with similar properties.

6. **(d)** nitrogen dioxide

For compounds of two nonmetals, the symbol for the element that is to the left in the periodic table is written first and this element is named first. Thus, (c) dioxygen nitride is eliminated. Since NO_2 is an electrically neutral compound, (b) nitrite ion is eliminated. Roman numerals are used only in naming compounds containing metals that form more than one type of compound with the same nonmetal. They are not used in naming compounds of two nonmetals. Thus, (a) nitrogen(II) oxide is eliminated. The element that is written first in the formula is called by its name; therefore (e) must be eliminated. If there is no prefix before the name of the first element, mono- is understood. Thus, the correct name for NO_2 is (d) nitrogen dioxide.

7. **(e)** 1 Co, 2 N, 6 O

The subscripts in a chemical formula give the number of atoms of each element in one unit of the compound. A subscript following a parenthesis in a formula means that the number of atoms of each element inside the parentheses must be multiplied by that subscript.

8. (c) iii and iv

The element with atomic number 20 is calcium (Ca). This element is an alkaline earth metal in Group IIA of the periodic table. A calcium atom must lose two electrons to have the same number of electrons as the nearest noble gas, argon (atomic number 18). Therefore, calcium forms a 2+ ion, Ca^{2+}. Calcium is in the fourth period.

9. (e) K_2SO_4

Potassium (K) forms 1+ ions, and sulfate (SO_4^{2-}) ions carry a 2– charge. Only potassium ions and sulfate ions are present in potassium sulfate. Thus, K_2SO_3, K_2S, and $KHSO_4$ are all eliminated. $K(SO_4)_2$ is eliminated because it would have a 3– charge, which is impossible for a compound.

10. (e) $3Fe(s) + 4H_2O(g) \rightarrow Fe_3O_4(s) + 4H_2(g)$

Equation (a) is not corrrect because it does not indicate that hydrogen gas is a diatomic molecule, $H_2(g)$. Equation (b) incorrectly shows the reactant, H_2O, to be a liquid. Steam is gaseous water. Equation (c) has coefficients that are not the simplest whole number ratios of one another. (The coefficients can be divided by 2.) Equation (d) incorrectly shows the formula for steam to be $H_2O_2(g)$.

11. (b) a solid

(s) following a substance in a chemical equation always represents a solid.

12. (b) 3

Step 1:	carbon disulfide	+	chlorine gas	$\rightarrow$	carbon tetrachloride	+	disulfur dichloride
Step 2:	CS_2	+	Cl_2	$\rightarrow$	CCl_4	+	S_2Cl_2
Steps 3 and 4:	CS_2	+	$3Cl_2$	$\rightarrow$	CCl_4	+	S_2Cl_2

13. (c) H_2S

O_2 is an element rather than a compound. KI, $Fe(NO_3)_3$, and Na_2S are all compounds between metals and nonmetals or between metals and polyatomic anions. Thus, they are ionic. H_2S is a compound of two nonmetals. Such compounds are molecular.

14. (d) 35 protons and 36 electrons

The bromide ion (Br^-) has a 1– charge, indicating that it has one more electron than proton. From the table inside the front cover, the atomic number for bromine (Br) is 35. Thus, a bromine atom has 35 protons. Bromide ion, then, has 35 protons and 36 electrons.

15. (b) The gas is chlorine.

(a), (c), (d), and (e) are all observations which would not change if the experiment were repeated. The statement that the gas is chlorine, however, is not an observation and could change if more observations were taken.

Putting Things Together

1.91 (a) A law is a summary of many observations. Some examples from this chapter are the law of constant composition and the law of definite proportions.

(b) A theory is an explanation of a law or of a series of observations. The atomic theory is an example.

1.93 (a) The formula for sodium ion is Na^+ because a sodium atom, atomic number 11, must lose one electron to have 10 electrons like neon. The formula for the carbonate ion is CO_3^{2-} (Table 1.2). For the compound sodium carbonate to have zero net charge, two sodium ions must combine with one carbonate ion, and the formula is Na_2CO_3.

(b) The formula for the hydrogen carbonate ion is HCO_3^- and the formula for sodium hydrogen carbonate is $NaHCO_3$.

(c) The formula for potassium ion is K^+ because a potassium atom, atomic number 19, must lose one electron to have18 electrons like argon. From Table 1.2, the formula for the nitrate ion is NO_3^-. By analogy to sulfate, SO_4^{2-}, and sulfite, SO_3^{2-}, ions, the formula for the nitrite ion is NO_2^- with the same charge as the nitrate ion but one less oxygen. Thus, the formula for potassium nitrite is KNO_2.

(d) The formula for magnesium ion is Mg^{2+} because a magnesium atom, atomic number 12, must lose two electrons to have 10 electrons like neon. The formula for oxide ion is O^{2-} because an oxygen atom, atomic number 8, must gain two electrons to have 10 electrons like neon. The formula for magnesium oxide is MgO.

(e) The formula for calcium ion is Ca^{2+} because a calcium atom, atomic number 20, must lose two electrons to have 18 electrons like argon. The formula for chloride ion is Cl^- because a chlorine atom, atomic number 17, must gain one electron to have 18 electrons like argon. The formula for calcium chloride is $CaCl_2$.

(f) The formula for the vanadium ion in vanadium(II) sulfide is V^{2+}. The formula for sulfide ion is S^{2-} because a sulfur atom, atomic number 16, must gain two electrons to have 18 electrons like argon. The formula for vanadium(II) sulfide is VS.

(g) The formula for the iron ion in iron(III) sulfate is Fe^{3+}. The formula for the sulfate ion is SO_4^{2-} (Table 1.2). The formula for iron(III) sulfate is $Fe_2(SO_4)_3$.

1.95 Combination is a reaction in which substances combine to form more complex substances. Decomposition is a reaction in which substances break down to form simpler substances. Reaction (e), $P_4(s) + 5O_2(g) \rightarrow P_4O_{10}(s)$, is combination and reaction (c), $2H_2O_2(aq) \rightarrow 2H_2O(l) + O_2(g)$, is decomposition.

1.97 In the formula of an element, there is only one kind of atom. In the formulas of compounds, there are more than one kind of atom. Compounds of metals and nonmetals are ionic; compounds of nonmetals are molecular. Organic compounds contain carbon, C. Thus P_4 represents an element, MgO and $Ni(NO_3)_2$ represent ionic compounds, and HBr and C_2H_4 represent molecular compounds; C_2H_4 is organic.

1.99 The atomic number of chlorine is 17. The atomic number is the number of protons in the nucleus of an atom and the number of electrons outside the nucleus.

(a) An atom of chlorine has 17 electrons.

(b) A molecule of chlorine, Cl_2, is composed of two chlorine atoms combined and has a total of 2(17) = 34 electrons.

(c) To form a chloride ion Cl^-, a chlorine atom gains one electron, making 18 electrons in a chloride ion.

(d) A perchlorate ion, ClO_4^- is composed of one chlorine atom, four oxygen atoms, and an extra electron (to account for the 1– charge). The atomic number of oxygen is 8. Therefore, a ClO_4^- ion has a total of 1(17) + 4(8) + 1 = 50 electrons.

1.101 **(a)** From the list of elements inside the front cover, the atomic number of tin is 50.

(b) From the periodic table inside the front cover or Figure 1.10, tin is in Group IVA.

(c) Tin is in the fifth period.

(d) and (e) The fact that the atomic number of tin is 50 tells us that there are 50 protons in the nucleus of a tin atom and 50 electrons outside the nucleus.

(f) The fact that the charge on an Sn^{2+} ion is 2+ tells us that the tin atom has lost two electrons in forming the ion. Therefore, there are (50 – 2) = 48 electrons left in the Sn^{2+} ion.

1.103 **Step 1.** The reactants are methane gas and oxygen gas. Carbon dioxide gas and gaseous water are the products. Heat and light are given off. The word equation is

$$\text{methane gas} + \text{oxygen gas} \rightarrow \text{carbon dioxide gas} + \text{gaseous water} + \text{heat} + \text{light}$$

Step 2. The formulas are

$$CH_4 + O_2 \rightarrow CO_2 + H_2O$$

Step 3. Balanced, the expression is

$$CH_4 + 2O_2 \rightarrow CO_2 + 2H_2O$$

Step 4. Check. There are one C, four H, and four O on each side.

Step 5. Add symbols. The equation for the reaction is

$$CH_4(g) + 2O_2(g) \rightarrow CO_2(g) + 2H_2O(l) \text{ or } (g)$$

1.105

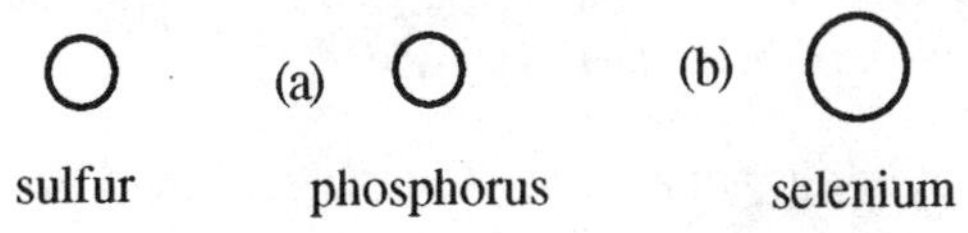

Phosphorus and sulfur are next to each other in the same period. From Figure 1.8 you can see that atoms of carbon, nitrogen, oxygen, and fluorine, which are next to each other in the same period, are all about the same size. Selenium is in the same group as sulfur, but one period lower. From Figure 1.8 you can see that bromine, which is in the same group as chlorine but one period lower, is larger than chlorine.

1.107 According to the index at the back of the book:

(a) Atomic number is discussed on pp 8–9, 57, 218, 277.

(b) Nomenclature of inorganic compounds is discussed on pp 19–22, A1–A5.

(c) Solutions are discussed on pp 106–140, 460–489.

Applications

1.109 Physical properties do not involve a substance changing into other substances. Chemical properties do involve a substance changing into other substances. The failure of gold to dissolve in water, its bright color, and its softness are physical properties. The failure of gold to form many compounds or to tarnish, which results in the fact that gold is mostly found as the metal (rather than in the form of compounds), are chemical properties.

1.111 nitrogen dioxide, NO_2; chlorine dioxide, ClO_2; ozone, O_3

1.113 **(a)** The term "inorganic" refers to compounds that do not contain carbon. (A few simple carbon–containing compounds, such as calcium carbonate, calcium hydrogen carbonate, carbon monoxide, and carbon dioxide, are usually considered to be inorganic.)

(b) Chemists call chalcocite, Cu_2S, copper(I) sulfide. The formula for sulfide ion is S^{2-} because a sulfur atom, atomic number 16, must gain two electrons to have 18 electrons like argon. Because two copper ions are combined with one sulfide ion in Cu_2S, each copper ion must have a 1+ charge for the combination to be electrically neutral. Therefore, the name of this compound is copper(I) sulfide.

(c) Hematite, Fe_2O_3, is called iron(III) oxide. The formula for oxide ion is O^{2-} because an oxygen atom, atomic number 8, must gain two electrons to have ten electrons like neon. For Fe_2O_3 to have a net charge of zero, each iron ion must have a 3+ charge.

(d) Siderite, $FeCO_3$, is iron(II) carbonate. From Table 1.2, the group "CO_3" is a carbonate ion, CO_3^{2-}. Therefore the charge on the iron in $FeCO_3$ is 2+.

(e) Corundum, Al_2O_3, is aluminum oxide. Aluminum only forms one compound with oxygen so no Roman numeral is necessary.

(f) Molybdenite, MoS_2, is called molybdenum(IV) sulfide. The formula for sulfide ion is S^{2-} because a sulfur atom, atomic number 16, must gain two electrons to have 18 electrons like argon. Because one molybdenum ion is combined with two sulfide ions in MoS_2, each molybdenum ion must have a 4+ charge for the combination to be electrically neutral. Therefore, the name of this compound is molybdenum(IV) sulfide.

(g) Quartz, SiO_2, is silicon dioxide. Silicon is a semimetal, that is, it is not a metal. Therefore, a Greek prefix is used in naming SiO_2.

(h) Calcite, $CaCO_3$, is called calcium carbonate. No Roman numeral is necessary because calcium is in Group IIA and forms only one carbonate.

(i) Corundum, Al_2O_3, is a compound of a reactive metal with a reactive nonmetal and is ionic. Corundum is composed of Al^{3+} and O^{2-} ions, which are held together by the electrostatic attraction between the positive and negative charges. No molecules of Al_2O_3 exist.

CHAPTER 2: Measurement

Practice Problems

2.1 **(a)** 6.39 kg

According to the SI rules you should use a dot on the baseline for the decimal point, use only the singular form of units, and should *not* put a period after the symbol.

(b) 0.000 015 K

If a number is less than one, you should write a zero to the left of the decimal point, group digits in threes around the decimal point, omit the degree sign when the kelvin scale is used for temperature, and separate the symbol from the number by a space.

(c) 299 792 458 m/s

You should *not* use commas to space digits in numbers. Use spaces to group digits in threes.

2.2 **(a)** From Table 2.1, the SI base unit for mass is the kilogram, kg.

(b) The SI base unit for length is the meter, m. Area has the dimensions $(\text{length})^2$; therefore the SI derived unit for area is m^2 or square meter.

2.3 **(a)** From Table 2.2, the prefix M means mega–. From Table 2.1, s is the symbol for second. A Ms is a megasecond.

(b) The prefix p means pico– and mol is the symbol for mole. Thus, a pmol is a picomole.

(c) The prefix c means centi–, A is the symbol for ampere, and a cA is a centiampere.

2.4 (a) From Table 2.2 the symbol for milli– is m, and from Tables 2.1 and 2.2 the symbol for gram is g. The symbol for milligram is mg.

(b) The symbol for nano– is n and the symbol for second is s. The symbol for nanosecond is ns.

(c) The symbol for kilo– is k and the symbol for kelvin is K. The symbol for kilokelvin is kK.

2.5 (a) From Table 2.2, the prefix mega– means 10^6. One megagram equals 10^6 grams.

(b) 1 Mg = 10^6 g or one million grams (1 000 000 g)

2.6 (a) From Table 2.2, the prefix milli– means 10^{-3} and 1 mm = 10^{-3} m. Division of both sides of this equation by 10^{-3} gives

$$\frac{1\text{ mm}}{10^{-3}} = 1\text{ m} = 10^3\text{ mm}$$

There are 10^3 mm, or 1000 mm, in a meter.

(b) A millimeter is smaller than a meter. There are one thousand millimeters in a meter.

(c) A nanometer is smaller than a micrometer. A nanometer is a smaller fraction of a meter (10^{-9} m) than a micrometer (10^{-6} m).

(d) From Table 2.2, the prefix nano– means 10^{-9} and 1 nm = 10^{-9} m. Division of both sides of this equation by 10^{-9} gives

$$\frac{1\text{ nm}}{10^{-9}} = 1\text{ m}$$

There are 10^9 nm, or 1 000 000 000 nm, in a meter.

2.7 From Table 2.2, 1 mg = 10^{-3} g. To convert mg to g, multiply by

$$\frac{10^{-3}\text{ g}}{1\text{ mg}}$$

so that mg cancel.

2.8 (a) 6.4×10^{-9} or 0.000 000 0064

$$(6.4\,\cancel{\text{ns}})\left(\frac{10^{-9}\text{ s}}{\cancel{\text{ns}}}\right) = 6.4 \times 10^{-9}\text{ s}$$

(b) 0.872

$$(872\,\cancel{\text{g}})\left(\frac{1\text{ kg}}{10^3\,\cancel{\text{g}}}\right) = 0.872\text{ kg}$$

2.9 $\left(\dfrac{10^{-6}\text{ g}}{\mu\text{g}}\right)$ and $\left(\dfrac{1\text{ kg}}{10^{3}\text{ g}}\right)$

2.10 **(a)** 4.9×10^{5}

$$(4.9\ \cancel{\text{km}})\left(\frac{10^{3}\ \cancel{\text{m}}}{\cancel{\text{km}}}\right)\left(\frac{1\text{ cm}}{10^{-2}\ \cancel{\text{m}}}\right) = 4.9 \times 10^{5}\text{ cm}$$

(b) 2.648×10^{-6}

$$(2648\ \cancel{\text{mg}})\left(\frac{10^{-3}\ \cancel{\text{g}}}{\cancel{\text{mg}}}\right)\left(\frac{1\text{ Mg}}{10^{6}\ \cancel{\text{g}}}\right) = 2.648 \times 10^{-6}\text{ Mg}$$

(c) 3.41×10^{-4}

$$(341\text{ mm}^2)\left(\frac{10^{-3}\text{ m}}{\text{mm}}\right)^2 = (341\ \cancel{\text{mm}^2})\left(\frac{10^{-6}\text{ m}^2}{\cancel{\text{mm}^2}}\right) = 3.41 \times 10^{-4}\text{ m}^2$$

2.11 **(a)** 2.89×10^{4}

$$(63.8\ \cancel{\text{lb}})\left(\frac{453.6\text{ g}}{\cancel{\text{lb}}}\right) = 2.89 \times 10^{4}\text{ g}$$

(b) 27.8

$$(44.7\ \cancel{\text{km}})\left(\frac{1\text{ mi}}{1.609\ \cancel{\text{km}}}\right) = 27.8\text{ mi}$$

2.12 **(a)** 4.4, 4.5, or 4.6 cm

Ruler (a) is divided into centimeters; therefore, tenths of a centimeter should be estimated in reading ruler (a). The left end of the rectangle is at 0.0 cm and the right end is at 4.5 cm. The rectangle is 4.5 cm long. If you estimate that the right end is at 4.4 or 4.6 cm and the rectangle is 4.4 or 4.6 cm long, your answer is also correct. The last digit is estimated and therefore uncertain.

(b) 3.88 or 3.89 cm **(c)** 3.0 cm **(d)** 2.4, 2.5, or 2.6 cm

2.13 **(a)** ruler (b)

(b)

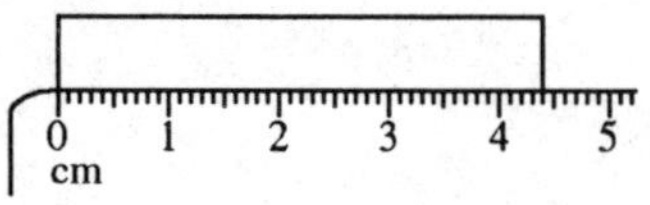

2.14 **(a)** 3.21×10^{5} to three significant figures

(b) 3.210×10^{5} to four significant figures

(c) $3.210\,00 \times 10^5$ to six significant figures

Exponential notation must be used to show the number of significant figures in numbers like 321 000.

2.15 (a) 4

All nonzero numbers are significant.

(b) 3

Zeros between nonzero numbers are significant.

(c) 5

Zeros at the end of a number to the right of the decimal point are significant.

(d) 2

There are just two significant figures in this measurement. The zeros only show the position of the decimal point, that is, $0.0065 = 6.5 \times 10^{-3}$ but $0.065 = 6.5 \times 10^{-2}$.

(e) 2 or 3

The zero may or may not be significant. It is impossible to tell when the measurement is written in this way.

(f) 2

There are just two significant figures in 4.8×10^{-9} m. The exponential part of the number tells the magnitude of the number but is disregarded in counting the number of significant figures.

2.16 (a) exactly

The number of students in a classroom can be counted.

(b) some uncertainty

The number of people in the United States cannot be counted exactly. The census takes a significant amount of time to complete. During that time many people are born and many die. These factors combined with the fact that some people aren't counted and others are counted more than once lead to a degree of uncertainty in the measurement.

(c) some uncertainty

Height is measured with some sort of ruler. In addition, the way you stand is not always exactly the same.

2.17 (a) 62.4 cm

62.4 cm is the median because it is the middle measurement in an odd number of measurements.

(b) 62.1 cm

$$\frac{(61.4\text{ cm} + 62.4\text{ cm} + 62.6\text{ cm})}{3} = 62.1\text{ cm}$$

(c) The median value. The median is less affected by the one lower reading.

(d) 62.5 m

62.5 m is the average of the two middle measurements in an even set of measurements.

2.18 Chances of measurements greater than the median = chances of measurements less than the median. Normal distribution is symmetrical about the mean.

2.19 **(a)** 3.43

In 3.432, the first digit to be removed (2) is less than 5. The unwanted digit is simply ignored.

(b) 3.44

In 3.438, the first digit to be removed (8) is greater than 5. The preceding digit is increased by one.

(c) 3.44

In 3.435, the first digit to be removed (5) is 5. The preceding digit is increased by one.

2.20 **(a)** 3

$$\frac{9.546}{2.31} = 4.13$$ ← only 3 significant figures

A quotient of measurements can have no more significant figures than the measurement with the least number of significant figures.

(b) 4

```
  53.674
 - 7.26    ← don't know this digit
  46.41    ← can't know this digit
```

The answer to a subtraction of measurements can have no more decimal places than the measurement with the least number of decimal places.

2.21 (b) wood (d) rain water (e) a dime

Matter occupies space and has mass.

2.22 **(a)** extensive **(b)** intensive **(c)** extensive **(d)** intensive **(e)** intensive

Extensive properties depend on sample size. Intensive properties do not depend on sample size.

2.23 **(a)** 11.3 g/cm^3

$$\text{density} = \frac{\text{mass}}{\text{volume}} = \frac{864\ \text{g}}{76.2\ \text{cm}^3} = 11.3\ \text{g/cm}^3$$

(b) 1.13×10^4 kg/m^3

$$\left(\frac{864\ \text{g}}{76.2\ \text{cm}^3}\right)\left(\frac{1\ \text{kg}}{10^3\ \text{g}}\right)\left(\frac{1\ \text{cm}}{10^{-2}\ \text{m}}\right)^3 = 1.13 \times 10^4\ \text{kg/m}^3$$

Note that the numerical value of density in SI units is 10^3 times that of the more commonly used units, g/cm^3.

2.24 61 cm^3

From Table 2.5, the density of ice is 0.92 g/cm^3.

$$(56.3\ \text{g ice})\left(\frac{1\ \text{cm}^3\ \text{ice}}{0.92\ \text{g ice}}\right) = 61\ \text{cm}^3\ \text{ice}$$

2.25 34 g

From Table 2.5, the density of iron is 7.9 g/cm^3

$$(4.36\ \text{cm}^3\ \text{iron})\left(\frac{7.9\ \text{g iron}}{\text{cm}^3\ \text{iron}}\right) = 34\ \text{g iron}$$

2.26 **(a)** 26 °C

$$\left(\frac{1\ °\text{C}}{1\ \text{K}}\right)(299\ \text{K}) - 273.15\ °\text{C} = 26\ °\text{C}$$

(b) 292.7 K

$$T\ \text{K} = (19.5\ °\text{C} + 273.15\ °\text{C})\left(\frac{1\ \text{K}}{1\ °\text{C}}\right)$$
$$= 292.7\ \text{K}$$

2.27 Celsius, 100 °C = 180 °F

100 Celsius degrees cover the same temperature range as 180 Fahrenheit degrees. Therefore, each Celsius degree is larger than a Fahrenheit degree.

2.28 **(a)** 39 °C

To convert temperature in °F to temperature in °C, solve equation 2.4 for t °C and substitute.

$$t_F\ °F = \left(\frac{9\ °F}{5\ °C}\right)(t\ °C) + 32\ °F$$

$$t_F\ °F - 32\ °F = \left(\frac{9\ °F}{5\ °C}\right)(t\ °C)$$

$$t\ °C = \left(\frac{5\ °C}{9\ °F}\right)(t_F\ °F - 32\ °F) = \left(\frac{5\ °C}{9\ \cancel{°F}}\right)(103\ \cancel{°F} - 32\ \cancel{°F}) = \left(\frac{5\ °C}{9\ \cancel{°F}}\right)(71\ \cancel{°F}) = 39\ °C$$

(b) 126 °F

To convert temperature in °C to temperature in °F, substitute in equation 2.4.

$$t_F = \left(\frac{9\ °F}{5\ °C}\right)(t\ °C) + 32\ °F = \left(\frac{9\ °F}{5\ \cancel{°C}}\right)(52\ \cancel{°C}) + 32\ °F = 94\ °F + 32\ °F = 126\ °F$$

2.29 **(a)** 86 400 or 8.6400×10^4

$$(1\ \cancel{\text{day}})\left(\frac{24\ \cancel{\text{h}}}{\cancel{\text{day}}}\right)\left(\frac{60\ \cancel{\text{min}}}{\cancel{\text{h}}}\right)\left(\frac{60\ \text{s}}{\cancel{\text{min}}}\right) = 86\ 400\ \text{s or } 8.6400 \times 10^4\ \text{s (exactly)}$$

(b) 2.7×10^3

$$(2.7\ \cancel{\text{ns}})\left(\frac{10^{-9}\ \cancel{\text{s}}}{\cancel{\text{ns}}}\right)\left(\frac{1\ \text{ps}}{10^{-12}\ \cancel{\text{s}}}\right) = 2.7 \times 10^3\ \text{ps}$$

(c) 6.0×10^{10}

$$(1\ \cancel{\text{min}})\left(\frac{60\ \cancel{\text{s}}}{\cancel{\text{min}}}\right)\left(\frac{1\ \text{ns}}{10^{-9}\ \cancel{\text{s}}}\right) = 6.0 \times 10^{10}\ \text{ns}$$

2.30 **(a)** 2 h, 50 min, 53 s

Note that in subtracting time units, we can "borrow" units just as we "borrow" in other types of subtraction problems. To find out how much time has passed, subtract the earlier time from the later time:

```
  12:13:27
 - 9:22:34
```

If the number of seconds in the earlier time is greater than the number of seconds in the later time, borrow one minute (1 min = 60 s) from the minutes:

12:13:27 = 12:12:87

Similarly, if the number of minutes in the earlier time is greater than the number of minutes in the later time, borrow one hour (1 h = 60 min) from the hours.

12:12:87 = 11:72:87

You are now ready to subtract.

$$\begin{array}{r} 11{:}72{:}87 \\ -\ \ 9{:}22{:}34 \\ \hline 2{:}50{:}53 \end{array}$$

It was 2 h, 50 min, and 53 s between the two events.

(b) 1.0253×10^4 s

$$\left[(2\,\cancel{h})\left(\frac{60\,\cancel{min}}{\cancel{h}}\right)\left(\frac{60\text{ s}}{\cancel{min}}\right)\right] + \left[(50\,\cancel{min})\left(\frac{60\text{ s}}{\cancel{min}}\right)\right] + 53\text{ s} = 10\,253\text{ s}$$

2.31 **(a)** 50

The atomic number of tin is 50. Therefore, the nucleus of the atom contains 50 protons.

(b) 68

From equation 2.5, mass number (A) = number of protons (Z) + number of neutrons (N)
number of neutrons = mass number – number of protons = 118 – 50 = 68

(c) $^{118}_{50}\text{Sn}$

In nuclide symbols, mass number is shown as a left superscript and atomic number is a left subscript.

(d) 48

The number of electrons in a neutral atom of tin equals the number of protons (50). In the Sn^{2+} ion, the tin atom has lost two electrons. Thus, Sn^{2+} has 48 electrons.

2.32 **(a)** 17

The atomic number of Cl is 17.

(b) 18

number of neutrons = mass number – number of protons = 35 – 17 = 18

(c) $^{35}_{17}\text{Cl}$

(d) 18

A neutral atom of chlorine contains the same number of electrons as protons (17). In the Cl^- ion, one additional electron has been added to give a 1– charge and a total of 18 electrons.

(e) 34

$$(2\ \text{atoms of chlorine})\left(\frac{17\ \text{electrons}}{1\ \text{atom of chlorine}}\right) = 34\ \text{electrons}$$

2.33 50% bromine–79, 50% bromine–81

The atomic mass of naturally occurring bromine (rounded to one decimal place) is 79.9 u, which is very near 80. Because 80 is half–way between 79 and 81, the isotopic composition of naturally occurring bromine must be about 50% bromine–79 and 50% bromine–81.

2.34 28.1 u

$$\left(\frac{92.2}{100}\right)(27.976\,927\ \text{u}) + \left(\frac{4.67}{100}\right)(28.976\,495\ \text{u}) + \left(\frac{3.10}{100}\right)(29.973\,770\ \text{u})$$

$$25.8\ \text{u} + 1.35\ \text{u} + 0.929\ \text{u} = 28.1\ \text{u}$$

2.35 **(a)** The mass number is normally the nearest whole number to the atomic mass of an isotope.

The mass number represents a whole number of protons and neutrons in the nucleus of a specific isotope. The atomic mass of an isotope is the mass of that isotope relative to the mass of carbon–12 isotope, which is taken to be exactly 12. The mass number is usually the nearest whole number to the atomic mass.

(b) 14

2.36 Naturally occurring samples of carbon contain other isotopes that have higher atomic masses than carbon-12 as well as carbon-12, causing the atomic mass to be greater than 12.000 u.

2.37 **(a)** 199.886 u

1 Ca	@ 40.078 u	=	40.078 u
2 Br	@ 79.904 u	=	159.808 u
	$1CaBr_2$	=	199.886 u

(b) 101.9613 u

2 Al	@ 26.981 539 u	=	53.963 078 u
3 O	@ 15.999 4 u	=	47.998 2 u
	$1Al_2O_3$	=	101.9613 u

(c) 68.143 u

2 N	@ 14.006 74 u	=	28.013 48 u
8 H	@ 1.007 94 u	=	8.063 52 u
1 S	@ 32.066 u	=	32.066 u
	$1(NH_4)_2S$	=	68.143 u

(d) 153.822 u

1 C	@ 12.011 u	=	12.011 u
4 Cl	@ 35.4527 u	=	141.8108 u
$1CCl_4$		=	153.822 u

2.38 **(a)** 219.075 u

1 Ca	@ 40.078 u	=	40.078 u
2 Cl	@ 35.4527 u	=	70.9054 u
12 H	@ 1.007 94 u	=	12.09528 u
6 O	@ 15.9994 u	=	95.9964 u
$1CaCl_2 \cdot 6H_2O$		=	219.075 u

(b) hexahydrate

(c) $CaCl_2 \cdot H_2O$ and $CaCl_2$

2.39 **(a)** 197.0 g

$$(1.000\ \cancel{\text{mol Au}})\left(\frac{196.97\text{ g Au}}{\cancel{\text{mol Au}}}\right) = 197.0\text{ g Au}$$

(b) 6.022×10^{23} atoms

$$(1.000\ \cancel{\text{mol Au}})\left(\frac{6.0221 \times 10^{23}\text{ atoms Au}}{\cancel{\text{mol Au}}}\right) = 6.022 \times 10^{23}\text{ atoms Au}$$

(c) 1.20×10^{24} atoms

$$(2.00\ \cancel{\text{mol Au}})\left(\frac{6.022 \times 10^{23}\text{ atoms Au}}{\cancel{\text{mol Au}}}\right) = 1.20 \times 10^{24}\text{ atoms Au}$$

(d) 3.01×10^{23} atoms

$$(0.500\ \cancel{\text{mol Au}})\left(\frac{6.022 \times 10^{23}\text{ atoms Au}}{\cancel{\text{mol Au}}}\right) = 3.01 \times 10^{23}\text{ atoms Au}$$

2.40 **(a)** 123.9 g

4 P @ 30.973 762 u = 123.895 048 u

$$(1.000\ \cancel{\text{mol } P_4})\left(\frac{123.90\text{ g } P_4}{\cancel{\text{mol } P_4}}\right) = 123.9\text{ g } P_4$$

(b) 6.022×10^{23} molecules

(c) 2.409×10^{24} atoms

$$(1.000\ \cancel{\text{mol P}_4})\left(\frac{6.0221 \times 10^{23}\ \cancel{\text{molecules P}_4}}{\cancel{\text{mol P}_4}}\right)\left(\frac{4\ \text{atoms P}}{\cancel{\text{molecule P}_4}}\right) = 2.409 \times 10^{24}\ \text{atoms P}$$

2.41 **(a)** 1.0000 mol

$$(17.031\ \cancel{\text{g NH}_3})\left(\frac{1\ \text{mol NH}_3}{17.0306\ \cancel{\text{g NH}_3}}\right) = 1.0000\ \text{mol NH}_3$$

(b) 6.0223×10^{23} atoms N; 1.8067×10^{24} atoms H

$$(17.031\ \cancel{\text{g NH}_3})\left(\frac{1\ \cancel{\text{mol NH}_3}}{17.0306\ \cancel{\text{g NH}_3}}\right)\left(\frac{6.022\ 14 \times 10^{23}\ \cancel{\text{molecules NH}_3}}{\cancel{\text{mol NH}_3}}\right)\left(\frac{1\ \text{atom N}}{\cancel{\text{molecule NH}_3}}\right) = 6.0223 \times 10^{23}\ \text{atoms N}$$

$$(17.031\ \cancel{\text{g NH}_3})\left(\frac{1\ \cancel{\text{mol NH}_3}}{17.0306\ \cancel{\text{g NH}_3}}\right)\left(\frac{6.022\ 14 \times 10^{23}\ \cancel{\text{molecules NH}_3}}{\cancel{\text{mol NH}_3}}\right)\left(\frac{3\ \text{atoms H}}{\cancel{\text{molecule NH}_3}}\right) = 1.8067 \times 10^{24}\ \text{atoms H}$$

2.42 1.2044×10^{24} ions

There are two nitrate ions in each unit of $Ca(NO_3)_2$. Therefore, 1.0000 mol $Ca(NO_3)_2$ contains 2.0000 mol nitrate ion.

$$(2.0000\ \cancel{\text{mol nitrate ions}})\left(\frac{6.022\ 14 \times 10^{23}\ \text{nitrate ions}}{\cancel{\text{mol nitrate ions}}}\right) = 1.2044 \times 10^{24}\ \text{nitrate ions}$$

2.43 53 g

$$(0.54\ \cancel{\text{mol H}_2\text{SO}_4})\left(\frac{98.1\ \text{g H}_2\text{SO}_4}{\cancel{\text{mol H}_2\text{SO}_4}}\right) = 53\ \text{g H}_2\text{SO}_4$$

(The formula mass of H_2SO_4 is given in the table inside the back cover.)

2.44 0.2280 mol

$$(22.36\ \cancel{\text{g H}_2\text{SO}_4})\left(\frac{1\ \text{mol H}_2\text{SO}_4}{98.079\ \cancel{\text{g H}_2\text{SO}_4}}\right) = 0.2280\ \text{mol H}_2\text{SO}_4$$

2.45 **(a)** 9.721×10^{23} atoms P

$$(5.000\ \cancel{\text{g P}_4})\left(\frac{1\ \text{mol P}_4}{123.90\ \cancel{\text{g P}_4}}\right) = 4.036 \times 10^{-2}\ \text{mol P}_4$$

$$(4.036 \times 10^{-2}\ \cancel{\text{mol P}_4})\left(\frac{4\ \text{mol P}}{1\ \cancel{\text{mol P}_4}}\right) = 1.614 \times 10^{-1}\ \text{mol P}$$

$$(1.614 \times 10^{-1}\ \cancel{\text{mol P}})\left(\frac{6.0221 \times 10^{23}\ \text{atoms P}}{1\ \cancel{\text{mol P}}}\right) = 9.720 \times 10^{22}\ \text{atoms P}$$

or, combining steps,

$$(5.000 \cancel{\text{g P}_4})\left(\frac{1 \cancel{\text{mol P}_4}}{123.90 \cancel{\text{g P}_4}}\right)\left(\frac{4 \cancel{\text{mol P}}}{1 \cancel{\text{mol P}_4}}\right)\left(\frac{6.0221 \times 10^{23} \text{ atoms P}}{1 \cancel{\text{mol P}}}\right) = 9.721 \times 10^{22} \text{ atoms P}_4$$

The answer obtained without intermediate round off is the better answer.

(b) 5.143×10^{-23} g

$$(1 \cancel{\text{atom P}})\left(\frac{1 \cancel{\text{mol P}}}{6.0221 \times 10^{23} \cancel{\text{atoms P}}}\right)\left(\frac{30.97 \text{ g P}}{1 \cancel{\text{mol P}}}\right) = 5.143 \times 10^{-23} \text{ g P}$$

Additional Practice Problems

2.97 There are no neutrons in its nucleus.

2.99 $1.2 \text{ g} \cdot \text{mL}^{-1}$

$$d = \frac{m}{V} = \frac{7.50 \text{ g}}{(26.0 \text{ mL} - 19.7 \text{ mL})} = \frac{7.50 \text{ g}}{6.3 \text{ mL}} = 1.2 \text{ g} \cdot \text{mL}^{-1}$$

2.101 5.49×10^{-4} g

$$(1.00 \cancel{\text{mol electron}})\left(\frac{6.022 \times 10^{23} \cancel{\text{electrons}}}{\cancel{\text{mol electron}}}\right)\left(\frac{9.11 \times 10^{-28} \text{ g}}{\cancel{\text{electron}}}\right) = 5.49 \times 10^{-4} \text{ g}$$

2.103 Mg-24 = 79%, Mg-25 = 10%, and Mg-26 = 11%

From the graph, the relative abundances of the isotopes are Mg-24 = 100, Mg-25 = 13, and Mg-26 = 14, which yield a total abundance of 127.

The percent abundance of Mg-24 is calculated as follows: $\frac{100}{(100 + 13 + 14)} \times 100 = 79\%$

Similarly, the percent abundance of Mg-25 = $\frac{13}{127}$ = 10% and the percent abundance of Mg-26 = $\frac{14}{127}$ = 11%

Stop and Test Yourself

1. **(c)** kg (Table 2.1)

2. **(a)** 10^{-3} meter (Table 2.2)

3. **(b)** 4727

$$1 \text{ h, } 18 \text{ min, } 47 \text{ s} = (1 \cancel{\text{h}})\left(\frac{60 \cancel{\text{min}}}{\cancel{\text{h}}}\right)\left(\frac{60 \text{ s}}{\cancel{\text{min}}}\right) + (18 \cancel{\text{min}})\left(\frac{60 \text{ s}}{\cancel{\text{min}}}\right) + 47 \text{ s} = 3.6 \times 10^3 \text{ (exact) s} + 1.08 \times 10^3 \text{ (exact) s} + 47 \text{ s}$$

$$= 4727 \text{ s}$$

4. (d) 4.3×10^6

$$(4.3\ \cancel{\text{Mm}})\left(\frac{10^6\ \text{m}}{\cancel{\text{Mm}}}\right) = 4.3 \times 10^6\ \text{m}$$

5. (c) 3.24×10^{-1}

$$(324\ \cancel{\text{ng}})\left(\frac{10^{-9}\ \cancel{\text{g}}}{\cancel{\text{ng}}}\right)\left(\frac{1\ \mu\text{g}}{10^{-6}\ \cancel{\text{g}}}\right) = 324 \times 10^{-3}\ \mu\text{g} = 3.24 \times 10^{-1}\ \mu\text{g}$$

6. (e) 5.50 cm

The ruler reads to 0.1 cm. This means that the second decimal place is an estimate and is, therefore, uncertain. The left end of the block of wood is at 1.00 cm. The right end is at 6.50 cm.

6.50 cm – 1.00 cm = 5.50 cm to two decimal places.

7. (c) 4

Neither of the two zeros on the left are significant. The nonzero digits, the zero between, and the zero on the right are significant.

8. (e) 4.17×10^8

To add numbers expressed in exponential notation, it is necessary to first change them to the same power of 10.

$4.62 \times 10^7 = 0.462 \times 10^8$

$$\begin{array}{r} 3.71 \times 10^8 \\ \underline{+\ 0.462 \times 10^8} \\ 4.17 \times 10^8 \end{array}$$

There can be only two digits to the right of the decimal point in the answer because there are only two digits to the right of the decimal point in 3.71×10^8.

9. **(d)** 3.00 cm^3

Using density to write a unit conversion factor between mass and volume gives

$$(7.44\ \not{g})\left(\frac{1\ cm^3}{2.48\ \not{g}}\right) = 3.00\ cm^3$$

If you prefer to use the definition of density and solve for volume,

$$\text{volume} = \frac{\text{mass}}{\text{density}} = \frac{7.44\ \not{g}}{\frac{2.48\ \not{g}}{1\ cm^3}} = 3.00\ cm^3$$

Answer (a) is incorrect due to the wrong number of significant figures. Answers (b), (c), and (e) are all incorrect because they result from multiplying mass and density, rather than dividing the mass by the density.

10. **(d)** 261.2

$$T\ \text{K} = (t\ °\text{C} + 273.15\ °\text{C})\left(\frac{1\ \text{K}}{1\ °\text{C}}\right) = (-12.0\ \not{°\text{C}} + 273.15\ \not{°\text{C}})\left(\frac{1\ \text{K}}{1\ \not{°\text{C}}}\right) = 261.2\ \text{K}$$

In the addition, only one decimal place is allowed because the temperature given only has one decimal place.

11. **(a)** 42 protons, 56 neutrons, 42 electrons

The atomic number (Z) of molybdenum is 42. Therefore, the number of protons is 42.
The mass number (A) of molybdenum–98 is 98
The number of neutrons (N) = mass number (A) – number of protons (Z) = 98 – 42 = 56.
In a neutral atom of molybdenum, the number of electrons = the number of protons = 42.

12. **(d)** ${}^{52}_{24}Cr^{3+}$

The ion containing 24 protons has an atomic number of 24. The element is chromium (Cr). This ion has 3 more protons than electrons so it has a 3+ charge. The mass number is the sum of the number of protons and the number of neutrons: 24 + 28 = 52. In the nuclide symbol, the number of protons is shown as the left subscript. The charge is shown as the right superscript and the mass number is a left superscript. The nuclide symbol is ${}^{52}_{24}Cr^{3+}$.

13. **(b)** 132.17 u

2 N	@	14.01 u	=	28.02 u
8 H	@	1.01 u	=	8.08 u
1 S	@	32.07 u	=	32.07 u
4 O	@	16.00 u	=	64.00 u
1$(NH_4)_2SO_4$			=	132.17 u

Note that the actual answer that you obtain for a problem of this type will depend on how you round off the atomic masses used in the calculation. If your answer disagrees with the possible solution only in the last digit, it is acceptable, since the last digit is understood to be uncertain.

14. (c) 1.204×10^{24} ammonium ions

$$(1.000 \text{ mol } (NH_4)_2SO_4)\left(\frac{2 \text{ mol } NH_4^+ \text{ ions}}{\text{mol } (NH_4)_2SO_4}\right)\left(\frac{6.0221 \times 10^{23} \, NH_4^+ \text{ ions}}{\text{mol } NH_4^+ \text{ ions}}\right) = 1.204 \times 10^{24} \, NH_4^+ \text{ ions}$$

$(NH_4)_2SO_4$ is an ionic compound and does not contain molecules. This eliminates answer (a). $(NH_4)_2SO_4$ contains 1.204×10^{24} nitrogen atoms. This eliminates (d). $(NH_4)_2SO_4$ contains 4.82×10^{24} hydrogen atoms. This eliminates (e). Choice (b) results from lowering the exponent by 1 rather than raising it.

15. (c) 10.666 g

$$(0.333\,33 \text{ mol } O_2)\left(\frac{31.9988 \text{ g } O_2}{\text{mol } O_2}\right) = 10.666 \text{ g } O_2$$

Putting Things Together

2.105 No. The fact that volume can only be measured to three significant figures limits the precision of the density to three significant figures.

2.107 (b) CO_2 (c) H_2O (e) CH_4. These are molecular compounds.

The units for these three compounds are molecules, since they are all combinations of nonmetals. Therefore, they are appropriately termed "molecular compounds" and they each have a "molecular mass".

For (a) NaCl and (d) $Ca(NO_3)_2$, the term "molecular mass" is inappropriate because the units of these compounds are ions, not molecules.

2.109 The chemical properties of the elements are more related to the number of electrons and protons than to the number of neutrons. Carbon–12 only differs from carbon–13 by one neutron. Therefore, their chemical properties would be expected to be very similar.

2.111 More moles of ammonia were produced.

For H_2SO_4, $(80.31 \times 10^9 \text{ lb})\left(\frac{453.6 \text{ g}}{\text{lb}}\right)\left(\frac{1 \text{ mol } H_2SO_4}{98.079 \text{ g}}\right) = 3.714 \times 10^{11} \text{ mol } H_2SO_4$

For NH_3, $(34.50 \times 10^9 \text{ lb})\left(\frac{453.6 \text{ g}}{\text{lb}}\right)\left(\frac{1 \text{ mol } NH_3}{17.031 \text{ g}}\right) = 9.189 \times 10^{11} \text{ mol } NH_3$

2.113 (a) volume (b) density (c) area (d) amount of substance (e) mass

2.115 (a) If two elements form more than one compound, the masses of one element that combine with a fixed mass of the other element are in ratios of small whole numbers. (This generalization was first proposed by Dalton early in the 19th century and is called the law of multiple proportions.)

(b) Possible pairs are CO and CO_2, H_2O and H_2O_2, $FeCl_2$ and $FeCl_3$, and SnO and SnO_2.

(c) Yes, because, according to Dalton's atomic theory, compounds are combinations of indivisible units called atoms.

Applications

2.117 7×10^6 - 9×10^6 km^2

One uncertain digit should be shown. Since the uncertainty is in the first digit, only that digit is shown.

2.119 73 000 g/mol

The molecular mass is 73 000 u (see below), which makes the molar mass 73 000 g/mol.

$$(73\ \cancel{kD})\left(\frac{1000\ \cancel{D}}{1\ \cancel{kD}}\right)\left(\frac{1\ u}{1\ \cancel{D}}\right) = 73\ 000\ u$$

2.121 **(a)** 1.0, 1.0, 1.2, 1.3, 1.5, 1.6

Mass number (A) = Atomic number (Z) + Neutron number (N)
Therefore, N = A – Z

Species	A	Z	N = A – Z	N/Z
carbon–12	12	6	6	6/6 = 1.0
$^{32}_{16}S$	32	16	16	16/16 = 1.0
$^{56}_{26}Fe$	56	26	30	30/26 = 1.2
$^{107}_{47}Ag$	107	47	60	60/47 = 1.3
$^{197}_{79}Au$	197	79	118	118/79 = 1.5
$^{238}_{92}U$	238	92	146	146/92 = 1.6

(b) The ratio of neutrons to protons increases (beyond sulfur).

2.123 3×10^3 y

The error is 365.2425 days·y^{-1} - 365.242 1934 days·y^{-1} = 0.0003 days·y^{-1} or

$$\frac{1\ y}{0.0003\ days} = 3 \times 10^3\ y{\cdot}day^{-1}$$

2.125 Empty, wash, and dry the can. Weigh. Fill the can with water. Be sure the outside of the can is dry. Reweigh. Measure the temperature of the water in the can. From the table of density of water at different temperatures given in the Appendix, find the density of water. Use the mass of the water and its density to calculate the volume of the can. Subtract 355 mL to obtain the volume of the empty space.

2.127 38, 3.8×10^3, 0.48

The relationships given in the problem give us the necessary conversion factors to work the problem.

$$(3.8 \text{ furlongs})\left(\frac{10 \text{ chains}}{\text{furlong}}\right) = 38 \text{ chains}$$

$$(38 \text{ chains})\left(\frac{100 \text{ links}}{\text{chain}}\right) = 3800 \text{ links or } 3.8 \times 10^3 \text{ links (to two significant figures)}$$

$$(38 \text{ chains})\left(\frac{1 \text{ mi}}{80 \text{ chains}}\right) = 0.48 \text{ mi}$$

2.129 **(a)** 1273 K, 671 mi/h

$$T\text{ K} = (t\text{ °C} + 273.15\text{ °C})\left(\frac{1\text{ K}}{1\text{ °C}}\right) = (1000\text{ °C} + 273.15\text{ °C})\left(\frac{1\text{ K}}{1\text{ °C}}\right) = 1273\text{ K}$$

$$\left(\frac{300 \text{ m}}{\text{s}}\right)\left(\frac{60 \text{ s}}{1 \text{ min}}\right)\left(\frac{60 \text{ min}}{1 \text{ h}}\right)\left(\frac{1 \text{ km}}{1000 \text{ m}}\right)\left(\frac{1 \text{ mi}}{1.609 \text{ km}}\right) = 671 \text{ mi/h}$$

(b) $2.1 \times 10^{11} - 4.2 \times 10^{11}$ mol SO_2

$$(15 \text{ x } 10^6 \text{ ton } SO_2)\left(\frac{2000 \text{ lb } SO_2}{1 \text{ ton } SO_2}\right)\left(\frac{454 \text{ g } SO_2}{1 \text{ lb } SO_2}\right)\left(\frac{1 \text{ mol } SO_2}{64.1 \text{ g } SO_2}\right) = 2.1 \times 10^{11} \text{ mol } SO_2$$

$$(30 \text{ x } 10^6 \text{ ton } SO_2)\left(\frac{2000 \text{ lb } SO_2}{1 \text{ ton } SO_2}\right)\left(\frac{454 \text{ g } SO_2}{1 \text{ lb } SO_2}\right)\left(\frac{1 \text{ mol } SO_2}{64.1 \text{ g } SO_2}\right) = 4.2 \times 10^{11} \text{ mol } SO_2$$

(c) You need to know the world populations for these time periods.

2.131 **(a)** 6×10^3 molecules

$$(1 \times 10^{-20} \text{ mol})\left(\frac{6.0 \times 10^{23} \text{ molecules}}{\text{mol}}\right) = 6 \times 10^3 \text{ molecules}$$

(b) 3×10^3 atoms

$$(0.5 \text{ µm})\left(\frac{10^{-6} \text{ m}}{\text{µm}}\right)\left(\frac{1 \text{ pm}}{10^{-12} \text{ m}}\right) = 5 \times 10^5 \text{ pm}$$

The diameter of a carbon atom is twice the radius, or 154 pm.

$$(5 \text{ x } 10^5 \text{ pm})\left(\frac{1 \text{ C atom}}{154 \text{ pm}}\right) = 3 \times 10^3 \text{ C atoms}$$

CHAPTER 3: Stoichiometry

Practice Problems

3.1 (a) $4HNO_3(aq) \rightarrow 4NO_2(aq) + 2H_2O(l) + O_2(g)$

$HNO_3(aq) \rightarrow NO_2(aq) + H_2O(l) + O_2(g)$

Step 1: There are two hydrogen atoms on the right. Multiply $HNO_3(aq)$ by two to get two hydrogen atoms on the left:

$2HNO_3(aq) \rightarrow NO_2(aq) + H_2O(l) + O_2(g)$

Step 2: Now there are two nitrogen atoms on the left and only one on the right. Multiply $NO_2(aq)$ by two to get two nitrogen atoms on the right:

$2HNO_3(aq) \rightarrow 2NO_2(aq) + H_2O(l) + O_2(g)$

Step 3: Now there are six oxygen atoms on the left but seven on the right. Solve this dilemma by multiplying $O_2(g)$ by $\frac{1}{2}$ and then multiplying the whole equation by two:

$2HNO_3(aq) \rightarrow 2NO_2(aq) + H_2O(l) + \frac{1}{2}O_2(g)$

$4HNO_3(aq) \rightarrow 4NO_2(aq) + 2H_2O(l) + O_2(g)$

Note that there are often more than one set of steps possible for balancing equations by inspection depending on where you begin. Any set of steps that leads to the correct equation is all right.

(b) $B_2O_3(s) + 3H_2O(l) \rightarrow 2H_3BO_3(aq)$

$B_2O_3(s) + H_2O(l) \rightarrow H_3BO_3(aq)$
$B_2O_3(s) + H_2O(l) \rightarrow 2H_3BO_3(aq)$
$B_2O_3(s) + 3H_2O(l) \rightarrow 2H_3BO_3(aq)$

(c) $Cu(s) + 2AgNO_3(aq) \rightarrow 2Ag(s) + Cu(NO_3)_2(aq)$

$Cu(s) + AgNO_3(aq) \rightarrow Ag(s) + Cu(NO_3)_2(aq)$
$Cu(s) + 2AgNO_3(aq) \rightarrow Ag(s) + Cu(NO_3)_2(aq)$
$Cu(s) + 2AgNO_3(aq) \rightarrow 2Ag(s) + Cu(NO_3)_2(aq)$

(d) $4NH_3(g) + 5O_2(g) \rightarrow 4NO(g) + 6H_2O(l)$

$NH_3(g) + O_2(g) \rightarrow NO(g) + H_2O(l)$
$2NH_3(g) + O_2(g) \rightarrow NO(g) + 3H_2O(l)$
$2NH_3(g) + O_2(g) \rightarrow 2NO(g) + 3H_2O(l)$
$2NH_3(g) + \frac{5}{2}O_2(g) \rightarrow 2NO(g) + 3H_2O(l)$
$4NH_3(g) + 5O_2(g) \rightarrow 4NO(g) + 6H_2O(l)$

(e) $C_3H_6O(l) + 4O_2(g) \rightarrow 3CO_2(g) + 3H_2O(l)$

$C_3H_6O(l) + O_2(g) \rightarrow CO_2(g) + H_2O(l)$
$C_3H_6O(l) + O_2(g) \rightarrow 3CO_2(g) + 3H_2O(l)$
$C_3H_6O(l) + 4O_2(g) \rightarrow 3CO_2(g) + 3H_2O(l)$

3.2 **(a)** 2 mol

$3Mg(s) + N_2(g) \xrightarrow{\text{heat}} Mg_3N_2(s)$

3 mol Mg(s) ~ 1 mol $Mg_3N_2(s)$

$$(6\ \cancel{\text{mol Mg}})\left(\frac{1\ \text{mol}\ Mg_3N_2}{3\ \cancel{\text{mol Mg}}}\right) = 2\ \text{mol}\ Mg_3N_2$$

(b) 0.3 mol

3 mol Mg(s) ~ 1 mol $N_2(g)$

$$(1\ \cancel{\text{mol Mg}})\left(\frac{1\ \text{mol}\ N_2}{3\ \cancel{\text{mol Mg}}}\right) = 0.3\ \text{mol}\ N_2$$

(c) 9 mol

1 mol Mg_3N_2 ~ 3 mol Mg

$$(3\ \cancel{\text{mol}\ Mg_3N_2})\left(\frac{3\ \text{mol Mg}}{\cancel{\text{mol}\ Mg_3N_2}}\right) = 9\ \text{mol Mg}$$

3.3 **(a)** 28.5 g

Problem	63.5 g			? g
Equation	3Na(s)	+	P(s)	$\rightarrow Na_3P(s)$
Formula mass, u	22.99		30.97	
Recipe, mol	3			1

$$(63.5\ \cancel{g\ Na})\left(\frac{1\ \cancel{mol\ Na}}{22.99\ \cancel{g\ Na}}\right)\left(\frac{1\ \cancel{mol\ P}}{3\ \cancel{mol\ Na}}\right)\left(\frac{30.97\ g\ P}{\cancel{mol\ P}}\right) = 28.5\ g\ P$$

(If you have forgotten how to calculate formula masses, review Section 2.12.)

(b) 92.0 g

Problem	63.5 g				? g
Equation	3Na(s)	+	P(s)	$\rightarrow$	$Na_3P(s)$
Formula mass, u	22.99				99.94
Recipe, mol	3				1

$$(63.5\ \cancel{g\ Na})\left(\frac{1\ \cancel{mol\ Na}}{22.99\ \cancel{g\ Na}}\right)\left(\frac{1\ \cancel{mol\ Na_3P}}{3\ \cancel{mol\ Na}}\right)\left(\frac{99.94\ g\ Na_3P}{\cancel{mol\ Na_3P}}\right) = 92.0\ g\ Na_3P$$

3.4 **(a)** 1.93×10^{-2} mol

Problem	3.91 g			? mol
Equation	$2KNO_3(s) \xrightarrow{heat} 2KNO_2(s)$		+	$O_2(g)$
Formula mass, u	101.1			32.00
Recipe, mol	2			1

$$(3.91\ \cancel{g\ KNO_3})\left(\frac{1\ \cancel{mol\ KNO_3}}{101.1\ \cancel{g\ KNO_3}}\right)\left(\frac{1\ mol\ O_2}{2\ \cancel{mol\ KNO_3}}\right) = 1.93 \times 10^{-2}\ mol\ O_2$$

(b) 4.056 g

Problem	25.63 g			? g
Equation	$2KNO_3(s) \xrightarrow{heat} 2KNO_2(s)$		+	$O_2(g)$
Formula mass, u	101.10			31.999
Recipe, mol	2			1

$$(25.63\ \cancel{g\ KNO_3})\left(\frac{1\ \cancel{mol\ KNO_3}}{101.10\ \cancel{g\ KNO_3}}\right)\left(\frac{1\ \cancel{mol\ O_2}}{2\ \cancel{mol\ KNO_3}}\right)\left(\frac{31.999\ g\ O_2}{1\ \cancel{mol\ O_2}}\right) = 4.056\ g\ O_2$$

3.5 **(a)** 2.60×10^3 g

Problem	487 g		? g	
Equation	C(s)	+	$2S(g) \xrightarrow{high\ temp.}$	$CS_2(g)$
Formula mass, u	12.01		32.07	
Recipe, mol	1		2	

$$(487\ \cancel{g\ C})\left(\frac{1\ \cancel{mol\ C}}{12.01\ \cancel{g\ C}}\right)\left(\frac{2\ \cancel{mol\ S}}{1\ \cancel{mol\ C}}\right)\left(\frac{32.07\ g\ S}{1\ \cancel{mol\ S}}\right) = 2.60 \times 10^3\ g\ S$$

(b) 199 g

Problem		? g	236 g
Equation	C(s) +	2S(g) $\xrightarrow{\text{high temp.}}$	$CS_2(g)$
Formula mass, u		32.07	76.14
Recipe, mol		2	1

$$(236\ \cancel{g\ CS_2})\left(\frac{1\ \cancel{mol\ CS_2}}{76.14\ \cancel{g\ CS_2}}\right)\left(\frac{2\ \cancel{mol\ S}}{1\ \cancel{mol\ CS_2}}\right)\left(\frac{32.07\ g\ S}{1\ \cancel{mol\ S}}\right) = 199\ g\ S$$

(c) 2.60×10^3 tons

We found in part (a) that 2.60×10^3 g S are needed to react with 487 g of carbon. Because atomic masses are relative masses, different units of mass can be used in stoichiometric calculations without changing the arithmetic as long as the same units are used throughout a calculation. Therefore, 2.60×10^3 tons sulfur are needed to react with 487 tons of carbon.

(d) 199 lb

We found in part (b) that 199 g S are needed to make 236 g CS_2. Therefore, 199 lb S are needed to make 236 lb CS_2. Alternatively:

$$(236\ \cancel{lb\ CS_2})\left(\frac{453.6\ \cancel{g\ CS_2}}{1\ \cancel{lb\ CS_2}}\right)\left(\frac{1\ \cancel{mol\ CS_2}}{76.14\ \cancel{g\ CS_2}}\right)\left(\frac{2\ \cancel{mol\ S}}{1\ \cancel{mol\ CS_2}}\right)\left(\frac{32.07\ \cancel{g\ S}}{1\ \cancel{mol\ S}}\right)\left(\frac{1\ lb\ S}{453.6\ \cancel{g\ S}}\right) = 199\ lb\ S$$

3.6

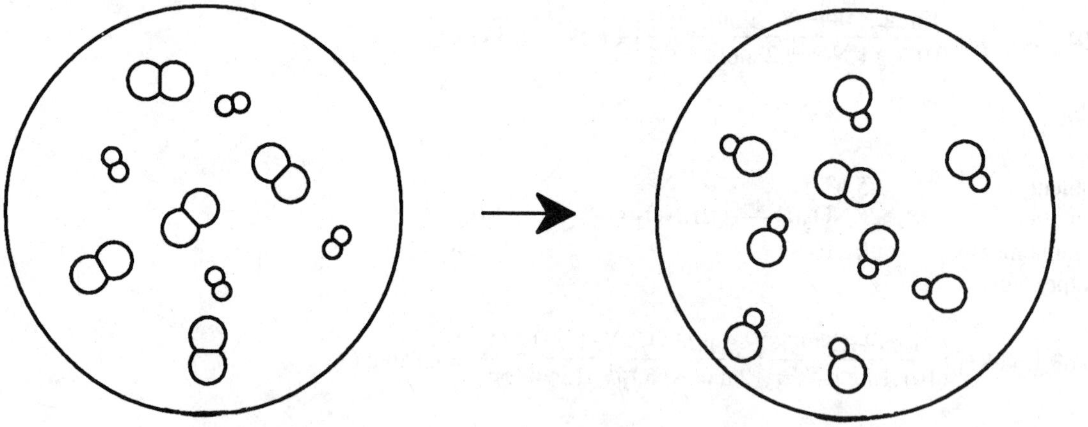

The reaction between hydrogen and chlorine to form hydrogen chloride has the equation $H_2\ (g) + Cl_2\ (g) \rightarrow 2HCl\ (g)$. There are four H_2 molecules and five Cl_2 molecules. Since the reactants combine in a 1:1 ratio, there will be one Cl_2 molecule left over unreacted.

3.7 **(a)** H_2SO_4

Problem	22.7 g		54.8 g		? g
Equation	$2NH_3(g)$	+	$H_2SO_4(aq)$	→	$(NH_4)_2SO_4(aq)$
Formula mass, u	17.03		98.08		132.1
Recipe, mol	2		1		1

Step 1: Determine the moles of $NH_3(g)$ available:

$$(22.7\ \cancel{g\ NH_3})\left(\frac{1\ mol\ NH_3}{17.03\ \cancel{g\ NH_3}}\right) = 1.33\ mol\ NH_3$$

Step 2: Determine the moles of $H_2SO_4(aq)$ available:

$$(54.8\ \cancel{g\ H_2SO_4})\left(\frac{1\ mol\ H_2SO_4}{98.08\ \cancel{g\ H_2SO_4}}\right) = 0.559\ mol\ H_2SO_4$$

Step 3: Determine which reactant is limiting:

$$\frac{1.33\ \cancel{mol\ NH_3}}{2\ \cancel{mol\ NH_3}} = 0.665\ \text{of a recipe} \qquad \frac{0.559\ \cancel{mol\ H_2SO_4}}{1\ \cancel{mol\ H_2SO_4}} = 0.559\ \text{of a recipe}$$

Therefore, $NH_3(g)$ is in excess (0.665 > 0.559), and H_2SO_4 is the limiting reactant.

(b) 73.8 g

$$(54.8\ \cancel{g\ H_2SO_4})\left(\frac{1\ \cancel{mol\ H_2SO_4}}{98.08\ \cancel{g\ H_2SO_4}}\right)\left(\frac{1\ \cancel{mol\ (NH_4)_2SO_4}}{1\ \cancel{mol\ H_2SO_4}}\right)\left(\frac{132.1\ g\ (NH_4)_2SO_4}{1\ \cancel{mol\ (NH_4)_2SO_4}}\right) = 73.8\ g\ (NH_4)_2SO_4$$

(c) 3.7 g $NH_3(g)$

$$(54.8\ \cancel{g\ H_2SO_4})\left(\frac{1\ \cancel{mol\ H_2SO_4}}{98.08\ \cancel{g\ H_2SO_4}}\right)\left(\frac{2\ \cancel{mol\ NH_3}}{1\ \cancel{mol\ H_2SO_4}}\right)\left(\frac{17.03\ g\ NH_3}{1\ \cancel{mol\ NH_3}}\right) = 19.0\ g\ NH_3$$

22.7 g NH_3 available – 19.0 g NH_3 reacted = 3.7 g NH_3 left over.

3.8 **(a)** 9.59 g

Problem	5.67 g	excess		? g
Equation	$C_2H_2(g)$ +	$H_2O(l)$	$\xrightarrow{HgSO_4,\ H_2SO_4}$	$C_2H_4O\ (g)$
Formula mass, u	26.04			44.05
Recipe, mol	1			1

$$(5.67\ \cancel{g\ C_2H_2})\left(\frac{1\ \cancel{mol\ C_2H_2}}{26.04\ \cancel{g\ C_2H_2}}\right)\left(\frac{1\ \cancel{mol\ C_2H_4O}}{1\ \cancel{mol\ C_2H_2}}\right)\left(\frac{44.05\ g\ C_2H_4O}{1\ \cancel{mol\ C_2H_4O}}\right) = 9.59\ g\ C_2H_4O$$

(b) 87.4%

$$\% \text{ yield} = \left(\frac{\text{Actual Yield}}{\text{Theoretical Yield}}\right)(100) = \left(\frac{8.38 \cancel{\text{g } C_2H_4O}}{9.59 \cancel{\text{g } C_2H_4O}}\right)(100) = 87.4\%$$

3.9 **(a)** 92.61%

$$\left(\frac{0.7732 \text{ g Hg}}{0.8349 \text{ g oxide}}\right)(100) = 92.61\%$$

(b) 7.39%

0.8349 g oxide – 0.7732 g Hg = 0.0617 g oxygen

$$\left(\frac{0.0617 \text{ g O}}{0.8349 \text{ g oxide}}\right)(100) = 7.39\%$$

To check: % Hg + % O must = 100%
92.61% + 7.39% = 100%

3.10 **(a)** 64.80%

$$(0.044\ 38 \cancel{\text{g } CO_2})\left(\frac{12.011 \text{ g C}}{44.010 \cancel{\text{g } CO_2}}\right) = 0.012\ 11 \text{ g C}$$

$$\left(\frac{0.012\ 11 \text{ g C}}{0.018\ 69 \text{ g sample}}\right)(100) = 64.79\% \text{ C}$$

(Best answer, without intermediate round off, is 64.80%)

(b) 13.60%

$$(0.022\ 72 \cancel{\text{g } H_2O})\left(\frac{2.0159 \text{ g H}}{18.015 \cancel{\text{g } H_2O}}\right) = 2.542 \times 10^{-3} \text{ g H}$$

$$\left(\frac{2.542 \times 10^{-3} \text{ g H}}{0.018\ 69 \text{ g sample}}\right)(100) = 13.60\% \text{ H}$$

(c) 21.60%

100.00% – (64.80% + 13.60%) = 21.60% O

3.11 C_2H_5

The empirical formula shows the lowest whole number ratio of atoms. This is obtained for C_4H_{10} by dividing four and ten by their lowest common denominator, two. This gives C_2H_5.

3.12 CH_2

The lowest common denominator of three and six is three. Dividing both subscripts in C_3H_6 by three gives CH_2.

3.13 KH_2PO_4

Assume 100.00–g sample.

$$(28.73\ \cancel{g\ K})\left(\frac{1\ mol\ K}{39.098\ \cancel{g\ K}}\right) = 0.7348\ mol\ K$$

$$(1.48\ \cancel{g\ H})\left(\frac{1\ mol\ H}{1.008\ \cancel{g\ H}}\right) = 1.47\ mol\ H$$

$$(22.76\ \cancel{g\ P})\left(\frac{1\ mol\ P}{30.974\ \cancel{g\ P}}\right) = 0.7348\ mol\ P$$

$$(47.03\ \cancel{g\ O})\left(\frac{1\ mol\ O}{15.999\ \cancel{g\ O}}\right) = 2.940\ mol\ O$$

$K_{0.7348}H_{1.47}P_{0.7348}O_{2.940}$

Divide each subscript by the smallest one (0.7348):

$\frac{0.7348}{0.7348} = 1.000$; $\frac{1.47}{0.7348} = 2.00$; $\frac{0.7348}{0.7348} = 1.000$; $\frac{2.940}{0.7348} = 4.001$ (4.000 without intermediate round off)

KH_2PO_4 is the empirical formula.

3.14 $Na_2Cr_2O_7$

Assume 100.00–g sample.

$$(17.55\ \cancel{g\ Na})\left(\frac{1\ mol\ Na}{22.990\ \cancel{g\ Na}}\right) = 0.7634\ mol\ Na$$

$$(39.70\ \cancel{g\ Cr})\left(\frac{1\ mol\ Cr}{51.996\ \cancel{g\ Cr}}\right) = 0.7635\ mol\ Cr$$

$$(42.75\ \cancel{g\ O})\left(\frac{1\ mol\ O}{15.999\ \cancel{g\ O}}\right) = 2.672\ mol\ O$$

$\frac{0.7634}{0.7634} = 1.000$; $\frac{0.7635}{0.7634} = 1.000$; $\frac{2.672}{0.7634} = 3.500$

This gives us $NaCrO_{3.500}$. Multiply each subscript by 2 to get whole numbers. $Na_2Cr_2O_7$ is the empirical formula.

3.15 $C_3Cl_6O_3$

1 C @ 12.011 u	=	12.011 u
2 Cl @ 35.4527 u	=	70.9054 u
1 O @ 15.9994 u	=	15.9994 u
formula mass for CCl_2O	=	98.916 u

molecular mass = n(formula mass)

$$n = \frac{\text{molecular mass}}{\text{formula mass}} = \frac{297\,\cancel{u}}{98.92\,\cancel{u}} = 3.00$$

$$3.00(CCl_2O) = C_3Cl_6O_3$$

3.16

(a)

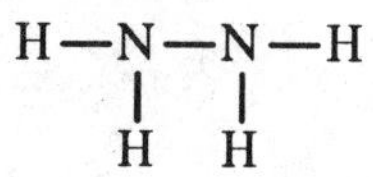

(b)

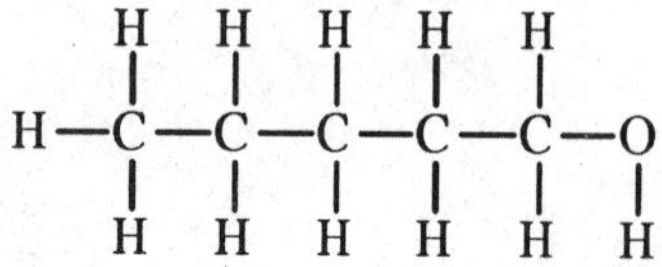

(c)

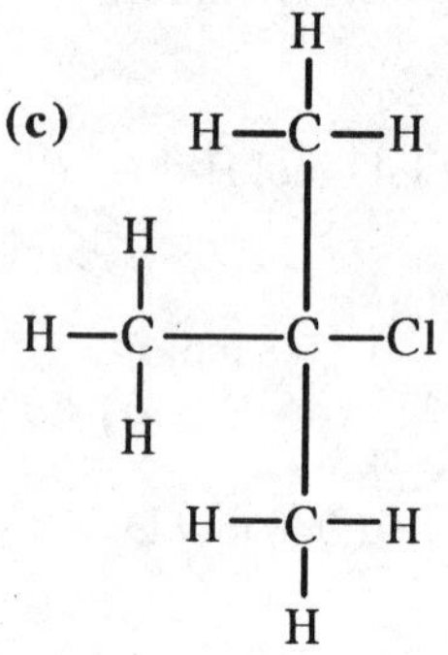

3.17 **(a)** C_6H_{14} **(b)** C_3H_7

The lowest common denominator of six and fourteen is two. Dividing both subscripts in C_6H_{14} by two gives C_3H_7.

3.18 **(a)** 12.6%

2 N	@	14.0067 u	=	28.0134 u
4 H	@	1.007 94 u	=	4.03176 u
	N_2H_4		=	32.0452 u

$$\left(\frac{4.032\,\cancel{u}}{32.05\,\cancel{u}}\right)(100) = 12.6\%\ \text{H (rounded to one decimal place)}$$

(b) 87.4%

$$\left(\frac{28.01\,\cancel{u}}{32.05\,\cancel{u}}\right)(100) = 87.4\%\ \text{N (rounded to one decimal place)}$$

(c) No. The molecular formula is a whole number multiple of the empirical formula, and thus the ratio of N to H does not change.

3.19 Cr_2O_3

Assume 100.00–g sample.

$$(68.42\,\cancel{\text{g Cr}})\left(\frac{1\ \text{mol Cr}}{51.996\,\cancel{\text{g Cr}}}\right) = 1.316\ \text{mol Cr}$$

$$(31.58\,\cancel{\text{g O}})\left(\frac{1\ \text{mol O}}{15.999\,\cancel{\text{g O}}}\right) = 1.974\ \text{mol O}$$

$$\frac{1.316}{1.316} = 1.000; \quad \frac{1.974}{1.316} = 1.500$$

$CrO_{1.500}$ – multiply subscripts by 2 to get whole numbers.

$2(CrO_{1.500}) = Cr_2O_3$

Additional Practice Problems

3.53 **(a)** $3A + B \rightarrow 2C$

There are 8A's and 2B's on the reactant side and 4C's and 2A's on the product side:

$$8A + 2B \rightarrow 4C + 2A$$

Since 2A's did not change, i.e., did not react, they are subtracted from both sides:

$$6A + 2B \rightarrow 4C$$

Chemical equations are written with the smallest whole number set of coefficients; therefore, each side is divided by two to yield the final balanced equation:

$$3A + B \rightarrow 2C$$

(b) B is limiting.

The limiting reactant controls or limits the yield of product. In this example, all of reactant B was consumed, but not all of A. A was present in excess, and B limited the yield of C.

3.55 CCl_4

The empirical formula is simply the lowest whole number molar ratio of the elements in a compound. To find this ratio, divide both molar amounts by the lower molar amount:

$$\frac{0.27\ \cancel{mol}}{0.27\ \cancel{mol}} = 1.0 \qquad \frac{1.08\ \cancel{mol}}{0.27\ \cancel{mol}} = 4.0$$

The empirical formula is CCl_4.

3.57 60.12% H_2O

3 Mg	@	24.31 u	=	72.93 u
2 P	@	30.97 u	=	61.94 u
8 O	@	15.999 u	=	127.99 u
22 H_2O	@	18.015 u	=	396.33 u
$1Mg_3(PO_4)_2 \cdot 22H_2O$			=	659.19 u

$$\frac{396.33\ \cancel{u}}{659.19\ \cancel{u}} \times 100 = 60.12\%\ H_2O \text{ to four significant figures}$$

3.59 **(a)** 0.759 mol HCl

Problem	24.8 g		? mol			? mol
Equation	Zn(s)	+	2HCl(aq)	→	$ZnCl_2$(aq) +	H_2(g)
Formula mass, u	65.39				136.29	
Recipe, mol	1		2		1	1

$$(24.8\ \cancel{\text{g Zn}})\left(\frac{1\ \cancel{\text{mol Zn}}}{65.39\ \cancel{\text{g Zn}}}\right)\left(\frac{2\ \text{mol HCl}}{1\ \cancel{\text{mol Zn}}}\right) = 0.759\ \text{mol HCl}$$

(b) 0.379 mol H_2

$$(24.8\ \cancel{\text{g Zn}})\left(\frac{1\ \cancel{\text{mol Zn}}}{65.39\ \cancel{\text{g Zn}}}\right)\left(\frac{1\ \text{mol H}_2}{1\ \cancel{\text{mol Zn}}}\right) = 0.379\ \text{mol H}_2$$

(c) 2.2×10^2 g Zn

$$(3.4\ \cancel{\text{mol H}_2})\left(\frac{1\ \cancel{\text{mol Zn}}}{1\ \cancel{\text{mol H}_2}}\right)\left(\frac{65.4\ \text{g Zn}}{\cancel{\text{mol Zn}}}\right) = 2.2 \times 10^2\ \text{g Zn}$$

(d) 4.6×10^2 g $ZnCl_2$

$$(3.4\ \cancel{\text{mol H}_2})\left(\frac{1\ \cancel{\text{mol ZnCl}_2}}{1\ \cancel{\text{mol H}_2}}\right)\left(\frac{136\ \text{g ZnCl}_2}{\cancel{\text{mol ZnCl}_2}}\right) = 4.6 \times 10^2\ \text{g ZnCl}_2$$

3.61 The determination of the limiting reagent depends on the molar ratio of reactants in the balanced chemical equation as well as the actual number of moles of each reactant.

3.63 14 g

The mass of the metal oxide, 3.48 g, is equal to the sum of the masses of the metal, 3.24 g, and oxygen. The mass of oxygen is, therefore,

$$3.48\ \text{g metal oxide} - 3.24\ \text{g metal} = 0.24\ \text{g oxygen}$$

This shows the mass relationship between metal and oxygen as 3.24 g metal ≈ 0.24 g oxygen, from which two conversion factors can be written:

$$\frac{3.24\ \text{g metal}}{0.24\ \text{g oxygen}} \text{ and } \frac{0.24\ \text{g oxygen}}{3.24\ \text{g metal}}$$

The first conversion factor can be used to convert 1.00 g oxygen to grams of metal:

$$(1.00\ \cancel{\text{g O}})\left(\frac{3.24\ \text{g metal}}{0.24\ \cancel{\text{g O}}}\right) = 14\ \text{g metal}$$

3.65 86.0 tons

Problem			? ton		153 tons		
Equation	CaO(s)	+	3C(s) →		CaC_2(l)	+	CO(g)
Formula mass, u			12.01		64.09		
Recipe, mol			3		1		

Since formula masses are relative, the problem can be solved in grams. The ratio in tons will be the same.

$$(153 \cancel{\text{g CaC}_2})\left(\frac{1 \cancel{\text{mol CaC}_2}}{64.09 \cancel{\text{g CaC}_2}}\right)\left(\frac{3 \cancel{\text{mol C}}}{1 \cancel{\text{mol CaC}_2}}\right)\left(\frac{12.01 \text{ g C}}{1 \cancel{\text{mol C}}}\right) = 86.0 \text{ g C}$$

If 86.0 g C are needed to prepare 153 g CaC_2, then 86.0 tons C are needed to prepare 153 tons CaC_2.

3.67 **(a)** 81%

$$(90)\left(\frac{90}{100}\right) = 81\%$$

(b) 73%

$$(90)\left(\frac{81}{100}\right) = 73\%$$

(c) The overall yield decreases as the number of steps in a synthesis increases.

(d) In order to maximize the overall yield, keep the number of steps as small as possible.

3.69 **(a)** $(CH_3)_3CH$ or $(CH_3)CH(CH_3)_2$

(b) C_4H_{10}

(c) C_2H_5

Stop and Test Yourself

1. **(e)** 6

$C_4H_{10}O(l) + O_2(g) \rightarrow CO_2(g) + H_2O(l)$

$C_4H_{10}O(l) + O_2(g) \rightarrow 4CO_2(g) + 5H_2O(l)$ (balance C and H)

$C_4H_{10}O(l) + 6O_2(g) \rightarrow 4CO_2(g) + 5H_2O(l)$ (balance O)

2. **(e)** 6

$Na_3PO_4(aq) + BaCl_2(aq) \rightarrow Ba_3(PO_4)_2(s) + NaCl(aq)$

$2Na_3PO_4(aq) + BaCl_2(aq) \rightarrow Ba_3(PO_4)_2(s) + 6NaCl$ (balance Na, P, and O)

$2Na_3PO_4(aq) + 3BaCl_2(aq) \rightarrow Ba_3(PO_4)_2(s) + 6NaCl(aq)$ (balance Ba and Cl)

3. **(e)** $3A + B \rightarrow 2C$

The diagram shows $6A + 4B \rightarrow 4C + 2B$. Before determining the correct equation for the problem, you must eliminate the molecules of B that did not change, and therefore did not react. This leaves the following change: $6A + 2B \rightarrow 4C$. Thus answers (a), (b), and (c) may be eliminated as they do not have the correct mole ratio of reactants. While answer (d) matches the diagram, it does not use the smallest whole number coefficients. Therefore, it is not the best equation. Answer (e) uses both the correct ratios and the smallest whole number set of coefficients, making it the best equation.

4. **(c)** 20.4

Problem	25.0 g		? g		? g				
Equation	$MnO_2(s)$	+	$4HCl(aq)$	$\rightarrow$	$Cl_2(g)$	+	$MnCl_2(aq)$	+	$2H_2O(l)$
Formula mass, u	86.9		36.5		70.9				
Recipe, mol	1		4		1				

$$(25.0\ \cancel{g\ MnO_2})\left(\frac{1\ \cancel{mol\ MnO_2}}{86.9\ \cancel{g\ MnO_2}}\right)\left(\frac{1\ \cancel{mol\ Cl_2}}{1\ \cancel{mol\ MnO_2}}\right)\left(\frac{70.9\ g\ Cl_2}{\cancel{mol\ Cl_2}}\right) = 20.4\ g\ Cl_2$$

5. **(a)** 1.15

(Refer to problem set–up in problem 4 above.)

$$(25.0\ \cancel{g\ MnO_2})\left(\frac{1\ \cancel{mol\ MnO_2}}{86.9\ \cancel{g\ MnO_2}}\right)\left(\frac{4\ mol\ HCl}{1\ \cancel{mol\ MnO_2}}\right) = 1.15\ mol\ HCl$$

6. **(e)** 0.50

(Refer to problem set–up in problem 4 above.)

First determine the limiting reagent.

$$\frac{0.75\ \cancel{mol\ MnO_2}}{1\ \cancel{mol\ MnO_2}} = 0.75 \text{ of a recipe} \qquad \frac{2.0\ \cancel{mol\ HCl}}{4\ \cancel{mol\ HCl}} = 0.50 \text{ of a recipe}$$

Since HCl has the smaller fraction of the reaction recipe, it is the limiting reactant and will determine the amount of products formed.

$$(2.0\ \cancel{mol\ HCl})\left(\frac{1\ mol\ Cl_2}{4\ \cancel{mol\ HCl}}\right) = 0.50\ mol\ Cl_2$$

7. **(d)** 0.25 mol MnO_2

(Refer to problem set–up in problem 4 above.)

Since HCl is the limiting reactant, some MnO_2 is left over. To determine how much, calculate how much reacts with 2.0 mol HCl and subtract that amount from the amount available, 0.75 mol:

$$(2.0\ \cancel{\text{mol HCl}})\left(\frac{1\ \text{mol } MnO_2}{4\ \cancel{\text{mol HCl}}}\right) = 0.50\ \text{mol } MnO_2$$

0.75 mol MnO_2 available – 0.50 mol MnO_2 reacted = 0.25 mol MnO_2 left over

8. **(a)** 0.75

First write the balanced equation:

$$2Na(s) + Cl_2(g) \rightarrow 2NaCl(s)$$

Then set up the stoichiometry problem:

Problem		? mol		1.5 mol
Equation	2Na(s) +	$Cl_2(g)$	→	2NaCl(s)
Formula mass, u	(not needed)			
Recipe, mol	2	1		2

$$(1.5\ \cancel{\text{mol NaCl}})\left(\frac{1\ \text{mol } Cl_2}{2\ \cancel{\text{mol NaCl}}}\right) = 0.75\ \text{mol } Cl_2$$

9. **(d)** 85.4

To calculate the percent yield, determine the theoretical yield from the balanced chemical equation, divide that into the actual yield, and multiply by 100.

Problem	762 g		? g (theoretical) 2048 g (actual)	
Equation	$CH_4(g) + Cl_2(g)$	$\xrightarrow{\text{light}}$	$CH_3Cl(g)$ +	HCl(g)
Formula mass, u	16.04		50.49	
Recipe, mol	1		1	

$$(762\ \cancel{\text{g } CH_4})\left(\frac{1\ \cancel{\text{mol } CH_4}}{16.04\ \cancel{\text{g } CH_4}}\right)\left(\frac{1\ \cancel{\text{mol } CH_3Cl}}{1\ \cancel{\text{mol } CH_4}}\right)\left(\frac{50.49\ \text{g } CH_3Cl}{\cancel{\text{mol } CH_3Cl}}\right) = 2.40 \times 10^3\ \text{g } CH_3Cl$$

Theoretical yield = 2.40×10^3 g

Actual yield = 2048 g

$$\%\ \text{yield} = \frac{\text{Actual yield}}{\text{Theoretical yield}} \times 100 = \frac{2048\ \cancel{\text{g}}}{2.40 \times 10^3\ \cancel{\text{g}}} \times 100 = 85.3\%$$

(Best answer, without intermediate round off, is 85.4%.)

10. (c) 40.00

Determine the formula mass of $C_6H_{12}O_6$.

6 C	@	12.011 u	=	72.066 u
12 H	@	1.0079 u	=	12.095 u
6 O	@	15.999 u	=	95.994 u
		$1C_6H_{12}O_6$	=	180.155 u

$$\frac{72.066 \text{ u C}}{180.16 \text{ u } C_6H_{12}O_6} \times 100 = 40.00\%$$

11. (c) 12.84

1 Ca	@	40.08 u	=	40.08 u
2 N	@	14.01 u	=	28.02 u
9 O	@	15.999 u	=	143.99 u
6 H	@	1.01 u	=	6.06 u
$1Ca(NO_3)_2 \cdot 3H_2O$			=	218.15 u

$$\frac{28.02 \text{ u N}}{218.15 \text{ u } Ca(NO_3)_2 \cdot 3H_2O} \times 100 = 12.84\%$$

12. (d) $C_{13}H_6Cl_6O_2$

When these three percents are added, they do not add up to 100%. This indicates that another element is present. Assume that this element is oxygen. Determine the percent oxygen by difference:

100.00% – (38.37% + 1.49% + 52.28%) = 100% – 92.14% = 7.86% O

Assume 100.00 g of sample. The compound then contains 38.37 g C, 1.49 g H, 52.28 g Cl, and 7.86 g O. From these masses, determine the moles of each element present:

$$(38.37 \cancel{\text{g C}})\left(\frac{1 \text{ mol C}}{12.011 \cancel{\text{g C}}}\right) = 3.195 \text{ mol C}$$

$$(1.49 \cancel{\text{g H}})\left(\frac{1 \text{ mol H}}{1.008 \cancel{\text{g H}}}\right) = 1.48 \text{ mol H}$$

$$(52.28 \cancel{\text{g Cl}})\left(\frac{1 \text{ mol Cl}}{35.453 \cancel{\text{g Cl}}}\right) = 1.475 \text{ mol Cl}$$

$$(7.86 \cancel{\text{g O}})\left(\frac{1 \text{ mol O}}{16.00 \cancel{\text{g O}}}\right) = 0.491 \text{ mol O}$$

The empirical formula is the simplest whole number molar ratio of the elements. To calculate the subscripts for the formula, divide each number of moles by the smallest, 0.491:

$\frac{3.195 \text{ mol C}}{0.491} = 6.51 \text{ mol C}$ $\frac{1.48 \text{ mol H}}{0.491} = 3.01 \text{ mol H}$

$\frac{1.475 \text{ mol Cl}}{0.491} = 3.00 \text{ mol Cl}$ $\frac{0.491 \text{ mol O}}{0.491} = 1.00 \text{ mol O}$

This gives the following formula: $C_{6.51}H_{3.01}Cl_{3.00}O_{1.00}$

Since 6.51 is not close to a whole number, multiply all of the subscripts by 2 to make it a whole number.

$2(C_{6.51}H_{3.01}Cl_{3.00}O_{1.00}) = C_{13.0}H_{6.02}Cl_{6.00}O_{2.00}$ or $C_{13}H_6Cl_6O_2$

13. **(d)** $C_4H_2Cl_4$

The molecular formula is a whole number multiple of the empirical formula. Therefore, (a), (b), (c), and (e) are all eliminated. The possible molecular formula is (d) $C_4H_2Cl_4$, where the whole number multiplier is 2.

$2(C_2HCl_2) = C_4H_2Cl_4$

14. **(e)** C_5H_{10}

Since the molecular formula is a whole number multiple of the empirical formula, if the molecular mass is known and the formula mass is known, the whole number multiplier can be determined.

The formula mass for CH_2 = 12 u + 2(1.01 u) = 12 u + 2.02 u = 14 u

$n(14 \text{ u}) = 70 \text{ u}$

$n = \frac{70 \cancel{\text{u}}}{14 \cancel{\text{u}}} = 5.0$

$5.0(CH_2) = C_5H_{10}$

15. **(a)** 10.1% C, 0.847% H

All of the H in the sample is incorporated into H_2O. All of the C in the sample is incorporated into CO_2.

Formula mass for H_2O = 18.0153 u
Formula mass for CO_2 = 44.010 u

From these formula masses and the atomic masses of hydrogen and carbon, conversion factors can be written that can be used to determine the mass of hydrogen in the water obtained and the mass of carbon in the CO_2 obtained:

$(0.001\,16 \cancel{\text{g H}_2\text{O}})\left(\frac{2.016 \text{ g H}}{18.02 \cancel{\text{g H}_2\text{O}}}\right) = 1.30 \times 10^{-4} \text{ g H}$

$(0.005\,65 \cancel{\text{g CO}_2})\left(\frac{12.01 \text{ g C}}{44.01 \cancel{\text{g CO}_2}}\right) = 1.54 \times 10^{-3} \text{ g C}$

$$\frac{1.30 \times 10^{-4}\text{ g H}}{0.015\ 32\text{ g sample}} \times 100 = 0.849\%\text{ H}$$

$$\frac{1.54 \times 10^{-3}\text{ g C}}{0.015\ 32\text{ g sample}} \times 100 = 10.1\%\text{ C}$$

(Best answers, without intermediate round off, are 0.847% H and 10.1% C.)

Putting Things Together

3.71 **(a)** magnesium

Problem	36.5 g		27.0 g		? g
Equation	2Mg(s)	+	O_2(g)	→	2MgO(s)
Formula mass, u	24.31		32.00		40.30
Recipe, mol	2		1		2

$$(36.5\ \cancel{\text{g Mg}})\left(\frac{1\text{ mol Mg}}{24.31\ \cancel{\text{g Mg}}}\right) = 1.50\text{ mol Mg}$$

$$(27.0\ \cancel{\text{g O}_2})\left(\frac{1\text{ mol O}_2}{32.00\ \cancel{\text{g O}_2}}\right) = 0.844\text{ mol O}_2$$

$$\frac{1.50\ \cancel{\text{mol Mg}}}{2\ \cancel{\text{mol Mg}}} = 0.750\text{ of a recipe}$$

$$\frac{0.844\ \cancel{\text{mol O}_2}}{1\ \cancel{\text{mol O}_2}} = 0.844\text{ of a recipe}$$

Since Mg has the smaller fraction of the recipe, it is the limiting reactant.

(b) 3.0 g O_2

Oxygen is in excess. To calculate grams oxygen left over, you must first calculate grams oxygen consumed by reaction with 36.5 g Mg.

$$(36.5\ \cancel{\text{g Mg}})\left(\frac{1\ \cancel{\text{mol Mg}}}{24.31\ \cancel{\text{g Mg}}}\right)\left(\frac{1\ \cancel{\text{mol O}_2}}{2\ \cancel{\text{mol Mg}}}\right)\left(\frac{32.00\text{ g O}_2}{\cancel{\text{mol O}_2}}\right) = 24.0\ \text{ g O}_2$$

Subtract the grams O_2 consumed from the grams O_2 available to obtain the grams O_2 left over: 27.0 - 24.0 = 3.0

(c) 60.5 g

$$(36.5\ \cancel{\text{g Mg}})\left(\frac{1\ \cancel{\text{mol Mg}}}{24.31\ \cancel{\text{g Mg}}}\right)\left(\frac{2\ \cancel{\text{mol MgO}}}{2\ \cancel{\text{mol Mg}}}\right)\left(\frac{40.30\text{ g MgO}}{\cancel{\text{mol MgO}}}\right) = 60.5\ \text{ g MgO}$$

(d) 97.7%

$$\%\ \text{yield} = \frac{\text{Actual yield}}{\text{Theoretical yield}} \times 100 = \frac{59.1\ \cancel{\text{g MgO}}}{60.5\ \cancel{\text{g MgO}}} \times 100 = 97.7\%$$

3.73 $n = 2$

Write the equation for the dehydration of $CaCl_2 \cdot nH_2O$:

$$CaCl_2 \cdot nH_2O(s) \xrightarrow{\text{heat}} CaCl_2(s) + nH_2O(l)$$

The mass of the hydrate, 6.38 g, is equal to the sum of the masses of $CaCl_2$ and H_2O, since no mass can be lost or gained during the reaction. This allows the calculation of the mass of water given off.

mass $CaCl_2 \cdot nH_2O$ = mass $CaCl_2$ + mass H_2O

mass H_2O = mass $CaCl_2 \cdot nH_2O$ – mass $CaCl_2$ = 6.38 g – 4.82 g = 1.56 g

Using the formula masses of $CaCl_2$ and water, calculate the moles of each present after dehydration:

$$(4.82\ \cancel{\text{g } CaCl_2})\left(\frac{1\text{ mol } CaCl_2}{111.0\ \cancel{\text{g } CaCl_2}}\right) = 0.0434\text{ mol } CaCl_2$$

$$(1.56\ \cancel{\text{g } H_2O})\left(\frac{1\text{ mol } H_2O}{18.02\ \cancel{\text{g } H_2O}}\right) = 0.866\text{ mol } H_2O$$

Determine the relative molar ratio of $CaCl_2$ to water:

$$\frac{0.0434}{0.0434} = 1.00 \qquad \frac{0.0866}{0.0434} = 2.00$$

Thus, the ratio is 1 : 2 and the formula is $CaCl_2 \cdot 2H_2O$

3.75 97.2% Fe

Problem	? g		excess				9.54×10^{-3} mol
Equation	Fe(s)	+	2HCl(aq)	→	$FeCl_2$(aq)	+	H_2(g)
Formula mass, u	55.85						
Recipe, mol	1						1

First determine the mass of Fe actually consumed in the reaction:

$$(9.54 \times 10^{-3}\ \cancel{\text{mol } H_2})\left(\frac{1\ \cancel{\text{mol Fe}}}{1\ \cancel{\text{mol } H_2}}\right)\left(\frac{55.85\text{ g Fe}}{\cancel{\text{mol Fe}}}\right) = 0.533\text{ g Fe}$$

The percent iron in the original sample is the actual mass of iron consumed divided by the mass of the sample times 100:

$$\%\ \text{iron} = \frac{0.533\text{ g Fe}}{0.5483\text{ g sample}} \times 100 = 97.2\%$$

3.77 4

Assume that 100.00 g of sample are present. Then 28.96 g Cl are present from which the moles of Cl can be determined.

$$(28.96\ \cancel{g\ Cl})\left(\frac{1\ mol\ Cl}{35.453\ \cancel{g\ Cl}}\right) = 0.8169\ mol\ Cl$$

The formula, $NaClO_n$, shows that the moles of Na are equivalent to the moles of Cl.

$$(0.8169\ \cancel{mol\ Cl})\left(\frac{1\ mol\ Na}{1\ \cancel{mol\ Cl}}\right) = 0.8169\ mol\ Na$$

From moles Na, calculate grams of Na:

$$(0.8169\ \cancel{mol\ Na})\left(\frac{22.990\ g\ Na}{\cancel{mol\ Na}}\right) = 18.78\ g\ Na$$

In a 100-g sample, 18.78% is Na.

Next find the % oxygen from the percentages of Cl and Na:

%O = 100.00% sample – (28.96% Cl + 18.78% Na) = 52.26%

A 100-g sample contains 52.26 grams of O, from which the number of moles can be determined:

$$(52.26\ \cancel{g\ O})\left(\frac{1\ mol\ O}{15.999\ \cancel{g\ O}}\right) = 3.266\ mol\ O$$

From the moles of each element, we can calculate the empirical formula:

$Na_{0.8169}Cl_{0.8169}O_{3.266}$ Divide each subscript by 0.8169:

$Na_{0.8169/0.8169}Cl_{0.8169/0.8169}O_{3.266/0.8169} = Na_{1.000}Cl_{1.000}O_{3.998}$

To the nearest whole numbers, the empirical formula is $NaClO_4$. Thus, $n = 4$.

3.79 **(a)** C_5H_{12}

$$(0.049\ 77\ \cancel{g\ CO_2})\left(\frac{12.011\ g\ C}{44.010\ \cancel{g\ CO_2}}\right) = 0.013\ 58\ g\ C$$

$$(0.024\ 44\ \cancel{g\ H_2O})\left(\frac{2.0159\ g\ H}{18.015\ \cancel{g\ H_2O}}\right) = 0.002\ 735\ g\ H$$

$$(0.013\ 58\ \cancel{g\ C})\left(\frac{1\ mol\ C}{12.011\ \cancel{g\ C}}\right) = 0.001\ 131\ mol\ C$$

$$(0.002\ 735\ \cancel{g\ H})\left(\frac{1\ mol\ H}{1.0079\ \cancel{g\ H}}\right) = 0.002\ 714\ mol\ H$$

$C_{0.001131/0.001131}H_{0.002714/0.001131} = CH_{2.400}$

To get whole numbers, multiply the formula by 5:

$5(CH_{2.400}) = C_5H_{12}$

(b) C_5H_{12}

$$n = \frac{72\ \cancel{u}}{72.150\ \cancel{u}} = 1.0$$

$1.0(C_5H_{12}) = C_5H_{12}$

(c) 0.016 32 g

0.013 58 g C + 0.002 735 g H = 0.016 32 g sample

3.81 **(a)** 0.2332 g oxygen, 3.973 g Cu/g O

The mass of oxygen reacted is determined by subtracting the mass of copper from the mass of the oxide:

$$\begin{array}{r} 1.1596 \text{ g oxide} \\ \underline{-\,0.9264 \text{ g copper}} \\ 0.2332 \text{ g oxygen} \end{array}$$

$$\frac{0.9264 \text{ g copper}}{0.2332 \text{ g oxygen}} = 3.973 \text{ g Cu/g O}$$

(b) 0.2203 g oxygen, 7.943 g Cu/g O

$$\begin{array}{r} 1.9701 \text{ g oxide} \\ \underline{-\,1.7498 \text{ g copper}} \\ 0.2203 \text{ g oxygen} \end{array}$$

$$\frac{1.7498 \text{ g copper}}{0.2203 \text{ g oxygen}} = 7.943 \text{ g Cu/g O}$$

(c) Yes.

$$\frac{7.943\ \cancel{\text{g Cu/g O}}}{3.973\ \cancel{\text{g Cu/g O}}} = 1.999 \text{ (2.00 to 3 significant figures)}$$

(d) CuO, copper(II) oxide and Cu_2O, copper(I) oxide

$$\%\text{ Cu in oxide (a)} = \frac{0.9264 \text{ g Cu}}{1.1596 \text{ g oxide}} \times 100 = 79.89\%\text{ Cu}$$

$$\%\text{ O in oxide (a)} = \frac{0.2332 \text{ g O}}{1.1596 \text{ g oxide}} \times 100 = 20.11\%\text{ O}$$

Assume 100.00 g of oxide. Then the oxide contains 79.89 g Cu and 20.11 g O.

$$(79.89 \text{ g Cu})\left(\frac{1 \text{ mol Cu}}{63.546 \text{ g Cu}}\right) = 1.257 \text{ mol Cu}$$

$$(20.11 \text{ g O})\left(\frac{1 \text{ mol O}}{15.999 \text{ g O}}\right) = 1.257 \text{ mol O}$$

The empirical formula is CuO and the name is copper(II) oxide. Since the g Cu/g O ratio of oxide (a) to oxide (b) is 1.999 (2.00), oxide (a) has twice as much oxygen as oxide (b). Therefore, the empirical formula for oxide (b) is $CuO_{1/2}$ or Cu_2O, and the name is copper(I) oxide.

3.83 **(a)** 104 g

Determine the formula for molecular phosphorus by dividing the molecular mass by the atomic mass of phosphorus:

$$n(\text{atomic mass P}) = \text{molecular mass P}$$

$$n = \frac{\text{molecular mass P}}{\text{atomic mass P}} = \frac{124 \text{ u}}{31 \text{ u}} = 4$$

The molecular formula for phosphorus is P_4. Now set up the problem table:

Problem	87.8 g		45.3 g		? g
Equation	$P_4(s)$	+	$3O_2(g)$	→	$P_4O_6(s)$
Formula mass, u	123.9		32.00		219.9
Recipe, mol	1		3		1

$$(87.8 \text{ g } P_4)\left(\frac{1 \text{ mol } P_4}{123.9 \text{ g } P_4}\right) = 0.709 \text{ mol } P_4$$

$$(45.3 \text{ g } O_2)\left(\frac{1 \text{ mol } O_2}{32.00 \text{ g } O_2}\right) = 1.42 \text{ mol } O_2$$

$$\frac{0.709 \text{ mol } P_4}{1 \text{ mol}} = 0.709 \text{ of a recipe}$$

$$\frac{1.42 \text{ mol } O_2}{3 \text{ mol}} = 0.473 \text{ of a recipe}$$

O_2 is the limiting reactant.

$$(45.3 \text{ g } O_2)\left(\frac{1 \text{ mol } O_2}{32.00 \text{ g } O_2}\right)\left(\frac{1 \text{ mol } P_4O_6}{3 \text{ mol } O_2}\right)\left(\frac{219.9 \text{ g } P_4O_6}{\text{mol } P_4O_6}\right) = 104 \text{ g } P_4O_6$$

(b) 48.8%

$$\%\text{yield} = \frac{\text{Actual yield}}{\text{Theoretical yield}} \times 100 = \frac{50.8 \text{ g}}{104 \text{ g}} \times 100 = 48.8\%$$

(c) 29.3 g P_4

$$(45.3\ \cancel{g\ O_2})\left(\frac{1\ \cancel{mol\ O_2}}{32.00\ \cancel{g\ O_2}}\right)\left(\frac{1\ \cancel{mol\ P_4}}{3\ \cancel{mol\ O_2}}\right)\left(\frac{123.9\ g\ P_4}{\cancel{mol\ P_4}}\right) = 58.5\ g\ P_4\ \text{reacted}$$

87.8 g P_4 available – 58.5 g P_4 reacted = 29.3 g P_4 left over

3.85 (a) $NH_4NO_3(l) \xrightarrow{heat} N_2O(g) + 2H_2O(g)$

(b) side reactions

(c) 0.021 613 mol

Problem	1.7300 g		? mol		
Equation	$NH_4NO_3(l)$	$\xrightarrow{heat}$	$N_2O(g)$	+	$2H_2O(g)$
Formula mass, u	80.0434				
Recipe, mol	1		1		2

$$(1.7300\ \cancel{g\ NH_4NO_3})\left(\frac{1\ \cancel{mol\ NH_4NO_3}}{80.0434\ \cancel{g\ NH_4NO_3}}\right)\left(\frac{1\ mol\ N_2O}{1\ \cancel{mol\ NH_4NO_3}}\right) = 0.021\ 613\ mol\ N_2O$$

(d) 84.30%

$$\%\ yield = \frac{\text{Actual yield}}{\text{Theoretical yield}} \times 100 = \frac{0.018\ 22\ \cancel{mol\ N_2O}}{0.021\ 613\ \cancel{mol\ N_2O}} \times 100 = 84.30\%$$

(e) The chloride ion increases the reaction rate of the side reaction without being used up. This reduces the yield of the main product. The catalyst is shown above the arrow in an equation.

3.87 0.6251% Ag

All of the Ag in the ore is converted to AgCl. Determine the mass of silver in the AgCl to find the mass of silver in the sample:

1 Ag @ 107.87 u	=	107.87 u
<u>1 Cl @ 35.453 u</u>	=	<u>35.453 u</u>
1 AgCl	=	143.32 u

$$(0.2940\ \cancel{g\ AgCl})\left(\frac{107.87\ g\ Ag}{143.32\ \cancel{g\ AgCl}}\right) = 0.2213\ g\ Ag$$

$$\frac{0.2213\ g\ Ag}{35.40\ g\ ore} \times 100 = 0.6251\%$$

Applications

3.89

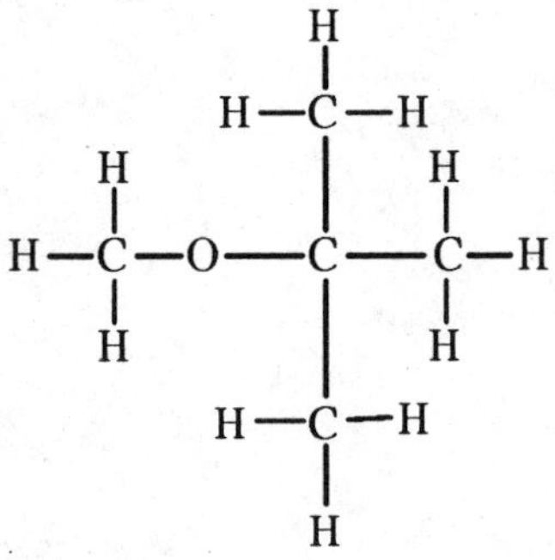

3.91 urea, CH_4N_2O

Find the % nitrogen in each compound:

NH_4NO_3	CH_4N_2O
$2(14.01 \text{ u N}) = 28.02 \text{ u N}$	$2(14.01 \text{ u N}) = 28.02 \text{ u N}$
$\dfrac{28.02 \text{ u N}}{80.04 \text{ u } NH_4NO_3} \times 100 = 35.01\% \text{ N}$	$\dfrac{28.02 \text{ u N}}{60.06 \text{ u urea}} \times 100 = 46.65\% \text{ N}$

Assume exactly 1 kg of each compound is available:

$$\left(\frac{1 \text{ kg } NH_4NO_3}{7.82 \text{ dollar}}\right)\left(\frac{35.01 \text{ kg N}}{100 \text{ kg } NH_4NO_3}\right) = \frac{0.0448 \text{ kg N}}{\text{dollar}} \qquad \left(\frac{1 \text{ kg urea}}{8.11 \text{ dollar}}\right)\left(\frac{46.65 \text{ kg N}}{100 \text{ kg urea}}\right) = \frac{0.0575 \text{ kg N}}{\text{dollar}}$$

Therefore, urea gives more nitrogen per dollar.

3.93 4 atoms Fe

Determine the mass of Fe atoms in 6.45×10^4 u of hemoglobin (one molecule):

$$(6.45 \times 10^4 \text{ u hemoglobin})\left(\frac{0.35 \text{ u Fe}}{100 \text{ u hemoglobin}}\right) = 2.3 \times 10^2 \text{ u Fe}$$

From the mass of Fe in one molecule and the atomic mass of Fe, determine the moles of iron and the atoms of iron.

$$(2.3 \times 10^2 \text{ u Fe})\left(\frac{1 \text{ mol Fe}}{55.8 \text{ u Fe}}\right) = 4.1 \text{ atoms Fe} \approx 4 \text{ atoms Fe}$$

3.95 No. The % C and % H in the sample don't match the percent composition of cocaine.

All the C in the unknown is converted to CO_2 and all the H is converted to water. Determine the % C and % H in the sample:

$$(0.048\ 04\ \cancel{g\ CO_2})\left(\frac{12.011\ g\ C}{44.010\ \cancel{g\ CO_2}}\right) = 0.013\ 11\ g\ C$$

$$(0.010\ 99\ \cancel{g\ H_2O})\left(\frac{2.0159\ g\ H}{18.015\ \cancel{g\ H_2O}}\right) = 1.230 \times 10^{-3}\ g\ H$$

$$\frac{0.013\ 11\ g\ C}{0.018\ 32\ g\ sample} \times 100 = 71.56\%\ C$$

$$\frac{1.230 \times 10^{-3}\ g\ H}{0.018\ 32\ g\ sample} \times 100 = 6.714\%\ H$$

Next calculate the % C and % H in cocaine:

17 C	@	12.011 u	=	204.19 u
21 H	@	1.0079 u	=	21.166 u
1 N	@	14.007 u	=	14.007 u
4 O	@	15.999 u	=	63.996 u
1 $C_{17}H_{21}NO_4$			=	303.36 u

$$\frac{204.19\ \cancel{u}}{303.36\ \cancel{u}} \times 100 = 67.31\%\ C$$

$$\frac{21.166\ \cancel{u}}{303.36\ \cancel{u}} \times 100 = 6.977\%\ H$$

$71.57\%\ C \neq 67.31\%\ C$
$6.714\%\ H \neq 6.977\%\ H$

Therefore, the unknown sample cannot be cocaine.

3.97 **(a)** 657 g

Write the chemical equations for each step of the process:

$$2SO_2(g) + O_2(g) \xrightarrow{V} 2SO_3(g)$$

$$SO_3(g) + H_2O(l) \rightarrow H_2SO_4(l)$$

Set up the stoichiometry problem for the first reaction and calculate the theoretical yield of SO_3:

Problem	454 g		excess		? g
Equation	$2SO_2(g)$	+	$O_2(g)$	$\xrightarrow{V}$	$2SO_3(g)$
Formula mass, u	64.07		32.00		80.06
Recipe, mol	2		1		2

$$(454\ \cancel{g\ SO_2})\left(\frac{1\ \cancel{mol\ SO_2}}{64.07\ \cancel{g\ SO_2}}\right)\left(\frac{2\ \cancel{mol\ SO_3}}{2\ \cancel{mol\ SO_2}}\right)\left(\frac{80.06\ g\ SO_3}{\cancel{mol\ SO_3}}\right) = 567\ g\ SO_3$$

Determine actual yield of SO_3 from percent yield given in the problem and theoretical yield calculated:

$$\text{percent yield} = \frac{\text{Actual yield}}{\text{Theoretical yield}} \times 100$$

$$\text{Actual yield} = \frac{(\text{Theoretical yield})(\text{percent yield})}{100} = \frac{(567\ g\ SO_3)(96.5)}{100} = 547\ g\ SO_3$$

Use the actual yield of SO_3 to set up the stoichiometry problem for the second reaction and calculate the theoretical yield of H_2SO_4:

Problem	547 g				? g
Equation	$SO_3(g)$	+	$H_2O(l)$	→	$H_2SO_4(l)$
Formula mass, u	80.06				98.08
Recipe, mol	1				1

$$(547\ \cancel{g\ SO_3})\left(\frac{1\ \cancel{mol\ SO_3}}{80.06\ \cancel{g\ SO_3}}\right)\left(\frac{1\ \cancel{mol\ H_2SO_4}}{1\ \cancel{mol\ SO_3}}\right)\left(\frac{98.08\ g\ H_2SO_4}{\cancel{mol\ H_2SO_4}}\right) = 6.70 \times 10^2\ g\ H_2SO_4$$

Determine the actual yield of H_2SO_4 from the percent yield given in the problem and the theoretical yield calculated:

$$\text{Actual yield} = \frac{(6.70 \times 10^2\ g\ H_2SO_4)(98.0)}{100} = 657\ g\ H_2SO_4$$

(b) 357 mL

$$(657\ \cancel{g\ H_2SO_4})\left(\frac{1\ mL\ H_2SO_4}{1.84\ \cancel{g\ H_2SO_4}}\right) = 357\ mL\ H_2SO_4$$

3.99 **(a)** (i) $C_{15}H_{32}(l) + 23O_2(g) \rightarrow 15CO_2(g) + 16H_2O(l)$

(ii) $2C_{15}H_{32}(l) + 31O_2(g) \rightarrow 30CO(g) + 32H_2O(l)$

(iii) $C_{15}H_{32}(l) + 8O_2(g) \rightarrow 15C(s) + 16H_2O(l)$

(b) Formation of C requires the smallest number of oxygen per molecule of $C_{15}H_{32}$. When more oxygen is present, more of the $C_{15}H_{32}$ can be converted to carbon monoxide and carbon dioxide; therefore, less smoke is formed.

3.101 **(a)** 2 tablets

1000 mg Ca - 650 mg Ca = 350 mg Ca needed

$$(350\ \cancel{mg\ Ca})\left(\frac{100\ \cancel{mg\ CaCO_3}}{40.1\ \cancel{mg\ Ca}}\right)\left(\frac{1\ \text{tablet}}{500\ \cancel{mg\ CaCO_3}}\right) = 1.75\ \text{tablets}$$

Since the tablets are not divisible, take two.

(b) Yes. Each 100 mg $CaCO_3$ contains 60 mg carbonate but only 40 mg calcium. Since the composition of a compound is always the same, you cannot obtain more calcium from calcium carbonate than carbonate. In order to have more calcium than carbonate in your diet, you will need another source of calcium that does not contain carbonate.

3.103 0.250 kg NaOH, 0.221 kg Cl_2, 0.006 30 kg H_2

Problem	0.365 kg	0.635 kg	? kg	? kg		? kg
Equation	2NaCl(aq) +	$2H_2O(l) \xrightarrow{\text{electrolysis}}$	2NaOH(aq) +	$Cl_2(g)$	+	$H_2(g)$
Formula mass, u	58.44	18.02	40.00	70.91		2.016
Recipe, mol	2	2	2	1		1

1.000 kg of salt solution contains 0.365 kg NaCl and 0.635 kg H_2O. Determine which reactant is limiting:

$$(0.365\ \cancel{\text{kg NaCl}})\left(\frac{1\ \text{kg mol NaCl}}{58.44\ \cancel{\text{kg NaCl}}}\right) = 0.006\ 25\ \text{kg mol NaCl}$$

$$(0.006\ 25\ \cancel{\text{kg mol NaCl}})\left(\frac{1\ \text{recipe NaCl}}{2\ \cancel{\text{kg mol NaCl}}}\right) = 0.003\ 13\ \text{recipe NaCl}$$

$$(0.635\ \cancel{\text{kg H}_2\text{O}})\left(\frac{1\ \text{kg mol H}_2\text{O}}{18.02\ \cancel{\text{kg H}_2\text{O}}}\right) = 0.0352\ \text{kg mol H}_2\text{O}$$

$$\frac{0.0352\ \cancel{\text{kg mol H}_2\text{O}}}{2\ \cancel{\text{kg mol H}_2\text{O}}} = 0.0176\ \text{recipe H}_2\text{O}$$

Since 0.0176 is greater than 0.003 13, NaCl is the limiting reactant.

$$(0.365\ \cancel{\text{kg NaCl}})\left(\frac{1\ \cancel{\text{kg mol NaCl}}}{58.44\ \cancel{\text{kg NaCl}}}\right)\left(\frac{2\ \cancel{\text{kg mol NaOH}}}{2\ \cancel{\text{kg mol NaCl}}}\right)\left(\frac{40.00\ \text{kg NaOH}}{\cancel{\text{kg mol NaOH}}}\right) = 0.250\ \text{kg NaOH}$$

$$(0.365\ \cancel{\text{kg NaCl}})\left(\frac{1\ \cancel{\text{kg mol NaCl}}}{58.44\ \cancel{\text{kg NaCl}}}\right)\left(\frac{1\ \cancel{\text{kg mol Cl}_2}}{2\ \cancel{\text{kg mol NaCl}}}\right)\left(\frac{70.91\ \text{kg Cl}_2}{\cancel{\text{kg mol Cl}_2}}\right) = 0.221\ \text{kg Cl}_2$$

$$(0.365\ \cancel{\text{kg NaCl}})\left(\frac{1\ \cancel{\text{kg mol NaCl}}}{58.44\ \cancel{\text{kg NaCl}}}\right)\left(\frac{1\ \cancel{\text{kg mol H}_2}}{2\ \cancel{\text{kg mol NaCl}}}\right)\left(\frac{2.016\ \text{kg H}_2}{\cancel{\text{kg mol H}_2}}\right) = 0.006\ 30\ \text{kg H}_2$$

3.105 **(a)** $N_2(g) + O_2(g) \xrightarrow{\text{heat}} 2NO(g)$

(b) $2NO(g) + O_2(g) \rightarrow 2NO_2(g)$

(c) $2CO(g) + O_2(g) \xrightarrow{\text{catalyst}} 2CO_2(g)$

(d) $2C_8H_{18}(l) + 25O_2(g) \xrightarrow{\text{catalyst}} 16CO_2(g) + 18H_2O(l)$

Step 2: $C_8H_{18} + O_2 \xrightarrow{\text{catalyst}} CO_2 + H_2O$

Steps 3 & 4: $C_8H_{18} + O_2 \xrightarrow{\text{catalyst}} 8CO_2 + 9H_2O$

$$C_8H_{18} + \frac{25}{2}O_2 \xrightarrow{\text{catalyst}} 8CO_2 + 9H_2O$$

$$2C_8H_{18} + 25O_2 \xrightarrow{\text{catalyst}} 16CO_2 + 18H_2O$$

Step 5: $2C_8H_{18}(l) + 25O_2(g) \xrightarrow{\text{catalyst}} 16CO_2(g) + 18H_2O(l)$

(e) $2SO_2(g) + O_2(g) \xrightarrow{\text{catalyst}} 2SO_3(g)$

CHAPTER 4:
Reactions in Solution

Practice Problems

4.1 **(a)** water **(b)** sugar

Since the solution is in the liquid state and water is a liquid, water is the solvent and sugar (a solid) is the solute.

4.2 **(a)** alcohol **(b)** water

Since both substances are liquids, alcohol, the substance present in greater quantity, is the solvent.

4.3 **(a)** more soluble

For all four solids for which solubility data are given in Table 4.1, solubility is greater at the higher temperature.

(b) decreases

For all three gases for which solubility data are given in Table 4.1, solubility is lower at the higher temperature.

4.4 **(a)** saturated

The solubility of $HgBr_2$ is 0.61 g/100 mL water at 25 °C. Therefore, a solution that contains 0.61g/100 mL of water at 25 °C is a saturated solution.

(b) unsaturated

Because 0.46 g < 0.61 g, the solution is unsaturated.

(c) supersaturated

Because 0.62 g > 0.61 g, the solution is supersaturated.

(d) unsaturated

One hundred milliliters of a solution that contains 0.61 g $HgBr_2$/200 mL water contains

$$(100\,\cancel{mL})\left(\frac{0.61\text{ g }HgBr_2}{200\,\cancel{mL}}\right) = 0.31\text{ g }HgBr_2$$

Because 0.31 g < 0.61 g, the solution is unsaturated.

4.5 precipitation

As the temperature is decreased from 100 °C to 25 °C, $HgBr_2(s)$ will probably begin to precipitate out of solution owing to its decreasing solubility with decreasing temperature. Only 0.61 g $HgBr_2$ will remain in solution at 25 °C. However, $HgBr_2$ *may* remain in solution to form a supersaturated solution.

4.6 (a) saturated

$$\frac{0.000\,0445\text{ g AgCl}}{50\text{ mL }H_2O} = \frac{0.000\,0890\text{ g AgCl}}{100\text{ mL }H_2O}$$

Because, according to Table 4.1, the solubility of AgCl in water at 10 °C is 0.000 089 g/100 mL H_2O, the solution is saturated.

(b) dilute

The concentration of the solute, AgCl, is very low.

4.7 (a) 3 (b) 1 (c) 2 (d) 0

Hydrogens that form ions in solution are written at the left of formulas. Therefore, H_3PO_4 has three hydrogens that form hydrogen ions in aqueous solution, $HC_7H_5O_2$ has one, $H_2C_8H_4O_4$ has two, and CH_4O has none.

4.8 hydrobromic acid

Bromine is in the same group in the periodic table as chlorine (VIIA). Therefore, the names of bromine compounds are similar to the names of chlorine compounds.

4.9 If you have forgotten the formulas for the common polyatomic ions, review Table 1.2.

(a) nitric acid

HNO_3 is recognized as an acid since its formula begins with an ionizable hydrogen.

(b) magnesium perchlorate

$Mg(ClO_4)_2$ results from the combination of a metal ion, Mg^{2+}, with a polyatomic ion, ClO_4^-.

(c) potassium hydrogen carbonate

$KHCO_3$ results from the combination of a potassium ion with the polyatomic ion formed when carbonic acid loses one hydrogen ion.

(d) phosphoric acid

H_3PO_4 is recognized as an acid since its formula begins with ionizable hydrogens.

4.10 **(a)** CaS

The calcium ion has a 2+ charge, and the sulfide ion has a 2– charge. Therefore, the formula is CaS.

(b) H_2SO_4

Sulfuric acid has a formula beginning with hydrogen. Since the sulfate ion has a 2– charge, it takes two hydrogens to balance the charge and yield a neutral molecule.

(c) H_2CO_3

Carbonic acid is similar to sulfuric acid, because the carbonate ion also has a 2– charge.

(d) $Al_2(SO_4)_3$

The aluminum ion has a 3+ charge and the sulfate ion has a 2– charge. Therefore, the formula of the compound does not involve a 1-to-1 ratio of ions. It takes two aluminum ions to balance the charge of three sulfate ions.

4.11 (a) KOH (c) SrO

Bases usually contain the hydroxide ion, as in KOH. Oxides are also usually classified as bases, as in SrO.

4.12 (a) KI (d) NH_4NO_3 (e) Ag_2SO_4

Salts are compounds of metals (or polyatomic cations such as the NH_4^+ ion) and nonmetals (including polyatomic anions such as NO_3^- and SO_4^{2-} ions). However, oxides and hydroxides are *not* classified as salts. According to this definition, KI, NH_4NO_3, and Ag_2SO_4 are salts.

4.13 **(a)** $KClO_4$—potassium perchlorate

The formula for potassium ion is K^+ because a potassium atom, atomic number 19, must lose one electron to have 18 electrons like argon. The formula for the perchlorate ion is ClO_4^-. The potassium salt of perchloric acid is a compound and must have zero net charge. Therefore, the ratio of K^+ ions to ClO_4^- ions must be 1:1, and the formula of the salt is $KClO_4$. The name of this compound is potassium perchlorate.

(b) $Ca(NO_3)_2$—calcium nitrate

The formula for calcium ion is Ca^{2+}, the formula for nitrate ion is NO_3^-, and the calcium salt of nitric acid is $Ca(NO_3)_2$, calcium nitrate.

(c) $KHCO_3$—potassium hydrogen carbonate and K_2CO_3—potassium carbonate

Carbonic acid, H_2CO_3, has two ionizable hydrogens and can form two potassium salts, $KHCO_3$ and K_2CO_3. The names of these salts are potassium hydrogen carbonate and potassium carbonate, respectively.

(d) $Ca(HCO_3)_2$—calcium hydrogen carbonate and $CaCO_3$—calcium carbonate

The calcium salts of carbonic acid are $Ca(HCO_3)_2$, calcium hydrogen carbonate, and $CaCO_3$, calcium carbonate.

(e) $AlCl_3$—aluminum chloride

The aluminum salt of hydrochloric acid is $AlCl_3$, aluminum chloride.

4.14 (a) $H_2SO_4(aq) + 2NaOH(aq) \rightarrow Na_2SO_4(aq) + 2H_2O(l)$

Remember and apply the steps for writing equations:

Step 1: Identify reactants and products and write a word equation. The products of the neutralization of an acid with a base are a salt and water. According to the information given in this problem, the salt is sodium sulfate. The word equation is

aqueous sulfuric acid + aqueous sodium hydroxide → aqueous sodium sulfate + water

Step 2: Write symbols and formulas for the substances involved in the reaction.

$$H_2SO_4 + NaOH \rightarrow Na_2SO_4 + H_2O$$

Step 3: Balance by changing coefficients.

$$H_2SO_4 + 2NaOH \rightarrow Na_2SO_4 + 2H_2O$$

Step 4: Simplify if possible and check. The coefficients have no common denominator; simplification is not possible. There are 4 H, 1 S, 6 O, and 2 Na on each side.

Step 5: Add symbols showing states.

$$H_2SO_4(aq) + 2NaOH(aq) \rightarrow Na_2SO_4(aq) + 2H_2O(l)$$

(b) $2HNO_3(aq) + Ba(OH)_2(aq) \rightarrow Ba(NO_3)_2(aq) + 2H_2O(l)$

The problem states that both hydroxide ions of barium hydroxide, $Ba(OH)_2$, react. Two H_2O's and the salt $Ba(NO_3)_2$ must be formed. The equation is

$$2HNO_3(aq) + Ba(OH)_2(aq) \rightarrow Ba(NO_3)_2(aq) + 2H_2O(l)$$

4.15 Use Table 4.3.

(a) weak electrolyte

The compound HF is an acid because it has a H written on the left side. According to Table 4.3, the acids HCl, HBr, HI, $HClO_4$, HNO_3, and H_2SO_4 are strong electrolytes. All other common acids are weak electrolytes. Therefore, HF is a weak electrolyte.

(b) strong electrolyte

The compound KOH is the hydroxide of a Group IA element. Therefore, it is a strong electrolyte.

(c) strong electrolyte

The compound $MgCl_2$ is a salt. Therefore, it is a strong electrolyte.

(d) nonelectrolyte

According to Table 4.3, the compound C_3H_6O is neither a strong nor a weak electrolyte. Therefore, it is a nonelectrolyte.

(e) strong electrolyte

The compound $(NH_4)_2SO_4$ is a salt and therefore is a strong electrolyte.

4.16 Use Table 4.4.

(a) slightly soluble

According to Table 4.4, Ag_2SO_4 is slightly soluble.

(b) insoluble

All carbonates are insoluble except those of the Group IA metals and the ammonium ion. Calcium is a Group IIA metal. Therefore, $CaCO_3$ is insoluble.

(c) soluble

All chlorides are soluble except AgCl and Hg_2Cl_2; $PbCl_2$ is only slightly soluble. Therefore, $CoCl_2$ is soluble.

(d) soluble

The compound HNO_3 is an acid; all acids are soluble.

(e) insoluble

All sulfides are insoluble except those of the Group IA and IIA metals and ammonium sulfide. Since Pb is in Group IVA, PbS is insoluble.

4.17 **(a)** Yes. $2NaOH(aq) + MgCl_2(aq) \rightarrow Mg(OH)_2(s) + 2NaCl(aq)$

A solution of sodium hydroxide contains Na^+ and OH^- ions. A solution of magnesium chloride contains Mg^{2+} and Cl^- ions. Ions present are

Na^+ OH^-

Mg^{2+} Cl^-

Possible new combinations of ions are

Na^+ OH^-

Mg^{2+} Cl^-

The rule says that if either of the possible products is insoluble or a weak or nonelectrolyte, a reaction will take place. $Mg(OH)_2$ is insoluble; therefore, a reaction will take place. A precipitate of $Mg(OH)_2$ will be formed. Sodium chloride is soluble and the equation for the reaction that will take place is

$$2NaOH(aq) + MgCl_2(aq) \rightarrow Mg(OH)_2(s) + 2NaCl(aq)$$

(b) Yes. $K_2SO_3(aq) + 2HCl(aq) \rightarrow SO_2(g) + H_2O(l) + 2KCl(aq)$

Possible new combinations of ions are

K^+ SO_3^{2-}

H^+ Cl^-

Sulfurous acid, $H_2SO_3(aq)$, which decomposes to $SO_2(g)$ and $H_2O(l)$, is a weak electrolyte. A reaction will take place. The equation is

$$K_2SO_3(aq) + 2HCl(aq) \rightarrow SO_2(g) + H_2O(l) + 2KCl(aq)$$

(c) no reaction

Possible new combinations of ions are

NH_4^+ I^-

Zn^{2+} Cl^-

Both NH_4Cl and ZnI_2 are soluble, strong electrolytes. No reaction will take place.

(d) Yes. $NaC_2H_3O_2(aq) + HBr(aq) \rightarrow HC_2H_3O_2(aq) + NaBr(aq)$

Possible new combinations of ions are

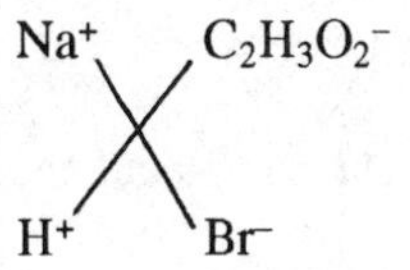

Acetic acid, $HC_2H_3O_2$, is a weak electrolyte. A reaction will take place and the equation is

$$NaC_2H_3O_2(aq) + HBr(aq) \rightarrow HC_2H_3O_2(aq) + NaBr(aq)$$

(e) Yes. $Na_2CO_3(aq) + CaCl_2(aq) \rightarrow CaCO_3(s) + 2NaCl(aq)$

Possible new combinations of ions are

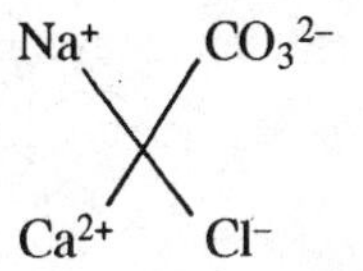

Calcium carbonate, $CaCO_3$, is insoluble. A reaction will take place and the equation is

$$Na_2CO_3(aq) + CaCl_2(aq) \rightarrow CaCO_3(s) + 2NaCl(aq)$$

4.18 In ionic equations, compounds that exist mostly as ions in solution (strong electrolytes) are shown as ions. Molecular formulas are used for weak electrolytes and insoluble compounds.

(a) $CaCO_3(s)$

Calcium carbonate is insoluble and is written $CaCO_3(s)$.

(b) $H^+ + NO_3^-$ or $H^+(aq) + NO_3^-(aq)$

Nitric acid is a strong electrolyte and is written $H^+ + NO_3^-$.

(c) $HC_2H_3O_2(aq)$

Acetic acid is a weak acid and is written $HC_2H_3O_2(aq)$.

(d) $Mg^{2+} + 2Cl^-$ or $Mg^{2+}(aq) + 2Cl^-(aq)$

Magnesium chloride is a strong electrolyte and is written $Mg^{2+} + 2Cl^-$.

(e) $Fe(OH)_2(s)$

Iron(II) hydroxide is insoluble and is written $Fe(OH)_2(s)$.

4.19 Hydrogen sulfide is a weak electrolyte and is written $H_2S(g)$ in ionic equations. Zinc nitrate and nitric acid are strong electrolytes and are written $Zn^{2+} + 2NO_3^-$ and $H^+ + NO_3^-$ in ionic equations. Zinc sulfide is insoluble and is written $ZnS(s)$.

(a) $H_2S(g) + Zn^{2+} + 2NO_3^- \rightarrow ZnS(s) + 2H^+ + 2NO_3^-$

The complete ionic equation shows all ions present. The complete ionic equation is

$$H_2S(g) + Zn^{2+} + 2NO_3^- \rightarrow ZnS(s) + 2H^+ + 2NO_3^-$$

(b) NO_3^-

Ions that are present in the solution both before and after reaction takes place are called spectator ions because they do not take part in the reaction. The nitrate ions, NO_3^-, are spectator ions.

(c) $H_2S(g) + Zn^{2+} \rightarrow ZnS(s) + 2H^+$

The net ionic equation shows only the species that take part in the reaction. It is obtained from the complete ionic equation by canceling spectator ions:

$$H_2S(g) + Zn^{2+} + 2NO_3^- \rightarrow ZnS(s) + 2H^+ + 2NO_3^-$$

The net ionic equation is

$$H_2S(g) + Zn^{2+} \rightarrow ZnS(s) + 2H^+$$

Net ionic equations must show the same charge as well as the same number of atoms of each kind on both sides. This net ionic equation shows 2 H, 1 S, 1 Zn, and a 2+ charge on each side.

(d) Yes. Zinc(II) chloride

Zinc(II) chloride would give the same net reaction.

4.20 **(a)** Complete ionic equation:

$$2Na^+ + 2OH^- + Mg^{2+} + 2Cl^- \rightarrow Mg(OH)_2(s) + 2Na^+ + 2Cl^-$$

Net ionic equation: $Mg^{2+} + 2OH^- \rightarrow Mg(OH)_2(s)$

(b) Complete ionic equation:

$$2K^+ + SO_3^{2-} + 2H^+ + 2Cl^- \rightarrow SO_2(g) + H_2O(l) + 2K^+ + 2Cl^-$$

Net ionic equation: $2H^+ + SO_3^{2-} \rightarrow SO_2(g) + H_2O(l)$

(c) Complete ionic equation:

$$NH_4^+ + I^- + Zn^{2+} + 2Cl^- \rightarrow NH_4^+ + I^- + Zn^{2+} + 2Cl^-$$

Net ionic equation: No net ionic equation (no reaction)

(d) Complete ionic equation:

$$Na^+ + C_2H_3O_2^- + H^+ + Br^- \rightarrow HC_2H_3O_2(aq) + Na^+ + Br^-$$

Net ionic equation: $H^+ + C_2H_3O_2^- \rightarrow HC_2H_3O_2(aq)$

(e) Complete ionic equation:

$$2Na^+ + CO_3^{2-} + Ca^{2+} + 2Cl^- \rightarrow CaCO_3(s) + 2Na^+ + 2Cl^-$$

Net ionic equation: $Ca^{2+} + CO_3^{2-} \rightarrow CaCO_3(s)$

4.21 Use the activity series for metals in Table 4.5 and the periodic table to make predictions.

(a) no reaction

Aluminum does not react with water.

(b) Yes. Molecular equation: $Sr(s) + 2H_2O(l) \rightarrow Sr(OH)_2(aq) + H_2(g)$

Complete ionic equation:

$$Sr(s) + 2H_2O(l) \rightarrow Sr^{2+} + 2OH^- + H_2(g)$$

There are no spectator ions, so the net ionic equation is the same as the complete ionic equation.

Strontium metal is below calcium metal in Group IIA and is therefore more reactive toward water than calcium. According to Table 4.5, Ca reacts with water; therefore, Sr reacts with water.

(c) Yes. Molecular equation: $2Al(s) + 6HCl(aq) \rightarrow 2AlCl_3(aq) + 3H_2(g)$

Complete ionic equation:

$$2Al(s) + 6H^+ + 6Cl^- \rightarrow 2Al^{3+} + 6Cl^- + 3H_2(g)$$

Net ionic equation: $2Al(s) + 6H^+ \rightarrow 2Al^{3+} + 3H_2(g)$

According to Table 4.5, aluminum will react with dilute acid.

(d) no reaction

Gold metal does not react with dilute acids.

(e) Yes. Molecular equation: $2Cr(s) + 3Cu(NO_3)_2(aq) \rightarrow 2Cr(NO_3)_3(aq) + 3Cu(s)$

Complete ionic equation:

$$2Cr(s) + 3Cu^{2+} + 6NO_3^- \rightarrow 2Cr^{3+} + 6NO_3^- + 3Cu(s)$$

Net ionic equation: $2Cr(s) + 3Cu^{2+} \rightarrow 2Cr^{3+} + 3Cu(s)$

Chromium metal will react with copper(II) nitrate solution because Cr is above Cu in the activity series.

(f) no reaction

Iron metal will not react with zinc chloride solution because iron is below zinc in the activity series.

4.22 Chlorine is more reactive than bromine and takes the place of bromine in NaBr.

Molecular equation: $Cl_2(aq) + 2NaBr(aq) \rightarrow 2NaCl(aq) + Br_2(aq)$

Complete ionic equation: $Cl_2(aq) + 2Na^+ + 2Br^- \rightarrow 2Na^+ + 2Cl^- + Br_2(aq)$

Net ionic equation: $Cl_2(aq) + 2Br^- \rightarrow 2Cl^- + Br_2(aq)$

4.23 Yes. A reaction will take place because bromine is more reactive than iodine.

Molecular equation: $Br_2(aq) + 2KI(aq) \rightarrow 2KBr(aq) + I_2(s)$

Complete ionic equation: $Br_2(aq) + 2K^+ + 2I^- \rightarrow 2K^+ + 2Br^- + I_2(s)$

Net ionic equation: $Br_2(aq) + 2I^- \rightarrow 2Br^- + I_2(s)$

4.24 **(a)** No. Both H_2SO_4 and $BaCl_2$ are soluble, strong electrolytes so $BaSO_4$ is *not* soluble in HCl(aq).

(b) Yes. H^+ from HCl(aq) will react with CO_3^{2-} from $BaCO_3$ to form the weak electrolyte $H_2CO_3(aq)$ or $CO_2(g) + H_2O(l)$. Therefore, $BaCO_3$ is soluble in HCl(aq).

4.25 **(a)** $MgCl_2(aq) + 2NaOH(aq) \rightarrow Mg(OH)_2(s) + 2NaCl(aq)$

Because $Mg(OH)_2$ is insoluble in water, it can be made by precipitation. Any combination of aqueous solutions of a soluble magnesium salt and a soluble strong base is suitable. If $MgCl_2(aq)$ and NaOH(aq) are used, the molecular equation for the reaction is

$$MgCl_2(aq) + 2NaOH(aq) \rightarrow Mg(OH)_2(s) + 2NaCl(aq)$$

(b) $\underset{\text{excess}}{Mg(s)} + 2HCl(aq) \rightarrow MgCl_2(aq) + H_2(g)$. Filter and evaporate.

Magnesium chloride is soluble. It can be made by treating magnesium metal with a limited quantity of hydrochloric acid, removing the excess magnesium by filtration, and evaporating water. The molecular equation for this method of preparation is

$$\underset{\text{excess}}{Mg(s)} + 2HCl(aq) \rightarrow MgCl_2(aq) + H_2(g)$$

The solid magnesium chloride left after evaporation of water will be hydrated. The water of hydration can be removed by heating in an oven. Magnesium chloride could also be made by treating an insoluble magnesium compound that is soluble in acid, such as magnesium hydroxide, with a limited quantity of hydrochloric acid, removing the excess magnesium compound by filtration, evaporating water, and drying. A molecular equation illustrating this method of preparation is

$$\underset{\text{excess}}{Mg(OH)_2(s)} + 2HCl(aq) \rightarrow MgCl_2(aq) + 2H_2O(l)$$

(c) $KOH(aq) + HI(aq) \rightarrow KI(aq) + H_2O(l)$. Use stoichiometric amounts. Evaporate.

Potassium iodide can be made by neutralizing a solution of potassium hydroxide with a solution of hydroiodic acid. The reactants should be mixed in the proportions shown by the equation for the reaction so that both are used up by the reaction.

$$KOH(aq) + HI(aq) \rightarrow KI(aq) + H_2O(l)$$

Potassium iodide does not form a hydrate.

4.26 Before writing directions, we need to calculate how many milliliters of ethyl alcohol to use. A 12.0% by volume solution of ethyl alcohol in water will contain 12.0 mL ethyl alcohol for every 100 mL of solution. Therefore, 12.0 mL ethyl alcohol ~ 100 mL of solution.

$$(5.00 \times 10^2 \cancel{\text{ mL solution}})\left(\frac{12.0 \text{ mL alcohol}}{100 \cancel{\text{ mL solution}}}\right) = 60.0 \text{ mL alcohol}$$

Use a 100–mL graduated cylinder to measure 60.0 mL of ethyl alcohol. Pour the ethyl alcohol into a 500–mL graduated cylinder and dilute to the 500–mL mark with deionized water. Mix thoroughly.

Another way to calculate the volume of ethyl alcohol (EA) that should be used is to apply the definition of percent by volume:

$$\text{volume \% EA} = \frac{\text{volume EA}}{\text{total volume of solution}} \times 100$$

Solution of the definition for volume EA and substitution of the information from the problem gives

$$\text{volume EA} = \frac{\text{volume \% EA (total volume of solution)}}{100} = \frac{12.0 \text{ (500 mL)}}{100} = 60.0 \text{ mL}$$

4.27 Before writing directions, we need to calculate how many grams of KI and water to use. An 8.2% by mass solution of KI in water will contain 8.2 g KI(s) per 100 g of solution. Therefore, 8.2 g KI ~ 100 g solution.

$$(75 \cancel{\text{ g solution}})\left(\frac{8.2 \text{ g KI}}{100 \cancel{\text{ g solution}}}\right) = 6.2 \text{ g KI}$$

The rest of the solution is water

$$75 \text{ g solution} - 6.2 \text{ g KI} = 69 \text{ g } H_2O$$

It is easier to measure the volume of a liquid than to measure its mass. Because the density of water at room temperature is 1.00 g/mL, 69 g H_2O = 69 mL H_2O. Measure 6.2 g KI(s) into a 150–mL beaker, add 69 mL $H_2O(l)$ from a 100–mL graduated cylinder, and stir until all the KI(s) has dissolved.

Another way to calculate the mass of KI(s) that should be used is to apply the definition of percent by mass:

$$\text{mass \% KI} = \frac{\text{mass KI}}{\text{total mass of solution}} \times 100$$

Solution of the definition for mass KI and substitution of the information from the problem gives

$$\text{mass KI} = \frac{\text{mass \% KI (total mass of solution)}}{100} = \frac{8.2\ (75\ \text{g})}{100} = 6.2\ \text{g}$$

4.28 330 ppb (volume)

$$\text{ppb(volume)} = \frac{\text{volume of solute}}{\text{volume of solution}} \times 10^9 = \left(\frac{3.30 \times 10^{-5}\ \text{mL}\ N_2O}{100.0\ \text{mL air}}\right) \times 10^9 = 330\ \text{ppb (volume)}$$

4.29 $$\text{parts per trillion (volume)} = \frac{\text{volume solute}}{\text{volume solution}} \times 10^{12}$$

Note that ppt can also mean "parts per thousand" or "precipitate." Thus, this abbreviation should be avoided.

4.30 0.6965 M

$$\left(\frac{57.81\ \cancel{\text{g KI}}}{500.0\ \cancel{\text{mL soln}}}\right)\left(\frac{1000\ \cancel{\text{mL soln}}}{1\ \text{L soln}}\right)\left(\frac{1\ \text{mol KI}}{166.00\ \cancel{\text{g KI}}}\right) = \frac{0.6965\ \text{mol KI}}{\text{L soln}} = 0.6965\ \text{M}$$

4.31 **(a)** 6.91 g

$$(250.0\ \cancel{\text{mL soln}})\left(\frac{1\ \cancel{\text{L soln}}}{10^3\ \cancel{\text{mL soln}}}\right)\left(\frac{0.325\ \cancel{\text{mol NaNO}_3}}{\cancel{\text{L soln}}}\right)\left(\frac{84.99\ \text{g NaNO}_3}{\cancel{\text{mol NaNO}_3}}\right) = 6.91\ \text{g NaNO}_3$$

(b) 1. Measure 6.91 g $NaNO_3$(s) using a balance.
2. Dissolve the $NaNO_3$ in a minimal amount of deionized water.
3. When the $NaNO_3$ is fully dissolved, transfer the solution to a 250–mL volumetric flask using a glass stirring rod and a funnel to reduce the risk of losing part of the solution.
4. Use a stream of water from a wash bottle containing deionized water to complete the transfer. If the solution does not fill two thirds to three quarters of the bulb of the flask, add more deionized water.
5. Swirl the mixture and let it stand long enough to come to room temperature.
6. Dilute the solution up to the mark on the neck of the flask using a dropper to add the last few drops.
7. Mix thoroughly by repeatedly inverting the stoppered flask.

4.32 231 mL

$$(0.0750\ \cancel{\text{mol NaOH}})\left(\frac{1\ \cancel{\text{L soln}}}{0.325\ \cancel{\text{mol NaOH}}}\right)\left(\frac{1000\ \text{mL soln}}{\cancel{\text{L soln}}}\right) = 231\ \text{mL soln}$$

4.33 Calculate the volume of concentrated solution required:

$$(1.000\ \text{L dil soln})\left(\frac{0.0500\ \text{mol NaOH}}{\text{L dil soln}}\right)\left(\frac{1\ \text{L conc soln}}{0.325\ \text{mol NaOH}}\right)\left(\frac{10^3\ \text{mL conc soln}}{1\ \text{L conc soln}}\right) = 154\ \text{mL concentrated solution}$$

Using a graduated cylinder, measure 154 mL concentrated solution (0.325 M) into about 500 mL of deionized water in a 1000–mL volumetric flask. Swirl until mixed. Fill to the mark with deionized water and mix thoroughly to make 1.000 L of dilute solution (0.0500 M).

4.34 11.6 M

$$\left(\frac{1.179\ \text{g soln}}{\text{mL soln}}\right)\left(\frac{1000\ \text{mL soln}}{\text{L soln}}\right)\left(\frac{36.0\ \text{g HCl}}{100\ \text{g soln}}\right)\left(\frac{1\ \text{mol HCl}}{36.46\ \text{g HCl}}\right) = \frac{11.6\ \text{mol HCl}}{\text{L soln}} = 11.6\ \text{M}$$

4.35 0.273 M

Step 1: Calculate moles NaOH from both solutions.

$$(247\ \text{mL soln})\left(\frac{1\ \text{L soln}}{10^3\ \text{mL soln}}\right)\left(\frac{0.325\ \text{mol NaOH}}{\text{L soln}}\right) = 8.03 \times 10^{-2}\ \text{mol NaOH}$$

$$(538\ \text{mL soln})\left(\frac{1\ \text{L soln}}{10^3\ \text{mL soln}}\right)\left(\frac{0.249\ \text{mol NaOH}}{\text{L soln}}\right) = 0.134\ \text{mol NaOH}$$

0.0803 mol NaOH + 0.134 mol NaOH = 0.214 mol NaOH

Step 2: Calculate the total volume in L.

$$(247\ \text{mL soln} + 538\ \text{mL soln})\left(\frac{1\ \text{L soln}}{10^3\ \text{mL soln}}\right) = 0.785\ \text{L soln}$$

Step 3: Calculate molarity.

$$\frac{0.214\ \text{mol NaOH}}{0.785\ \text{L soln}} = 0.273\ \text{M}$$

4.36

Vol HCl —(M_{HCl})→ Mol HCl —(Conversion factor from equation)→ Mol NaOH —(M_{NaOH})→ Vol NaOH

4.37 26.5 mL

Problem	24.96 mL, 0.1254 M	? mL, 0.236 M
Equation	$H_2SO_4(aq)$ +	$2KOH(aq) \rightarrow K_2SO_4(aq) + 2H_2O(l)$
Recipe, mol	1	2

$$(24.96\ \text{mL}\ H_2SO_4\ \text{soln})\left(\frac{1\ \text{L}\ H_2SO_4\ \text{soln}}{10^3\ \text{mL}\ H_2SO_4\ \text{soln}}\right)\left(\frac{0.1254\ \text{mol}\ H_2SO_4}{\text{L}\ H_2SO_4\ \text{soln}}\right)\left(\frac{2\ \text{mol KOH}}{\text{mol}\ H_2SO_4}\right)\left(\frac{1\ \text{L KOH soln}}{0.236\ \text{mol KOH}}\right)\left(\frac{10^3\ \text{mL KOH soln}}{\text{L KOH soln}}\right)$$

$$= 26.5\ \text{mL KOH soln}$$

4.38 3.97 g H_2 formed, 0.29 mol HCl left

Problem	35.4 g		721 mL, 5.86 M		? g
Equation	2Al(s)	+	6HCl(aq)	→	$3H_2(g) + 2AlCl_3(aq)$
Formula masses, u	26.98		36.46		2.016
Recipe, mol	2		6		3

$$(721 \text{ mL HCl soln})\left(\frac{1 \text{ L HCl soln}}{10^3 \text{ mL HCl soln}}\right)\left(\frac{5.86 \text{ mol HCl}}{\text{L HCl soln}}\right) = 4.23 \text{ mol HCl}$$

$$(35.4 \text{ g Al})\left(\frac{1 \text{ mol Al}}{26.98 \text{ g Al}}\right) = 1.31 \text{ mol Al}$$

$$\frac{1.31 \text{ mol Al}}{2 \text{ mol Al}} = 0.656 \text{ of a recipe} \qquad \frac{4.23 \text{ mol HCl}}{6 \text{ mol HCl}} = 0.705 \text{ of a recipe}$$

0.656 < 0.705, so Al is limiting.

$$(35.4 \text{ g Al})\left(\frac{1 \text{ mol Al}}{26.98 \text{ g Al}}\right)\left(\frac{3 \text{ mol } H_2}{2 \text{ mol Al}}\right)\left(\frac{2.016 \text{ g } H_2}{1 \text{ mol } H_2}\right) = 3.97 \text{ g } H_2(g)$$

$$(35.4 \text{ g Al})\left(\frac{1 \text{ mol Al}}{26.98 \text{ g Al}}\right)\left(\frac{6 \text{ mol HCl}}{2 \text{ mol Al}}\right) = 3.94 \text{ mol HCl}$$

4.23 mol HCl – 3.94 mol HCl = 0.29 mol HCl excess.

4.39 **(a)** 1.03 M

$$(24.62 \text{ mL soln})\left(\frac{1 \text{ L soln}}{10^3 \text{ mL soln}}\right)\left(\frac{0.250 \text{ mol NaOH}}{\text{L soln}}\right)\left(\frac{1 \text{ mol } H_2SO_4}{2 \text{ mol NaOH}}\right) = 3.08 \times 10^{-3} \text{ mol } H_2SO_4$$

$$\left(\frac{3.08 \times 10^{-3} \text{ mol } H_2SO_4}{3.00 \text{ mL soln}}\right)\left(\frac{10^3 \text{ mL soln}}{\text{L soln}}\right) = 1.03 \text{ M}$$

(b) 0.302 g

$$(3.08 \times 10^{-3} \text{ mol } H_2SO_4)\left(\frac{98.08 \text{ g } H_2SO_4}{\text{mol } H_2SO_4}\right) = 0.302 \text{ g } H_2SO_4$$

(c) 9.47%

$$\text{mass \% } H_2SO_4 = \left(\frac{\text{mass } H_2SO_4}{\text{mass soln}}\right)(100)$$

Mass of H_2SO_4 was calculated in part (b). Use the density of the solution to calculate the mass of the solution:

$$(3.00 \text{ mL soln})\left(\frac{1.0626 \text{ g soln}}{\text{mL soln}}\right) = 3.19 \text{ g soln}$$

Then substitute into the definition of mass %.

$$\left(\frac{0.302\text{ g } H_2SO_4}{3.19\text{ g soln}}\right)(100) = 9.47\text{ mass \% } H_2SO_4$$

Additional Practice Problems

4.75 Because the student has not labeled the numbers completely, he or she has not taken into account the fact that the solution is 6.00% by mass $BaCl_2$ nor that the density is in fact the density of the solution, not the density of $BaCl_2$. All quantities should be labeled completely. The correct solution is

$$\left(\frac{6.00\text{ }\cancel{\text{g } BaCl_2}}{100.0\text{ }\cancel{\text{g soln}}}\right)\left(\frac{1\text{ mol } BaCl_2}{208.2\text{ }\cancel{\text{g } BaCl_2}}\right)\left(\frac{1.063\text{ }\cancel{\text{g soln}}}{\cancel{\text{mL soln}}}\right)\left(\frac{1000\text{ }\cancel{\text{mL soln}}}{\text{L soln}}\right) = 0.306\text{ M}$$

4.77 **(a)** less than 0.200 M because part of what was used was water rather than LiBr.

(b) less than 0.200 M because less LiBr is in solution than was intended.

(c) less than 0.200 M because as the solution warms up, its volume will increase. This will cause the molarity to be lower than expected.

(d) 0.200 M because once it is made, every part of the solution has the same concentration. The part that was spilled had a concentration of 0.200 M, and the part remaining had a concentration of 0.200 M.

(e) less than 0.200 M because the water on the inside of the bottle dilutes the solution.

4.79 Zn^{2+} because zinc sulfide is insoluble. According to Table 4.4, all sulfides are insoluble except those of Group IA, Group IIA, and ammonium. Na^+ and Ca^{2+} are in these groups.

Stop and Test Yourself

1. **(d)** unsaturated but concentrated

A solution of 25 g of potassium carbonate (K_2CO_3) in 25 mL of water at 20 °C is equivalent in concentration to a solution containing 100 g K_2CO_3 in 100 mL of water. Therefore, since the solubility of K_2CO_3 in water at 20 °C is 112 g/100 mL, the solution is unsaturated. Since it has a relatively high amount of K_2CO_3, it is concentrated.

2. **(e)** 11

$$\left(\frac{2.3\text{ mol } NH_3}{1\text{ }\cancel{\text{L soln}}}\right)(4.6\text{ }\cancel{\text{L soln}}) = 11\text{ mol } NH_3$$

3. **(b)** 0.0706

$$(4.21\text{ }\cancel{\text{g } NH_3})\left(\frac{1\text{ mol } NH_3}{17.03\text{ }\cancel{\text{g } NH_3}}\right) = 0.247\text{ mol } NH_3 \qquad \frac{0.247\text{ mol } NH_3}{3.50\text{ L soln}} = 0.0706\text{ M}$$

4. **(d)** 14

$$\left(\frac{0.52 \text{ mol HCl}}{1\ \cancel{\text{L soln}}}\right)(0.748\ \cancel{\text{L soln}}) = 0.39 \text{ mol HCl}$$

$$(0.39\ \cancel{\text{mol HCl}})\left(\frac{36.5 \text{ g HCl}}{\cancel{\text{mol HCl}}}\right) = 14 \text{ g HCl}$$

5. **(a)** 5.00×10^{-2} L

$$(250.0\ \cancel{\text{mL dil soln}})\left(\frac{1\ \cancel{\text{L dil soln}}}{1000\ \cancel{\text{mL dil soln}}}\right)\left(\frac{0.200\ \cancel{\text{mol KNO}_3}}{1\ \cancel{\text{L dil soln}}}\right)\left(\frac{1 \text{ L conc soln}}{1.000\ \cancel{\text{mol KNO}_3}}\right) = 5.00 \times 10^{-2} \text{ L conc soln}$$

6. **(c)** 29 mL

Problem	8.65 g		? mL 6.0 M
Equation	$CaCO_3(s)$	+	$2HCl(aq) \rightarrow CaCl_2(aq) + H_2O(l) + CO_2(g)$
Formula mass, u	100.1		
Recipe, mol	1		2

$$(8.65\ \cancel{\text{g CaCO}_3})\left(\frac{1\ \cancel{\text{mol CaCO}_3}}{100.1\ \cancel{\text{g CaCO}_3}}\right)\left(\frac{2\ \cancel{\text{mol HCl}}}{1\ \cancel{\text{mol CaCO}_3}}\right)\left(\frac{1\ \cancel{\text{L soln}}}{6.0\ \cancel{\text{mol HCl}}}\right)\left(\frac{1000 \text{ mL soln}}{1\ \cancel{\text{L soln}}}\right) = 29 \text{ mL soln}$$

7. **(b)** 8.6

Problem	23.32 g		65.3 mL 6.0 M						? g
Equation	$CaCO_3(s)$	+	$2HCl(aq)$	$\rightarrow$	$CaCl_2(aq)$	+	$H_2O(l)$	+	$CO_2(g)$
Formula mass, u	100.09								44.010
Recipe, mol	1		2						1

$$(23.32\ \cancel{\text{g CaCO}_3})\left(\frac{1 \text{ mol CaCO}_3}{100.09\ \cancel{\text{g CaCO}_3}}\right) = 0.2330 \text{ mol CaCO}_3$$

$$(65.3\ \cancel{\text{mL HCl}})\left(\frac{1\ \cancel{\text{L HCl}}}{10^3\ \cancel{\text{mL HCl}}}\right)\left(\frac{6.0 \text{ mol HCl}}{\cancel{\text{L HCl}}}\right) = 0.39 \text{ mol HCl}$$

$$\frac{0.2330\ \cancel{\text{mol CaCO}_3}}{1\ \cancel{\text{mol CaCO}_3}} = 0.233 \text{ of a recipe} \qquad \frac{0.39\ \cancel{\text{mol HCl}}}{2\ \cancel{\text{mol HCl}}} = 0.20 \text{ of a recipe}$$

$0.20 < 0.233$ Therefore, HCl is limiting.

$$(0.39\ \cancel{\text{mol HCl}})\left(\frac{1\ \cancel{\text{mol CO}_2}}{2\ \cancel{\text{mol HCl}}}\right)\left(\frac{44.0 \text{ g CO}_2}{\cancel{\text{mol CO}_2}}\right) = 8.6 \text{ g CO}_2$$

8. **(d)** 3.7 g $CaCO_3$

Note that the quantities given in this problem are the same as in the previous problem, except that the molarity of the hydrochloric acid is given to three significant figures.

$$(65.3\ \cancel{\text{mL HCl}})\left(\frac{1\ \cancel{\text{L HCl}}}{10^3\ \cancel{\text{mL HCl}}}\right)\left(\frac{6.00\ \cancel{\text{mol HCl}}}{1\ \cancel{\text{L HCl}}}\right)\left(\frac{1\ \cancel{\text{mol CaCO}_3}}{2\ \cancel{\text{mol HCl}}}\right)\left(\frac{100.1\text{ g CaCO}_3}{\cancel{\text{mol CaCO}_3}}\right) = 19.6\text{ g CaCO}_3 \text{ are needed to react with the HCl}$$

23.32 g $CaCO_3$ present – 19.6 g $CaCO_3$ reacted = 3.7 g $CaCO_3$ left

9. **(d)** both cations and anions

Electrons are not present in the free state in aqueous solutions. Both positive and negative charges must move through solutions.

10. **(e)** $HC_2H_3O_2$

According to Table 4.3, H_2SO_4, HNO_3, and HCl are all strong electrolytes, and NH_3 is a base rather than an acid. Therefore, $HC_2H_3O_2$ is a weak electrolyte.

11. **(a)** NiS

According to Table 4.4, $NiSO_4$, Na_3PO_4, and $(NH_4)_2CO_3$ are all soluble. Both NiS and $Fe(OH)_3$ are insoluble but $Fe(OH)_3$ is a *base*, not a salt.

12. **(b)** $NaNO_3$

HNO_3 will react to form the weak electrolyte water. NH_4Cl will react to form NH_3(aq), a weak electrolyte. $NiCl_2$ will react to form insoluble $Ni(OH)_2$. The potential products of the reaction of $NaNO_3$ with KOH are NaOH and KNO_3, both of which are soluble, strong electrolytes.

13. **(c)** $Cl_2 + 2I^- \rightarrow 2Cl^- + I_2$

Answer (a) has 2 Cl on the left and only 1 Cl on the right. It also has 1 I on the left and 2 I on the right. Answer (b) has 2(1–) = 2– charge on the left and zero charge on the right. Answer (d) is a molecular equation, while (e) is a complete ionic equation.

14. **(d)** $MnCl_2$ and Na_2S

To be used to carry out a reaction in aqueous solution, water–soluble salts of manganese(II) and sulfide must be used. Both $MnCO_3$ and ZnS are insoluble, eliminating answers (b) and (c). The compound $MgCl_2$ is magnesium chloride, not manganese(II) chloride; Na_2SO_4 is a sulfate, not a sulfide. Thus, (a) and (e) are eliminated.

15. **(e)** single replacement

In the reaction $Zn(s) + 2HCl(aq) \rightarrow ZnCl_2(aq) + H_2(g)$, Zn replaces H.

Putting Things Together

4.81 **(a)** Add $AgNO_3(aq)$ to the liquid. If a white precipitate of AgCl forms, the liquid is HCl(aq).

(b) Add $AgNO_3(aq)$ to an aqueous solution of the white solid. If a white precipitate of AgCl forms, the white solid is NaCl.

(c) Add $AgNO_3(aq)$ to an aqueous solution of the white solid. If a white precipitate of AgCl forms, the white solid is $CaCl_2$.

(d) Add a concentrated aqueous solution of NaOH to the white solid. If the mixture has the odor of ammonia, the solid is NH_4Cl.

(e) Attempt to dissolve the white solid in water. If it dissolves, it is Na_2CO_3.

4.83 **(a)** CuS

The copper (II) ion has a 2+ charge and the sulfide ion has a 2– charge. Therefore, the formula is CuS.

(b) $FePO_4 \cdot 2H_2O$

The iron (III) ion has a 3+ charge and the phosphate ion has a 3– charge. So the formula of the anhydrous compound is $FePO_4$. The dihydrate has 2 water molecules as part of the compound.

(c) $MgBr_2 \cdot 6H_2O$

The magnesium ion is in Group IIA and has a 2+ charge. Bromine is in Group VIIA, so the bromide ion has a charge of 1–. Thus, the formula of the anhydrous compound is $MgBr_2$. A hexahydrate has 6 water molecules as part of the formula.

(d) $Sn(C_2H_3O_2)_2$

The tin (II) ion has a 2+ charge and the acetate ion has a 1– charge. Therefore, the formula of the compound is $Sn(C_2H_3O_2)_2$.

(e) $Ba(H_2PO_4)_2$

Barium is in Group IIA, so its ion has a 2+ charge. The dihydrogen phosphate ion has a 1– charge. Therefore, the formula for the compound is $Ba(H_2PO_4)_2$.

4.85 71 ppt

$$\text{ppt} = \frac{\text{part}}{\text{whole}} \times 10^{12} = \left(\frac{0.000\,000\,0020}{27.976\,926\,5324}\right)(10^{12}) = 71\ \text{ppt}$$

4.87 (a) 6.0 M

The molecular equation is

$Zn(s) + 2HCl(aq) \rightarrow H_2(g) + ZnCl_2(aq)$

and the complete ionic equation is

$Zn(s) + 2H^+ + 2Cl^- \rightarrow H_2(g) + Zn^{2+} + 2Cl^-$

Chloride ions are spectator ions; the concentration of Cl^- does not change.

(b) 4.8 M H^+

Set up a table of information.

Problem	2.00 g		50.0 mL 6.0 M
Equation	$Zn(s)$	+	$2HCl(aq) \rightarrow H_2(g) + ZnCl_2(aq)$
Formula mass, u	65.39		
Recipe, moles	1		2

The wording of the question suggests that an excess of HCl was used.

To determine the amount of H^+ present, determine the amount of HCl(aq) in excess.

$$(50.0\ \cancel{\text{mL HCl}})\left(\frac{1\ \cancel{\text{L HCl}}}{10^3\ \cancel{\text{mL HCl}}}\right)\left(\frac{6.0\ \text{mol HCl}}{\cancel{\text{L HCl}}}\right) = 0.30\ \text{mol HCl present at beginning}$$

$$(2.00\ \cancel{\text{g Zn}})\left(\frac{1\ \cancel{\text{mol Zn}}}{65.39\ \cancel{\text{g Zn}}}\right)\left(\frac{2\ \text{mol HCl}}{1\ \cancel{\text{mol Zn}}}\right) = 0.0612\ \text{mol HCl needed to dissolve Zn}$$

0.30 mol HCl available – 0.0612 mol HCl reacted = 0.24 mol HCl left over

$$(0.24\ \cancel{\text{mol HCl}})\left(\frac{1\ \text{mol H}^+}{1\ \cancel{\text{mol HCl}}}\right) = 0.24\ \text{mol H}^+$$

$$\left(\frac{0.24\ \text{mol H}^+}{50.2\ \cancel{\text{mL soln}}}\right)\left(\frac{10^3\ \cancel{\text{mL soln}}}{1\ \text{L soln}}\right) = 4.8\ \text{M H}^+$$

4.89 1.953×10^{-1} M

Set up a table of information.

Problem	? M		31.24 mL 0.1250 M
Equation	$Ba(OH)_2 \cdot 8H_2O(aq)$	+	$2HCl(aq) \rightarrow BaCl_2(aq) + 10H_2O(l)$
Formula mass, u			
Recipe, mol	1		2

Calculate the moles of $Ba(OH)_2 \cdot 8H_2O$ in the supernatant liquid.

$$(31.24 \text{ mL HCl soln})\left(\frac{1 \text{ L HCl soln}}{10^3 \text{ mL HCl soln}}\right)\left(\frac{0.1250 \text{ mol HCl}}{1 \text{ L HCl soln}}\right)\left(\frac{1 \text{ mol Ba(OH)}_2 \cdot 8\text{H}_2\text{O}}{2 \text{ mol HCl}}\right)$$

$$= 1.953 \times 10^{-3} \text{ mol Ba(OH)}_2 \cdot 8\text{H}_2\text{O}$$

Using the known volume of sample and the moles, calculate the molarity.

$$\left(\frac{1.953 \times 10^{-3} \text{ mol Ba(OH)}_2 \cdot 8\text{H}_2\text{O}}{10.00 \text{ mL soln}}\right)\left(\frac{10^3 \text{ mL soln}}{1 \text{ L soln}}\right) = 1.953 \times 10^{-1} \text{ M Ba(OH)}_2 \cdot 8\text{H}_2\text{O}$$

4.91 4.593×10^{-2} M NaOH

Set up a table of information.

Problem	0.1285 g	22.91 mL ? M
Equation	$HC_7H_5O_2(s)$ +	$NaOH(aq) \rightarrow H_2O(l) + NaC_7H_5O_2(aq)$
Formula mass, u	122.12	
Recipe, mol	1	1

$$(0.1285 \text{ g BA})\left(\frac{1 \text{ mol BA}}{122.12 \text{ g BA}}\right)\left(\frac{1 \text{ mol NaOH}}{1 \text{ mol BA}}\right) = 1.052 \times 10^{-3} \text{ mol NaOH}$$

From the moles of NaOH and its volume, calculate the molarity.

$$\left(\frac{1.052 \times 10^{-3} \text{ mol NaOH}}{22.91 \text{ mL NaOH soln}}\right)\left(\frac{10^3 \text{ mL NaOH soln}}{1 \text{ L NaOH soln}}\right) = 4.593 \times 10^{-2} \text{ M NaOH}$$

4.93 Yes. At the endpoint, the ions in solution have essentially been removed.

The complete ionic equation is

$$Ba^{2+} + 2OH^- + 2H^+ + SO_4^{2-} \rightarrow BaSO_4(s) + 2H_2O(l)$$

As the titration progresses, the ions in solution are removed as $BaSO_4(s)$ and $H_2O(l)$ until at the endpoint the quantity of ions in solution is negligible. Past the endpoint, the conductivity increases because H_2SO_4 is a strong acid.

4.95 Six moles of HCl contain 3.613×10^{24} molecules of HCl. 6 M HCl refers to a solution containing 6 mol of HCl in each liter of solution.

$$(6 \text{ mol HCl})\left(\frac{6.022 \times 10^{23} \text{ molecules HCl}}{\text{mol HCl}}\right) = 3.613 \times 10^{24} \text{ molecules HCl}$$

4.97 **(a)** ammonia (NH_3), water (H_2O)

(b) hydroxide ion (OH^-)

(c) ammonium ion (NH_4^+)

4.99 6.02×10^{16}

$$(1.00\ \text{mL soln})\left(\frac{1\ \text{L soln}}{10^3\ \text{mL soln}}\right)\left(\frac{1.00 \times 10^{-4}\ \text{mol } CH_4O}{\text{L soln}}\right)\left(\frac{6.022 \times 10^{23}\ \text{molecules } CH_4O}{\text{mol } CH_4O}\right) = 6.02 \times 10^{16}\ \text{molecules } CH_4O$$

4.101 **(a)** Molecular equation: $Na_2CO_3(aq) + 2HC_2H_3O_2(aq) \rightarrow 2NaC_2H_3O_2(aq) + CO_2(g) + H_2O(l)$

Complete ionic equation: $2Na^+ + CO_3^{2-} + 2HC_2H_3O_2(aq) \rightarrow 2Na^+ + 2C_2H_3O_2^- + CO_2(g) + H_2O(l)$

Net ionic equation: $CO_3^{2-} + 2HC_2H_3O_2(aq) \rightarrow 2C_2H_3O_2^- + CO_2(g) + H_2O(l)$

(b) sodium acetate, carbon dioxide, and water

4.103 **(a)** $Cu(s) + 2AgNO_3(aq) \rightarrow Cu(NO_3)_2(aq) + 2Ag(s)$ is one example. There are others.

The pure metal reactant in a single replacement reaction must be listed above the pure metal product in the Activity Series of Metals (Table 4.5).

(b) $CH_4(g) + 2O_2(g) \rightarrow CO_2(g) + 2H_2O(g)$ is one example. There are others.

Combustion reactions typically involve the reaction of a compound containing carbon and hydrogen with oxygen to form carbon dioxide and water. Methane, CH_4, is natural gas and is combusted in gas furnaces.

(c) All the oxidation-reduction reactions involve an element as either a reactant or a product.

Applications

4.105 1.65 ppm

To calculate ppm, first calculate the number of seconds in a week:

$$(1\ \text{week})\left(\frac{7\ \text{days}}{1\ \text{week}}\right)\left(\frac{24\ \text{h}}{1\ \text{day}}\right)\left(\frac{60\ \text{min}}{1\ \text{h}}\right)\left(\frac{60\ \text{s}}{1\ \text{min}}\right) = 6.048 \times 10^5\ \text{s}$$

$$\text{ppm (time)} = \left(\frac{1\ \text{s}}{6.048 \times 10^5\ \text{s}}\right)(10^6) = 1.65\ \text{ppm}$$

4.107 **(a)** 460 ppt and 40 ppt

$460\ \text{pg} = 460 \times 10^{-12}\ \text{g}$

$$\text{ppt (mass)} = \frac{\text{mass lead}}{\text{mass wine}} \times 10^{12} = \left(\frac{460 \times 10^{-12}\ \text{g Pb}}{1\ \text{g wine}}\right)(10^{12}) = 460\ \text{ppt}$$

Similarly, $\left(\frac{40 \times 10^{-12}\ \text{g Pb}}{1\ \text{g wine}}\right)(10^{12}) = 40\ \text{ppt}$

(b) 23.2 ppm

$$\left(\frac{21\,530\ \cancel{\mu g\ Pb}}{1\ \cancel{L\ brandy}}\right)\left(\frac{10^{-6}\ g\ Pb}{1\ \cancel{\mu g\ Pb}}\right)\left(\frac{1\ \cancel{L\ brandy}}{1000\ \cancel{mL\ brandy}}\right)\left(\frac{1\ \cancel{mL\ brandy}}{0.930\ g\ brandy}\right) = \frac{2.32\times10^{-5}\ g\ Pb}{g\ brandy}$$

To obtain a value between 1 and 999, multiply by 10^6, which gives parts per million.

$$\text{ppm (mass)} = \frac{\text{mass Pb}}{\text{mass brandy}} \times 10^6 = \left(\frac{2.32\times10^{-5}\ g\ Pb}{1\ g\ brandy}\right)(10^6) = 23.2\ \text{ppm}$$

4.109 Tapwater in the bathtub, swimming–pool water, lake water, and seawater contain dissolved electrolytes that make the water conduct. Only pure water is a nonelectrolyte.

4.111 $Ca^{2+} + SO_4^{2-} \rightarrow CaSO_4(s)$
$Ba^{2+} + SO_4^{2-} \rightarrow BaSO_4(s)$
$Sr^{2+} + SO_4^{2-} \rightarrow SrSO_4(s)$

4.113 **(a)** $Mg(s) + Ag_2S(s) \rightarrow MgS(aq) + 2Ag(s)$

(b) In this process, Ag^+ in Ag_2S is converted to Ag(s), and Mg(s) to Mg^{2+}. MgS is a soluble ionic compound. Mg^{2+} ions and S^{2-} ions, therefore, remain in solution. The statement is inaccurate in that "tarnish" (Ag_2S) is not transferred to the magnesium bar. Rather the S^{2-} ion in tarnish is brought into solution by the reaction with Mg(s). The bar of Mg(s) does not "tarnish."

(c) No. The bar will be consumed during repeated uses when sufficient Ag_2S has been encountered to react with all of it.

4.115 As carbonated beverages warm up, the solubility of CO_2 in them decreases, causing CO_2 to escape from solution. This causes the drink to taste "flat."

4.117 **(a)** In a 1.00–L volumetric flask, dissolve 56 g $C_6H_{12}O_6$ in about 500 mL deionized water. Add enough deionized water to this solution to fill the flask to the mark. Mix.

Calculate the mass of glucose in 0.31 mol glucose.

$$(0.31\ \cancel{mol\ C_6H_{12}O_6})\left(\frac{180\ g\ C_6H_{12}O_6}{1\ \cancel{mol\ C_6H_{12}O_6}}\right) = 56\ g\ C_6H_{12}O_6$$

Dissolve 56 g $C_6H_{12}O_6$ in about 500 mL deionized water. Add additional deionized water to make 1.00 L of solution. Mix.

(b) 0.16 M NaCl

Each mole of NaCl contains 2 mol of ions. Thus, the concentration of NaCl need only be $\frac{1}{2}$ that of the glucose concentration.

4.119 All of the major components of seawater are soluble ionic compounds that are strong electrolytes. Therefore, they all exist in a fully ionized state in water.

4.121 **(a)** 0.526 M Cl^-

$Ag^+ + Cl^- \rightarrow AgCl(s)$

$$(26.35\ \cancel{mL\ AgNO_3\ soln})\left(\frac{1\ \cancel{L\ AgNO_3\ soln}}{10^3\ \cancel{mL\ AgNO_3\ soln}}\right)\left(\frac{0.0998\ mol\ AgNO_3}{1\ \cancel{L\ AgNO_3\ soln}}\right) = 2.63 \times 10^{-3}\ mol\ AgNO_3$$

$$(2.63 \times 10^{-3}\ \cancel{mol\ AgNO_3})\left(\frac{1\ \cancel{mol\ Ag^+}}{1\ \cancel{mol\ AgNO_3}}\right)\left(\frac{1\ mol\ Cl^-}{1\ \cancel{mol\ Ag^+}}\right) = 2.63 \times 10^{-3}\ mol\ Cl^-$$

$$\left(\frac{2.63 \times 10^{-3}\ mol\ Cl^-}{5.00\ \cancel{mL\ soln}}\right)\left(\frac{10^3\ \cancel{mL\ soln}}{1\ L\ soln}\right) = 0.526\ M\ Cl^-$$

(b) 1.84%

$$(2.63 \times 10^{-3}\ \cancel{mol\ Cl^-})\left(\frac{35.45\ g\ Cl^-}{1\ \cancel{mol\ Cl^-}}\right) = 0.0932\ g\ Cl^-$$

$$(5.00\ \cancel{mL\ sample})\left(\frac{1.0116\ g\ sample}{1\ \cancel{mL\ sample}}\right) = 5.06\ g\ sample$$

$$\left(\frac{0.0932\ g\ Cl^-}{5.06\ g\ sample}\right)(100) = 1.84\%\ Cl^-$$

4.123 **(a)** $CO(g)$ and $ZrO_2(s)$

(b) carbon monoxide and zirconium(IV) oxide

4.125 $Fe(s) + Cu^{2+} \rightarrow Fe^{2+} + Cu(s)$ or $2Fe(s) + 3Cu^{2+} \rightarrow 2Fe^{3+} + 3Cu(s)$

4.127 **(a)** Barium ion is toxic. The chloride ion cannot be toxic, since it is a component of table salt, which is not toxic.

(b) Barium sulfate can safely be taken before intestinal X-rays because it not soluble.

4.129 **(a)** $Zn(s) + 2H^+ \rightarrow Zn^{2+} + H_2(g)$

(b) The "boiling" is bubbles of hydrogen gas. Hydrogen gas forms explosive mixtures with air. Smoking should be prohibited in the area and nearby electric motors should be spark-proofed or relocated.

4.131
1. Carefully pipet 60.00 mL of the stock solution into a 100.0-mL volumetric flask.
2. Dilute the solution up to the mark with deionized water using a dropper to add the last few drops.
3. Mix thoroughly.

Calculate the volume of stock solution needed as follows:

$$(100.0\ \cancel{mL\ dil\ soln})\left(\frac{1\ \cancel{dL\ dil\ soln}}{100\ \cancel{mL\ dil\ soln}}\right)\left(\frac{150.0\ \cancel{mg\ glucose}}{1\ \cancel{dL\ dil\ soln}}\right)\left(\frac{1\ g\ glucose}{1000\ \cancel{mg\ glucose}}\right) = 0.1500\ g\ glucose$$

$$(0.1500\ \cancel{g\ glucose})\left(\frac{1\ L\ stock\ soln}{2.5000\ \cancel{g\ glucose}}\right) = 0.060\ 00\ L\ or\ 60.00\ mL\ stock\ soln$$

CHAPTER 5:
Gases

Practice Problems

5.1 **(a)** 1 Pa

There are $\frac{1.013\,25 \times 10^5 \text{ Pa}}{760 \text{ mmHg}} = 133.3$ Pa for every 1 mmHg.

(b) larger

Pascals are smaller than mmHg.

5.2 **(a)** 473.1 torr

$$(473.1 \cancel{\text{mmHg}})\left(\frac{1 \text{ torr}}{1 \cancel{\text{mmHg}}}\right) = 473.1 \text{ torr}$$

(b) 6.149×10^{-1} atm

$$(467.3 \cancel{\text{mmHg}})\left(\frac{1 \text{ atm}}{760 \cancel{\text{mmHg}}}\right) = 6.149 \times 10^{-1} \text{ atm}$$

(c) 3.67×10^3 mmHg

$$(4.83 \cancel{\text{atm}})\left(\frac{760 \text{ mmHg}}{\cancel{\text{atm}}}\right) = 3.67 \times 10^3 \text{ mmHg}$$

(d) 6.307×10^4 Pa

$$(473.1 \text{ mmHg})\left(\frac{1.013\ 25 \times 10^5 \text{ Pa}}{760 \text{ mmHg}}\right) = 6.307 \times 10^4 \text{ Pa}$$

(e) 2.3×10^5 Pa

$$(2.3 \text{ atm})\left(\frac{1.013\ 25 \times 10^5 \text{ Pa}}{\text{atm}}\right) = 2.3 \times 10^5 \text{ Pa}$$

5.3 volume increases

At constant temperature, as pressure decreases, volume increases.

5.4 less than 5.0 mL; data in Table 5.1(a) show that volume decreases as pressure increases.

5.5 about 7 mL

Find 58.0 cmHg on the vertical axis (y–axis) and lightly draw a line parallel to the horizontal axis (x–axis) from 58.0 cmHg on the y–axis to the curve. From the point where the line intersects the curve, drop a perpendicular to the x–axis and read the volume from the x–axis.

5.6 Volume is reduced to $\frac{1}{2}$ its original volume when pressure is doubled at constant temperature.

5.7 Pressure is reduced to $\frac{1}{3}$ its original pressure when volume is increased threefold at constant temperature.

5.8 638 mL

	Volume, mL	Pressure, mmHg
Old	325	538
New	?	274

The new pressure is less than the old pressure, which at constant temperature will cause the volume to become larger. For the volume to become larger, the old volume must be multiplied by a factor greater than 1.

$$(325 \text{ mL})\left(\frac{538 \text{ mmHg}}{274 \text{ mmHg}}\right) = 638 \text{ mL}$$

Check: The pressure was decreased by about $\frac{1}{2}$. Therefore, the volume is expected to about double. It does.

5.9 **(a)** 714 Pa

	Volume, mL	Pressure, Pa
Old	9.80	437
New	6.00	?

The new volume is less than the old volume, which at constant temperature requires that the pressure be increased. For the pressure to increase, the old pressure must be multiplied by a factor greater than 1.

$$(437\ \text{Pa})\left(\frac{9.80\ \cancel{\text{mL}}}{6.00\ \cancel{\text{mL}}}\right) = 714\ \text{Pa}$$

(b) 714 Pa

For ideal gases, the identity of the gas has no effect on its pressure–volume behavior.

5.10 **(a)** 273.2 K

$$(0.0\ \cancel{°\text{C}} + 273.2\ \cancel{°\text{C}})\left(\frac{1\ \text{K}}{1\ \cancel{°\text{C}}}\right) = 273.2\ \text{K}$$

(b) 313.2 K

$$(40.0\ \cancel{°\text{C}} + 273.2\ \cancel{°\text{C}})\left(\frac{1\ \text{K}}{1\ \cancel{°\text{C}}}\right) = 313.2\ \text{K}$$

(c) 235.8 K

$$(-37.4\ \cancel{°\text{C}} + 273.2\ \cancel{°\text{C}})\left(\frac{1\ \text{K}}{1\ \cancel{°\text{C}}}\right) = 235.8\ \text{K}$$

(d) 92 °C

$$(365\ \cancel{\text{K}})\left(\frac{1\ °\text{C}}{1\ \cancel{\text{K}}}\right) - 273\ °\text{C} = 92\ °\text{C}$$

(e) –68 °C

$$(205\ \cancel{\text{K}})\left(\frac{1\ °\text{C}}{1\ \cancel{\text{K}}}\right) - 273\ °\text{C} = -68\ °\text{C}$$

5.11 Volume will decrease to $\frac{1}{2}$ the original volume when Kelvin temperature is decreased by one–half at constant pressure.

At constant pressure, the volume of a gas decreases with decreasing Kelvin temperature.

5.12 Temperature will increase to fourfold the original temperature.

The volume of a sample of gas at constant pressure is increased fourfold by increasing the Kelvin temperature fourfold.

5.13 3.2 L

$$(40.0\ \cancel{°C} + 273.2\ \cancel{°C})\left(\frac{1\ K}{1\ \cancel{°C}}\right) = 313.2\ K$$

$$(0.0\ \cancel{°C} + 273.2\ \cancel{°C})\left(\frac{1\ K}{1\ \cancel{°C}}\right) = 273.2\ K$$

	Volume, L	Temperature, °C	Temperature, K
Old	3.7	40.0	313.2
New	?	0.0	273.2

The new temperature is less than the old temperature, which at constant pressure will cause the volume to decrease. For the volume to decrease, the old volume must be multiplied by a factor less than 1.

$$(3.7\ L)\left(\frac{273.2\ \cancel{K}}{313.2\ \cancel{K}}\right) = 3.2\ L$$

5.14 158 °C

$$(21.0\ \cancel{°C} + 273.2\ \cancel{°C})\left(\frac{1\ K}{1\ \cancel{°C}}\right) = 294.2\ K$$

	Volume, mL	Temperature, °C	Temperature, K
Old	6.82	21.0	294.2
New	10.00	?	?

The new volume is higher than the old volume, which at constant pressure requires the temperature to increase. For the temperature to increase, the old temperature must be multiplied by a factor greater than 1.

$$(294.2\ K)\left(\frac{10.00\ \cancel{mL}}{6.82\ \cancel{mL}}\right) = 431\ K$$

$$(431\ \cancel{K})\left(\frac{1\ °C}{1\ \cancel{K}}\right) - 273\ °C = 158\ °C$$

5.15 393 mL

$$(19.8\ \cancel{°C} + 273.2\ \cancel{°C})\left(\frac{1\ K}{1\ \cancel{°C}}\right) = 293.0\ K$$

$$(0.0\ \cancel{°C} + 273.2\ \cancel{°C})\left(\frac{1\ K}{1\ \cancel{°C}}\right) = 273.2\ K$$

	Volume, mL	Pressure, mmHg	Temperature, °C	Temperature, K
Old	419	765.4	19.8	293.0
New	?	760.0	0.0	273.2

The new pressure is lower than the old pressure, which will make the volume increase. For the volume to increase, the old volume must be multiplied by a factor greater than 1. This factor is $\frac{765.4\ mmHg}{760.0\ mmHg}$.

The new temperature is lower than the old temperature, which will make the volume decrease. For the volume to decrease, the old volume must be multiplied by a factor less than 1. This factor is $\frac{273.2\text{ K}}{293.0\text{ K}}$.

$$(419\text{ mL})\left(\frac{765.4\text{ mmHg}}{760.0\text{ mmHg}}\right)\left(\frac{273.2\text{ K}}{293.0\text{ K}}\right) = 393\text{ mL}$$

5.16 302 mL

$$(0.0\ ^\circ\text{C} + 273.2\ ^\circ\text{C})\left(\frac{1\text{ K}}{1\ ^\circ\text{C}}\right) = 273.2\text{ K}$$

$$(27.3\ ^\circ\text{C} + 273.2\ ^\circ\text{C})\left(\frac{1\text{ K}}{1\ ^\circ\text{C}}\right) = 300.5\text{ K}$$

	Volume, mL	Pressure, mmHg	Temperature, °C	Temperature, K
Old	274	760.0	0.0	273.2
New	?	758.4	27.3	300.5

The new pressure is lower than the old pressure, which will make the volume increase. For the volume to increase, the old volume must be multiplied by a factor greater than 1. That factor is $\frac{760.0\text{ mmHg}}{758.4\text{ mmHg}}$.

The new temperature is higher than the old temperature, which will make the volume increase. For the volume to increase, the old volume must be multiplied by a factor greater than 1. That factor is $\frac{300.5\text{ K}}{273.2\text{ K}}$.

$$(274\text{ mL})\left(\frac{760.0\text{ mmHg}}{758.4\text{ mmHg}}\right)\left(\frac{300.5\text{ K}}{273.2\text{ K}}\right) = 302\text{ mL}$$

5.17 **(a)** 50.6 L

At constant temperature and pressure, the same number of molecules of any gas occupies the same volume.

(b) 101.2 L (or 101 L)

Twice as many molecules will occupy twice the volume at constant temperature and pressure.

(Note: Answers with different numbers of significant figures are obtained depending upon which mathematical operation, addition or multiplication, is used. This happens because the rules for significant figures used in this book are simplified for ease of use. Better, but more complicated, rules for significant figures exist that lead to a single answer regardless of mathematical operation.)

5.18 **(a)** 1.748 L (or 1.75 L)

Problem	0.874 L		? L		? L		? L
Equation	$CH_4(g)$	+	$2O_2(g)$	→	$2H_2O(g)$	+	$CO_2(g)$
Recipe, mol	1		2		2		1

$$(0.874 \text{ L } \cancel{CH_4})\left(\frac{2 \text{ L } O_2}{1 \text{ L } \cancel{CH_4}}\right) = 1.748 \text{ L } O_2 \text{ or } 1.75 \text{ L } O_2$$

(Note: Answers with different numbers of significant figures are obtained depending upon which mathematical operation, addition or multiplication, is used. This happens because the rules for significant figures used in this book are simplified for ease of use. Better, but more complicated, rules for significant figures exist that lead to a single answer regardless of mathematical operation.)

(b) 1.748 L H_2O (or 1.75 L), 0.874 L CO_2

$$(0.874 \cancel{\text{ L } CH_4})\left(\frac{2 \text{ L } H_2O}{1 \cancel{\text{ L } CH_4}}\right) = 1.748 \text{ L } H_2O \text{ or } 1.75 \text{ L } H_2O$$

$$(0.874 \cancel{\text{ L } CH_4})\left(\frac{1 \text{ L } CO_2}{1 \cancel{\text{ L } CH_4}}\right) = 0.874 \text{ L } CO_2$$

(Note: Answers with different numbers of significant figures are obtained depending upon which mathematical operation, addition or multiplication, is used. This happens because the rules for significant figures used in this book are simplified for ease of use. Better, but more complicated, rules for significant figures exist that lead to a single answer regardless of mathematical operation.)

5.19 **(a)** 10.8 L

Problem	21.6 L		? L		? L
Equation	2CO(g)	+	O_2(g)	→	$2CO_2$(g)
Recipe, mol	2		1		2

$$(21.6 \cancel{\text{ L } CO})\left(\frac{1 \text{ L } O_2}{2 \cancel{\text{ L } CO}}\right) = 10.8 \text{ L } O_2$$

(b) 21.6 L

$$(21.6 \cancel{\text{ L } CO})\left(\frac{2 \text{ L } CO_2}{2 \cancel{\text{ L } CO}}\right) = 21.6 \text{ L } CO_2$$

5.20 68 L

$$T = (22.1 \cancel{°C} + 273.2 \cancel{°C})\left(\frac{1 \text{ K}}{1 \cancel{°C}}\right) = 295.3 \text{ K}$$

$$PV = nRT$$

$$V = \frac{nRT}{P} = \frac{(2.8 \cancel{\text{mol}})\left(62.4 \frac{\cancel{\text{mmHg}} \cdot \text{L}}{\cancel{\text{mol}} \cdot \cancel{\text{K}}}\right)(295.3 \cancel{\text{K}})}{753.4 \cancel{\text{mmHg}}} = 68 \text{ L}$$

5.21 0.23 mol

$$T = (25.9\ \text{°C} + 273.2\ \text{°C})\left(\frac{1\ \text{K}}{1\ \text{°C}}\right) = 299.1\ \text{K}$$

$$PV = nRT$$

$$n = \frac{PV}{RT} = \frac{(759.4\ \text{mmHg})(5.7\ \text{L})}{\left(62.4\ \frac{\text{mmHg} \cdot \text{L}}{\text{mol} \cdot \text{K}}\right)(299.1\ \text{K})} = 0.23\ \text{mol}$$

5.22 1.520 L

Problem	27.65 g			18.7 °C, 764.3 mmHg, ? L
Equation	$2HgO(s) \xrightarrow{\text{heat}}$	$2Hg(l)$	+	$O_2(g)$
Formula mass, u	216.59			
Recipe, mol	2			1

$$(18.7\ \text{°C} + 273.2\ \text{°C})\left(\frac{1\ \text{K}}{1\ \text{°C}}\right) = 291.9\ \text{K} = T$$

$$(27.65\ \text{g HgO})\left(\frac{1\ \text{mol HgO}}{216.59\ \text{g HgO}}\right)\left(\frac{1\ \text{mol}\ O_2}{2\ \text{mol HgO}}\right) = 0.063\ 83\ \text{mol}\ O_2 = n$$

$$PV = nRT$$

$$V = \frac{nRT}{P} = \frac{(6.383 \times 10^{-2}\ \text{mol})\left(62.36\ \frac{\text{mmHg} \cdot \text{L}}{\text{mol} \cdot \text{K}}\right)(291.9\ \text{K})}{764.3\ \text{mmHg}} = 1.520\ \text{L}$$

5.23 9.51 g/L

$$(22.4\ \text{°C} + 273.2\ \text{°C})\left(\frac{1\ \text{K}}{1\ \text{°C}}\right) = 295.6\ \text{K}.$$ Assume 1 mol He.

$$PV = nRT$$

$$V = \frac{nRT}{P} = \frac{(1\ \text{mol})\left(0.082\ 06\ \frac{\text{atm} \cdot \text{L}}{\text{mol} \cdot \text{K}}\right)(295.6\ \text{K})}{57.6\ \text{atm}} = 0.421\ \text{L He}$$

$$(1\ \text{mol He})\left(\frac{4.003\ \text{g He}}{\text{mol He}}\right) = 4.003\ \text{g He}$$

$$d = \frac{m}{V} = \frac{4.003\ \text{g He}}{0.421\ \text{L He}} = 9.51\ \text{g/L}$$

5.24 Chlorine gas has a higher density owing to its higher formula mass.

If we consider the same number of moles of each gas, then at the same temperature and pressure, fluorine gas and chlorine gas will have the same number of molecules in the same volume. However, chlorine molecules have a higher mass than fluorine molecules, giving chlorine gas a higher total mass in the same volume, or a higher density.

5.25 28.1 u

$$(24.7\ \cancel{°C} + 273.2\ \cancel{°C})\left(\frac{1\text{ K}}{1\ \cancel{°C}}\right) = 297.9\text{ K}$$

$$PV = nRT$$

$$n = \frac{PV}{RT} = \frac{(758.3\ \cancel{\text{mmHg}})(2.87\ \cancel{\text{L}})}{\left(62.36\ \frac{\cancel{\text{mmHg}} \cdot \cancel{\text{L}}}{\text{mol} \cdot \cancel{\text{K}}}\right)(297.9\ \cancel{\text{K}})} = 0.117\text{ mol}$$

$$\frac{3.29\text{ g sample}}{0.117\text{ mol sample}} = 28.1\text{ g/mol.}$$ The molecular mass is 28.1 u.

5.26 0.641 atm

$$P_{\text{total}} = p_{O_2} + p_{N_2} = 0.362\text{ atm} + 0.279\text{ atm} = 0.641\text{ atm}$$

5.27 **(a)** $p_{N_2} = 61$ atm, $p_{H_2} = 76$ atm

$$p_{N_2}V = nRT$$

$$p_{N_2} = \frac{nRT}{V} = \frac{(3.7\ \cancel{\text{mol } N_2})\left(0.0821\ \frac{\text{atm} \cdot \cancel{\text{L}}}{\cancel{\text{mol}} \cdot \cancel{\text{K}}}\right)(5.00 \times 10^2\ \cancel{\text{K}})}{2.5\ \cancel{\text{L}}} = 61\text{ atm}$$

$$p_{H_2}V = nRT$$

$$p_{H_2} = \frac{nRT}{V} = \frac{(4.6\ \cancel{\text{mol } H_2})\left(0.0821\ \frac{\text{atm} \cdot \cancel{\text{L}}}{\cancel{\text{mol}} \cdot \cancel{\text{K}}}\right)(5.00 \times 10^2\ \cancel{\text{K}})}{2.5\ \cancel{\text{L}}} = 76\text{ atm}$$

(b) 136 atm

$$P_{\text{total}} = p_{N_2} + p_{H_2} = 61\text{ atm} + 76\text{ atm} = 137\text{ atm}$$

(Best answer, without intermediate round off in part **a**, is 136 atm.)

5.28 141 atm

$$P_{\text{total}} = \frac{nRT}{V} = \frac{(5.83\ \cancel{\text{mol}} + 2.76\ \cancel{\text{mol}})(0.082\ 06\text{ atm} \cdot \cancel{\text{L}} \cdot \cancel{\text{mol}^{-1}} \cdot \cancel{\text{K}^{-1}})(752\ \cancel{\text{K}})}{3.75\ \cancel{\text{L}}} = 141\text{ atm}$$

5.29 **(a)** 13.634 mmHg

From Table 5.4, the vapor pressure of water is 13.634 mmHg at 16.0 °C.

(b) 17.212 mmHg

From Table 5.4, the vapor pressure of water is 17.105 mmHg at 19.6 °C and 17.319 mmHg at 19.8 °C.

$$\frac{(17.105\text{ mmHg} + 17.319\text{ mmHg})}{2} = 17.212\text{ mmHg}$$

5.30 0.550 L

$Zn(s) + 2HCl(aq) \rightarrow ZnCl_2(aq) + H_2(g)$

$753.8\text{ mmHg} = P_{total}$
$0.567\text{ L} = V_{wet}$
$22.922\text{ mmHg} = p_{H_2O(g)}$ (Table 5.4)

$$P_{total} = p_{H_2(g)} + p_{H_2O(g)}$$

$$p_{H_2(g)} = P_{total} - p_{H_2O(g)} = 753.8\text{ mmHg} - 22.922\text{ mmHg} = 730.9\text{ mmHg}$$

	Volume, L	Pressure, mmHg	Temperature, °C
$H_2(g)$, wet	0.567	730.9	24.4
$H_2(g)$, dry	?	753.8	24.4

The dry $H_2(g)$ pressure is greater than the wet $H_2(g)$ pressure, which at constant temperature would make the volume decrease. For the volume to decrease, it must be multiplied by a factor less than 1.

$$(0.567\text{ L }H_2)\left(\frac{730.9\text{ }\cancel{\text{mmHg}}}{753.8\text{ }\cancel{\text{mmHg}}}\right) = 0.550\text{ L }H_2(g)\text{, dry}$$

5.31 $CO(g)$ (molecular mass = 28 u) will diffuse faster than $CO_2(g)$ (molecular mass = 44 u) because gases having low molecular masses diffuse faster than gases having higher molecular masses.

5.32 Charles's law states that at constant pressure, the volume of a gas is directly proportional to the Kelvin temperature. According to the kinetic–molecular theory, as the Kelvin temperature is increased, the average kinetic energy of the molecules is increased. This causes the molecules to have higher speeds and to have more frequent and forceful collisions with the walls of the container. If the pressure is held constant, the volume will increase. If the volume is held constant, the pressure of the gas will increase.

5.33 **(a)** intermolecular attraction; compressibility factor < 1.0

According to Figure 5.17(a), at 200 atm the compressibility factor for CH_4 is less than 1.0, indicating the importance of intermolecular attractions. Attractions between molecules cause a reduction in pressure.

(b) molecular volume; compressibility factor > 1.0

At 800 atm, the compressibility factor for CH_4 is significantly greater than 1.0, showing that molecular volume has a greater effect than intermolecular attractions. However, even at 200 atm, the molecular volume has begun to have an influence, as indicated by the fact that the minimum in the curve is to the left of 200 atm for CH_4 in Figure 5.17(a).

5.34 low temperatures

According to Figure 5.17(c), at 100 atm the difference between the compressibility factor for the real gas nitrogen and an ideal gas is greater at low temperatures. As temperature increases, the curves are closer to the ideal gas line.

Additional Practice Problems

5.67 $p_{Ne} = 1.2 \times 10^2$ mmHg, $p_{Kr} = 1.9 \times 10^2$ mmHg, $P_{total} = 3.0 \times 10^2$ mmHg

Since the temperature is held constant during this experiment, it will have no effect on the pressure of the system. To determine the partial pressures of Ne and Kr, treat them as totally independent of one another.

Ne:	Volume, L	Pressure, mmHg
Old	5.0	237
New	10.0	?

$$p_{Ne} = (237\text{ mmHg})\left(\frac{5.0\cancel{L}}{10.0\cancel{L}}\right) = 1.2 \times 10^2\text{ mmHg}$$

Kr:	Volume, L	Pressure, mmHg
Old	5.0	372
New	10.0	?

$$p_{Kr} = (372\text{ mmHg})\left(\frac{5.0\cancel{L}}{10.0\cancel{L}}\right) = 1.9 \times 10^2\text{ mmHg}$$

$$P_{total} = p_{Ne} + p_{Kr} = 1.2 \times 10^2\text{ mmHg} + 1.9 \times 10^2\text{ mmHg} = 3.1 \times 10^2\text{ mmHg}$$

(Best answer, without intermediate round off, is 3.0×10^2 mmHg.)

5.69 The water within the glass tube will be pushed down about six inches, but no farther.

The pressure at the bottom of the glass tube, which is at the bottom of the graduated cylinder, is equal to atmospheric pressure plus about 20 inches of water (see below). The pressure within the gas line is not large enough to push all the water out of the tube.

$$(50\,\cancel{cm})\left(\frac{1\text{ in.}}{2.54\,\cancel{cm}}\right) = 20\text{ in.}$$

5.71 **(a)** near zero atmospheres

As the pressure on a real gas is reduced, the molecules move farther apart, allowing for significantly less interaction between them.

(b) 800 atm

(c) As pressure increases, the compressibility factors of real gases decrease due to stronger attractions between molecules. They also increase with increasing pressure due to the actual volume of the molecules becoming a significant part of the total gas volume. At this particular pressure, the two effects exactly offset each other.

5.73 The same. By increasing the diameter of the tube, the force exerted downward by the increased amount of mercury would increase. However, the area over which this force would be exerted would also increase, meaning that the pressure (force/area) would not change. In order for the height of the mercury to change, the pressure must change.

5.75 775.3 mmHg

At constant volume, the pressure of the neon will increase with increasing temperature.

	Pressure, mmHg	Temperature, °C	Temperature, K
Old	742.3	16.8	290.0
New	?	29.7	302.9

$$(16.8\ \cancel{°C} + 273.2\ \cancel{°C})\left(\frac{1\ K}{1\ \cancel{°C}}\right) = 290.0\ K$$

$$(29.7\ \cancel{°C} + 273.2\ \cancel{°C})\left(\frac{1\ K}{1\ \cancel{°C}}\right) = 302.9\ K$$

$$(742.3\ mmHg)\left(\frac{302.9\ \cancel{K}}{290.0\ \cancel{K}}\right) = 775.3\ mmHg$$

5.77 2.60×10^2 mL

The total pressure of the gas collected over water will be the barometric pressure, 762.3 mmHg. According to Dalton's law, this pressure will be equal to the sum of the partial pressure of the nitrogen and the partial pressure of water vapor. According to Table 5.4, the vapor pressure of water at 20.0 °C is 17.535 mmHg.

$$P_{atm} = p_{N_2} + p_{H_2O\ vapor}$$

$$P_{atm} - p_{H_2O\ vapor} = p_{N_2} \quad = 762.3\ mmHg - 17.535\ mmHg$$
$$= 744.8\ mmHg$$

	Volume, mL	Pressure, mmHg	Temperature, °C
dry N_2	254	762.3	20.0
wet N_2	?	744.8	20.0

$$(254\ mL)\left(\frac{762.3\ \cancel{mmHg}}{744.8\ \cancel{mmHg}}\right) = 260\ mL$$

5.79 **(a)** The volume will decrease by a factor of $\frac{373\ \cancel{K}}{473\ \cancel{K}} = 0.789$.

$$(200\ \cancel{°C} + 273\ \cancel{°C})\left(\frac{1\ K}{1\ \cancel{°C}}\right) = 473\ K; \qquad (100\ \cancel{°C} + 273\ \cancel{°C})\left(\frac{1\ K}{1\ \cancel{°C}}\right) = 373\ K$$

By Charles's law, the volume is directly proportional to the Kelvin temperature. Therefore, the volume will be reduced as the temperature is reduced by a factor of $\frac{373\ \cancel{K}}{473\ \cancel{K}}$.

(b) The volume will decrease by a factor of $\frac{323\ \cancel{K}}{373\ \cancel{K}} = 0.866$.

Stop and Test Yourself

1. **(b)** 9.93×10^4

$$(745\ \cancel{mmHg})\left(\frac{1.013 \times 10^5\ Pa}{760\ \cancel{mmHg}}\right) = 9.93 \times 10^4\ Pa$$

2. **(d)** 0.299 atm

	Volume, cm^3	Pressure, atm
Old	39.0	0.568
New	74.0	?

The volume increased, which at constant temperature means the pressure must decrease. For the pressure to decrease, the old pressure must be multiplied by a factor less than one.

$$(0.568\ atm)\left(\frac{39.0\ \cancel{cm^3}}{74.0\ \cancel{cm^3}}\right) = 0.299\ atm$$

3. **(d)** $24.5\ cm^3$

$$(125\ \cancel{°C} + 273\ \cancel{°C})\left(\frac{1\ K}{1\ \cancel{°C}}\right) = 398\ K; \qquad (-23\ \cancel{°C} + 273\ \cancel{°C})\left(\frac{1\ K}{1\ \cancel{°C}}\right) = 250\ K$$

	Volume, cm^3	Temperature, °C	Temperature, K
Old	39.0	125	398
New	?	–23	250

For the temperature to decrease at constant pressure, the volume must decrease. Therefore, the old volume must be multiplied by a factor less than one.

$$(39.0\ cm^3)\left(\frac{250\ \cancel{K}}{398\ \cancel{K}}\right) = 24.5\ cm^3$$

4. **(a)** 6.80 L

$(67\ \cancel{°C} + 273\ \cancel{°C})\left(\frac{1\ K}{1\ \cancel{°C}}\right) = 340\ K$; $(0\ \cancel{°C} + 273\ \cancel{°C})\left(\frac{1\ K}{1\ \cancel{°C}}\right) = 273\ K$

	Volume, liters	Pressure, mmHg	Temperature, °C	Temperature, K
Old	8.60	748	67	340
New	?	760	0	273

The pressure increase will cause the volume of the gas to decrease. Therefore, the old volume must be multiplied by a factor less than one. The temperature decrease will also cause the volume to decrease; therefore, the volume must be multiplied by a factor less than one.

$$(8.60\ L)\left(\frac{748\ \cancel{mmHg}}{760\ \cancel{mmHg}}\right)\left(\frac{273\ \cancel{K}}{340\ \cancel{K}}\right) = 6.80\ L$$

5. **(d)** Both contain the same number of molecules of gas.

Avogadro stated that equal volumes of all gases at the same temperature and pressure contain the same number of molecules.

(a) is incorrect because $NH_3(g)$ has a higher molecular mass than $H_2(g)$.

(b) is incorrect because every molecule of $NH_3(g)$ contains 50% more hydrogen atoms than any $H_2(g)$ molecule.

(c) is incorrect because, in fact, the volume of the molecules of a gas is very small compared to the volume of the container.

6. **(a)** 4.9 L

Problem	9.8 L		? L		
Equation	$2SO_2(g)$	+	$O_2(g)$	→	$2SO_3(g)$
Recipe, mol	2		1		

$$(9.8\ \cancel{L\ SO_2})\left(\frac{1\ L\ O_2}{2\ \cancel{L\ SO_2}}\right) = 4.9\ L\ O_2$$

(b) and (c) are incorrect because they would result from improperly balancing the equation.

(d) is incorrect because this is the volume that one mole of any ideal gas occupies at STP.

7. **(c)** 0.602 g/L

$PV = nRT$ Therefore, $\frac{P}{RT} = \frac{n}{V}$

$$\frac{745 \text{ mmHg}}{(62.36 \text{ mmHg} \cdot \text{L} \cdot \text{mol}^{-1} \cdot \text{K}^{-1})(338 \text{ K})} = 0.0353 \text{ mol/L}$$

$$\left(\frac{0.0353 \text{ mol NH}_3}{\text{L}}\right)\left(\frac{17.03 \text{ g NH}_3}{\text{mol NH}_3}\right) = 0.601 \text{ g/L}$$

(Best answer, without intermediate round, off is 0.602 g/L.)

8. **(d)** 30.1

$(0\ °\text{C} + 273\ °\text{C})\left(\frac{1 \text{ K}}{1\ °\text{C}}\right) = 273 \text{ K};$ $(448 \text{ mL})\left(\frac{1 \text{ L}}{10^3 \text{ mL}}\right) = 0.448 \text{ L}$

$PV = nRT$ Therefore, $n = \frac{PV}{RT} = \frac{(760 \text{ mmHg})(0.448 \text{ L})}{(62.36 \text{ mmHg} \cdot \text{L} \cdot \text{mol}^{-1} \cdot \text{K}^{-1})(273 \text{ K})} = 0.0200 \text{ mol}$

$$\text{molecular mass} = \frac{0.602 \text{ g sample}}{0.0200 \text{ mol sample}} = 30.1 \text{ g/mol}$$

9. **(b)** 728.7 mmHg

According to Table 5.4, the vapor pressure of water at 22.1 °C is:

(19.827 + 20.070)/2 mmHg = 19.949 mmHg

$P_{total} = p_{gas} + p_{water\ vapor}$ and $p_{gas} = p_{total} - p_{H_2O} = 748.6 \text{ mmHg} - 19.949 \text{ mmHg} = 728.7 \text{ mmHg}$

(a) and (c) both represent round–off errors.

(d) results from adding the pressure of the water vapor rather than subtracting it from the barometric pressure, P_{total}.

10. **(c)** $F_2(g)$

At a given temperature and pressure, the molecules of all gases have the same average kinetic energy. At the same average kinetic energy, mass and speed are inversely related. $F_2(g)$ has a lower molecular mass than any other halogen gas. Therefore, at the same temperature and pressure, its molecules will have a higher average speed.

11. **(c)** The molecules of a gas are attracted to the walls of the container.

12. Curve (a)

This curve has the lowest average molecular speed and the lowest "most probable" molecular speed.

13. (c) One at 1 atm and 50 °C

At lower pressures and higher temperatures, gases most closely approximate ideal behavior. This is because under these conditions the molecules are much farther apart on the average and have a higher average kinetic energy, reducing the intermolecular interactions.

(a) and (b) represent lower temperatures, at which the molecules have a lower average kinetic energy. This allows for higher intermolecular interactions. (d) is incorrect because at higher pressures, the molecules are closer together on the average, causing stronger intermolecular interactions.

14. (c) iii and iv

The theory of ideal gases is based on the assumption that ideal gas molecules have no volume and that they have no intermolecular interactions. Molecules of real gases do have both volume and intermolecular interactions.
(a) and (b) are not correct because molecules of ideal gases are also in constant motion and they collide with the walls of the container.

15. (c) $\frac{P_1}{T_1} = \frac{P_2}{T_2}$, $P = CT$

When the volume is constant, $\frac{P_1 V}{T_1} = \frac{P_2 V}{T_2}$.

Dividing both sides of the equation by V, $\frac{P_1}{T_1} = \frac{P_2}{T_2}$.

This expression shows that $\frac{P}{T}$ = constant or $P = CT$.

Putting Things Together

5.81 0.232 atm

$$(4.98\ \cancel{\text{kPa}})\left(\frac{10^3\ \cancel{\text{Pa}}}{\cancel{\text{kPa}}}\right)\left(\frac{1\ \text{atm}}{1.013\ 25 \times 10^5\ \cancel{\text{Pa}}}\right) = 4.91 \times 10^{-2}\ \text{atm}$$

$$(768\ \cancel{\text{mL}})\left(\frac{1\ \text{L}}{10^3\ \cancel{\text{mL}}}\right) = 0.768\ \text{L}$$

	Volume, L	Pressure, atm
Old	3.62	4.91×10^{-2}
New	0.768	?

$$(4.91 \times 10^{-2}\ \text{atm})\left(\frac{3.62\ \cancel{\text{L}}}{0.768\ \cancel{\text{L}}}\right) = 0.231\ \text{atm}$$

(Best answer, without intermediate round off, is 0.232 atm.)

5.83 **(a)** $Cl_2(g)$ at 35 °C; molecules move faster at higher temperatures.

Gas molecules move faster at higher temperatures because their average kinetic energy increases with temperature.

(b) through a vacuum; it would not be slowed down by collisions with N_2 and O_2 molecules in air.

5.85 3.00×10^2 mL

Problem	0.783 g		excess		? mL over water at 751.8 mmHg and 21.2 °C
Equation	Zn(s)	+	$2HCl(aq) \rightarrow ZnCl_2(aq)$	+	$H_2(g)$
Formula mass, u	65.39				2.016
Recipe, mol	1				1

Calculate moles of $H_2(g)$ formed.

$$(0.783\ \cancel{g\ Zn})\left(\frac{1\ \cancel{mol\ Zn}}{65.39\ \cancel{g\ Zn}}\right)\left(\frac{1\ mol\ H_2}{1\ \cancel{mol\ Zn}}\right) = 0.0120\ mol\ H_2$$

Convert Celsius temperature to Kelvin.

$$(21.2\ \cancel{°C} + 273.2\ \cancel{°C})\left(\frac{1\ K}{1\ \cancel{°C}}\right) = 294.4\ K$$

Determine the pressure of water vapor over water at 21.2 °C using Table 5.4: 18.880 mmHg.

Determine the pressure of dry H_2 from the barometric pressure and the water vapor pressure.

$$P_{atm} = p_{H_2} + p_{H_2O\ vapor}; \qquad p_{H_2} = P_{atm} - p_{H_2O\ vapor}$$

$$p_{H_2} = 751.8\ mmHg - 18.88\ mmHg = 732.9\ mmHg$$

Using the ideal gas equation, solve for volume.

$$V = \frac{nRT}{P} = \frac{(0.0120\ \cancel{mol})(62.36\ \cancel{mmHg} \cdot L \cdot \cancel{mol^{-1}} \cdot \cancel{K^{-1}})(294.4\ \cancel{K})}{732.9\ \cancel{mmHg}}$$

$$= (0.301\ \cancel{L})\left(\frac{10^3\ mL}{\cancel{L}}\right) = 301\ mL$$

(Best answer, without intermediate round off, is 3.00×10^2 mL.)

5.87 **(a)** 17

362 mL – 345 mL = 17 mL

Since equal volumes of H_2 and Cl_2 will react according to Gay–Lussac's law of combining volumes, only 345 mL H_2 will react with 345 mL of Cl_2.

(b) 0

The Cl_2 is the limiting reactant and will be used up.

(c) 6.90×10^2

$$(345\ \text{mL}\ Cl_2)\left(\frac{2\ \text{mL HCl}}{\text{mL}\ Cl_2}\right) = 690\ \text{mL HCl}$$

Every volume of Cl_2 produces two volumes of HCl.

5.89 **(a)** 1.06×10^{22}

$$(22.4\ °\text{C} + 273.2\ °\text{C})\left(\frac{1\ \text{K}}{1\ °\text{C}}\right) = 295.6\ \text{K}$$

$$(438\ \text{mL})\left(\frac{1\ \text{L}}{10^3\ \text{mL}}\right) = 0.438\ \text{L}$$

$PV = nRT$

$$n = \frac{PV}{RT} = \frac{(739.5\ \text{mmHg})(0.438\ \text{L})}{(62.36\ \text{mmHg}\cdot\text{L}\cdot\text{mol}^{-1}\cdot\text{K}^{-1})(295.6\ \text{K})} = 0.0176\ \text{mol}$$

$$(0.0176\ \text{mol})\left(\frac{6.022 \times 10^{23}\ \text{molecules}}{\text{mol}}\right) = 1.06 \times 10^{22}\ \text{molecules}$$

(b) 2.12×10^{22}

$$(1.06 \times 10^{22}\ \text{molecules}\ CO_2)\left(\frac{2\ \text{atoms O}}{1\ \text{molecule}\ CO_2}\right) = 2.12 \times 10^{22}\ \text{atoms O}$$

5.91 C_2H_6

80.0% C + 20.0% H = 100.0%. Compound does not contain any other elements.

Determine the empirical formula by first assuming 100.0 g of sample.

$$(80.0\ \text{g C})\left(\frac{1\ \text{mol C}}{12.01\ \text{g C}}\right) = 6.66\ \text{mol C}$$

$$(20.0\ \text{g H})\left(\frac{1\ \text{mol H}}{1.008\ \text{g H}}\right) = 19.8\ \text{mol H}$$

$$C_{6.66/6.66}H_{19.8/6.66} = CH_3$$

Using the ideal gas equation, determine the molar mass of the compound.

$$(22\ \text{°C} + 273\ \text{°C})\left(\frac{1\ \text{K}}{1\ \text{°C}}\right) = 295\ \text{K}$$

$$PV = nRT = \left(\frac{\text{mass}}{\text{molar mass}}\right)RT$$

$$\text{molar mass} = \frac{(\text{mass})RT}{PV} = \frac{(0.2367\ \text{g})(62.36\ \text{mmHg} \cdot \text{L} \cdot \text{mol}^{-1} \cdot \text{K}^{-1})(295\ \text{K})}{(756.8\ \text{mmHg})(0.1917\ \text{L})} = 30.0\ \text{g} \cdot \text{mol}^{-1}$$

Finally, find the molecular formula.

Calculate the formula mass for CH_3.

1 C @ 12.01 u	=	12.01 u
3 H @ 1.008 u	=	3.024 u
1 CH_3	=	15.03 u

$$\frac{30.0\ \text{g} \cdot \text{mol}^{-1}}{15.03\ \text{g} \cdot \text{mol}^{-1}} \approx 2$$

$2(CH_3) = C_2H_6$

5.93 (a) 2.37 L

Problem	5.68 g		6.46 mL of 17.8 M			? L, 24.3 °C, 762.4 mmHg
Equation	NaCl(s)	+	H_2SO_4(aq) $\xrightarrow{\text{heat}}$ $NaHSO_4$(s)		+	HCl(g)
Formula mass, u	58.44					
Recipe, mol	1		1			1

Find the limiting reactant.

$$(5.68\ \text{g NaCl})\left(\frac{1\ \text{mol NaCl}}{58.44\ \text{g NaCl}}\right) = 0.0972\ \text{mol NaCl}$$

$$(6.46\ \text{mL soln})\left(\frac{1\ \text{L soln}}{10^3\ \text{mL soln}}\right)\left(\frac{17.8\ \text{mol}\ H_2SO_4}{\text{L soln}}\right) = 0.115\ \text{mol}\ H_2SO_4$$

$$\frac{0.0972\ \text{mol NaCl}}{1\ \text{mol NaCl}} = 0.0972\ \text{of a recipe} \qquad \frac{0.115\ \text{mol}\ H_2SO_4}{1\ \text{mol}\ H_2SO_4} = 0.115\ \text{of a recipe}$$

$0.0972 < 0.115$. NaCl is the limiting reactant and H_2SO_4 is in excess.

$$(0.0972\ \text{mol NaCl})\left(\frac{1\ \text{mol HCl}}{1\ \text{mol NaCl}}\right) = 0.0972\ \text{mol HCl}$$

$$(24.3\ \text{°C} + 273.2\ \text{°C})\left(\frac{1\ \text{K}}{1\ \text{°C}}\right) = 297.5\ \text{K}$$

$$V = \frac{nRT}{P} = \frac{(0.0972\ \cancel{\text{mol}})(62.36\ \cancel{\text{mmHg}} \cdot \text{L} \cdot \cancel{\text{mol}^{-1}} \cdot \cancel{\text{K}^{-1}})(297.5\ \cancel{\text{K}})}{762.4\ \cancel{\text{mmHg}}} = 2.37\ \text{L}$$

(b) No. HCl is quite soluble in water.

5.95 8.9 L

			3.5×10^2 °C, 3.0×10^2 atm		0.7914 g/mL
Problem			? L		1.00 L
Equation	CO(g)	+	$2H_2(g)$	→	$CH_3OH(l)$
Formula mass, u			2.016		32.04
Recipe, mol			2		1

Determine the actual yield of CH_3OH from the volume and density.

$$(1.00\ \cancel{\text{L}})\left(\frac{10^3\ \cancel{\text{mL}}}{\cancel{\text{L}}}\right)\left(\frac{0.7914\ \text{g}}{\cancel{\text{mL}}}\right) = 791\ \text{g}\ CH_3OH$$

From the actual yield and the percent yield, determine the theoretical yield needed.

$$\frac{\text{Actual yield}}{\text{Theoretical yield}}(100) = \text{Percent yield}$$

$$\text{Theoretical yield} = \frac{\text{Actual yield (100)}}{\text{Percent yield}} = \frac{791\ \text{g (100)}}{95\%} = 8.3 \times 10^2\ \text{g}$$

Calculate the moles of H_2 required.

$$(8.3 \times 10^2\ \cancel{\text{g}\ CH_3OH})\left(\frac{1\ \cancel{\text{mol}\ CH_3OH}}{32.0\ \cancel{\text{g}\ CH_3OH}}\right)\left(\frac{2\ \text{mol}\ H_2}{1\ \cancel{\text{mol}\ CH_3OH}}\right) = 52\ \text{mol}\ H_2$$

$$(3.5 \times 10^2\ \cancel{°\text{C}} + 273\ \cancel{°\text{C}})\left(\frac{1\ \text{K}}{1\ \cancel{°\text{C}}}\right) = 6.2 \times 10^2\ \text{K}$$

$$V = \frac{nRT}{P} = \frac{(52\ \cancel{\text{mol}})(0.0821\ \text{L} \cdot \cancel{\text{atm}} \cdot \cancel{\text{mol}^{-1}} \cdot \cancel{\text{K}^{-1}})(6.2 \times 10^2\ \cancel{\text{K}})}{3.0 \times 10^2\ \cancel{\text{atm}}} = 8.8\ \text{L}$$

(Best answer, without intermediate round off, is 8.9 L.)

5.97 (b), (c), and (e). (b) and (e) are compounds of nonmetals and are molecular; (c) is a triatomic molecule of a nonmetal. (a) is an ionic compound and is a solid. (d) is also a solid under ordinary conditions (see Figure 2.16).

5.99 Yes. Under ordinary conditions, all gaseous mixtures are solutions (homogeneous mixtures). Gases are mostly empty space. In the gaseous mixture, water molecules are in the space between the pentane molecules, and pentane molecules are in the space between water molecules.

Applications

5.101 **(a)** 16.0 kPa, 11 kPa (or 16 kPa, 1×10^1 kPa if zeros in data are not significant)

$$(120\ \text{mmHg})\left(\frac{1.013 \times 10^5\ \text{Pa}}{760\ \text{mmHg}}\right)\left(\frac{1\ \text{kPa}}{10^3\ \text{Pa}}\right) = 16.0\ \text{kPa}$$

$$(80\ \text{mmHg})\left(\frac{1.01 \times 10^5\ \text{Pa}}{760\ \text{mmHg}}\right)\left(\frac{1\ \text{kPa}}{10^3\ \text{Pa}}\right) = 11\ \text{kPa}$$

(b) 6.0×10^{-3} atm

$$(6.1\ \text{mbar})\left(\frac{10^{-3}\ \text{bar}}{\text{mbar}}\right)\left(\frac{1\ \text{atm}}{1.01\ \text{bar}}\right) = 6.0 \times 10^{-3}\ \text{atm}$$

(c) 5×10^4 mmHg

$$(6\ \text{MPa})\left(\frac{10^6\ \text{Pa}}{\text{MPa}}\right)\left(\frac{760\ \text{mmHg}}{1.0 \times 10^5\ \text{Pa}}\right) = 5 \times 10^4\ \text{mmHg}$$

5.103 **(a)** The likely reason for having difficulty in opening the can is that the pressure of the air above the liquid in the can is lower than the surrounding atmospheric pressure. By warming the can, the pressure of the air in the can, which is confined in a constant volume, will be increased, reducing the difference in pressure.

(b) No. The reason the screw cap won't open is related to friction of the cap to the bottle rather than pressure differences. If pressure inside the screw–top container becomes greater than atmospheric pressure, the top will probably be even harder to unscrew than before the container was heated.

5.105 **(a)** The reduced pressure on the outside of the tennis ball causes the air inside the ball, which is at nearly twice the atmospheric pressure, to slowly leak out.

(b) 1. Coat the inside of the ball with a flexible but impervious coating to reduce the air leakage.

2. Use a gas that effuses slower than air, such as carbon dioxide.

3. Place a higher pressure of air in the ball to allow it to remain at an increased level for a longer time.

(c) $SF_6(g)$ has a larger formula mass (146 u) than air (~29 u) and therefore diffuses out of the tennis ball more slowly.

5.107 **(a)** ammonia, NH_3; argon, Ar; boron trichloride, BCl_3; carbon tetrafluoride, CF_4; chlorine, Cl_2; hydrogen bromide, HBr; hydrogen iodide, HI; hydrogen selenide, H_2Se; nitric oxide, NO; nitrogen dioxide, NO_2; nitrous oxide, N_2O; phosphorus pentafluoride, PF_5; silicon tetrachloride, $SiCl_4$; sulfur hexafluoride, SF_6

(b) hydrogen selenide (H_2Se), nitrogen dioxide (NO_2), and nitrous oxide (N_2O)

Triatomic molecules contain three atoms.

(c) All the elements that compose these gases are nonmetals.

5.109 2.46×10^{19} molecules/cm^3

Assume the pressure is 760 mmHg.

$$(25\ ^\circ\text{C} + 273\ ^\circ\text{C})\left(\frac{1\ \text{K}}{1\ ^\circ\text{C}}\right) = 298\ \text{K}$$

$$(1.00\ \text{cm}^3)\left(\frac{1\ \text{mL}}{\text{cm}^3}\right)\left(\frac{10^{-3}\ \text{L}}{\text{mL}}\right) = 1.00 \times 10^{-3}\ \text{L}$$

$$n = \frac{PV}{RT} = \frac{(760\ \text{mmHg})(1.00 \times 10^{-3}\ \text{L})}{(62.36\ \text{L} \cdot \text{mmHg} \cdot \text{mol}^{-1} \cdot \text{K}^{-1})(298\ \text{K})} = 4.09 \times 10^{-5}\ \text{mol air}$$

$$(4.09 \times 10^{-5}\ \text{mol air})\left(\frac{6.022 \times 10^{23}\ \text{molecules air}}{\text{mol air}}\right) = 2.46 \times 10^{19}\ \text{molecules air}$$

5.111 2.6×10^{21} molecules

$$(20\ ^\circ\text{C} + 273\ ^\circ\text{C})\left(\frac{1\ \text{K}}{1\ ^\circ\text{C}}\right) = 293\ \text{K}; \qquad (500\ \text{mL})\left(\frac{10^{-3}\ \text{L}}{\text{mL}}\right) = 0.500\ \text{L}$$

If the oxygen is 21% by volume, it would occupy only 21% of the volume of the air under the same pressure and temperature.

$$(0.500\ \text{L air})\left(\frac{21\ \text{L}\ O_2}{100\ \text{L air}}\right) = 0.11\ \text{L}\ O_2$$

$$n = \frac{PV}{RT} = \frac{(1.0\ \text{atm})(0.11\ \text{L})}{(0.0821\ \text{L} \cdot \text{atm} \cdot \text{mol}^{-1} \cdot \text{K}^{-1})(293\ \text{K})} = 4.6 \times 10^{-3}\ \text{mol}\ O_2$$

From the moles of O_2, calculate the molecules of oxygen.

$$(4.6 \times 10^{-3}\ \text{mol}\ O_2)\left(\frac{6.02 \times 10^{23}\ \text{molecules}\ O_2}{\text{mol}\ O_2}\right) = 2.8 \times 10^{21}\ \text{molecules}\ O_2$$

(Best answer, without intermediate round off, is 2.6×10^{21} molecules.)

5.113 The units are "inches of Hg" or in. of Hg.

$$\frac{763.3\ \text{mmHg}}{30.05\ ?} = 25.40\ \frac{\text{mmHg}}{?}.$$

From the conversions on the inside back cover of the text, $1\ \text{in} = (2.54\ \text{cm})\left(\frac{10\ \text{mm}}{1\ \text{cm}}\right) = 25.40\ \text{mm}$

Thus, the units are "inches of Hg" or in. of Hg.

5.115 9.9×10^4 atm

$$(10\ \text{GPa})\left(\frac{10^9\ \text{Pa}}{1\ \text{GPa}}\right)\left(\frac{1\ \text{atm}}{1.01 \times 10^5\ \text{Pa}}\right) = 9.9 \times 10^4\ \text{atm}$$

5.117 12.4 L

The volume of 6.022×10^{23} molecules of any gas is 22.41 L at STP.

$$(3.34 \times 10^{23} \cancel{\text{molecules}})\left(\frac{22.41 \text{ L}}{6.022 \times 10^{23} \cancel{\text{molecules}}}\right) = 12.4 \text{ L}$$

5.119 7×10^1 °C

Convert the temperature to Kelvins.

$$(21 \cancel{°C} + 273 \cancel{°C})\left(\frac{1 \text{ K}}{1 \cancel{°C}}\right) = 294 \text{ K}$$

25 psi + 14.7 psi = 40 psi and 31 psi + 14.7 psi = 46 psi actual tire pressure, since the gauge reads zero at atmospheric pressure.

	Pressure, psi	Temperature, °C	Temperature, K
Old	40	21	294
New	46	?	?

$$(294 \text{ K})\left(\frac{46 \cancel{\text{psi}}}{40 \cancel{\text{psi}}}\right) = 3.4 \times 10^2 \text{ K}$$

$$(3.4 \times 10^2 \cancel{\text{K}} - 273 \cancel{\text{K}})\left(\frac{1 \text{ °C}}{1 \cancel{\text{K}}}\right) = 7 \times 10^1 \text{ °C}$$

5.121 **(a)** –460 °F

Convert 0 K or –273.15 °C to Fahrenheit degrees to obtain 0 °R.

$$t_F \text{ °F} = \left(\frac{9 \text{ °F}}{5 \text{ °C}}\right) t_C \text{ °C} + 32 \text{ °F} = \left(\frac{9 \text{ °F}}{5 \cancel{°C}}\right)(-273.2 \cancel{°C}) + 32 \text{ °F} = -460 \text{ °F}$$

(b) 0.023 65 psi · ft^3 · mol^{-1} · °R^{-1}

Solving the ideal gas law for R gives

$$R = \frac{PV}{nT}$$

The volume of one mole of an ideal gas at STP is 22.4141 L. Convert this volume to cubic feet.

$$(22.4141 \cancel{\text{L}})\left(\frac{10^3 \cancel{\text{cm}^3}}{\cancel{\text{L}}}\right)\left(\frac{1 \cancel{\text{in.}}}{2.54 \cancel{\text{cm}}}\right)^{\cancel{3}}\left(\frac{1 \text{ ft}}{12 \cancel{\text{in.}}}\right)^3 = 0.791\ 55 \text{ ft}^3$$

Convert temperature from Kelvin to Rankine

$$(273.15 \cancel{\text{K}})\left(\frac{9 \text{ °R}}{5 \cancel{\text{K}}}\right) = 491.67 \text{ °R}$$

Thus

$$R = \frac{(14.69\text{ psi})(0.791\,55\text{ ft}^3)}{(1\text{ mol})(491.67\text{ °R})} = 0.023\,65\text{ psi} \cdot \text{ft}^3 \cdot \text{mol}^{-1} \cdot \text{°R}^{-1}$$

5.123 **(a)** 19.7 L

			255 °C, 1.00 atm		255 °C, 1.00 atm
Problem	36.4 g		? L		? L
Equation	$NH_4NO_3(l)$	$\xrightarrow{\text{heat}}$	$N_2O(g)$	+	$2H_2O(g)$
Formula mass, u	80.04				
Recipe, mol	1		1		2

Calculate moles of N_2O.

$$(36.4\text{ g } NH_4NO_3)\left(\frac{1\text{ mol } NH_4NO_3}{80.04\text{ g } NH_4NO_3}\right)\left(\frac{1\text{ mol } N_2O}{1\text{ mol } NH_4NO_3}\right) = 0.455\text{ mol } N_2O$$

Calculate the temperature in K.

$$(255\text{ °C} + 273\text{ °C})\left(\frac{1\text{ K}}{1\text{ °C}}\right) = 528\text{ K}$$

Using the ideal gas equation, determine the volume in liters.

$$V = \frac{nRT}{P} = \frac{(0.455\text{ mol})(0.082\,06\text{ L} \cdot \text{atm} \cdot \text{mol}^{-1} \cdot \text{K}^{-1})(528\text{ K})}{1.00\text{ atm}} = 19.7\text{ L}$$

(b) 39.4 L

Using Gay-Lussac's law of combining volumes, the volume of water vapor will be twice that of N_2O, since two moles of water vapor are generated for every one mole of N_2O.

2(19.7 L) = 39.4 L

5.125 CO_2, 0.32 atm; N_2, 0.16 atm; He, 2.5 atm

There are 19 mol (2.0 mol + 1.0 mol + 16 mol) of gas in the laser discharge cell. Using the ideal gas equation,

$$P_{\text{total}} = 3.0\text{ atm} = \frac{(19\text{ mol})RT}{V}$$

Thus, $\frac{RT}{V} = \frac{3.0\text{ atm}}{19\text{ mol}} = 0.16\,\frac{\text{atm}}{\text{mol}}$

The partial pressures can now be calculated using the ideal gas equation and the fact that $RT/V = 0.16$ atm/mol:

$$p_{CO_2} = \frac{(2.0\ \text{mol})RT}{V}$$

$$(2.0\ \cancel{\text{mol}})\left(0.16\ \frac{\text{atm}}{\cancel{\text{mol}}}\right) = 0.32\ \text{atm}$$

$$p_{N_2} = (1.0\ \cancel{\text{mol}})\left(0.16\ \frac{\text{atm}}{\cancel{\text{mol}}}\right) = 0.16\ \text{atm}$$

$$p_{He} = (16\ \cancel{\text{mol}})\left(0.16\ \frac{\text{atm}}{\cancel{\text{mol}}}\right) = 2.6\ \text{atm}$$

(Best answer, without intermediate round off, is $p_{He} = 2.5$ atm.)

5.127 1.1 g/L

Assume 1.0 L of gaseous mixture. It will contain 0.4 L of CO_2 (40% by volume) and 0.6 L of CH_4 (60% by volume). Solve for the mass of each gas present in 1.0 L of the mixture.

$PV = nRT$; $n = \frac{PV}{RT}$; and 25 °C = 298 K

CO_2: $$n = \frac{(1\ \cancel{\text{atm}})(0.4\ \cancel{\text{L}})}{(0.082\ \cancel{\text{L}} \cdot \cancel{\text{atm}} \cdot \text{mol}^{-1} \cdot \cancel{\text{K}^{-1}})(298\ \cancel{\text{K}})} = 0.02\ \text{mol}$$

$$\text{mass} = (0.02\ \cancel{\text{mol}})\left(\frac{44\ \text{g}}{1\ \cancel{\text{mol}}}\right) = 0.9\ \text{g}$$

(Best answer, without intermediate round off, is 0.7 g.)

CH_4: $$n = \frac{(1\ \cancel{\text{atm}})(0.6\ \cancel{\text{L}})}{(0.082\ \cancel{\text{L}} \cdot \cancel{\text{atm}} \cdot \text{mol}^{-1} \cdot \cancel{\text{K}^{-1}})(298\ \cancel{\text{K}})} = 0.02\ \text{mol}$$

$$\text{mass} = (0.02\ \cancel{\text{mol}})\left(\frac{16\ \text{g}}{1\ \cancel{\text{mol}}}\right) = 0.3\ \text{g}$$

(Best answer, without intermediate round off, is 0.4 g)

Since the mass of 1.0 L of the gaseous mixture equals 1.2 g (0.9 g + 0.3 g), the density of the gaseous mixture is 1.2 g/L.

(Best answer, without intermediate round off, is 1.1 g/L.)

CHAPTER 6: Chemical Thermodynamics: Thermochemistry

Practice Problems

6.1 **(a)** spontaneous **(b)** nonspontaneous **(c)** spontaneous **(d)** spontaneous

Spontaneous changes are changes that take place naturally.

6.2 **(a)** endothermic **(b)** exothermic **(c)** endothermic

Changes that heat their surroundings are exothermic. Changes that cool their surroundings are endothermic.

6.3 **(a)** No. The reverse of a spontaneous change is nonspontaneous.

6.4 **(a)** 98.3

$$(23.5\ \text{cal})\left(\frac{4.184\ \text{J}}{\text{cal}}\right) = 9.83 \times 10^1\ \text{J}$$

(b) 0.642

$$(642\ \text{J})\left(\frac{1\ \text{kJ}}{10^3\ \text{J}}\right) = 0.642\ \text{kJ}$$

(c) 3.26×10^3

$$(778\ \text{kcal})\left(\frac{4.184\ \text{kJ}}{\text{kcal}}\right) = 3.26 \times 10^3\ \text{kJ}$$

(d) 3.86×10^3

$$(3.86\ \text{kJ})\left(\frac{10^3\ \text{J}}{\text{kJ}}\right) = 3.86 \times 10^3\ \text{J}$$

6.5 1.31×10^3 J must be added.

$$(7.92\ \text{g CH}_3\text{CH}_2\text{OH})\left(\frac{2.42\ \text{J}}{\text{g CH}_3\text{CH}_2\text{OH} \cdot {}^\circ\text{C}}\right)(75.2\ {}^\circ\text{C} - 6.8\ {}^\circ\text{C}) = 1.31 \times 10^3\ \text{J}$$

6.6 2.53×10^4 J must be removed.

$$(287\ \text{g CH}_3\text{CH}_2\text{OH})\left(\frac{2.42\ \text{J}}{\text{g CH}_3\text{CH}_2\text{OH} \cdot {}^\circ\text{C}}\right)(19.8\ {}^\circ\text{C} - 56.2\ {}^\circ\text{C}) = -2.53 \times 10^4\ \text{J}$$

6.7 **(a)** –53 kJ/mol

25.8 °C – 19.5 °C = 6.3 °C

$$(400.0\ \text{mL soln})\left(\frac{1.00\ \text{g soln}}{\text{mL soln}}\right) = 4.00 \times 10^2\ \text{g soln}$$

thermal energy given off(–) = thermal energy gained by solution(+)

$$\text{thermal energy gained by solution} = (4.00 \times 10^2\ \text{g soln})\left(\frac{4.18\ \text{J}}{\text{g soln} \cdot {}^\circ\text{C}}\right)(6.3\ {}^\circ\text{C})$$

$$= (1.1 \times 10^4\ \text{J})\left(\frac{1\ \text{kJ}}{10^3\ \text{J}}\right) = 11\ \text{kJ}$$

thermal energy given off = –11 kJ

$$(200.0\ \text{mL soln})\left(\frac{1\ \text{L soln}}{10^3\ \text{mL soln}}\right)\left(\frac{1.00\ \text{mol HCl}}{\text{L soln}}\right) = 0.200\ \text{mol HCl}$$

$$\text{molar heat of neutralization} = \frac{-11\ \text{kJ}}{0.200\ \text{mol HCl}} = -55\ \text{kJ/mol}$$

(Best answer, without intermediate round off, is –53 kJ/mol.)

(b) molar heat of neutraliztion HCl = $\frac{1}{2}$ molar heat of neutralization H_2SO_4

H_2SO_4 provides two moles of H^+ ion per mole. HCl provides only one.

6.8 1.77×10^4 kJ

$$2(\text{CH}_3)_2\text{CHCH}_2\text{C(CH}_3)_3(\text{l}) + 25\text{O}_2(\text{g}) \rightarrow 16\text{CO}_2(\text{g}) + 18\text{H}_2\text{O(l)} \quad \Delta H_{rxn} = -10\ 930.9\ \text{kJ}$$

$$(369\ \text{g isooctane})\left(\frac{1\ \text{mol isooctane}}{114.2\ \text{g isooctane}}\right)\left(\frac{1.093 \times 10^4\ \text{kJ}}{2\ \text{mol isooctane}}\right) = 1.77 \times 10^4\ \text{kJ}$$

6.9 **(a)** endothermic

ΔH_{rxn} is +.

(b) exothermic

Thermal energy is a product of the reaction.

(c) exothermic

ΔH_{rxn} is –.

6.10 (a) $CaCO_3(s) + 178.3 \text{ kJ} \rightarrow CaO(s) + CO_2(g)$

(c) $NH_3(g) + HCl(g) \rightarrow NH_4Cl(s) + 176.01 \text{ kJ}$

6.11 **(a)** hot; reaction is exothermic.

Part of the thermal energy given off will go to heat the container.

(b) –184.62 kJ

6.12 **(a)** $2HgO(s) \rightarrow 2Hg(l) + O_2(g) \quad \Delta H_{rxn} = +181.66 \text{ kJ}$

(b) $Hg(l) + \frac{1}{2}O_2(g) \rightarrow HgO(s) \quad \Delta H_{rxn} = -90.83 \text{ kJ or } -90.830 \text{ kJ}$

6.13 –349.8 kJ

$Sn(s) + Cl_2(g) \rightarrow SnCl_2(s) \qquad \Delta H_{rxn} = ?$

$$\begin{array}{lll} & SnCl_4(l) \rightarrow SnCl_2(s) + Cl_2(g) & \Delta H_{rxn} = +195.4 \text{ kJ} \\ + & \underline{Sn(s) + 2Cl_2(g) \rightarrow SnCl_4(l)} & + \ \underline{\Delta H_{rxn} = -545.2 \text{ kJ}} \\ & Sn(s) + Cl_2(g) \rightarrow SnCl_2(s) & \Delta H_{rxn} = -349.8 \text{ kJ} \end{array}$$

6.14 **(a)** $Na(s) + \frac{1}{2}Cl_2(g) + \frac{3}{2}O_2(g) \rightarrow NaClO_3(s)$

(b) $\frac{1}{2}H_2(g) + \frac{1}{2}N_2(g) + \frac{3}{2}O_2(g) \rightarrow HNO_3(l)$

(c) $2C(\text{graphite}) + 3H_2(g) + \frac{1}{2}O_2(g) \rightarrow CH_3CH_2OH(l)$

6.15 **(a)** −104.86 kJ

Equation	$2NaClO_3(s)$ →	$2NaCl(s)$ +	$3O_2(g)$	ΔH^0_{rxn} = ?
ΔH_f^0, kJ/mol	−358.69	−411.12	0	

$$\Delta H^0_{rxn} = \left[(2\ \cancel{\text{mol NaCl}})\left(\frac{-411.12\ \text{kJ}}{\cancel{\text{mol NaCl}}}\right) + (3\ \cancel{\text{mol } O_2})\left(\frac{0\ \text{kJ}}{\cancel{\text{mol } O_2}}\right)\right] - \left[(2\ \cancel{\text{mol NaClO}_3})\left(\frac{-358.69\ \text{kJ}}{\cancel{\text{mol NaClO}_3}}\right)\right]$$

$$= (-822.24\ \text{kJ} + 0\ \text{kJ}) - (-717.38\ \text{kJ}) = -104.86\ \text{kJ}$$

(b) −71.7 kJ

Equation	$3NO_2(g)$ +	$H_2O(l)$ →	$2HNO_3(l)$ +	$NO(g)$	ΔH^0_{rxn} = ?
ΔH_f^0, kJ/mol	33.2	−285.83	−174.1	90.25	

$$\Delta H^0_{rxn} = \left[(2\ \cancel{\text{mol HNO}_3})\left(\frac{-174.1\ \text{kJ}}{\cancel{\text{mol HNO}_3}}\right) + (1\ \cancel{\text{mol NO}})\left(\frac{90.25\ \text{kJ}}{\cancel{\text{mol NO}}}\right)\right]$$
$$- \left[(3\ \cancel{\text{mol NO}_2})\left(\frac{33.2\ \text{kJ}}{\cancel{\text{mol NO}_2}}\right) + (1\ \cancel{\text{mol H}_2\text{O}})\left(\frac{-285.83\ \text{kJ}}{\cancel{\text{mol H}_2\text{O}}}\right)\right]$$

$$= (-348.2\ \text{kJ} + 90.25\ \text{kJ}) - [99.6\ \text{kJ} + (-285.83\ \text{kJ})] = -71.7\ \text{kJ}$$

(c) −1409.2 kJ

Equation	$CH_3CH_2OH(g)$ +	$3O_2(g)$ →	$2CO_2(g)$ +	$3H_2O(l)$
ΔH_f^0, kJ/mol	−235.3	0	−393.51	−285.83

$$\Delta H^0_{rxn} = \left[(2\ \cancel{\text{mol CO}_2})\left(\frac{-393.51\ \text{kJ}}{\cancel{\text{mol CO}_2}}\right) + (3\ \cancel{\text{mol H}_2\text{O}})\left(\frac{-285.83\ \text{kJ}}{\cancel{\text{mol H}_2\text{O}}}\right)\right]$$

$$- \left[(1\ \cancel{\text{mol CH}_3\text{CH}_2\text{OH}})\left(\frac{-235.3\ \text{kJ}}{\cancel{\text{mol CH}_3\text{CH}_2\text{OH}}}\right) + (3\ \cancel{\text{mol O}_2})\left(\frac{0\ \text{kJ}}{\cancel{\text{mol O}_2}}\right)\right]$$

$$= [-787.02\ \text{kJ} + (-857.49\ \text{kJ})] - (-235.3\ \text{kJ} + 0\ \text{kJ}) = -1409.2\ \text{kJ}$$

6.16 −487.0 kJ/mol

Use the molecular formula for CH_3COOH, $C_2H_4O_2$, to make balancing the equation for the combustion reaction easier.

Equation:	$C_2H_4O_2(l)$ +	$2O_2(g)$ →	$2CO_2(g)$ +	$2H_2O(l)$	ΔH_{rxn} = −871.7 kJ/mol
ΔH_f^0, kJ/mol	?	0	−393.51	−285.83	

$$\Delta H^0_{rxn} = [2\ \text{mol } CO_2 \cdot \Delta H_f^0\ (CO_2) + 2\ \text{mol } H_2O \cdot \Delta H_f^0\ (H_2O)] - [1\ \text{mol } C_2H_4O_2 \cdot \Delta H_f^0\ (C_2H_4O_2) + 2\ \text{mol } O_2 \cdot \Delta H_f^0\ (O_2)]$$

Solving this equation for $\Delta H_f^0\ (C_2H_4O_2)$ gives

$$\Delta H_f^0\ (C_2H_4O_2) = \frac{[2\ \text{mol } CO_2 \cdot \Delta H_f^0\ (CO_2) + 2\ \text{mol } H_2O \cdot \Delta H_f^0\ (H_2O)] - [2\ \text{mol } O_2 \cdot \Delta H_f^0\ (O_2)] - \Delta H^0_{rxn}}{1\ \text{mol } C_2H_4O_2}$$

Substitution gives

$$= \frac{[(-787.02\text{ kJ}) + (-571.66\text{ kJ})] - (0\text{ kJ}) - (-871.7\text{ kJ})}{\text{mol } C_2H_4O_2}$$

$$= -487.0\text{ kJ/mol } C_2H_4O_2$$

6.17 Formation reactions are reactions in which compounds are formed from elements. Combustion is the rapid reaction of materials with oxygen. As you have no way of judging whether the reaction given is fast or slow, you will have to assume that reactions with oxygen may be combustion.

(a) neither formation nor combustion

(b) formation

The compound, HBr, is formed from the elements, H_2 and Br_2.

(c) combustion

Note that this reaction is *not* formation because one of the reactants is a compound.

(d) neither formation nor combustion (It is decomposition.)

(e) formation and combustion

(f) formation and combustion

6.18 greater; the piston would fall, adding energy to the system

If the number of moles of gaseous products is less than the number of moles of gaseous reactants, the final volume will be less than the initial volume (the piston will fall) under constant pressure conditions. This means that work will be done on the system that will cause the amount of thermal energy given off to be greater than that given off under constant volume conditions.

Additional Practice Problems

6.57 **(a)** neither

The reaction does not involve the formation of a compound from its elements nor does it involve a reaction with oxygen.

(b) both

$NaNO_3$ is being formed from its elements, and the reaction involves oxygen.

(c) both

P_4O_{10} is being formed from its elements and P_4 is reacting with O_2.

(d) combustion

CO is reacting with O_2. CO is a compound; the reaction is *not* formation.

(e) neither

The reaction does not involve the formation of a compound from its elements nor does it involve a reaction with oxygen.

6.59 **(a)** (i) Changing a liquid to a gas is an endothermic process.

Equation (ii) involves liquid CH_3OH, which must be vaporized before combustion takes place. This requires thermal energy to be added to the system, decreasing the net release of thermal energy.

(b) more negative

Exothermic reactions involve a loss of thermal energy by the system to the surroundings. A more exothermic reaction gives off more thermal energy and has, thus, a more negative value of ΔH than the value of ΔH for the less exothermic reaction.

6.61 **(a)** $N_2(g) + 3H_2(g) \rightarrow 2NH_3(g)$

$2SO_2(g) + O_2(g) \rightarrow 2SO_3(g)$

$2HgO(s) \rightarrow 2Hg(l) + O_2(g)$

Under constant pressure conditions, each of these reactions will experience a change in volume due to the number of moles of gas changing in going from reactants to products. This means that work will be done by the system or on the system.

(b) The difference can usually be neglected because the difference is generally small compared with the magnitude of ΔH.

6.63 **(a)** –1214.66 kJ

Equation	$2CH_4(g)$	+	$3O_2(g)$	→	$2CO(g)$	+	$4H_2O(l)$
ΔH_f^0, kJ/mol	–74.85		0		–110.52		–285.83

$$\Delta H^0_{rxn} = \left[(2\ \cancel{mol\ CO})\left(\frac{-110.52\ kJ}{\cancel{mol\ CO}}\right) + (4\ \cancel{mol\ H_2O})\left(\frac{-285.83\ kJ}{\cancel{mol\ H_2O}}\right)\right]$$

$$-\left[(2\ \cancel{mol\ CH_4})\left(\frac{-74.85\ kJ}{\cancel{mol\ CH_4}}\right) + (3\ \cancel{mol\ O_2})\left(\frac{0\ kJ}{\cancel{mol\ O_2}}\right)\right]$$

$$= [-221.04\ kJ + (-1143.32\ kJ)] - (-149.70\ kJ + 0\ kJ) = -1214.66\ kJ$$

(b) 252 kJ

Equation	$CH_4(g)$	+	$NH_3(g)$	→	$HCN(g)$	+	$3H_2(g)$
ΔH_f^0, kJ/mol	–74.85		–46.11		131		0

$$\Delta H^0_{rxn} = \left[(1\ \text{mol HCN})\left(\frac{131\ \text{kJ}}{\text{mol HCN}}\right) + (3\ \text{mol H}_2)\left(\frac{0\ \text{kJ}}{\text{mol H}_2}\right)\right]$$

$$- \left[(1\ \text{mol CH}_4)\left(\frac{-74.85\ \text{kJ}}{\text{mol CH}_4}\right) + (1\ \text{mol NH}_3)\left(\frac{-46.11\ \text{kJ}}{\text{mol NH}_3}\right)\right]$$

$$= (131\ \text{kJ} + 0\ \text{kJ}) - [-74.85\ \text{kJ} + (-46.11\ \text{kJ})] = 252\ \text{kJ}$$

(c) –1169.54 kJ

Equation	$4NH_3(g)$	+	$5O_2(g)$	→	$4NO(g)$	+	$6H_2O(l)$
ΔH_f^0, kJ/mol	–46.11		0		90.25		–285.83

$$\Delta H^0_{rxn} = \left[(4\ \text{mol NO})\left(\frac{90.25\ \text{kJ}}{\text{mol NO}}\right) + (6\ \text{mol H}_2\text{O})\left(\frac{-285.83\ \text{kJ}}{\text{mol H}_2\text{O}}\right)\right]$$

$$- \left[(4\ \text{mol NH}_3)\left(\frac{-46.11\ \text{kJ}}{\text{mol NH}_3}\right) + (5\ \text{mol O}_2)\left(\frac{0\ \text{kJ}}{\text{mol O}_2}\right)\right]$$

$$= [361.00\ \text{kJ} + (-1714.98\ \text{kJ})] - (-184.44\ \text{kJ} + 0\ \text{kJ}) = -1169.54\ \text{kJ}$$

(d) 945.40 kJ

Equation	$N_2(g)$	→	$2N(g)$
ΔH_f^0, kJ/mol	0		472.70

$$\Delta H^0_{rxn} = (2\ \text{mol N})\left(\frac{472.70\ \text{kJ}}{\text{mol N}}\right) - (1\ \text{mol N}_2)\left(\frac{0\ \text{kJ}}{\text{mol N}_2}\right) = 945.40\ \text{kJ}$$

(e) 126.4 kJ

Equation	$NaHCO_3(s)$	→	$CO_2(g)$	+	$NaOH(s)$
ΔH_f^0, kJ/mol	–947.7		–393.51		–427.77

Note that the values of ΔH_f^0 for $NaHCO_3(s)$ and $NaOH(s)$ are found in Appendix E.

$$\Delta H^0_{rxn} = \left[(1\ \text{mol CO}_2)\left(\frac{-393.51\ \text{kJ}}{\text{mol CO}_2}\right) + (1\ \text{mol NaOH})\left(\frac{-427.77\ \text{kJ}}{\text{mol NaOH}}\right)\right]$$

$$- \left[(1\ \text{mol NaHCO}_3)\left(\frac{-947.7\ \text{kJ}}{\text{mol NaHCO}_3}\right)\right]$$

$$= [-393.51\ \text{kJ} + (-427.77\ \text{kJ})] - (-947.7\ \text{kJ}) = 126.4\ \text{kJ}$$

6.65 –454 kJ

$S(s) + O_2(g) \rightarrow SO_2(g) \quad \Delta H = -296.83\ \text{kJ}$

$$(49.1\ \text{g S})\left(\frac{1\ \text{mol S}}{32.07\ \text{g S}}\right)\left(\frac{-296.83\ \text{kJ}}{\text{mol S}}\right) = -454\ \text{kJ}$$

Stop and Test Yourself

1. **(b)** 0.346

$$(346\ \cancel{J})\left(\frac{1\ kJ}{10^3\ \cancel{J}}\right) = 0.346\ kJ$$

2. **(c)** 63.6

$$(0.266\ \cancel{kJ})\left(\frac{10^3\ \cancel{J}}{\cancel{kJ}}\right)\left(\frac{1\ cal}{4.184\ \cancel{J}}\right) = 63.6\ cal$$

3. **(e)** 71.7 kJ

Equation	$2HNO_3(l)$	+	$NO(g)$	→	$3NO_2(g)$	+	$H_2O(l)$
ΔH_f^0, kJ/mol	–174.1		90.25		33.2		–285.83

$$\Delta H^0_{rxn} = \left[(3\ \cancel{mol\ NO_2})\left(\frac{33.2\ kJ}{\cancel{mol\ NO_2}}\right) + (1\ \cancel{mol\ H_2O})\left(\frac{-285.83\ kJ}{\cancel{mol\ H_2O}}\right)\right]$$

$$-\left[(2\ \cancel{mol\ HNO_3})\left(\frac{-174.1\ kJ}{\cancel{mol\ HNO_3}}\right) + (1\ \cancel{mol\ NO})\left(\frac{90.25\ kJ}{\cancel{mol\ NO}}\right)\right]$$

$$= [99.6\ kJ + (-285.83\ kJ)] - (-348.2\ kJ + 90.25\ kJ) = 71.7\ kJ$$

4. **(a)** –1560 kJ

Equation	$CH_3CH_3(g)$	+	$\frac{7}{2}O_2(g)$	→	$2CO_2(g)$	+	$3H_2O(l)$
ΔH_f^0, kJ/mol	–84.67		0		–393.51		–285.83

$$\Delta H^0_{rxn} = \left[(2\ \cancel{mol\ CO_2})\left(\frac{-393.51\ kJ}{\cancel{mol\ CO_2}}\right) + (3\ \cancel{mol\ H_2O})\left(\frac{-285.83\ kJ}{\cancel{mol\ H_2O}}\right)\right]$$

$$-\left[(1\ \cancel{mol\ CH_3CH_3})\left(\frac{-84.67\ kJ}{\cancel{mol\ CH_3CH_3}}\right) + \left(\frac{7}{2}\ \cancel{mol\ O_2}\right)\left(\frac{0\ kJ}{\cancel{mol\ O_2}}\right)\right]$$

$$= [-787.02\ kJ + (-857.49\ kJ)] - (-84.67\ kJ + 0\ kJ) = -1559.84\ kJ = 1560\ kJ \text{ to the nearest kilojoule}$$

5. **(d)** $Cl_2(g) + \frac{1}{2}O_2(g) \rightarrow Cl_2O(g)$

Only (d) is a formation reaction. (e) involves a compound, H_2O, as reactant, so it is not a formation reaction. (a) and (c) are the reverse of formation reactions. (b) is also decomposition.

6. **(a)** endothermic

All ΔH^0_{rxn} have positive signs.

7. **(a)** cool

Thermal energy is absorbed by the system from its surroundings (your hand).

8. **(d)** 545.2 kJ

$$SnCl_4(l) \rightarrow SnCl_2(s) + Cl_2(g) \quad \Delta H^0_{rxn} = 195.4 \text{ kJ}$$
$$\underline{SnCl_2(s) \rightarrow Sn(s) + Cl_2(g) \quad \Delta H^0_{rxn} = 349.8 \text{ kJ}}$$
$$SnCl_4(l) \rightarrow Sn(s) + 2Cl_2(g) \quad \Delta H^0_{rxn} = 545.2 \text{ kJ}$$

9. **(a)** –160.6 kJ

$$Cl_2(g) + \frac{1}{2}O_2(g) \rightarrow Cl_2O(g) \quad \Delta H^0_{rxn} = 80.3 \text{ kJ}$$

$$Cl_2O(g) \rightarrow Cl_2(g) + \frac{1}{2}O_2(g) \quad \Delta H^0_{rxn} = -80.3 \text{ kJ}$$

$$2Cl_2O(g) \rightarrow 2Cl_2(g) + O_2(g) \quad \Delta H^0_{rxn} = -160.6 \text{ kJ}$$

10. **(c)** 62.95

Problem	? g	3169 kJ
Equation	$C_3H_8(g) + 5O_2(g) \rightarrow 3CO_2(g) + 4H_2O(l)$	+ 2220.00 kJ
Formula mass, u	44.097	

$$(3169\ \cancel{kJ})\left(\frac{1\ \cancel{mol\ C_3H_8}}{2220.00\ \cancel{kJ}}\right)\left(\frac{44.097\text{ g } C_3H_8}{\cancel{mol\ C_3H_8}}\right) = 62.95\text{ g } C_3H_8$$

11. **(d)** –108 kJ/mol

Equation	$CH_4(g)$	+	$4Cl_2(g)$	→	$CCl_4(g)$	+	$4HCl(g)$	$\Delta H^0_{rxn} = -402$ kJ
ΔH_f^0, kJ/mol	–74.85		0		?		–92.31	

$$-402\text{ kJ} = \left[(4\ \cancel{mol\ HCl})\left(\frac{-92.31\text{ kJ}}{\cancel{mol\ HCl}}\right) + (1\text{ mol } CCl_4)\cdot \Delta H_f^0\,(CCl_4)\right]$$

$$-\left[(1\ \cancel{mol\ CH_4})\left(\frac{-74.85\text{ kJ}}{\cancel{mol\ CH_4}}\right) - (4\ \cancel{mol\ Cl_2})\left(\frac{0\text{ kJ}}{\cancel{mol\ Cl_2}}\right)\right]$$

$$\Delta H_f^0\,(CCl_4) = \frac{(-402\text{ kJ}) + (-74.85\text{ kJ}) - (-369.24\text{ kJ})}{1\text{ mol } CCl_4} = -108\text{ kJ/mol } CCl_4$$

12. **(d)** 20 kJ

Specific heat of Al is $0.90\text{ J} \cdot g^{-1} \cdot {}^\circ C^{-1}$.

$$(289\text{ g }\cancel{Al})\left(\frac{0.90\ \cancel{J}}{\cancel{g\ Al}\cdot \cancel{{}^\circ C}}\right)[(25.0 - 100.0)\ \cancel{{}^\circ C}]\left(\frac{1\text{ kJ}}{10^3\ \cancel{J}}\right) = -20\text{ kJ}$$

13. **(a)** 568 kJ

Problem	36.0 g		excess	? kJ
Equation	2Al(s)	+	$Fe_2O_3(s) \rightarrow 2Fe(s) + Al_2O_3(s)$	+ 851.4 kJ
Formula mass, u	26.98			

$$(36.0 \cancel{\text{g Al}})\left(\frac{1 \cancel{\text{mol Al}}}{26.98 \cancel{\text{g Al}}}\right)\left(\frac{851.4 \text{ kJ}}{2 \cancel{\text{mol Al}}}\right) = 568 \text{ kJ}$$

14. **(b)** It is exothermic.

The products are at a lower enthalpy than the reactants.

Putting Things Together

6.67

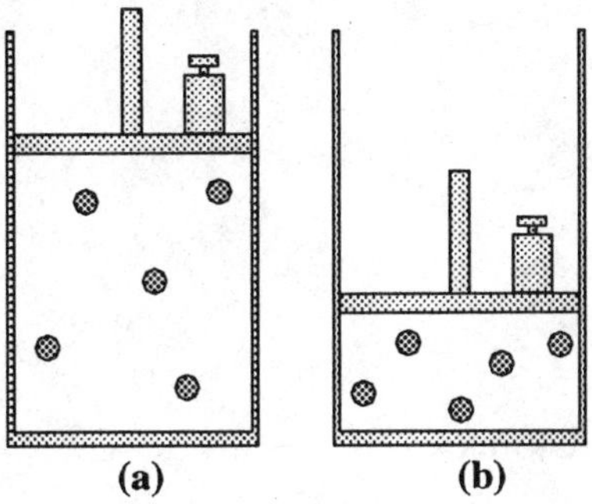

6.69 Real gases have intermolecular attractions. These intermolecular attractions must be overcome in order for the gas to expand. This requires energy from the surroundings, which results in a temperature decrease.

6.71 –24.8 kJ

$2Fe_2O_3(s) \rightarrow 3O_2(g) + 4Fe(s)$ $\quad \Delta H = 1648.4$ kJ

$\underline{6CO(g) + 3O_2(g) \rightarrow 6CO_2(g) \quad \Delta H = -1697.94 \text{ kJ}}$

$2Fe_2O_3(s) + 6CO(g) \rightarrow 4Fe(s) + 6CO_2(g)$ $\quad \Delta H = -49.5$ kJ

$Fe_2O_3(s) + 3CO(g) \rightarrow 2Fe(s) + 3CO_2(g)$ $\quad \Delta H = -49.5 \text{ kJ}/2 = -24.8$ kJ

6.73 $\frac{3}{2}NO_2(g) + \frac{1}{2}H_2O(l) \rightarrow HNO_3(l) + \frac{1}{2}NO(g) + 35.0$ kJ or $\frac{3}{2}NO_2(g) + \frac{1}{2}H_2O(l) \rightarrow HNO_3(l) + \frac{1}{2}NO(g) \quad \Delta H = -35.0$ kJ

Since thermal energy is given off in the reaction, the reaction is exothermic and the thermal energy can be written as a product of the reaction. ΔH is negative.

6.75 –177 kJ

Problem	1.25 g		11.00 g		
Equation	$H_2(g)$	+	$\frac{1}{2}O_2(g)$	→	$H_2O(l)$ $\Delta H_f^0 = -285.83$ kJ/mol
Formula mass, u	2.016		31.999		
Recipe, mol	1		$\frac{1}{2}$		1

$$(1.25\ \text{g H}_2)\left(\frac{1\ \text{mol H}_2}{2.016\ \text{g H}_2}\right) = 0.620\ \text{mol H}_2 \qquad (11.00\ \text{g O}_2)\left(\frac{1\ \text{mol O}_2}{31.999\ \text{g O}_2}\right) = 0.3438\ \text{mol O}_2$$

$$\frac{0.620\ \text{mol H}_2}{1\ \text{mol H}_2} = 0.620 \text{ of a recipe} \qquad \frac{0.3438\ \text{mol O}_2}{\frac{1}{2}\ \text{mol O}_2} = 0.6876 \text{ of a recipe}$$

0.620 < 0.6876. H_2 is limiting.

$$(0.620\ \text{mol H}_2)\left(\frac{-285.83\ \text{kJ}}{\text{mol H}_2}\right) = -177\ \text{kJ}$$

6.77 –53 kJ

Problem	3.6 L		2.4 L		
Equation	$\frac{1}{2}H_2(g)$	+	$\frac{1}{2}F_2(g)$	→	HF(g) $\Delta H_f^0 = -271.12$ kJ/mol
Recipe, mol	$\frac{1}{2}$		$\frac{1}{2}$		1

According to Gay–Lussac's law, one volume of H_2 reacts with one volume of F_2. Therefore, F_2 is limiting. The ideal gas equation can be used to calculate the number of moles of F_2:

$$n = \frac{PV}{RT} = \frac{(1.0\ \text{atm})(2.4\ \text{L})}{(0.0821\ \text{L}\cdot\text{atm}\cdot\text{mol}^{-1}\cdot\text{K}^{-1})(298\ \text{K})} = 9.8 \times 10^{-2}\ \text{mol F}_2$$

$$(9.8 \times 10^{-2}\ \text{mol F}_2)\left(\frac{1\ \text{mol HF}}{1/2\ \text{mol F}_2}\right)\left(\frac{-271.12\ \text{kJ}}{\text{mol HF}}\right) = -53\ \text{kJ}$$

Applications

6.79 49 kJ

$$(276\ \text{g Al})\left(\frac{0.90\ \text{J}}{\text{g Al}\cdot{}^\circ\text{C}}\right)[(218 - 22)\ {}^\circ\text{C}]\left(\frac{1\ \text{kJ}}{10^3\ \text{J}}\right) = 49\ \text{kJ}$$

6.81 As water freezes, it gives off thermal energy to the surroundings.

6.83 27.4 g NH_4NO_3

Thermal energy to be removed:

$$(117\ \cancel{mL\ H_2O})\left(\frac{1.00\ \cancel{g\ H_2O}}{1\ \cancel{mL\ H_2O}}\right)\left(\frac{4.18\ J}{\cancel{g\ H_2O\cdot °C}}\right)(5.0\ \cancel{°C} - 23.0\ \cancel{°C}) = -8.80 \times 10^3\ J$$

$$(8.80 \times 10^3\ \cancel{J})\left(\frac{1\ \cancel{mol\ NH_4NO_3}}{2.569 \times 10^4\ \cancel{J}}\right)\left(\frac{80.04\ g\ NH_4NO_3}{1\ \cancel{mol\ NH_4NO_3}}\right) = 27.4\ g\ NH_4NO_3$$

6.85 **(a)** –930 kJ

Equation	$2CH_4(g)$	+	$2NH_3(g)$	+	$3O_2(g)$	→	$2HCN(g)$	+	$6H_2O(g)$
ΔH_f^0, kJ/mol	–74.85		–46.11		0		131		–238.92

$$\Delta H^0_{rxn} = \left[(2\ \cancel{mol\ HCN})\left(\frac{131\ kJ}{\cancel{mol\ HCN}}\right) + (6\ \cancel{mol\ H_2O})\left(\frac{-238.92\ kJ}{\cancel{mol\ H_2O}}\right)\right]$$

$$-\left[(2\ \cancel{mol\ CH_4})\left(\frac{-74.85\ kJ}{\cancel{mol\ CH_4}}\right) + (2\ \cancel{mol\ NH_3})\left(\frac{-46.11\ kJ}{\cancel{mol\ NH_3}}\right) + (3\ \cancel{mol\ O_2})\left(\frac{0\ kJ}{\cancel{mol\ O_2}}\right)\right]$$

$$= [262\ kJ + (-1433.52\ kJ)] - [-149.70\ kJ + (-92.22\ kJ) + 0] = -930\ kJ$$

(b) –465 kJ

$$(1\ \cancel{mol\ HCN})\left(\frac{-930\ kJ}{2\ \cancel{mol\ HCN}}\right) = -465\ kJ$$

(c) exothermic

The sign of ΔH^0 is –.

6.87 **(a)** 7×10^1 kcal (one significant figure)

$$(7\ \cancel{g\ carbohydrate})\left(\frac{4\ kcal}{\cancel{g}}\right) + (4\ \cancel{g\ fat})\left(\frac{9\ kcal}{\cancel{g}}\right) + (1\ \cancel{g\ protein})\left(\frac{4\ kcal}{\cancel{g}}\right) = 68\ kcal$$

$= 7 \times 10^1$ kcal (to one significant figure)

(b) 70 calories

$$(7 \times 10^1\ \cancel{kcal})\left(\frac{1\ \text{nutritional calorie}}{\cancel{kcal}}\right) = 7 \times 10^1\ \text{calories} = 70\ \text{calories}$$

(c) 3×10^2 kJ

$$(7 \times 10^1\ \cancel{kcal})\left(\frac{4.2\ kJ}{\cancel{kcal}}\right) = 3 \times 10^2\ kJ$$

6.89 (a) 2×10^3 lb

$$(60\ \text{in})(84\ \text{in})(9\ \text{in})\left(\frac{2.54\ \text{cm}}{\text{in}}\right)^3\left(\frac{1.0\ \text{g}}{\text{cm}^3}\right)\left(\frac{1\ \text{lb}}{453.6\ \text{g}}\right) = 2 \times 10^3\ \text{lb}$$

(b) 1×10^4 kcal (using rounded value from part a)

$$\Delta T = (85 - 65)\ {}^\circ\text{F}\left(\frac{5\ {}^\circ\text{C}}{9\ {}^\circ\text{F}}\right) = 11\ {}^\circ\text{C}$$

The calorie was originally defined as the amount of energy required to heat one gram of water 1 °C (from 15 to 16 °C).

$$(2 \times 10^3\ \text{lb})\left(\frac{453.6\ \text{g}}{\text{lb}}\right)\left(\frac{1\ \text{cal}}{\text{g} \cdot {}^\circ\text{C}}\right)(11\ {}^\circ\text{C})\left(\frac{1\ \text{kcal}}{10^3\ \text{cal}}\right) = 1 \times 10^4\ \text{kcal}$$

(Best answer, without intermediate round off in part (a), is 8×10^3 kcal.)

6.91 5.7 °C

$$V_{air} = (12\ \text{ft} \times 11\ \text{ft} \times 8.0\ \text{ft})\left(\frac{12^3\ \text{in.}^3}{1\ \text{ft}^3}\right)\left(\frac{2.54^3\ \text{cm}^3}{1\ \text{in.}^3}\right)\left(\frac{1\ \text{L}}{1000\ \text{cm}^3}\right) = 3.0 \times 10^4\ \text{L}$$

$$\text{mass}_{air} = (3.0 \times 10^4\ \text{L})(1.2\ \text{g} \cdot \text{L}^{-1}) = 3.6 \times 10^4\ \text{g}$$

The two light bulbs provide equal light. The difference in power must be in the amount of thermal energy released, 75 W – 18 W = 57 W = 57 J/s.

$$(1.0\ \text{h})\left(\frac{60\ \text{min}}{1\ \text{h}}\right)\left(\frac{60\ \text{s}}{1\ \text{min}}\right)\left(\frac{57\ \text{J}}{1\ \text{s}}\right) = 2.1 \times 10^5\ \text{J of extra light}$$

Thermal energy gained by air = mass × specific heat × temperature change

$$\therefore\ \text{temperature change} = \frac{\text{thermal energy}}{\text{mass} \times \text{specific heat}} = \frac{2.1 \times 10^5\ \text{J}}{(3.6 \times 10^4\ \text{g}_{air})(1.007\ \text{J} \cdot \text{g}_{air}^{-1} \cdot {}^\circ\text{C}^{-1})} = 5.8\ {}^\circ\text{C}$$

(Best answer, without intermediate round off, is 5.7 °C.)

6.93 (a) 141.79 kJ

$$H_2(g) + \frac{1}{2}O_2(g) \rightarrow H_2O(l) \qquad \Delta H_f^0 = -285.83\ \text{kJ}$$

$$(1.0000\ \text{g}\ H_2)\left(\frac{1\ \text{mol}\ H_2}{2.015\ 88\ \text{g}\ H_2}\right)\left(\frac{-285.83\ \text{kJ}}{\text{mol}\ H_2}\right) = -141.79\ \text{kJ}$$

(b) 12.1 L

$PV = nRT$

$$V = \frac{nRT}{P} = \frac{(1.000\ \text{g H}_2)\left(\frac{1\ \text{mol H}_2}{2.016\ \text{g H}_2}\right)(0.082\ 06\ \text{L}\cdot\text{atm}\cdot\text{mol}^{-1}\cdot\text{K}^{-1})(298\ \text{K})}{1.000\ \text{atm}} = 12.1\ \text{L}$$

(c) $\Delta H^0{}_{\text{combustion}} = -5465.5$ kJ/mol, –47.846 kJ

Equation	$C_8H_{18}(l)$	+	$\frac{25}{2}O_2(g)$	→	$8CO_2(g)$	+	$9H_2O(l)$
$\Delta H_f{}^0$, kJ/mol	–255.1		0		–393.51		–285.83

$$\Delta H^0{}_{\text{combustion}} = \left[(9\ \text{mol H}_2\text{O})\left(\frac{-285.83\ \text{kJ}}{\text{mol H}_2\text{O}}\right) + (8\ \text{mol CO}_2)\left(\frac{-393.51\ \text{kJ}}{\text{mol CO}_2}\right)\right]$$

$$-\left[(1\ \text{mol C}_8\text{H}_{18})\left(\frac{-255.1\ \text{kJ}}{\text{mol C}_8\text{H}_{18}}\right) + \left(\frac{25}{2}\ \text{mol O}_2\right)\left(\frac{0\ \text{kJ}}{\text{mol O}_2}\right)\right]$$

$$= [-2572.47\ \text{kJ} + (-3148.08\ \text{kJ})] - (-255.1\ \text{kJ} + 0\ \text{kJ}) = -5465.5\ \text{kJ/mol C}_8\text{H}_{18}$$

$$(1.000\ 00\ \text{g C}_8\text{H}_{18})\left(\frac{1\ \text{mol C}_8\text{H}_{18}}{114.231\ \text{g C}_8\text{H}_{18}}\right)\left(\frac{-5465.5\ \text{kJ}}{\text{mol C}_8\text{H}_{18}}\right) = -47.846\ \text{kJ}$$

(d) 1.45 mL

$$(1.00\ \text{g C}_8\text{H}_{18})\left(\frac{1\ \text{mL C}_8\text{H}_{18}}{0.692\ \text{g C}_8\text{H}_{18}}\right) = 1.45\ \text{mL C}_8\text{H}_{18}$$

(e) Hydrogen gas takes up a lot of space! Also, hydrogen is difficult to store because of its tendency to diffuse out of containers. In addition, hydrogen's high heat of combustion would require more cooling capability than is required for a gasoline–powered engine.

6.95 **(a)** dimethylhydrazine

Hydrazine:

Equation	$H_2NNH_2(l)$	+	$O_2(g)$	→	$N_2(g)$	+	$2H_2O(l)$
$\Delta H_f{}^0$, kJ/mol	50.63		0		0		–285.83

$$\Delta H^0 = \left[(2\ \text{mol H}_2\text{O})\left(\frac{-285.83\ \text{kJ}}{\text{mol H}_2\text{O}}\right) + (1\ \text{mol N}_2)\left(\frac{0\ \text{kJ}}{\text{mol N}_2}\right)\right]$$

$$-\left[(1\ \text{mol O}_2)\left(\frac{0\ \text{kJ}}{\text{mol O}_2}\right) + (1\ \text{mol H}_2\text{NNH}_2)\left(\frac{50.63\ \text{kJ}}{\text{mol H}_2\text{NNH}_2}\right)\right]$$

$$= -622.29\ \text{kJ}$$

$$\left(\frac{-622.29\text{ kJ}}{\cancel{\text{mol}}}\right)\left(\frac{1\ \cancel{\text{mol}}}{32.0450\text{ g}}\right) = 19.419\text{ kJ/g}$$

Dimethylhydrazine:

Equation	$(CH_3)_2NNH_2(l)$	+	$4O_2(g)$	→	$N_2(g)$	+	$4H_2O(l)$	+	$2CO_2(g)$
ΔH_f^0, kJ/mol	48.9		0		0		−285.83		−393.51

$$\Delta H^0 = \left[(4\ \cancel{\text{mol H}_2\text{O}})\left(\frac{-285.83\text{ kJ}}{\cancel{\text{mol H}_2\text{O}}}\right) + (1\ \cancel{\text{mol N}_2})\left(\frac{0\text{ kJ}}{\cancel{\text{mol N}_2}}\right) + (2\ \cancel{\text{mol CO}_2})\left(\frac{-393.51\text{ kJ}}{\cancel{\text{mol CO}_2}}\right)\right]$$

$$-\left[(4\ \cancel{\text{mol O}_2})\left(\frac{0\text{ kJ}}{\cancel{\text{mol O}_2}}\right) + (1\ \cancel{\text{mol (CH}_3)_2\text{NNH}_2})\left(\frac{48.9\text{ kJ}}{\cancel{\text{mol (CH}_3)_2\text{NNH}_2}}\right)\right]$$

$$= -1979.2\text{ kJ}$$

$$\left(\frac{-1979.2\text{ kJ}}{\cancel{\text{mol}}}\right)\left(\frac{1\ \cancel{\text{mol}}}{60.099\text{ g}}\right) = 32.932\text{ kJ/g}$$

(b) Dimethylhydrazine is the better fuel because it yields more energy per gram of fuel.

6.97 **(a)** 6.36×10^2 kJ, 485 L

$$\left(\frac{-87.37\ \cancel{\text{kcal}}}{\text{mol}}\right)\left(\frac{4.184\text{ kJ}}{\cancel{\text{kcal}}}\right) = -365.6\ \frac{\text{kJ}}{\text{mol}}$$

Equation	$NH_4NO_3(s)$	→	$N_2(g)$	+	$\frac{1}{2}O_2(g)$	+	$2H_2O(g)$
ΔH_f^0, kJ/mol	−365.6		0		0		−238.92

$$\Delta H^0 = \left[(2\ \cancel{\text{mol H}_2\text{O}})\left(\frac{-238.92\text{ kJ}}{\cancel{\text{mol H}_2\text{O}}}\right) + \left(\frac{1}{2}\ \cancel{\text{mol O}_2}\right)\left(\frac{0\text{ kJ}}{\cancel{\text{mol O}_2}}\right) + (1\ \cancel{\text{mol N}_2})\left(\frac{0\text{ kJ}}{\cancel{\text{mol N}_2}}\right)\right]$$

$$-\left[(1\ \cancel{\text{mol NH}_4\text{NO}_3})\left(\frac{-365.6\text{ kJ}}{\cancel{\text{mol NH}_4\text{NO}_3}}\right)\right]$$

$$= -112.2\text{ kJ}$$

$$(454\ \cancel{\text{g NH}_4\text{NO}_3})\left(\frac{1\ \cancel{\text{mol NH}_4\text{NO}_3}}{80.04\ \cancel{\text{g NH}_4\text{NO}_3}}\right)\left(\frac{-112.2\text{ kJ}}{\cancel{\text{mol NH}_4\text{NO}_3}}\right) = -6.36 \times 10^2\text{ kJ}$$

$$(454\ \cancel{\text{g NH}_4\text{NO}_3})\left(\frac{1\ \cancel{\text{mol NH}_4\text{NO}_3}}{80.04\ \cancel{\text{g NH}_4\text{NO}_3}}\right)\left(\frac{3.5\text{ mol gas}}{\cancel{\text{mol NH}_4\text{NO}_3}}\right) = 19.9\text{ mol gas}$$

$$PV = nRT,\quad V = \frac{nRT}{P} = \frac{(19.9\ \cancel{\text{mol}})(0.082\,06\text{ L}\cdot\cancel{\text{atm}}\cdot\cancel{\text{mol}^{-1}}\cdot\cancel{\text{K}^{-1}})(298\ \cancel{\text{K}})}{1.000\ \cancel{\text{atm}}} = 487\text{ L}$$

(Best answer, without intermediate round off, is 485 L.)

(b) 485 L

$$(454\ \cancel{g})\left(\frac{1\ \text{cm}^3}{1.725\ \cancel{g}}\right) = 263\ \text{cm}^3 \text{ initial volume, } 485\ \text{L final volume}$$

$$\Delta V = 485\ \text{L} - (263\ \cancel{\text{cm}^3})\left(\frac{1\ \text{L}}{10^3\ \cancel{\text{cm}^3}}\right) = 485\ \text{L} - 0.263\ \text{L} = 485\ \text{L}$$

CHAPTER 7:
Atomic Structure

Practice Problems

7.1 **(a)** wavelength of 1 m

The shorter the wavelength, the higher the frequency.

(b) wavelength of 10^{-10} m

A wavelength of 10^{-7} m is longer than a wavelength of 10^{-10} m.

7.2

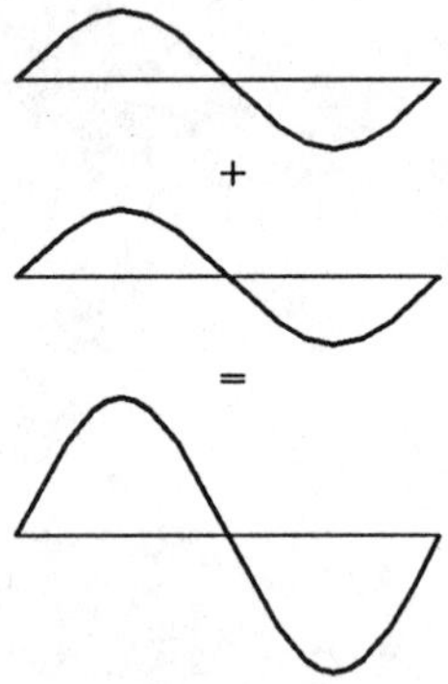

The amplitude of the new wave will be twice the amplitude of the original waves.

7.3 neither

All electromagnetic radiation travels through a vacuum with the same velocity, $2.998 \times 10^{8}\ m \cdot s^{-1}$.

7.4 10^3 s^{-1}

The lower the frequency, the longer the wavelength.

7.5 4.91×10^4 m

$\lambda \nu = c$

$$\lambda = \frac{c}{\nu} = \left(\frac{2.998 \times 10^8 \text{ m} \cdot \cancel{\text{s}^{-1}}}{6.10 \cancel{\text{kHz}}}\right)\left(\frac{1 \cancel{\text{kHz}}}{10^3 \cancel{\text{s}^{-1}}}\right) = 4.91 \times 10^4 \text{ m}$$

7.6 54 MHz

$\lambda \nu = c$

$$\nu = \frac{c}{\lambda} = \left(\frac{3.00 \times 10^8 \cancel{\text{m}} \cdot \cancel{\text{s}^{-1}}}{5.6 \cancel{\text{m}}}\right)\left(\frac{1 \cancel{\text{Hz}}}{\cancel{\text{s}^{-1}}}\right)\left(\frac{1 \text{ MHz}}{10^6 \cancel{\text{Hz}}}\right) = 54 \text{ MHz}$$

7.7 **(a)** yellow light

According to Figure 7.15 (b), the wavelength of blue light is $4.4–4.8 \times 10^{-7}$ m; the wavelength of yellow light is $5.3–5.9 \times 10^{-7}$ m.

(b) microwaves

According to Figure 7.15 (a), microwaves have a frequency of $10^{11}–10^{12}$ s^{-1}; radio waves have a frequency of $10^5–10^{11}$ s^{-1}.

7.8 violet or purple

The absorption of green, yellow, and orange light allows violet, blue, and red light to be transmitted and reflected.

7.9 246 kJ/mol

$$\text{energy of a photon} = \frac{hc}{\lambda} = \frac{(6.626 \times 10^{-34} \text{ J} \cdot \cancel{\text{s}})(2.998 \times 10^8 \cancel{\text{m}} \cdot \cancel{\text{s}^{-1}})}{(486 \cancel{\text{nm}})\left(\frac{10^{-9} \cancel{\text{m}}}{\cancel{\text{nm}}}\right)} = 4.09 \times 10^{-19} \text{ J}$$

For a mole of photons, the energy is

$$\left(\frac{4.09 \times 10^{-19} \cancel{\text{J}}}{\cancel{\text{photon}}}\right)\left(\frac{6.022 \times 10^{23} \cancel{\text{photon}}}{\text{mol}}\right)\left(\frac{1 \text{ kJ}}{10^3 \cancel{\text{J}}}\right) = 246 \text{ kJ/mol}$$

7.10 **(a)** 1.48×10^5 kJ/mol

$$\text{energy of a photon} = h\nu = (6.626 \times 10^{-34}\ \text{J}\cdot\text{s})(3.72 \times 10^{17}\ \text{s}^{-1})\left(\frac{1\ \text{kJ}}{10^3\ \text{J}}\right) = 2.46 \times 10^{-19}\ \text{kJ}$$

For a mole of photons, the energy is

$$\left(\frac{2.46 \times 10^{-19}\ \text{kJ}}{\text{photon}}\right)\left(\frac{6.022 \times 10^{23}\ \text{photon}}{\text{mol}}\right) = 1.48 \times 10^5\ \text{kJ/mol}$$

(b) X-rays

From Figure 7.15 (a), eletromagnetic radiation that has a frequency of $3.72 \times 10^{17}\ \text{s}^{-1}$ is in the X–ray region.

7.11 6.947×10^{-8} m

$$\left(\frac{1722\ \text{kJ}}{\text{mol}}\right)\left(\frac{10^3\ \text{J}}{\text{kJ}}\right)\left(\frac{1\ \text{mol}}{6.0221 \times 10^{23}\ \text{photon}}\right) = 2.859 \times 10^{-18}\ \text{J/photon}$$

energy of a photon = $\frac{hc}{\lambda}$, and

$$\lambda = \frac{hc}{E} = \frac{(6.6261 \times 10^{-34}\ \text{J}\cdot\text{s})(2.9979 \times 10^8\ \text{m}\cdot\text{s}^{-1})}{(2.859 \times 10^{-18}\ \text{J})} = 6.948 \times 10^{-8}\ \text{m}$$

(Best answer, without intermediate round-off, is 6.947×10^{-8} m.)

7.12 photons of gamma rays

Gamma rays have higher frequency and shorter wavelength than X–rays. Thus, they are more energetic.

7.13 486.2 nm

$$\frac{1}{\lambda} = 1.097 \times 10^7\ \text{m}^{-1}\left(\frac{1}{2^2} - \frac{1}{n^2}\right), \quad \text{where } n = 4$$

$$\lambda = \frac{1}{(1.097 \times 10^7\ \text{m}^{-1})\left(\frac{1}{2^2} - \frac{1}{4^2}\right)} = \frac{1}{(1.097 \times 10^7\ \text{m}^{-1})\left(\frac{1}{4} - \frac{1}{16}\right)}$$

$$= \frac{1}{(1.097 \times 10^7\ \text{m}^{-1})(0.1875)} = 4.862 \times 10^{-7}\ \text{m}$$

$$(4.862 \times 10^{-7}\ \text{m})\left(\frac{1\ \text{nm}}{10^{-9}\ \text{m}}\right) = 486.2\ \text{nm}$$

7.14 **(a)** 292 kJ/mol

$$\text{energy of a photon} = h\nu = \frac{hc}{\lambda} = \frac{(6.626 \times 10^{-34}\ \text{J}\cdot\text{s})(2.998 \times 10^{8}\ \text{m}\cdot\text{s}^{-1})}{(410\ \text{nm})\left(\frac{10^{-9}\ \text{m}}{\text{nm}}\right)} = 4.85 \times 10^{-19}\ \text{J/photon}$$

$$\left(\frac{4.85 \times 10^{-19}\ \text{J}}{\text{photon}}\right)\left(\frac{6.022 \times 10^{23}\ \text{photon}}{\text{mol}}\right)\left(\frac{1\ \text{kJ}}{10^{3}\ \text{J}}\right) = 292\ \text{kJ/mol}$$

(b)

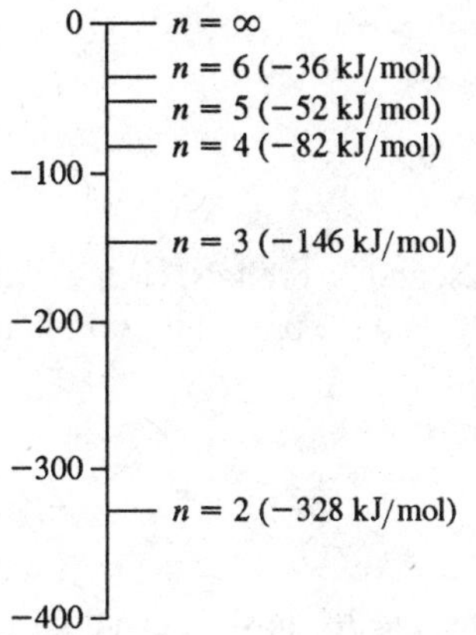

The $n = 6$ level is 292 kJ/mol above the $n = 2$ level, which is at –328 kJ/mol.

–328 kJ/mol + 292 kJ/mol = –36 kJ/mol

Thus, the $n = 6$ level is at –36 kJ/mol.

7.15 **(a)**

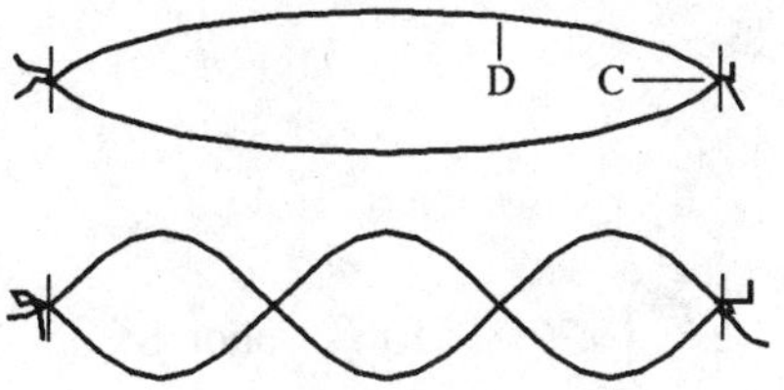

(b) two

There are two nodes in addition to the two at the ends.

(c) three

Three half–wavelengths are equal to the length of the box.

7.16

Principal quantum number, n (shell)	Angular momentum quantum number, l (subshell)	Subshell label	Magnetic quantum number, m_l	Number of orbitals in subshell
5	0	5*s*	0	1
	1	5*p*	−1, 0, +1	3
	2	5*d*	−2, −1, 0, +1, +2	5
	3	5*f*	−3, −2, −1, 0, +1, +2, +3	7
	4	5*g*	−4, −3, −2, −1, 0, +1, +2, +3, +4	9

7.17 **(a)** 3

The 4*s* orbital should have three nodes in keeping with the pattern of 1*s*, 2*s*, and 3*s* orbitals, each of which have one less node than the number of the principal energy level and one more node than the preceding *s* orbital.

(b) spherical

7.18 **(a)** 10

The 3*d* subshell contains 5 orbitals. Since each orbital can hold 2 electrons, the 3*d* subshell can hold 10 electrons.

(b) 18

The third shell has one 3*s* orbital, three 3*p* orbitals, and five 3*d* orbitals, for a total of nine orbitals. Therefore, the third shell can hold 18 electrons.

Additional Practice Problems

7.55 1.8×10^{27} photons

$$\text{energy of a photon} = h\nu = \left(\frac{6.626 \times 10^{-34}\ \text{J} \cdot \cancel{\text{s}}}{\text{photon}}\right)(4.56\ \cancel{\text{MHz}})\left(\frac{10^6\ \cancel{\text{Hz}}}{\cancel{\text{MHz}}}\right)\left(\frac{1\ \cancel{\text{s}^{-1}}}{\cancel{\text{Hz}}}\right) = 3.02 \times 10^{-27}\ \text{J/photon}$$

Since the total energy is 5.4 J, the number of photons can be calculated.

$$(5.4\ \cancel{\text{J}})\left(\frac{1\ \text{photon}}{3.02 \times 10^{-27}\ \cancel{\text{J}}}\right) = 1.8 \times 10^{27}\ \text{photons}$$

7.57 **(a)** green

Convert the frequency to wavelength and then use Figure 7.15(b) to determine the color.

$$\lambda\nu = c \qquad \lambda = \frac{c}{\nu} = \frac{2.998 \times 10^8\ \text{m} \cdot \cancel{\text{s}^{-1}}}{5.56 \times 10^{14}\ \cancel{\text{s}^{-1}}} = 5.39 \times 10^{-7}\ \text{m}$$

From Figure 7.15(b), green is the color of light of this wavelength.

(b) green, blue, violet

(c) yellow, orange, red

(d) 3.68×10^{-19} J

energy of a photon = $h\nu = (6.626 \times 10^{-34}\ \text{J} \cdot \cancel{\text{s}})(5.56 \times 10^{14}\ \cancel{\text{s}^{-1}}) = 3.68 \times 10^{-19}$ J

7.59 Ordinary uncertainty in measurements can be reduced by using more precise instruments. The uncertainty in measuring the position and speed of submicroscopic particles is unavoidable.

In order to identify the position and the speed of submicroscopic particles, light of extremely short wavelength and high energy is required. This energetic light will cause the particle to change speed. If lower energy, lower frequency light is used, the position of the particle cannot be accurately determined. In other words, the measuring "device" affects the measured value. This is not true for larger objects.

7.61 **(a)** No. Charges are not whole-number multiples of the smallest charge.

The smallest charge Millikan obtained in these experiments was 3.10×10^{-19}. This is *not* the unit of charge, since the other charges obtained are not whole number multiples of 3.10×10^{-19}. For example, 4.60×10^{-19} C/3.10×10^{-19} C = 1.5.

(b) 1.55×10^{-19} C

By dividing the smallest charge by 2, a factor is obtained that divides evenly into all of the recorded charges within experimental error.

(c) 3

$$(4.60 \times 10^{-19}\ \cancel{\text{C}})\left(\frac{1\ \text{unit charge}}{1.55 \times 10^{-19}\ \cancel{\text{C}}}\right) = 2.97$$

The answer is very close to 3.

Stop and Test Yourself

1. **(d)** 6.32 nm

The frequency of electromagnetic radiation is inversely proportional to the wavelength. Therefore, the electromagnetic radiation with the shortest wavelength, 6.32 nm, has the highest frequency.

2. **(e)** 1.091×10^{12} Hz

According to the equation $E = h\nu$, the energy of photons of electromagnetic radiation is directly proportional to the frequency. Therefore, the photons of electromagnetic radiation with the highest frequency, 1.091×10^{12} Hz, have the highest energy.

3. **(b)** 2.32×10^{-7}

$\lambda\nu = c$ $\lambda = \dfrac{c}{\nu} = \dfrac{2.998 \times 10^8 \text{ m} \cdot \cancel{s^{-1}}}{1.29 \times 10^{15} \cancel{s^{-1}}} = 2.32 \times 10^{-7} \text{ m}$

4. **(d)** 8.8×10^{13}

$\lambda\nu = c$ $\nu = \dfrac{c}{\lambda} = \left(\dfrac{3.00 \times 10^8 \cancel{m} \cdot s^{-1}}{3.4 \cancel{\mu m}}\right)\left(\dfrac{10^6 \cancel{\mu m}}{\cancel{m}}\right) = 8.8 \times 10^{13} \text{ s}^{-1}$

5. **(b)** 4.41×10^{-19}

energy of a photon $= \dfrac{hc}{\lambda} = \dfrac{(6.626 \times 10^{-34} \text{ J} \cdot \cancel{s})(2.998 \times 10^8 \cancel{m} \cdot \cancel{s^{-1}})}{4.50 \times 10^{-7} \cancel{m}} = 4.41 \times 10^{-19} \text{ J}$

6. **(c)** 33

energy of a photon $= h\nu = (6.626 \times 10^{-34} \text{ J} \cdot \cancel{s})(2.17 \times 10^{18} \cancel{Hz})\left(\dfrac{1 \cancel{s^{-1}}}{1 \cancel{Hz}}\right) = 1.44 \times 10^{-15} \text{ J}$

$(4.73 \times 10^{-14} \cancel{J})\left(\dfrac{1 \text{ photon}}{1.44 \times 10^{-15} \cancel{J}}\right) = 32.8$ photons = 33 photons because fractional photons do not exist.

7. **(c)** radio waves

See Figure 7.15 (a).

8. **(e)** has a lower frequency than radiation with wavelength 450 nm.

(a) is not true because longer wavelengths have lower frequencies.
(b) is not true because 550 nm is in the visible region of the spectrum.
(c) is not true because all electromagnetic radiation has the same velocity in vacuum.
(d) is not true because radiation with a shorter wavelength has a higher frequency and higher energy.

9. **(e)** proposed the theory that we call wave mechanics or quantum mechanics.

The other answers are attributable to the following people:
(a) J. J. Thomson
(b) Millikan
(c) Rutherford
(d) Bohr

10. **(c)** electrons move from a given energy level to a lower energy level.

When atoms are excited, electrons are often lifted to higher energy levels by the absorption of radiation of a specific wavelength. When these excited electrons fall back down to their original state, they emit energy.

11. **(a)** $2p_x$

See Figures 7.32 and 7.33.

12. **(c)** 10

Each of the five 3*d* orbitals can contain a maximum of two electrons.

13. **(d)** 8

The second shell has three *p* orbitals and one *s* orbital for a total of 4 orbitals, each of which can contain two electrons.

14. **(a)** principal

Only the principal quantum number, which designates the main shell, is shared by all of the orbitals in a shell.

Putting Things Together

7.63 **(a)** *Wavelength* is the distance between two similar points of a wave such as wave crests or wave troughs. The *frequency* is the number of cycles that pass a given point in one second.

(b) *Interference* occurs when two waves, which can be either in phase or out of phase with one another, are added together. *Diffraction* is the bending of waves around the edge of any object in their path that is about the same size or larger than the wavelength.

(c) *Traveling waves* move from one site to another, whereas *standing waves* are confined within a certain region of space.

(d) When atoms are in the lowest energy state they are said to be in their *ground state.* Any state of the atom that has a higher energy than the ground state is said to be an *excited state.*

(e) An *orbit* is a defined pathway along which an object (such as a planet) moves around another object (such as the Sun). An *orbital* is a region of space around an atom's nucleus where a relatively high probability of finding an electron exists.

7.65 1. What is the charge on the nucleus of an atom?

2. What keeps the electrons from falling into the nucleus?

7.67 **(a)** six

$n = 2$ to $n = 1$, $n = 3$ to $n = 1$, $n = 4$ to $n = 1$, $n = 3$ to $n = 2$, $n = 4$ to $n = 2$, $n = 4$ to $n = 3$

(b) $n = 4$ to $n = 1$

$n = 4$ to $n = 1$ is the largest energy transition.

(c) $n = 4$ to $n = 3$

$n = 4$ to $n = 3$ is the smallest energy transition.

7.69 **(a)** 984 kJ/mol

(Energy of $n = 2$ level) – (Energy of $n = 1$ level) = (–328 kJ/mol) – (–1312 kJ/mol) = 984 kJ/mol

(b) 1.22×10^{-7} m

$$\text{energy of photon} = \frac{hc}{\lambda}$$

$$\lambda = \frac{hc}{E} = \left(\frac{(6.626 \times 10^{-34}\,\cancel{J} \cdot \cancel{s} \cdot \cancel{photon^{-1}})(2.998 \times 10^{8}\,m \cdot \cancel{s^{-1}})}{(984\,\cancel{kJ} \cdot \cancel{mol^{-1}})(10^{3}\,\cancel{J} \cdot \cancel{kJ^{-1}})}\right)\left(\frac{6.022 \times 10^{23}\,\cancel{photon}}{\cancel{mol}}\right) = 1.22 \times 10^{-7}\,m$$

(c) $2.47 \times 10^{15}\,s^{-1}$

$$\lambda\nu = c \qquad \nu = \frac{c}{\lambda} = \frac{2.998 \times 10^{8}\,\cancel{m} \cdot s^{-1}}{1.22 \times 10^{-7}\,\cancel{m}} = 2.46 \times 10^{15}\,s^{-1}$$

(Best answer, without intermediate round off from part (b), is $2.47 \times 10^{15}\,s^{-1}$.)

(d) ultraviolet; radiation with frequencies between 10^{15} and $10^{16}\,s^{-1}$ is in the ultraviolet.

According to Figures 7.15(a) and (b), electromagnetic radiation with wavelengths 4×10^{-7} m and 10^{-7} m and with frequencies between $10^{15}\,s^{-1}$ and $10^{16}\,s^{-1}$ is in the ultraviolet region of the spectrum.

7.71 1312 kJ/mol

$$\frac{1}{\lambda} = 1.097 \times 10^{7}\,m^{-1}\left(\frac{1}{n_1^2} - \frac{1}{n^2}\right)$$

In this problem, $n_1 = 1$

But for ionization, $n = \infty$ and $\frac{1}{n^2} = 0$

Therefore,

$$\frac{1}{\lambda} = 1.097 \times 10^{7}\,m^{-1}\left(\frac{1}{1}\right) \text{ for ionization of the electron in hydrogen.}$$

$$\lambda = \frac{1}{1.097 \times 10^{7}\,m^{-1}} = 0.9116 \times 10^{-7}\,m$$

$$\text{ionization energy} = \frac{hc}{\lambda} = \frac{(6.6261 \times 10^{-34}\,J \cdot \cancel{s})(2.9979 \times 10^{8}\,\cancel{m} \cdot \cancel{s^{-1}})}{(0.9116 \times 10^{-7}\,\cancel{m})} = 2.179 \times 10^{-18}\,J$$

$$(2.179 \times 10^{-18}\,\cancel{J} \cdot \cancel{photon^{-1}})(6.0221 \times 10^{23}\,\cancel{photon} \cdot mol^{-1})\left(\frac{10^{-3}\,kJ}{1\,\cancel{J}}\right) = 1312\,kJ/mol$$

7.73 According to Heisenberg's uncertainty principle, the speed and the position of a subatomic particle cannot both be determined simultaneously with any degree of certainty. This is not true of large objects, such as baseballs, cars, and people. This difference is due to the fact that light is used to determine the position and speed of objects. If the wavelength of the light used is small compared with the object, the uncertainty in measurement is very small. However, if the wavelength of light is about the same order of magnitude or larger than the object, either the speed and/or the position of the object is uncertain.

7.75 The negative charge and the relatively low mass of the electrons would have caused most of the electrons to be reflected back rather than to penetrate the negatively charged electron clouds of the metal atoms. If an electron did have enough kinetic energy to penetrate, it would have been captured by the positively charged nuclei of the atoms.

7.77 **(a)** decreases

Li – 4.6
Na – 4.40
Rb – 3.46
Cs – 3.43

The work function decreases down a group

(b) increases

Na – 4.40 < Mg – 5.86 < Al – 6.86

The work function increases across a period (row) from left to right.

(c) 5.86×10^{-19} J

According to the table, 5.86×10^{-19} J of energy are required to knock one electron out of an atom of magnesium.

(d) ultraviolet

$$\text{energy of a photon} = \frac{hc}{\lambda}$$

$$\lambda = \frac{hc}{E} = \frac{(6.626 \times 10^{-34}\ \cancel{\text{J}} \cdot \cancel{\text{s}})(2.998 \times 10^{8}\ \text{m} \cdot \cancel{\text{s}^{-1}})}{5.86 \times 10^{-19}\ \cancel{\text{J}}} = 3.39 \times 10^{-7}\ \text{m}$$

According to Figure 7.15(b), this wavelength is in the ultraviolet region of the spectrum.

7.79 **(a)** 3×10^{15} g/cm^3

First, convert atomic mass units (u) to grams.

$$(1.0073\ \cancel{\text{u}})\left(\frac{1\ \text{g}}{6.022\ 14 \times 10^{23}\ \cancel{\text{u}}}\right) = 1.6727 \times 10^{-24}\ \text{g}$$

Next, calculate volume in cm^3.

$$V = \frac{\pi(d)^3}{6} = \frac{(3.1416)(1 \times 10^{-15}\ \cancel{\text{m}})^3\left(\frac{1\ \text{cm}}{10^{-2}\ \cancel{\text{m}}}\right)^3}{6} = 5 \times 10^{-40}\ \text{cm}^3$$

Finally, divide the mass in grams by the volume in cm³.

$$d = \frac{1.6727 \times 10^{-24}\ \text{g}}{5 \times 10^{-40}\ \text{cm}^3} = 3 \times 10^{15}\ \text{g/cm}^3$$

(b) 9.577×10^7 C/kg

$$\left(\frac{1.602 \times 10^{-19}\ \text{C}}{1.6727 \times 10^{-24}\ \cancel{\text{g}}}\right)\left(\frac{10^3\ \cancel{\text{g}}}{\text{kg}}\right) = 9.577 \times 10^7\ \text{C/kg}$$

(c) zero

The charge on a neutron is zero.

Applications

7.81 Because of the wave properties of an electron, the electron microscope is possible.

The wave nature of an electron allows it to be used in an electron microscope. The wavelength of an electron is about the same as that of X–rays. However, unlike X–rays, which are difficult to focus, the charge on the electron allows it to be easily focused using electric and/or magnetic fields.

7.83 461 m

$$\lambda\nu = c$$

$$\lambda = \frac{c}{\nu} = \frac{2.998 \times 10^8\ \text{m} \cdot \cancel{\text{s}^{-1}}}{(650\ \cancel{\text{kHz}})\left(\frac{10^3\ \cancel{\text{Hz}}}{\cancel{\text{kHz}}}\right)\left(\frac{1\ \cancel{\text{s}^{-1}}}{\cancel{\text{Hz}}}\right)} = 461\ \text{m}$$

7.85 (b) < (c) < (d) < (e) < (a)

From Figure 7.15(a) the order of decreasing wavelength (increasing energy per photon) is

microwave < red–orange < ultraviolet < X–rays < gamma rays.

7.87 **(a, b)**

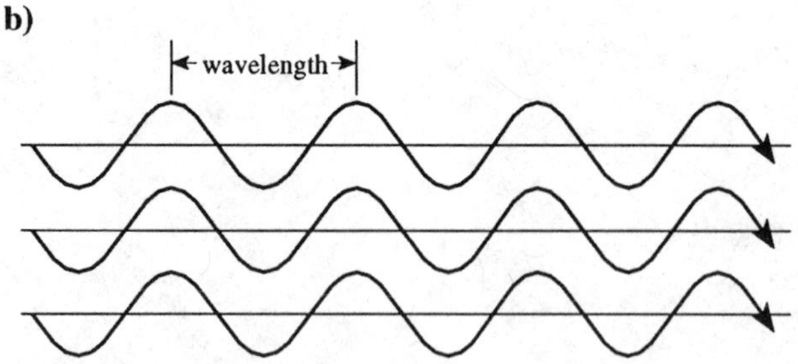

The photons are all of the same wavelength, they're in phase with each other, and they are parallel.

(c) The intensity of light is proportional to the amplitude of the wave (half the distance from the top of the wave to the bottom of the wave). Since each photon is in phase with the others and moving parallel to the others, significant constructive interference occurs, causing high–intensity light.

7.89 **(a)** (i) Both X–rays and visible light move with the same speed in a vacuum.

(ii) The wavelength of an X–ray is significantly shorter than the wavelength of visible light.

(iii) The frequency of an X–ray is significantly higher.

(iv) The energy per photon of an X–ray is significantly higher.

(v) The X–ray has a higher ability to penetrate flesh due to its significantly higher energy.

(b) Lead is a very dense material that can absorb electromagnetic radiation of X–ray frequency, thus transmitting little radiation.

7.91 12.9 kJ

$$\text{energy of a photon} = \frac{hc}{\lambda} = (6.6261 \times 10^{-34}\ \text{J} \cdot \cancel{\text{s}})\left(\frac{2.9979 \times 10^{8}\ \cancel{\text{m}} \cdot \cancel{\text{s}^{-1}}}{694.3\ \cancel{\text{nm}}}\right)\left(\frac{1\ \cancel{\text{nm}}}{10^{-9}\ \cancel{\text{m}}}\right) = 2.861 \times 10^{-19}\ \text{J}$$

$$(4.50 \times 10^{22}\ \cancel{\text{photon}})\left(\frac{2.861 \times 10^{-19}\ \cancel{\text{J}}}{\cancel{\text{photon}}}\right)\left(\frac{1\ \text{kJ}}{10^{3}\ \cancel{\text{J}}}\right) = 12.9\ \text{kJ}$$

7.93 from short radio waves and microwaves to X–rays

$$(20\ \cancel{\text{cm}})\left(\frac{10^{-2}\ \text{m}}{\cancel{\text{cm}}}\right) = 0.20\ \text{m} = 2 \times 10^{-1}\ \text{m}$$

$$(10^{-7}\ \cancel{\text{cm}})\left(\frac{10^{-2}\ \text{m}}{\cancel{\text{cm}}}\right) = 10^{-9}\ \text{m}$$

According to Figure 7.15(a), 20 cm to 10^{-7} cm corresponds to the region of the electromagnetic spectrum ranging from short radio waves and microwaves to X–rays.

7.95 1×10^{3} photon · s^{-1}

Calculate the area of the pupil of the eye.

$$A = \pi r^2 = (3.1)\left(\frac{0.6\ \cancel{\text{cm}}}{2}\right)^2\left(\frac{10^{-2}\ \text{m}}{\cancel{\text{cm}}}\right)^2 = 3 \times 10^{-5}\ \text{m}^2$$

Calculate the energy per photon of 550 nm light.

$$\text{energy of a photon} = \frac{hc}{\lambda} = (6.626 \times 10^{-34}\ \text{J} \cdot \cancel{\text{s}} \cdot \text{photon}^{-1})\left(\frac{2.998 \times 10^{8}\ \cancel{\text{m}} \cdot \cancel{\text{s}^{-1}}}{550\ \cancel{\text{nm}}}\right)\left(\frac{1\ \cancel{\text{nm}}}{10^{-9}\ \cancel{\text{m}}}\right) = 3.61 \times 10^{-19}\ \text{J/photon}$$

Using the given intensity of the light, the area of the pupil of the eye, and the energy per photon, calculate the photons per second entering the eye.

$$\left(\frac{1.5 \times 10^{-11}\ \text{J}}{\text{m}^2 \cdot \text{s}}\right)\left(\frac{1\ \text{photon}}{3.61 \times 10^{-19}\ \text{J}}\right)(3 \times 10^{-5}\ \text{m}^2) = 1 \times 10^3\ \text{photon} \cdot \text{s}^{-1}$$

7.97 **(a)** 1.661×10^{-19} J

$$\left(\frac{100.0\ \text{kJ}}{\text{mol}}\right)\left(\frac{1\ \text{mol}}{6.0221 \times 10^{23}\ \text{photon}}\right)\left(\frac{10^3\ \text{J}}{\text{kJ}}\right) = 1.661 \times 10^{-19}\ \text{J/photon}$$

(b) The energy per photon of visible light (2.8×10^{-19} J) is higher than the minimum required to darken silver bromide, whereas the energy per photon of the signal from a TV station (10^{-25} J) is not.

For signal from TV station,

$$\text{energy of photon} = h\nu_{TV} = (6.626 \times 10^{-34}\ \text{J} \cdot \text{s})(10^8\ \text{s}^{-1}) = 10^{-25}\ \text{J}$$

For lowest energy visible light,

$$\text{energy of a photon} = h\nu_{\text{visible}} = (6.626 \times 10^{-34}\ \text{J} \cdot \text{s})(4.3 \times 10^{14}\ \text{s}^{-1}) = 2.8 \times 10^{-19}\ \text{J}$$

$2.8 \times 10^{-19}\ \text{J} > 1.661 \times 10^{-19}\ \text{J} > 10^{-25}\ \text{J}$

7.99 **(a)** ultraviolet

$$(180.7\ \text{nm})\left(\frac{10^{-9}\ \text{m}}{\text{nm}}\right) = 1.807 \times 10^{-7}\ \text{m}$$

1.807×10^{-7} m is in the ultraviolet region according to Figure 7.15(a).

(b) 1.099×10^{-18} J

$$\text{energy of a photon} = \frac{hc}{\lambda} = (6.6261 \times 10^{-34}\ \text{J} \cdot \text{s})\left(\frac{2.9979 \times 10^8\ \text{m} \cdot \text{s}^{-1}}{1.807 \times 10^{-7}\ \text{m}}\right) = 1.099 \times 10^{-18}\ \text{J}$$

7.101 **(a)** hydrogen **(b)** helium

7.103 1.88×10^{27} photons

Calculate the thermal energy absorbed by the water in joules.

$$\text{thermal energy absorbed} = (\text{mass of } H_2O)(\text{specific heat})(\Delta t) = (250.0\ \text{g})\left(\frac{4.184\ \text{J}}{\text{g} \cdot {}^\circ\text{C}}\right)(100.0\ {}^\circ\text{C} - 10.6\ {}^\circ\text{C}) = 9.35 \times 10^4\ \text{J}$$

Calculate the energy of a photon of 4.00-mm radiation.

$$\text{energy of a photon} = \frac{hc}{\lambda} = \left(\frac{(6.626 \times 10^{-34}\ \text{J} \cdot \text{s})(2.998 \times 10^8\ \text{m} \cdot \text{s}^{-1})}{4.00\ \text{mm}}\right)\left(\frac{1\ \text{mm}}{10^{-3}\ \text{m}}\right) = 4.97 \times 10^{-23}\ \text{J}$$

Calculate the number of photons required to yield the total energy, 9.35×10^4 J.

$$(9.35 \times 10^4 \text{ J})\left(\frac{1 \text{ photon}}{4.97 \times 10^{-23} \text{ J}}\right) = 1.88 \times 10^{27} \text{ photons}$$

7.105 **(a)** 3.843×10^5 km

$$(2.564 \text{ s})(2.998 \times 10^8 \text{ m} \cdot \text{s}^{-1})\left(\frac{1 \text{ km}}{10^3 \text{ m}}\right) = 7.687 \times 10^5 \text{ km round trip}$$

$$\text{Distance to moon} = \frac{7.687 \times 10^5}{2} \text{ km} = 3.844 \times 10^5 \text{ km}$$

(Best answer, without intermediate round off, is 3.843×10^5 km.)

(b) 11 significant figures

$$(2.5 \text{ cm})\left(\frac{10^{-2} \text{ m}}{\text{cm}}\right)\left(\frac{1 \text{ km}}{10^3 \text{ m}}\right) = 0.000\,025 \text{ km}$$

The distance to the moon, therefore, can be measured to 11 significant figures.

384 300.000 00 km

(Note, however, that it is only known to four significant figures from the data in this problem.)

7.107 3.86×10^{-7} m

$$\left(\frac{310 \text{ kJ}}{\text{mol Ag}}\right)\left(\frac{1000 \text{ J}}{1 \text{ kJ}}\right)\left(\frac{1 \text{ mol Ag}}{6.022 \times 10^{23} \text{ atoms}}\right) = 5.15 \times 10^{-19} \text{ J/Ag atom}$$

$$\lambda = \frac{hc}{E} = \frac{(6.626 \times 10^{-34} \text{ J} \cdot \text{s})(2.998 \times 10^8 \text{ m} \cdot \text{s}^{-1})}{5.15 \times 10^{-19} \text{ J}} = 3.86 \times 10^{-7} \text{ m}$$

7.109 **(a)** green

(b) 6.14×10^{14} s^{-1} and 5.83×10^{14} s^{-1}

$$(488 \text{ nm})\left(\frac{10^{-9} \text{ m}}{\text{nm}}\right) = 4.88 \times 10^{-7} \text{ m} \qquad (514 \text{ nm})\left(\frac{10^{-9} \text{ m}}{\text{nm}}\right) = 5.14 \times 10^{-7} \text{ m}$$

$$\lambda \nu = c \qquad \nu = \frac{c}{\lambda}$$

$$\nu = \frac{c}{\lambda} = \frac{2.998 \times 10^8 \text{ m} \cdot \text{s}^{-1}}{4.88 \times 10^{-7} \text{ m}} = 6.14 \times 10^{14} \text{ s}^{-1} \qquad \nu = \frac{c}{\lambda} = \frac{2.998 \times 10^8 \text{ m} \cdot \text{s}^{-1}}{5.14 \times 10^{-7} \text{ m}} = 5.83 \times 10^{14} \text{ s}^{-1}$$

(c) 18 protons, 17 electrons

(d) 1.000 mmHg, 1.316×10^{-3} atm, 133.3 Pa

$$(1.000\ \cancel{\text{torr}})\left(\frac{1\ \text{atm}}{760\ \cancel{\text{torr}}}\right) = 1.316 \times 10^{-3}\ \text{atm}$$

$$(1.000\ \cancel{\text{torr}})\left(\frac{1\ \cancel{\text{atm}}}{760\ \cancel{\text{torr}}}\right)\left(\frac{1.103\ 25 \times 10^{5}\ \text{Pa}}{1\ \cancel{\text{atm}}}\right) = 133.3\ \text{Pa}$$

CHAPTER 8:
Electronic Structure and the Periodic Table

Practice Problems

8.1   There are two electrons in the s subshell of the third main shell or principal energy level.

8.2 $3s^1$ or $4s^1$, etc.

       An infinite number of excited states are possible.

8.3   **(a)**

$$\uparrow \quad \uparrow \quad \uparrow$$
$$2p$$

Increasing energy →

$$\uparrow\downarrow$$
$$2s$$

$$\uparrow\downarrow$$
$$1s$$

N

       (b) $1s^2 2s^2 2p^3$ or $1s^2 2s^2 2p_x^1 2p_y^1 2p_z^1$

8.4   **(a)**

$$\downarrow \quad \underline{\quad} \quad \underline{\quad}$$
$$2p$$

Increasing energy →

$$\uparrow$$
$$2s$$

$$\uparrow\downarrow$$
$$1s$$

Be

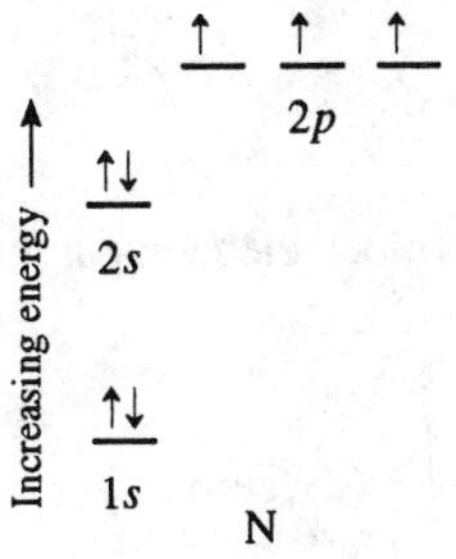

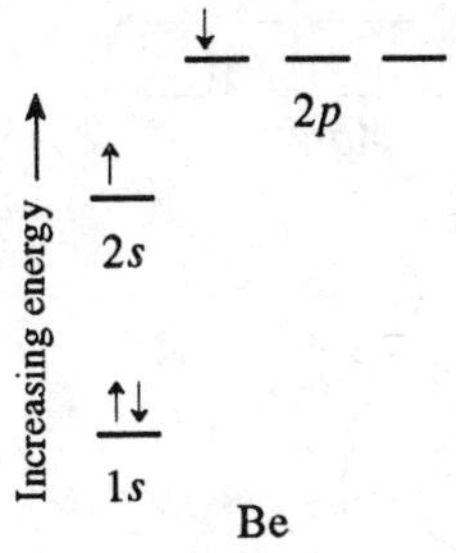

(b) $1s^22s^12p^1$ or $1s^22s^12p_x^1$

8.5 **(a)** $[He]2s^1$

(b) $[He]2s^22p^4$

(c) [Ne] or $[He]2s^22p^6$

8.6 **(a)** Li [↑↓] (1s) [↑] (2s) (one unpaired electron)

(b) O [↑↓] (1s) [↑↓] (2s) [↑↓][↑][↑] ($2p_x$ $2p_y$ $2p_z$) (two unpaired electrons)

(c) Ne [↑↓] (1s) [↑↓] (2s) [↑↓][↑↓][↑↓] ($2p_x$ $2p_y$ $2p_z$) (zero unpaired electrons)

8.7 C [He] [↑] (2s) [↑][↑][↓] ($2p_x$ $2p_y$ $2p_z$)

8.8 **(a)** $[Ar]4s^23d^9$

(b) $[Ar]4s^13d^{10}$

8.9 **(a)** 118 **(b)** 121

Elements 110–118 will complete the seventh period. Elements 119 and 120 will be *s*–block elements in the eighth period. Element 121 will be the first *g*–block element.

8.10 **(i)** **(a)** $1s^22s^22p^6$ **(b)** [Ne] or $[He]2s^22p^6$ **(c)** [Ne] or [He] [↑↓] (2s) [↑↓][↑↓][↑↓] (2p) **(d)** zero

(ii) **(a)** $1s^22s^22p^63s^23p^5$ **(b)** $[Ne]3s^23p^5$ **(c)** [Ne] [↑↓] (3s) [↑↓][↑↓][↑] (3p) **(d)** one

(iii) **(a)** $1s^22s^22p^63s^23p^64s^23d^{10}4p^65s^24d^2$ **(b)** $[Kr]5s^24d^2$ **(c)** [Kr] [↑↓] (5s) [↑][↑][ ][ ][ ] (4d)

(d) two

(iv) **(a)** $1s^22s^22p^63s^23p^64s^23d^{10}4p^65s^24d^{10}5p^66s^24f^5$ **(b)** $[Xe]6s^24f^5$

(c) [Xe] [↑↓] (6s) [↑][↑][↑][↑][↑][ ][ ] (4f) **(d)** five

8.11 **(a)** Cl^- $1s^22s^22p^63s^23p^6$

(b) Cl^- [Ne] [↑↓] [↑↓][↑↓][↑↓] or [Ar]
3*s* 3*p*

8.12 (a) Cu^+ $1s^22s^22p^63s^23p^63d^{10}$

Cu^{2+} $1s^22s^22p^63s^23p^63d^9$

(b) Cu^+ [Ar] [↑↓][↑↓][↑↓][↑↓][↑↓]
3*d*

Cu^{2+} [Ar] [↑↓][↑↓][↑↓][↑↓][↑]
3*d*

8.13 (a) Lithium and beryllium are both in the 2nd period of the periodic table. This means that their outer shell electrons are being added to the same principal energy level (main shell). Since beryllium has one more proton in the nucleus than lithium, the outer shell electrons in beryllium feel a stronger force of attraction and are pulled in toward the nucleus.

(b) Chlorine and fluorine are in the same main group (Group VIIA) but they are in different periods, chlorine being in the third period and fluorine in the second period. In a chlorine atom, the outer electrons are in the third shell; in a fluorine atom the outer electrons are in the second shell. Electrons in the third shell are farther from the nucleus on average and are shielded from the positive charge of the nucleus by an additional layer of electrons. Therefore, a chlorine atom is larger than a fluorine atom in spite of the chlorine atom's higher nuclear charge.

(c) Between Ag and Au, the 4*f* orbitals of the lanthanides have been filled and many additional protons have been added to the nuclei. Thus, the higher energy level effects on size (5th level vs. 4th level) are offset by the substantial increase in nuclear charge, causing Au to be about the same size as Ag. This effect is referred to as the lanthanide contraction.

8.14 (a) K

K has a larger radius than sodium because its outer shell electron is in a higher principal energy level.

(b) K

K has a larger radius because calcium has one additional proton and the outer shell electrons are in the same energy level as the outer shell electron in K. The greater force of attraction "shrinks" the calcium atom relative to K.

8.15 B < Be < Mg

$\longrightarrow$

increasing radius

A boron atom is smaller than a beryllium atom, since atomic radius decreases across rows in the periodic table. A magnesium atom is larger than a beryllium atom, since atomic radius increases down main groups.

8.16 (a) When a sodium atom is ionized to a Na^+ ion, the one outer shell electron ($3s^1$) is removed. Thus, in effect, the outer shell is removed. In addition, the remaining electrons are held even more tightly by the nucleus owing to the excess positive charge.

(b) When a chlorine atom forms a chloride ion, Cl^-, the extra electron is added to a shell that already contains 7 electrons. Thus, electronic repulsion is increased, and the size increases.

8.17 (a) Al^{3+} (b) F^- (c) Na^+

See Figure 8.9. Al^{3+} is smaller than Mg^{2+} as a result of its higher nuclear and ionic charges. F^- is smaller than O^{2-} as a result of its higher nuclear and lower ionic charges. Na^+ is smaller than F^- as a result of its positive charge.

8.18 K^+, Cl^-

Isoelectronic species have the same electron configurations. Positive ions that are isoelectronic with argon are K^+, Ca^{2+}, and Sc^{3+}. Negative ions that are isoelectronic with argon are P^{3-}, S^{2-}, and Cl^-.

8.19 (a) The 4*s* electron in K is in the 4th principal energy level and is farther from the nucleus than the 3*s* electron in Na. The 4*s* electron is also better shielded from the nucleus. Thus, the 4*s* electron in K is more easily removed than the 3*s* electron in Na.

(b) F has one more proton than O while having outer electrons in the same shell. Thus, F's outer shell electrons are held more tightly by the nuclear charge, giving F a higher first ionization energy.

(c) Oxygen has 2 paired electrons in the $2p_x$ orbital, while N has no paired electrons in the 2*p* sublevel. The pairing of electrons in O produces electron repulsion, making it easier to remove an electron.

(d) Au has a much higher effective nuclear charge than does Ag while being about the same size. The higher effective nuclear charge is due to an increase of 32 protons in the nucleus from Ag to Au. This higher effective nuclear charge results in a higher first ionization energy.

8.20 (a) He

Hydrogen and helium are in the same period; helium has one more proton in its nucleus. Also, helium is a noble gas and has a very stable electron configuration.

(b) Na

Potassium and sodium are in the same main group, but potassium is in the fourth period whereas sodium is in the third period. Sodium's outer electron is closer to the nucleus and not as well shielded.

(c) Hg

Cadmium and mercury are transition elements. The atoms are about the same size, but the mercury atom has 32 more protons in its nucleus.

8.21 The *y*–axis on which the ionization energies are plotted would be much too long to fit on the page. In a log plot, each unit on the *y*–axis represents 10 times the previous unit.

8.22 (a) The second electron must be removed from a positively charged species, Be^+.

(b) The 3rd ionization energy represents the energy required to remove a core electron from an electron configuration electronically identical to that of the noble gas He. This electron configuration is very stable.

8.23 21 kJ/mol

$He(g) + e^- \rightarrow He^- (g)$ $\Delta H = -$electron affinity for He
$\Delta H = -(-21$ kJ/mol He)
$\Delta H = 21$ kJ/mol

8.24 When an electron is added to an oxygen atom to form O^-, that electron must go into a relatively small energy level (shell) that already contains six other electrons. By contrast, the 3rd, 4th, 5th, and 6th energy levels in S, Se, Te, and Po, respectively, are much bigger (they have a higher volume), and electronic repulsion in these elements is much less. Also, S, Se, Te, and Po have higher effective nuclear charges than O, causing more energy to be liberated when an electron is added to them.

8.25 For magnesium to add an electron, the electron must go into a higher energy sublevel due to the fact that magnesium's *s* sublevel is filled. Magnesium does not have a large enough effective nuclear charge for the addition of another electron to be an exothermic or energetically favorable change.

8.26 **(a)** Exothermic. Electron affinity of Na is positive.

This reaction represents the electron affinity of sodium, which is positive. Since electron affinity is the negative of the enthalpy change associated with the reaction, the enthalpy change will be negative and the process is exothermic.

(b) Endothermic. Electron affinity of Mg is negative.

This reaction represents the electron affinity of magnesium, which is negative. Since electron affinity is the negative of the enthalpy change associated with the reaction, the enthalpy change will be positive and the process is endothermic.

8.27 **(a)** Exothermic. Electron attracted by 2+ charge.

When adding an electron to a Mg^{2+} ion, the electron will be strongly attracted by the 2+ charge. Thus, the process will be exothermic.

(b) Endothermic. Electron repelled by 1– charge.

The additional electron being added to a S^- ion will feel a strong repulsion from excess negative charge on the ion. Thus, the process will be endothermic.

8.28 **(a)** NaOH

(b) Na_2CO_3

(c) Na_2SO_4

(d) SO_2

(e) HCl

8.29 **(a)** sodium hydride

(b) xenon difluoride

(c) potassium hydroxide

(d) carbon dioxide

(e) magnesium nitride

8.30 Mg < Na < K

8.31 (a) $Ca(s) + H_2(g) \rightarrow CaH_2(s)$

(b) $2K(s) + 2H_2O(l) \rightarrow 2KOH(aq) + H_2(g)$

(c) $Mg(s) + 2H_2O(g) \rightarrow Mg(OH)_2(s) + H_2(g)$

(d) $3Ca(s) + N_2(g) \rightarrow Ca_3N_2(s)$

(e) no reaction

(f) $2K(s) + Cl_2(g) \rightarrow 2KCl(s)$

(g) $Ca(s) + Cl_2(g) \rightarrow CaCl_2(s)$

(h) $Cl_2(g \text{ or } aq) + 2NaBr(aq) \rightarrow Br_2(l \text{ or } aq) + 2NaCl(aq)$

(i) no reaction

(j) $CH_3CH_3(g) + 2H_2O(g) \rightarrow 5H_2(g) + 2CO(g)$

Additional Practice Problems

8.75 (a) The ground state of hydrogen has one *s* electron. Similarly, the alkali metals in Group IA each have a single *s* electron in their outermost occupied shell. However, by gaining one additional electron, a hydrogen atom in its ground state can get the electron configuration of the next noble gas. In this respect, hydrogen is like the halogens in Group VIIA, each of which, by adding one electron, has the electron configuration of a noble gas.

(b) Both of them have a half–filled outer shell.

Hydrogen has a half–filled first shell and carbon has a half–filled second shell.

8.77 (a) nitrogen (b) lanthanum (c) actinium

See Figure 8.4

8.79 (a) Across a row or period of the periodic table, the nuclear charge increases but the principal electron energy level stays the same. Thus, a contraction in size occurs. For the transition metals, inner orbitals are being filled that serve to shield the outer electrons from the attraction of the nucleus.

(b) Initially, the increased attraction of the nucleus for the electrons causes contraction. However, as more electrons are added to the *d* orbitals, electron–electron repulsion becomes important and a "swelling" of the atom takes place.

(c) This phenomenon is known as the lanthanide contraction. Between the second and third transition series a large number of protons are added to the nucleus, which causes significant contraction of the electron clouds.

8.81 Group VA

Elements in Group VA have three unpaired electrons in *p* orbitals. If one of these electrons is lost due to ionization, a singly charged cation is formed that has two unpaired electrons.

8.83 (a) Mn^{3+}

Mn loses two 4*s* electrons and one 3*d* electron when it forms a 3+ ion.

(b) Ag^{+}

Ag loses one 5*s* electron to form a 1+ ion.

(c) I^{-}

I gains one electron in the 5*p* subshell to form the 1– ion.

8.85 Metals are shiny when their surfaces are clean. With the exception of mercury, metals are solids under ordinary conditions. Metals conduct thermal energy and electricity. They can be rolled or hammered into sheets and pulled into wire. Their chemical properties vary from unreactive (e.g., gold and platinum) to very reactive (e.g., rubidium, which burns when exposed to air).

8.87 (a) No. They are similar in reactivity to magnesium, which is not found free in nature.

(b) Lanthanide ores are mixtures containing all of the lanthanides because the lanthanides have very similar chemical properties.

8.89 5140 kJ

From Table 8.1, the first three ionization energies for aluminum are 576 kJ/mol, 1814 kJ/mol, and 2750 kJ/mol. The sum of these three ionization energies (576 + 1814 + 2750 = 5140) equals the amount of energy required to remove three electrons from 1 mol of gaseous aluminum atoms.

8.91 (b), (c), and (e)

These compounds contain transition elements.

Stop and Test Yourself

1. (b) $1s^22s^22p^63s^23p^64s^23d^{10}4p^65s^24d^{10}5p^66s^24f^4$

2. **(c)** The number of unpaired electrons is 3 and the Co^{2+} ion is paramagnetic.

The two electrons removed to form the Co^{2+} cation are removed from the 4*s* orbital. This leaves 7 electrons in the outer shell of electrons: 4 paired and 3 unpaired.

3. **(c)** [Kr]

Krypton, Kr, is the noble gas in the row just above the row iodine, I, is in.

4. **(d)** $[Ne]3s^1 3p^6$

This represents an excited state of the element chlorine.

5. **(d)** $1s^2 2s^2 2p^6 3s^2 3p^6 3d^7 4s^2$

(a) has too many 3*p* electrons.
(b) has two electrons in a 3*p* orbital with the same spin, which violates the Pauli exclusion principle.
(c) has three electrons in one 2*p* orbital.
(e) has a 2*d* orbital, which does not exist.

6. **(a)** 4*s*

The 4*s* orbital penetrates closer to the nucleus than the other orbitals in the 4th principal energy level.

7. **(c)** 10

There are five orbitals, each of which can hold two electrons, in *d* subshells.

8. **(d)** $K > Na > Mg > Mg^{2+}$

K > Na because atomic radii increase down main groups. Na > Mg because atomic radii decrease from left to right across periods. Since two 3*s* electrons are lost from Mg in forming Mg^{2+}, an entire electron shell has been removed and Mg^{2+} is the smallest of the four species.

9. **(c)** $Cl < Br < Br^-$

Bromine has one more occupied shell than chlorine and is, thus, larger than Cl. The bromide ion is larger than a bromine atom because adding an extra electron to the outer shell causes a "swelling" of the shell due to repulsions between electrons.

10. **(e)** Cl > Br > I

As you go down a family, it becomes easier to remove an electron from the atom due to the fact that the electron is farther away from the nucleus and is partially shielded from the nucleus by other electrons in lower energy levels. (a) is backwards. This series should be Mg > Ca > Sr. Ionization energies decrease down main groups. (b) is also backwards. It should be B > Be > Li. Ionization energies increase across periods from left to right. (c) Ne is a noble gas and has the highest ionization energy of the three. (d) Mg > Na. Ionization energies increase across periods from left to right.

11. **(a)** B

A large jump in ionization energy occurs between I_3 and I_4, indicating that the atom has three outer electrons. Boron is the only one of the elements listed for which this is true.

12. **(c)** Cl

Ar is a noble gas and has a very low (negative) electron affinity. In general, electron affinities decrease down groups and increase across periods from left to right except for the noble gases. Thus Cl > Br, S > Se, and Cl > S. Cl has the highest electron affinity of all the elements according to Figure 8.13.

13. **(d)** Ne

No compounds have been observed for noble gases above Kr in the periodic table.

14. **(d)** Mg

All the alkali metals and the alkaline earth metals *below* Mg in the periodic table react with water at room temperature.

15. **(c)** *d*–block element

Vanadium is a transition metal. (See Figure 8.5.)

Putting Things Together

8.93 (a) 496 kJ/mol

The electron affinity for Na^+ is the reverse of the first ionization energy for Na, which is given in Figure 8.10.

$Na^+(g) + e^- \rightarrow Na(g)$ (electron affinity)
$Na(g) \rightarrow Na^+(g) + e^-$ (first ionization energy of Na, 496 kJ/mol)

(b) 1447 kJ/mol

The electron affinity for Mg^{2+} is the reverse of the second ionization energy for Mg, which is given in Table 8.1.

$Mg^{2+}(g) + e^- \rightarrow Mg^+(g)$ (electron affinity)
$Mg^+(g) \rightarrow Mg^{2+}(g) + e^-$ (second ionization energy of Mg, 1447 kJ/mol)

8.95 The properties of the elements are periodic functions of their atomic number.

8.97 (c) Cl

A chlorine atom has one unpaired electron in the 3*p* subshell.

8.99 **(a)** $1s^22s^22p^63s^23p^64s^2 \rightarrow 1s^22s^22p^63s^23p^64s^1$

$\rightarrow 1s^22s^22p^63s^23p^6 \rightarrow 1s^22s^22p^63s^23p^5$

(b) The second electron must be removed from a positively charged ion. Since it has a positive charge, the Ca^+ ion holds more tightly to the electrons that remain.

(c) The first two electrons are removed from the fourth principal energy level, which is farther away from the nucleus and less shielded than the third energy level. The third electron not only must be removed from an energy level closer to the nucleus and from an electric field with two extra positive charges, but it also must be removed from a very stable electronic configuration identical to that of the noble gas argon.

(d) For aluminum, the biggest increase in ionization energy is between the third and fourth ionization energies because the fourth electron must be removed from the core.

8.101 Magnesium has a higher first ionization energy than Na because Mg has a smaller radius and higher nuclear charge than Na. Magnesium's first ionization energy is also higher than that of Al because the first electron must be removed from an *s* energy level rather than from a *p* energy level.

8.103 **(a)** Atoms with high first ionization energies have a relatively large attraction for electrons. Thus, they also tend to have high electron affinities. Normally, by adding electrons, these atoms can attain a noble gas electron configuration.

(b) The noble gases have very high ionization energies because of their very stable electron configuration. For the same reason, they have no affinity for any additional electrons.

8.105 **(a)** Cl_2, H_2, and Ne

(b) none

(c) Ca and Na

8.107 In the hydrogen atom, the 2*p* orbital is degenerate with the 2*s* orbital, and the 3*p* and 3*d* orbitals are degenerate with the 3*s* orbital.

In hydrogen, all orbitals within a given principal energy level are degenerate, that is, have the same energy.

8.109 **(a)** K^+, potassium ion; Ca^{2+}, calcium ion

(b) Cl^-, chloride ion; S^{2-}, sulfide ion

(c) H^+, hydrogen ion; transition metal ions such as Fe^{3+}

8.111 **(a)** solid

The trend in physical state for the halogens from top to bottom in the periodic table is gas (F_2) to liquid (Br_2) to solid (I_2). Therefore, At_2, which is beneath I_2, would be expected to be a solid under ordinary conditions.

(b) 117, $[Rn]7s^2 5f^{14} 6d^{10} 7p^5$ or $[Rn]5f^{14} 6d^{10} 7s^2 7p^5$

8.113 **(a)** no reaction

(b) $Ba(s) + 2H_2O(l) \xrightarrow{\text{room temp.}} Ba(OH)_2(aq) + H_2(g)$

(c) $2K(s) + Br_2(g) \rightarrow 2KBr(s)$

(d) no reaction

(e) no reaction

8.115 (a) 24 cm^3/mol

$$\left(\frac{1\ cm^3}{0.97\ \cancel{g}}\right)\left(\frac{23.0\ \cancel{g}}{mol}\right) = \frac{24\ cm^3}{mol}$$

(b) No. At the same temperature and pressure, one mole of any gas (assuming ideal behavior) has the same volume as one mole of any other gas.

8.117 (a) 2.05 g/cm^3 Graph given data and extrapolate.

(b) 12 °C, liquid Graph given data and extrapolate.

(c) 285 pm Graph data from Figure 8.8 and extrapolate.

(d) Fr^+

(e) 185 pm Graph data from Figure 8.9 and extrapolate.

(f) 320 kJ/mol Graph data from Figure 8.10 and extrapolate.

(g) greater than that of Cs, which reacts exposively with water (see Table 1.1); FrOH and $H_2(g)$ should be formed.

(h) FrCl

The formula for sodium chloride is NaCl.

(i) soluble

All Na^+ and K^+ compounds are soluble (see Table 4.4).

8.119 (a) Potassium nitrate must be the least soluble of the four compounds.

(b) concentrated

According to Table 4.4, all of these salts are soluble.

(c) Molecular equation: $NaNO_3(aq) + KCl(aq) \rightarrow KNO_3(s) + NaCl(aq)$

Complete ionic equation: $Na^+ + NO_3^- + K^+ + Cl^- \rightarrow KNO_3(s) + Na^+ + Cl^-$

Net ionic equation: $NO_3^- + K^+ \rightarrow KNO_3(s)$

8.121 13.51 M

$$\left(\frac{77.67\,\text{g } H_2SO_4}{100\,\text{g soln}}\right)\left(\frac{1.706\,\text{g soln}}{\text{mL soln}}\right)\left(\frac{1000\,\text{mL soln}}{\text{L soln}}\right)\left(\frac{1\,\text{mol } H_2SO_4}{98.080\,\text{g } H_2SO_4}\right) = 13.51\,\frac{\text{mol } H_2SO_4}{\text{L soln}}$$

Applications

8.123 Strontium is an alkaline earth metal, as is calcium; thus, they have very similar properties. Strontium has an identical outer shell electron configuration, ns^2, and is more reactive, in general, than calcium. It also has nearly the same atomic and ionic radii (slightly larger) and the same ionic charge (2+). Since strontium is more reactive, it can displace magnesium from its compounds.

8.125 **(a)** $1s^22s^22p^63s^03p^1$ or $[Ne]3s^03p^1$

(b) 3.37×10^{-19} J

The difference in energy between $3p$ and $3s$ orbitals of the sodium atom is equal to the energy of a photon of radiation having a wavelength of 589 nm.

$$\text{energy of a photon} = \frac{hc}{\lambda} = \left(\frac{(6.626 \times 10^{-34}\,\text{J}\cdot\text{s})(2.998 \times 10^8\,\text{m}\cdot\text{s}^{-1})}{589\,\text{nm}}\right)\left(\frac{1\,\text{nm}}{10^{-9}\,\text{m}}\right) = 3.37 \times 10^{-19}\,\text{J}$$

8.127 **(a)** AlI_3, $BaBr_2$, CsI, $CoCl_2$, $ScCl_3$

(b) 5, the same number as the price; $2729.93

$$(1\,\text{mol } AlI_3)\left(\frac{407.695\,\text{g } AlI_3}{1\,\text{mol } AlI_3}\right)\left(\frac{\$167.40}{25.000\,\text{g } AlI_3}\right) = \$2729.93$$

In the real world, the price would be $2729.93 (6 figures), even though the last digit is uncertain.

(c) 0.01%

$$\frac{100}{1\,000\,000} \times 100 = 0.01\%$$

(d) The anhydrous halides are packaged in sealed ampules under argon in order to keep them out of contact with air, which usually contains water vapor.

8.129 **(a)** 349 kJ/mol

See Figure 8.13.

(b) 349 kJ/mol

$Cl^-(g) \rightarrow Cl(g) + e^-$ is the reverse of the electron affinity for Cl(g).

(c) 3.43×10^{-7} m, ultraviolet (UV)

$$\left(\frac{349\ \text{kJ}}{\text{mol Cl}^-}\right)\left(\frac{1\ \text{mol Cl}^-}{6.022\times10^{23}\ \text{Cl}^-}\right)\left(\frac{1000\ \text{J}}{\text{kJ}}\right) = 5.80\times10^{-19}\ \text{J}$$

$$\lambda = \frac{hc}{E} = \frac{(6.626\times10^{-34}\ \text{J}\cdot\text{s})(2.998\times10^{8}\ \text{m}\cdot\text{s}^{-1})}{5.80\times10^{-19}\ \text{J}} = 3.42\times10^{-7}\ \text{m}$$

(Best answer, without intermediate round off, is 3.43×10^{-7} m.)

8.131 **(a)** 0.59 kg

$$(1.73\ \text{kg})\left(\frac{2.7\ \text{g/cm}^3}{7.9\ \text{g/cm}^3}\right) = 0.59\ \text{kg}$$

or

$$V_{Fe} = (1.73\ \text{kg})\left(\frac{1000\ \text{g}}{\text{kg}}\right)\left(\frac{1\ \text{cm}^3}{7.9\ \text{g}}\right) = 219\ \text{cm}^3$$

$$m_{Al} = (219\ \text{cm}^3)\left(\frac{2.7\ \text{g}}{\text{cm}^3}\right) = 5.9\times10^2\ \text{g} = 0.59\ \text{kg}$$

(b) 34 %

$$\frac{0.59\ \text{kg}}{1.73\ \text{kg}}\times 100 = 34\ \%$$

(c) The aluminum part is lighter, which is especially important in transportation because less energy will be required to move the aluminum part.

8.133 **(a)** $2\,Mg(s) + O_2(g) \rightarrow 2MgO(s)$

(b) $Mg(s) + 2H_2O(l) \xrightarrow{\text{heat}} Mg(OH)_2(s) + H_2(g)$

$2H_2(g) + O_2(g) \xrightarrow{\text{heat}} 2H_2O(g)$

The reaction of H_2 and O_2, once initiated, takes place explosively.

(c) –116 kJ. The reaction is exothermic. Therefore, the fire will burn more vigorously if a carbon dioxide extinguisher is used.

Equation	$MgO(s)$	+	$CO_2(g)$	→	$MgCO_3(s)$
ΔH_f^0, kJ/mol	–601.5		–393.51		–1111

$$\Delta H^0_{rxn} = \left[(1\ \text{mol MgCO}_3)\left(\frac{-1111\ \text{kJ}}{\text{mol MgCO}_3}\right)\right]$$

$$-\left[(1\ \text{mol MgO})\left(\frac{-601.5\ \text{kJ}}{\text{mol MgO}}\right) + (1\ \text{mol CO}_2)\left(\frac{-393.51\ \text{kJ}}{\text{mol CO}_2}\right)\right]$$

$$= -116\ \text{kJ}$$

CHAPTER 9:
Chemical Bonds

Practice Problems

9.1 **(a)**

(i) $[Ar]4s^1 + [Kr]5s^2\,4d^{10}\,5p^5 \rightarrow [Ar] + [Kr]5s^2\,4d^{10}\,5p^6$ (or [Xe])

$K + I \rightarrow K^+ + I^-$

(ii) $[Ar]4s^2 + \begin{matrix}[Ar]4s^2\,3d^{10}\,4p^5\\ [Ar]4s^2\,3d^{10}\,4p^5\end{matrix} \rightarrow [Ar] + \begin{matrix}[Ar]4s^2\,3d^{10}\,4p^6 \text{ (or [Kr])}\\ [Ar]4s^2\,3d^{10}\,4p^6 \text{ (or [Kr])}\end{matrix}$

$Ca + 2Br \rightarrow Ca^{2+} + 2Br^-$

(iii) $\begin{matrix}[Ne]3s^1\\ [Ne]3s^1\end{matrix} + [Ne]3s^2\,3p^4 \rightarrow \begin{matrix}[Ne]\\ [Ne]\end{matrix} + [Ne]3s^2\,3p^6$ (or [Ar])

$2Na + S \rightarrow 2Na^+ + S^{2-}$

(b) KI, $CaBr_2$, Na_2S

9.2 **(a)** two

$[Ar]3d^8 = [Ar]$ [↑↓][↑↓][↑↓][↑][↑]
3*d*

(b) four

$[Ar]3d^6 4s^2 = [Ar]$ [↑↓][↑][↑][↑][↑] [↑↓]
3*d* 4*s*

(c) $[Ar]3d^8$

Because the nickel(II) ion has two unpaired electrons, its electron configuration must be $[Ar]3d^8$.

9.3 Both electron configurations have four unpaired electrons.

Magnetic measurements measure the number of unpaired electrons in an atom or ion. To determine the number of unpaired electrons present in these two possible electron configurations, use orbital diagrams:

$[Ar]3d^6$ = [Ar] [↑↓][↑][↑][↑][↑]
3*d*

$[Ar]3d^4 4s^2$ = [Ar] [↑][↑][↑][↑][ ] [↑↓]
3*d* 4*s*

The orbital diagrams show that in either possible electron configuration, the number of unpaired electrons is the same (4). Thus, magnetic measurements cannot be used to tell the difference between them.

9.4 two Fe^{3+} and three O^{2-}

For the compound iron(III) oxide to have zero net charge, there must be two Fe^{3+} ions and three O^{2-} ions. The formula for iron(III) oxide is Fe_2O_3.

9.5 **(a)** Al^{3+}, S^{2-}

The electron configuration of Al^{3+} is [Ne]; the electron configuration of S^{2-} is $[Ne]3s^2 3p^6$ or [Ar].

(b) Bi^{3+}, Cu^+, Hg^{2+}

The electron configuration of Bi^{3+} is $[Xe]6s^2 4f^{14} 5d^{10}$, that of Cu^+ is $[Ar]3d^{10}$, and that of Hg^{2+} is $[Xe]4f^{14} 5d^{10}$.

9.6 **(a)** $MgCl_2$. Mg^{2+} is smaller.

$MgCl_2$ has a higher lattice energy than $CaCl_2$ because Mg^{2+} and Ca^{2+} have the same charge but Mg^{2+} is smaller.

(b) CaO. Ca^{2+} has higher charge.

CaO has a higher lattice energy than NaF because while Na^+ and Ca^{2+} are almost the same size (Figure 8.9), Ca^{2+} has twice the positive charge. Also, while F^- is slighty smaller than O^{2-}, O^{2-} has twice the negative charge. Thus, the lattice energy in CaO will be larger.

(c) Fe_2O_3. Fe^{3+} is smaller and has higher charge.

Fe_2O_3 has a higher lattice energy than FeO because Fe^{3+} has a higher charge and is smaller than Fe^{2+}, creating a stronger attraction for O^{2-} ions.

(d) NaBr. Br^- is smaller.

NaBr has a higher lattice energy than NaI because the Br^- ion is smaller than the I^- ion.

9.7

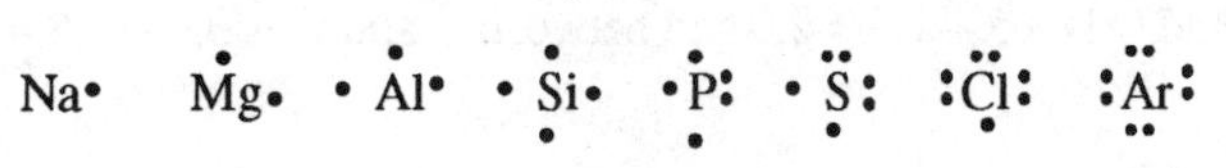
Na
Mg
Al
Si
P
S
Cl
Ar

9.8 (a) 3

The Lewis formula for ammonia is

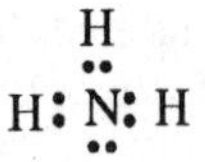

(b) 2

The Lewis formula for water is

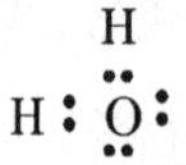

9.9

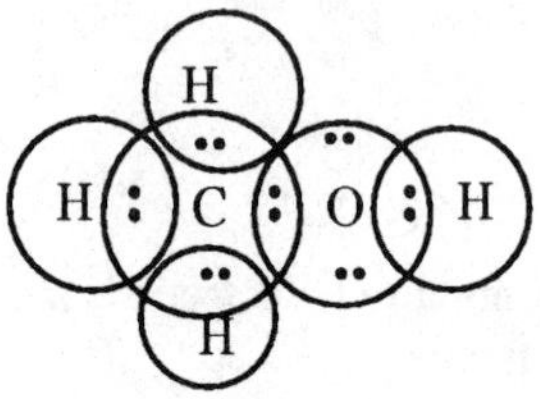

9.10

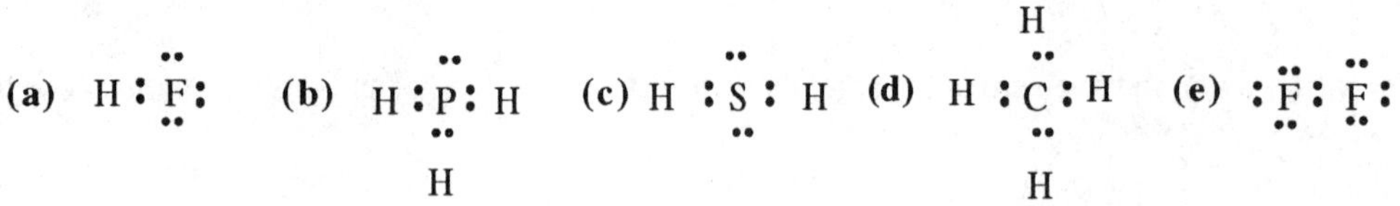

9.11

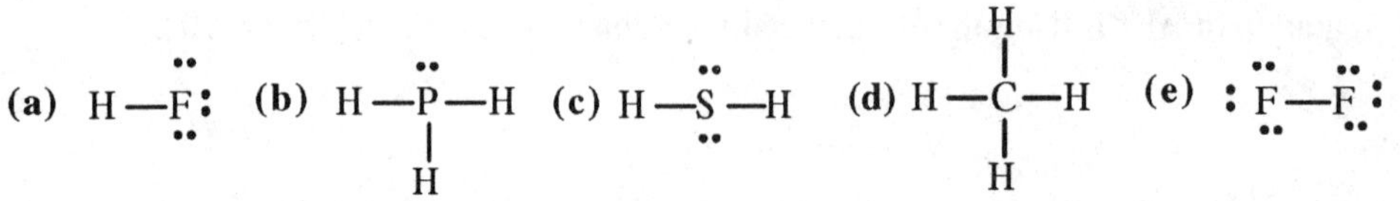

9.12

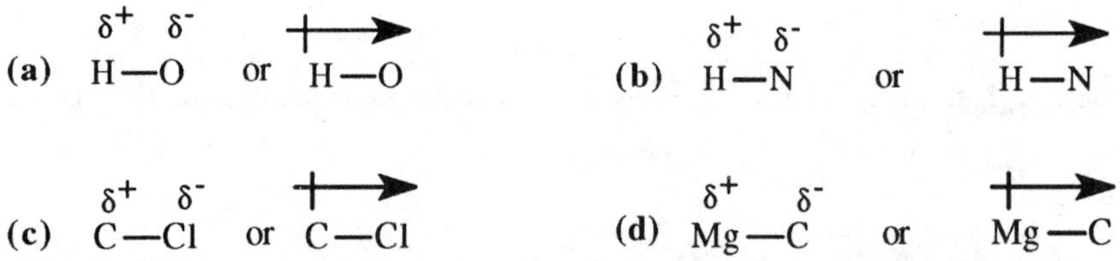

The negative end of each bond is the end having the more electronegative atom.

9.13 (a) O—O

The electronegativity difference between O and O in O—O is zero. Therefore, this bond is nonpolar.

(b) Ba—O

Ba—O is ionic, since it has the largest electronegativity difference (3.5 – 1.0 = 2.5).

(c) O—O < C—O < Be—O < Ba—O

The difference in electronegativity between Be and O is 3.5 – 1.5 = 2.0 and between C and O is 3.5 – 2.5 = 1.0.

9.14 **(a)** ionic

CaO is a metal oxide.

(b) ionic

NaI is a compound of a Group IA metal and a nonmetal.

(c) molecular

CS_2 is a compound of two nonmetals.

(d) ionic

AlF_3 is a metal fluoride.

(e) ionic

Ammonium compounds are ionic because they contain the polyatomic ion NH_4^+.

(f) molecular

NO is a compound of two nonmetals.

9.15 

9.16 **(a)** 12

2 C	@	4 valence electrons	=	8 valence electrons
4 H	@	1 valence electron	=	4 valence electrons
		Total	=	12 valence electrons

(b) 10

2 C	@	4 valence electrons	=	8 valence electrons
2 H	@	1 valence electron	=	2 valence electrons
		Total	=	10 valence electrons

(c) 24

1 N	@	5 valence electrons	=	5 valence electrons
3 O	@	6 valence electrons	=	18 valence electrons
1 extra electron for 1– charge			=	1 valence electron
		Total	=	24 valence electrons

(d) 32

4 C	@	4 valence electrons	=	16 valence electrons
12 H	@	1 valence electron	=	12 valence electrons
1 N	@	5 valence electrons	=	5 valence electrons
1 less electron for 1+ charge			=	–1 valence electron
		Total	=	32 valence electrons

9.17　(a)　H⦂C ⦂⦂C⦂H　or　H—C=C—H　　(b)　H⦂C ⫶⫶C⦂H　or　H—C≡C—H
　　　　　　　　H　H　　　　　　　　H　H

(c)
　　　　H　H　　　　　　　　H　H
　　H⦂C⦂C⦂Cl⦂　or　H—C—C—Cl⦂
　　　　H　H　　　　　　　　H　H

(d)
$$\left[\begin{array}{c} H \\ H⦂C⦂O⦂ \\ H \end{array} \right]^{-} \text{or} \left[\begin{array}{c} H \\ H—C—O⦂ \\ H \end{array} \right]^{-}$$

(e)　⦂F⦂O⦂F⦂　or　⦂F—O—F⦂

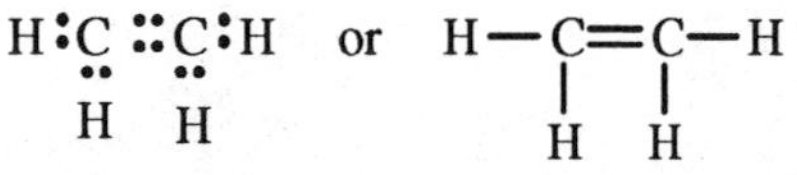

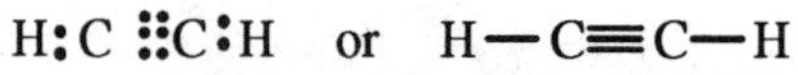

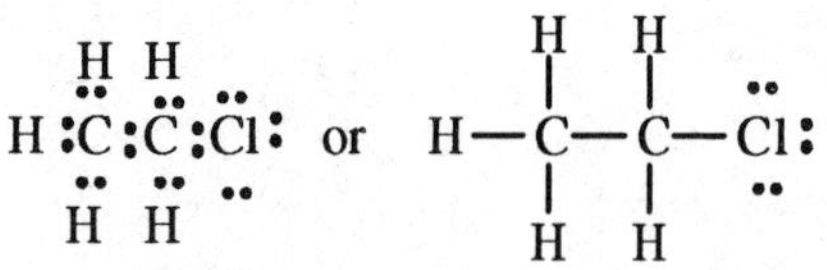

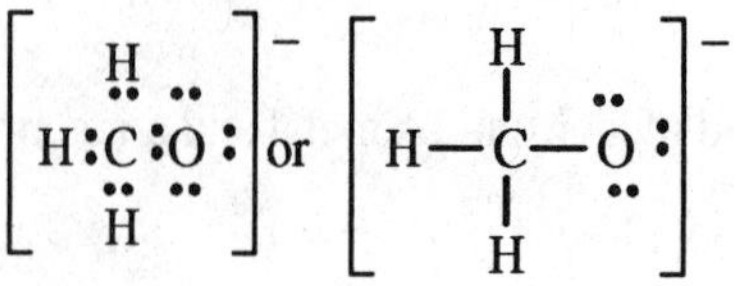

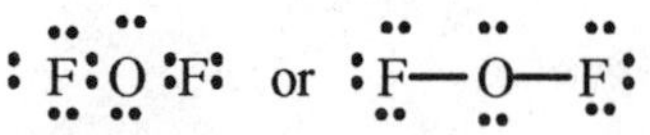

9.18　(a) the N—N single bond　(b) the triple bond　(c) the double bond

9.19　Formal charge on H $= 1 - \left(\frac{1}{2}(2) + 0 \right) = 0$

Formal charge on N $= 5 - \left(\frac{1}{2}(8) + 0 \right) = +1$

Sum of formal charges $= 4(0) + 1(+1) = +1$

9.20　oxygen

The Lewis formula for the hydroxide ion is

$$\left[H : \overset{\cdot\cdot}{\underset{\cdot\cdot}{O}} : \right]^{-} \quad \text{or} \quad \left[H — \overset{\cdot\cdot}{\underset{\cdot\cdot}{O}} : \right]^{-}$$

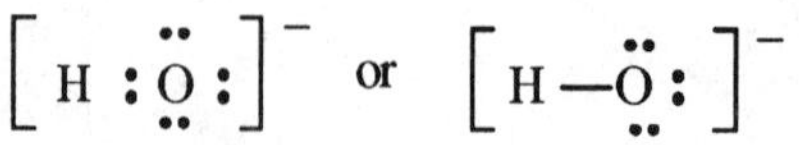

Formal charge on H $= 1 - \left(\frac{1}{2}(2) + 0 \right) = 0$

Formal charge on O $= 6 - \left(\frac{1}{2}(2) + 6 \right) = -1$

Sum of formal charges $= 0 + (-1) = -1$

9.21

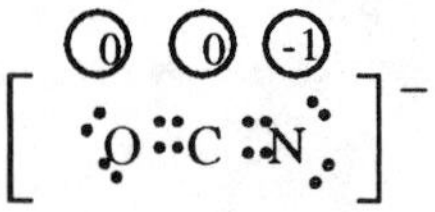

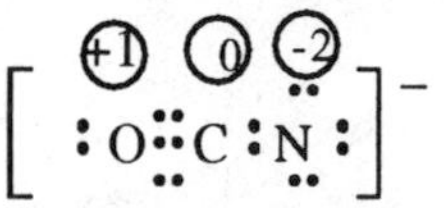

$$\left[\overset{\textcircled{0} \quad \textcircled{0} \quad \textcircled{-1}}{⦂O⫶⫶C⦂N⦂} \right]^{-}$$

$$\left[\overset{\textcircled{+1} \quad \textcircled{0} \quad \textcircled{-2}}{⦂O⫶⫶C⦂N⦂} \right]^{-}$$

Formal charge on O $= 6 - \left(\frac{1}{2}(6) + 2 \right) = +1$

Formal charge on C $= 4 - \left(\frac{1}{2}(8) + 0 \right) = 0$

Formal charge on N $= 5 - \left(\frac{1}{2}(2) + 6 \right) = -2$

Sum of formal charges $= 1(+1) + 1(0) + 1(-2) = -1$

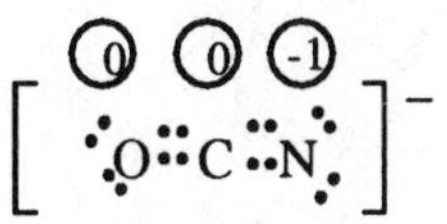

$$\left[\; \overset{\textstyle \overset{0}{}\;\overset{0}{}\;\overset{-1}{}}{\ddot{\text{O}} :: \text{C} :: \ddot{\text{N}}} \;\right]^{-}$$

Formal charge on O = $6 - \left(\dfrac{1}{2}(4) + 4\right) = 0$

Formal charge on C = $4 - \left(\dfrac{1}{2}(8) + 0\right) = 0$

Formal charge on N = $5 - \left(\dfrac{1}{2}(4) + 4\right) = -1$

Sum of formal charges = $1(0) + 1(0) + 1(-1) = -1$

The structure with a single formal charge of –1 is better.

9.22

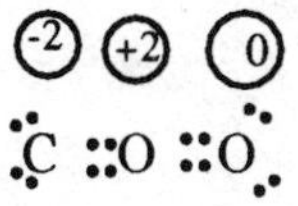

$$\overset{\textstyle \overset{-2}{}\;\overset{+2}{}\;\overset{0}{}}{\ddot{\text{C}} :: \text{O} :: \ddot{\text{O}}}$$

Formal charge on C = $4 - \left(\dfrac{1}{2}(4) + 4\right) = -2$

Formal charge on central O = $6 - \left(\dfrac{1}{2}(8) + 0\right) = +2$

Formal charge on terminal O = $6 - \left(\dfrac{1}{2}(4) + 4\right) = 0$

$$\overset{\textstyle \overset{-1}{}\;\overset{+2}{}\;\overset{-1}{}}{: \text{C} :: \text{O} \cdot\cdot \ddot{\text{O}} :}$$

Formal charge on C = $4 - \left(\dfrac{1}{2}(6) + 2\right) = -1$

Formal charge on central O = $6 - \left(\dfrac{1}{2}(8) + 0\right) = +2$

Formal charge on terminal O = $6 - \left(\dfrac{1}{2}(2) + 6\right) = -1$

$$\overset{\textstyle \overset{0}{}\;\overset{0}{}\;\overset{0}{}}{\ddot{\text{O}} :: \text{C} :: \ddot{\text{O}}}$$

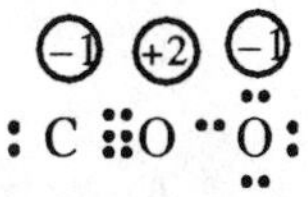

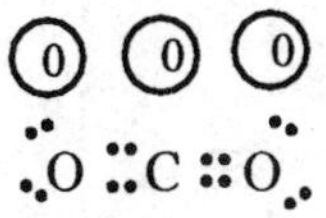

Formal charge on both O's = $6 - \left(\dfrac{1}{2}(4) + 4\right) = 0$

Formal charge on C = $4 - \left(\dfrac{1}{2}(8) + 0\right) = 0$

The top structures have higher formal charges than the bottom structure. Thus the best Lewis structure is: $\ddot{\text{O}} :: \text{C} :: \ddot{\text{O}}$

9.23 **(a)** $\left[\; : \text{C} \equiv \text{N} - \ddot{\text{O}} : \;\right]^{-}$ The central atom is N instead of C.

(b) $\left[\; \overset{\overset{-1}{}\;\overset{0}{}\;\overset{0}{}}{\ddot{\text{N}} = \text{C} = \ddot{\text{O}}} \;\right]^{-}$ has lower formal charge than $\left[\; \overset{\overset{-2}{}\;\overset{0}{}\;\overset{+1}{}}{\ddot{\text{N}} - \text{C} \equiv \text{O} :} \;\right]^{-}$

$\left[\; \overset{\overset{-1}{}\;\overset{0}{}\;\overset{0}{}}{\ddot{\text{N}} = \text{C} = \ddot{\text{O}}} \;\right]^{-}$

Formal charge on N = $5 - \left(\dfrac{1}{2}(4) + 4\right) = -1$

Formal charge on C = $4 - \left(\dfrac{1}{2}(8) + 0\right) = 0$

Formal charge on O = $6 - \left(\dfrac{1}{2}(4) + 4\right) = 0$

Total formal charge = –1

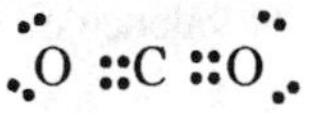

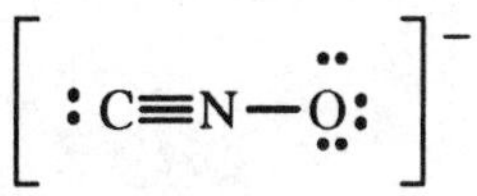

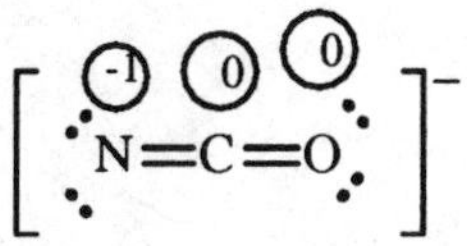

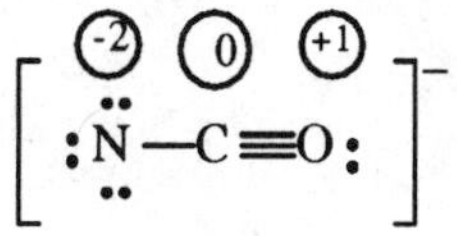

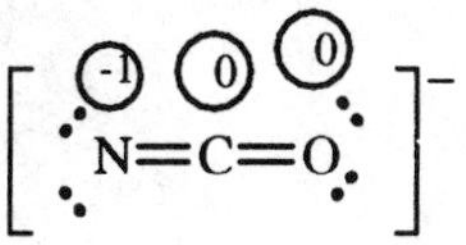

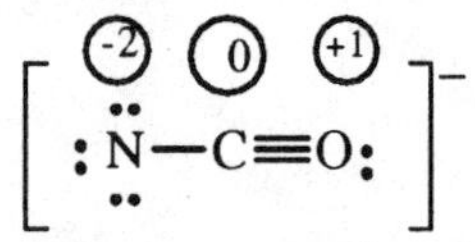

$$\text{Formal charge on N} = 5 - \left(\frac{1}{2}(2) + 6\right) = -2$$

$$\text{Formal charge on C} = 4 - \left(\frac{1}{2}(8) + 0\right) = 0$$

$$\text{Formal charge on O} = 6 - \left(\frac{1}{2}(6) + 2\right) = +1$$

Total formal charge = −1

The first structure is better because the formal charge is lower.

9.24 (a)

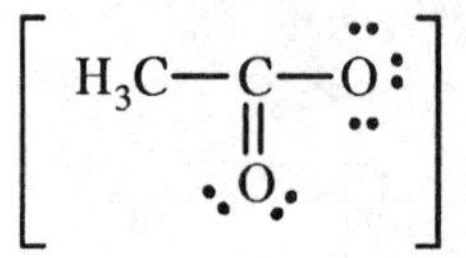

Resonance structures must have all atoms in the same positions. They differ in the arrangement of the electrons.

(b) 1.5

The second C—O bond in the acetate ion is shared equally by both oxygen atoms. Thus, the bond order for each is 1.5.

9.25 (a)

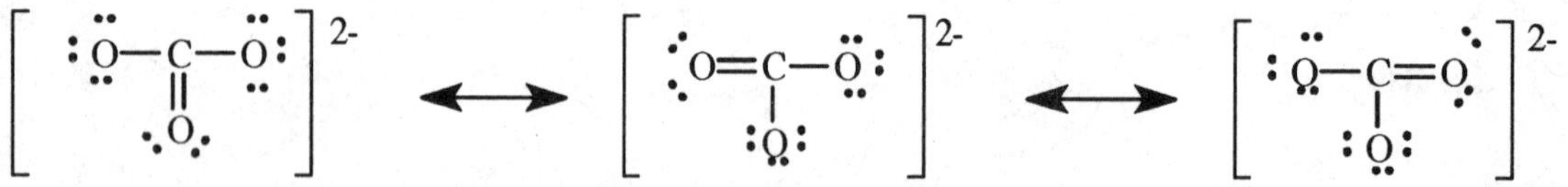

1 C @	4 valence electrons	=	4 valence electrons
3 O @	6 valence electrons	=	18 valence electrons
2 extra electrons for 2– charge		=	2 valence electrons
	Total	=	24 valence electrons

(b)

3 O @	6 valence electrons	=	18 valence electrons
	Total	=	18 valence electrons

9.26 (a)

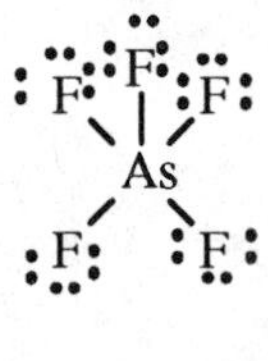

1 As @	5 valence electrons	=	5 valence electrons
5 F @	7 valence electrons	=	35 valence electrons
	Total	=	40 valence electrons

Consider As as the central atom:

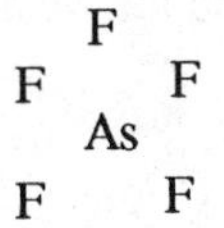

40 electrons are needed

Place one pair of electrons between each pair of bonded atoms:

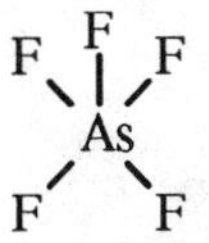

30 electrons remain

Beginning at the outside of the formula, place the remaining 30 electrons in pairs until each fluorine has eight electrons:

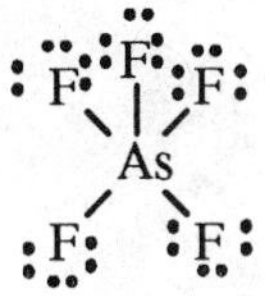

(b)

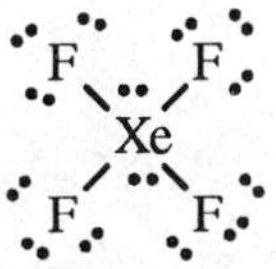

1 Xe @	8 valence electrons	=	8 valence electrons
4 F @	7 valence electrons	=	28 valence electrons
	Total	=	36 valence electrons

Consider Xe as the central atom:

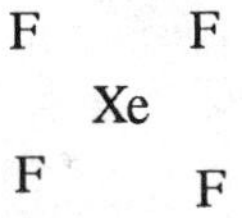

Place one pair of electrons between each pair of bonded atoms. Then begin at the outside of the formula and place eight electrons around each fluorine:

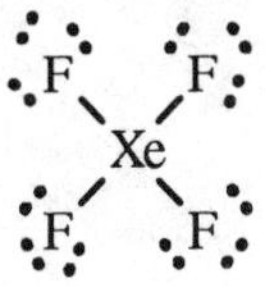

Place the four electrons that are left around the central atom:

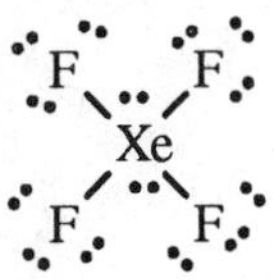

XeF_4 has two nonbonded pairs of electrons on the Xe atom and a total of 12 electrons around the Xe atom. Xe, being lower in the periodic table than period 2, can have more than an octet of electrons.

9.27 (a)

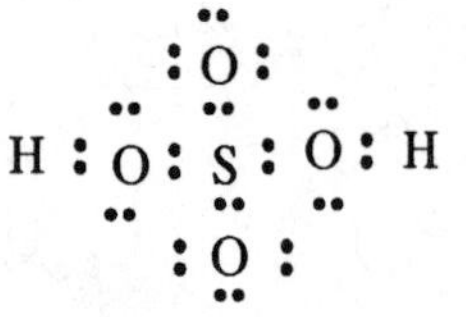

or

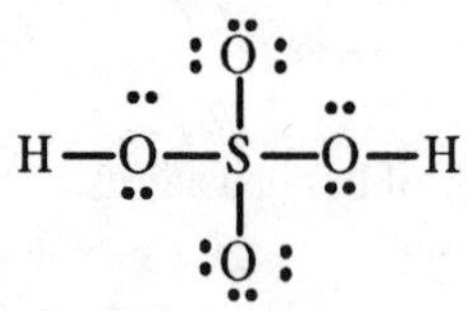

2 H	@	1 valence electron	=	2 valence electrons
1 S	@	6 valence electrons	=	6 valence electrons
4 O	@	6 valence electrons	=	24 valence electrons
		Total	=	32 valence electrons

Consider S as the central atom surrounded by four oxygens:

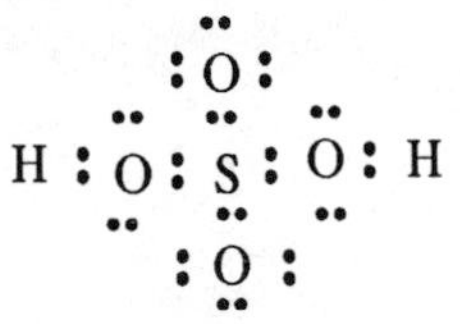

or

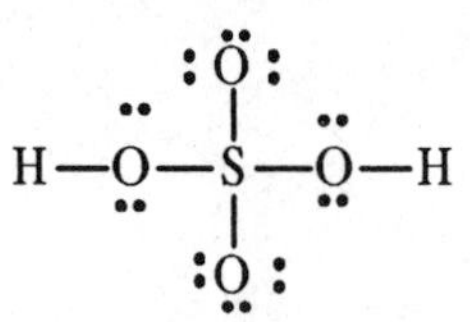

(b)

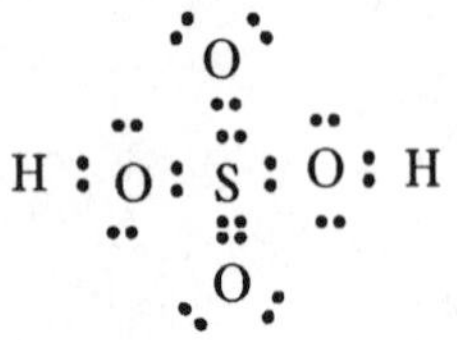

or

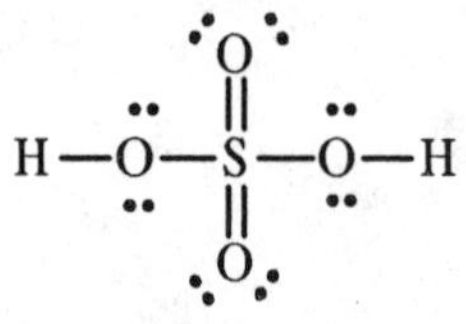

(c)

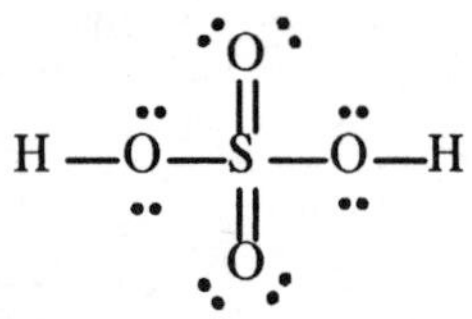

is better

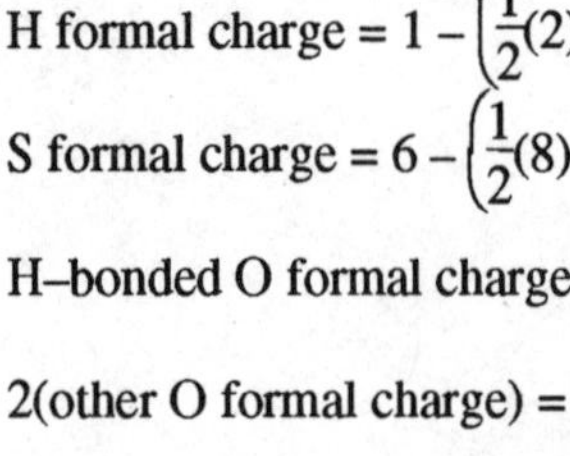

$$\text{H formal charge} = 1 - \left(\frac{1}{2}(2) + 0\right) = 0$$

$$\text{S formal charge} = 6 - \left(\frac{1}{2}(8) + 0\right) = +2$$

$$\text{H–bonded O formal charge} = 6 - \left(\frac{1}{2}(4) + 4\right) = 0$$

$$2(\text{other O formal charge}) = 2\left(6 - \left(\frac{1}{2}(2) + 6\right)\right) = 2(-1) = -2$$

Total = 0 formal charge

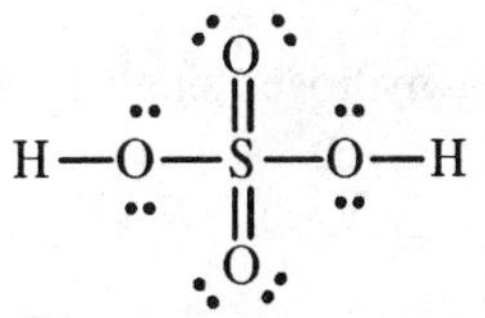

$$\text{H formal charge} = 1 - \left(\frac{1}{2}(2) + 0\right) = 0$$

$$\text{S formal charge} = 6 - \left(\frac{1}{2}(12) + 0\right) = 0$$

$$\text{H–bonded O formal charge} = 6 - \left(\frac{1}{2}(4) + 4\right) = 0$$

$$\text{other O formal charge} = 6 - \left(\frac{1}{2}(4) + 4\right) = 0$$

Total = 0 formal charge

The second formula is better because no formal charges are present on any of the atoms.

9.28 ClO_2

ClO_2	1 Cl @ 7 valence electrons	=	7 valence electrons
odd	2 O @ 6 valence electrons	=	12 valence electrons
	Total	=	19 valence electrons

Cl_2O	2 Cl @ 7	=	14		ClN_3	1 Cl @ 7	=	7
even	1 O @ 6	=	6		even	3 N @ 5	=	15
	Total	=	20			Total	=	22

Cl_2O_7	2 Cl @ 7	=	14		N_2O_5	2 N @ 5	=	10
even	7 O @ 6	=	42		even	5 O @ 6	=	30
	Total	=	56			Total	=	40

9.29

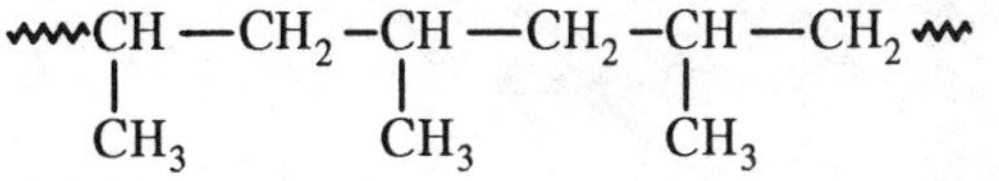

To form the polymer, break one bond of the carbon-carbon double bond. Attach each carbon by a single bond to a carbon atom from another monomer unit.

9.30 Any of the following is correct: CFCl═CF_2, ClFC═CF_2, FCCl═CF_2, CClF═CF_2, FClC═CF_2, ClCF═CF_2.

The repeat pattern features ~CFCl—CF_2~. So the monomer must contain a double bond between the two carbon atoms in that unit.

9.31 **(a)** $\Delta H_{rxn} = -8.0 \times 10^1$ kJ

Equation:	$N_2(g)$	+	$3H_2(g)$	→	$2NH_3(g)$
Lewis formulas:	:N≡N:		3 H—H		2 H—N(H)—H (with lone pair on N)
Bond energies, kJ/mol:	946		436		389

For this reaction to occur, one nitrogen–nitrogen triple bond and three hydrogen–hydrogen single bonds must be broken. Therefore,

946 kJ + 3(436) kJ = 2254 kJ of energy are required to break these bonds.

When six hydrogen–nitrogen bonds are formed during the formation of NH_3,

6(389) kJ = 2334 kJ of energy are released.

The difference between the energy required to break the bonds of the reactants and the energy released during bond formation for the products is the ΔH_{rxn}:

$$\Delta H_{rxn} = 2254 - 2334 \text{ kJ} = -8.0 \times 10^1 \text{ kJ}$$

(b) 12 kJ

$-80 \text{ kJ} - (-92.22 \text{ kJ}) = 12 \text{ kJ}$

(c) 13%

$$\left(\frac{12 \text{ kJ}}{92.22 \text{ kJ}}\right)(100) = 13\% \text{ error}$$

9.32 **(a)** Single bonds are weaker (compare C—O to C=O and C—C to C=C).

Carbon–carbon and carbon–oxygen bond energies given in Table 9.4 demonstrate that single bonds are weaker than double bonds:

	C—C	C=C
Bond energy, kJ/mol	347	615

	C—O	C=O
Bond energy, kJ/mol	360	728

(b) decreases (H—F > H—Cl > H—Br > H—I)

The hydrogen halides demonstrate this:

	H—F	H—Cl	H—Br	H—I
Bond energies, kJ/mol	568	432	366	298

A number of other examples from Table 9.4 also support the conclusion that bond strength usually decreases down a group: C—C (347 kJ/mol) > Si—Si (192 kJ/mol); O—H (464 kJ/mol) > S—H (368 kJ/mol); N—H (389 kJ/mol) > P—H (322 kJ/mol); C—H (414 kJ/mol) > Si—H (318 kJ/mol). However, the C—O (360 kJ/mol) and Si—O (464 kJ/mol) bonds are a notable exception.

Additional Practice Problems

9.69 Some of the common properties of ionic compounds are hardness, high melting and boiling points, ability to conduct electricity when melted but not when in the solid state, and ability to dissolve in water to form solutions that conduct electricity.

Lattice energy (the energy required to separate the ions of an ionic compound) is a measure of the force of attraction between ions. The fact that the lattice energies of ionic compounds are high accounts for the hardness and high melting and boiling points of ionic compounds. In the solid state, the ions are fixed in position; therefore, ionic solids do not conduct electricity. In the liquid state and in aqueous solution, the ions are free to move and conduct electricity.

9.71 (a) BaF_2 and (b) CaO. Ba^{2+}, F^-, Ca^{2+}, and O^{2-} ions all have noble gas electron configurations and are combined so that net charge is zero.

BaF_2 and CaO are stable ionic compounds because each consists of ions with stable noble gas electron configurations combined so that the net charge is zero.

Ba^{2+} is isoelectronic with Xe
F^- is isoelectronic with Ne
O^{2-} is isoelectronic with Ne
Ca^{2+} is isoelectronic with Ar

HCl is a polar covalent molecule. K_2Br and $RbCl_2$ are unstable. If the ions in K_2Br have their usual charges, the species would have a 1+ charge and be an ion, not a compound. For the combination of 2 K and 1 Br to have zero net charge, either K would have to have half a positive charge, which is impossible, or Br would have to have a 2– charge, which is unlikely energetically. If the ions in $RbCl_2$ have their usual charges, the species would have a 1– charge and be an ion, not a compound. For the combination of 1 Rb and 2 Cl to have a zero net charge, either Rb would have to have a 2+ charge, which is unlikely energetically, or Cl would have to have half a negative charge, which is impossible.

9.73 (a)

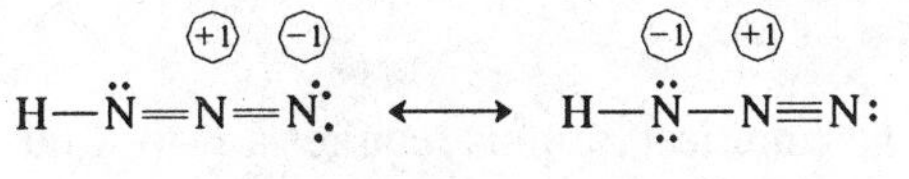

Calculation of number of valence electrons:

1 H @ 1 valence electron	=	1 valence electron
3 N @ 5 valence electrons	=	15 valence electrons
Total	=	16 valence electrons

Calculation of formal charges:

Left structure:

Formal charge on H = $1-\left(\frac{1}{2}(2)+0\right)=0$

Formal charge on left N = $5-\left(\frac{1}{2}(6)+2\right)=0$

Formal charge on middle N = $5-\left(\frac{1}{2}(8)+0\right)=+1$

Formal charge on right N = $5-\left(\frac{1}{2}(4)+4\right)=-1$

Right structure:

Formal charge on H = $1-\left(\frac{1}{2}(2)+0\right)=0$

Formal charge on left N = $5-\left(\frac{1}{2}(4)+4\right)=-1$

Formal charge on middle N = $5-\left(\frac{1}{2}(8)+0\right)=+1$

Formal charge on right N = $5-\left(\frac{1}{2}(6)+2\right)=0$

(b) lefthand bond, 1.5; righthand bond, 2.5

The lefthand bond has a bond order of 2 in the first structure and a bond order of 1 in the second structure. Therefore, it has a bond order of 1.5 in the molecule. Similarly, the bond order of the righthand bond is 2.5, the average of 2 and 3.

9.75 Group VA

There are 26 electrons in the structure. Since each fluorine atom contributes 7 electrons, *X* must contribute a total of 5 electrons [$26\ e^- - 3(7\ e^-) = 5\ e^-$]. Thus, *X* belongs to Group VA.

Stop and Test Yourself

1. **(a)** AlN

$$\underset{\text{Al}}{[\text{Ne}]3s^23p^1} + \underset{\text{N}}{[\text{He}]2s^22p^3} \rightarrow \underset{\text{Al}^{3+}}{[\text{Ne}]} + \underset{\text{N}^{3-}}{[\text{He}]2s^22p^6 \text{ or } [\text{Ne}]}$$

Since the ions have 3+ and 3– ionic charges, respectively, the empirical formula requires a 1–to–1 ratio for an electrically neutral compound.

2. **(c)** [Ar] [↑] [↑] [↑] [ ] [ ] (3*d*) [ ] (4*s*)

The orbital diagram for the chromium atom is

[Ar] [↑] [↑] [↑] [↑] [↑] (3*d*) [↑] (4*s*)

The one 4*s* and two 3*d* electrons are lost when a chromium(III) ion is formed. The orbital diagram for the chromium(III) ion is

[Ar] [↑] [↑] [↑] [ ] [ ] (3*d*) [ ] (4*s*)

3. **(b)** C

Carbon is unable to have more than eight electrons around it because its valence electrons are in the second energy level; in the $n = 2$ energy level, there are no 2*d* orbitals for expanding beyond the 2*s*2*p* eight–electron capacity.

4. (b) BCl_3

BCl_3 exhibits the most covalent (i.e., least ionic) chemical bonds because the electronegativity difference between boron and chlorine atoms (2.8 – 2.0 = 0.8) is smaller than the electronegativity differences between Al and Cl atoms (2.8 – 1.5 = 1.3), K and Cl atoms (2.8 – 0.9 = 1.9), Mg and Cl atoms (2.8 – 1.2 = 1.6), and Na and Cl atoms (2.8 – 1.0 = 1.8).

5. (c) 34

1 Cl @ 7 valence electrons	=	7 valence electrons
4 F @ 7 valence electrons	=	28 valence electrons
1 less electron for 1+ charge	=	–1 valence electron
Total	=	34 valence electrons

6. (c) CaO

Lattice energies increase with increasing ionic charges and decreasing ionic radii. CaO and CaS both have a 2+ cation and a 2– anion, but since the O^{2-} ion is smaller than the S^{2-} ion, CaO must have the higher lattice energy.

7. (c) BrF_3

BrF_3 has more than 8 valence electrons around the central bromine atom. The Lewis formula for BrF_3, which must account for 28 valence electrons, is shown to the right.

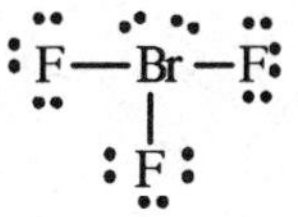

8. (c) 0 kJ

In the reaction

one O—H bond is broken and one O—H bond is formed. Therefore, $\Delta H_{rxn} = 0$.

9. (c) 10

$AsCl_3$ has 10 unshared pairs of valence electrons in its Lewis formula:

10. (d)

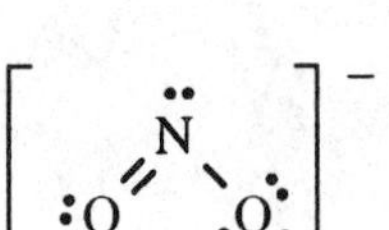

Structure (d) is a resonance structure for the structure to the right because it is the only Lewis structure that has all the atoms in the same positions but with different positions for the same number of electrons. Structure (a) is wrong because it only shows 17 valence electrons instead of 18.

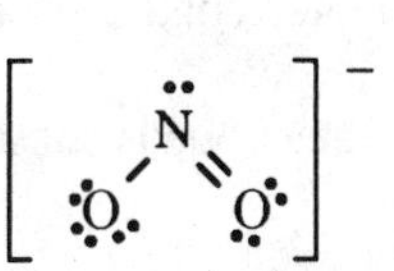

11.   **(d)** +1

$$\text{formal charge} = \text{\# of valence} - \left(\frac{1}{2}(\text{\# of valence electrons in shared pairs}) + \text{\# of valence electrons unshared}\right)$$
$$\text{electrons in}$$
$$\text{free atom}$$

$$\text{formal charge} = 6 - \left(\frac{1}{2}(6) + 2\right) = +1$$

12.   **(c)**

$$\left[\; \overset{..}{\underset{..}{S}} = C = \overset{..}{\underset{..}{N}} \;\right]^{-}$$

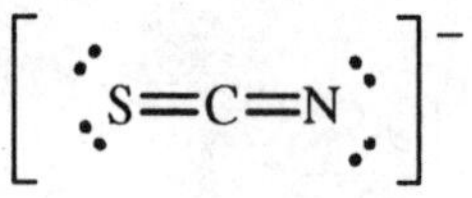

The Lewis formulas with formal charges are

(a)
$$\left[\; \overset{\boxed{-2}}{\overset{..}{\underset{..}{C}}} = \overset{\boxed{+1}}{N} = \overset{..}{\underset{..}{S}} \;\right]^{-}$$

(b)
$$\left[\; \overset{\boxed{-2}}{\overset{..}{\underset{..}{C}}} = \overset{\boxed{+2}}{S} = \overset{\boxed{-1}}{\overset{..}{\underset{..}{N}}} \;\right]^{-}$$

(c)
$$\left[\; \overset{..}{\underset{..}{S}} = C = \overset{\boxed{-1}}{\overset{..}{\underset{..}{N}}} \;\right]^{-}$$

(d)
$$\left[\; \overset{\boxed{-1}}{\overset{..}{\underset{..}{S}}} - C \equiv N\!: \;\right]^{-}$$

(e)
$$\left[\; :S \equiv \overset{\boxed{+1}}{C} - \overset{\boxed{-2}}{\overset{..}{\underset{..}{N}}}: \;\right]^{-}$$

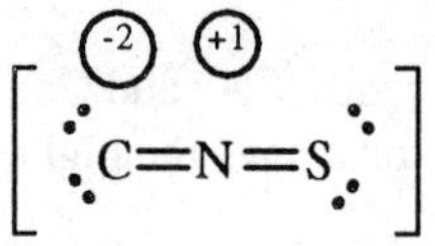

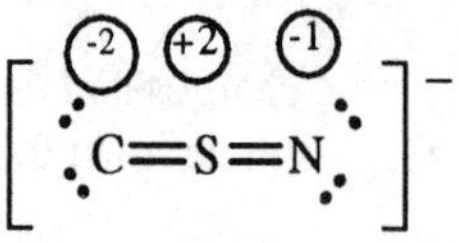

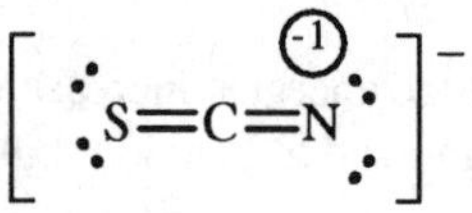

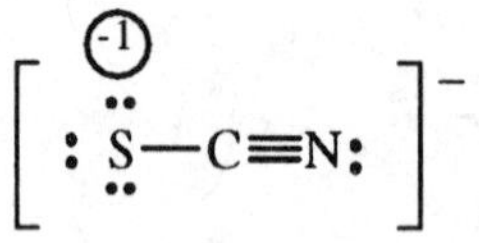

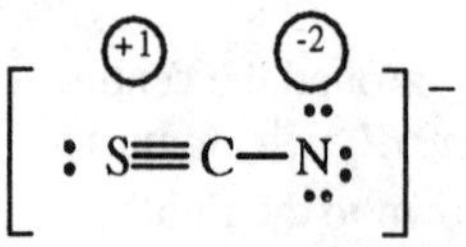

(a), (b), and (e) can be eliminated because they have higher formal charges than (c) and (d). Structure (c) is better than structure (d) because in (c) the –1 formal charge is on N rather than on S, which is less electronegative than N.

13.   **(e)** 5

The $C_5 - C_6$ carbon bond is the shortest and strongest of the carbon–carbon bonds listed because it is a triple bond. (The other carbon–carbon triple bonds are equally strong.)

14.   **(d)** $H_2C = C(CH_2CH_3)COOCH_3$

Monomers (c) and (e) are eliminated because they have CH_2CH_3 groups attached to the oxygen rather than CH_3 groups as shown in the polymer. Monomers (a), (b), and (c) can be eliminated because the other group is CH_3 rather than CH_2CH_3 as shown in the polymer. Monomers (a) and (c) can be further eliminated because the two attached groups are attached to different carbons, whereas both groups are attached to the same carbon in the polymer. Only monomer (d) meets the criteria of having the correct groups attached to the correct carbon.

Putting Things Together

9.77   **(a)** An ionic bond is the electrostatic attraction between positively and negatively charged ions.

(b) Covalent bonds consist of pairs of electrons shared between two atoms.

9.79 **(a)**

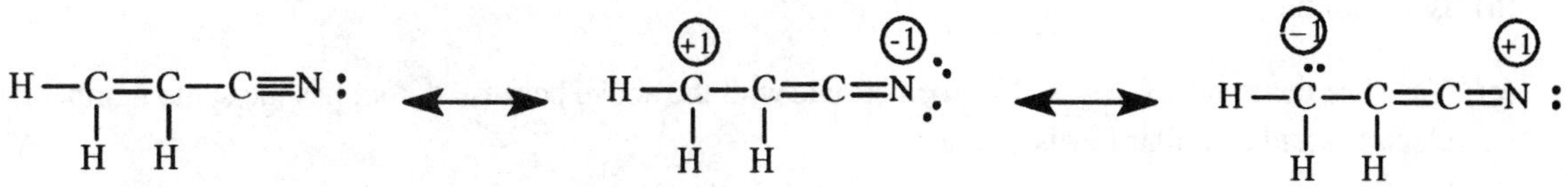

Lewis structure I has no atom with a formal charge.

(b) Each of the contributing Lewis structures is called a resonance structure.

(c) ↔ should be used between resonance structures.

(d) Structure I, in which all atoms have a formal charge of zero, contributes most to the physical properties.

(e) Structure III, in which the most electronegative atom has a positive formal charge, is the least important structure.

9.81 anion, 2–

The Lewis formula shows 26 electrons.

1 S	@	6 valence electrons	=	6 valence electrons
3 O	@	6 valence electrons	=	18 valence electrons
		Total	=	24 valence electrons

Since the species has 2 extra valence electrons beyond those from its atoms, it must have a 2– charge.

9.83 This results in both atoms having formal charges of zero.

The Lewis formula for NO has the odd electron on the nitrogen atom because this results in both atoms having formal charges of zero. If the odd electron were on the oxygen atom, nitrogen would have a formal charge of –1 and oxygen would have a formal charge of +1. The preferred Lewis formula is the one which has no atom with a formal charge other than zero.

9.85 *Ionic compounds* such as NaCl have high melting points, conduct electricity in the molten state, and conduct electricity in aqueous solutions. *Molecular compounds* such as sugar have moderately low melting points and do not conduct electricity in the molten state.

9.87 Acetylene is a molecular compound, so the molecular formula C_2H_2 is appropriate. Salt is an ionic compound consisting of positive Na^+ ions and negative Cl^- ions held together by the attractive forces between oppositely charged ions in a three–dimensional arrangement (remember Figure 1.18). No molecules of NaCl exist. Therefore, a molecular formula such as Na_2Cl_2 is inappropriate, and the empirical formula, NaCl, should be used for salt.

9.89 $[:C{\equiv}N:]^-$ $:C{\equiv}O:$ $:N{\equiv}N:$ $[:N{\equiv}O:]^+$

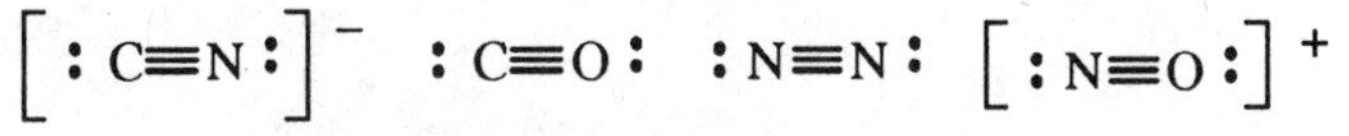

(a) All four Lewis formulas have the same number and arrangement of electrons.

(b) isoelectronic

(c) Yes. The compound C_2H_2 is also isoelectronic with the above because it, too, has the same number of valence electrons and a similar Lewis formula:

9.91 (a) BH_4^- is more stable than BH_3 because the B has an octet in BH_4^-:

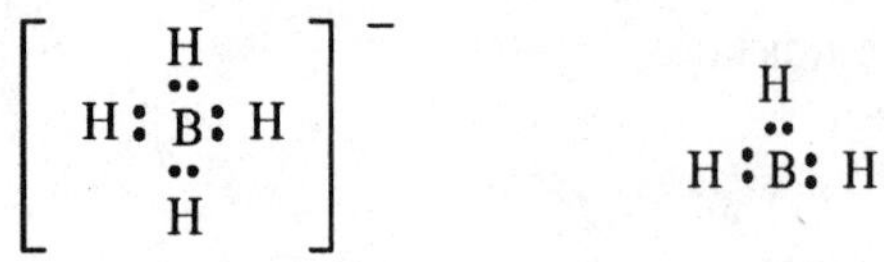

(b) CaO is more stable than CaF because CaF needs another F^- ion.

(c) KCl is more stable than KCl_2 because KCl_2 has one too many Cl^- ions.

(d) PF_5 is more stable than NF_5 because phosphorus is in the third period and can expand its octet. Nitrogen is in the second period and cannot expand its octet because it has no *d* orbitals available.

(e) The NO_2^- ion is more stable than the NO_2 molecule because the NO_2 molecule has an odd number of electrons.

9.93 (a)

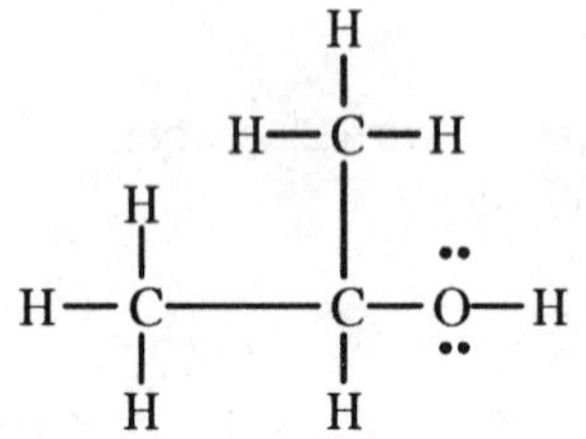

(b)

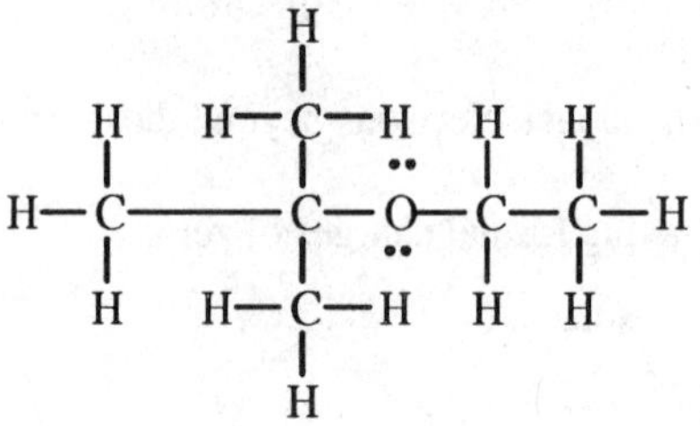

(c)

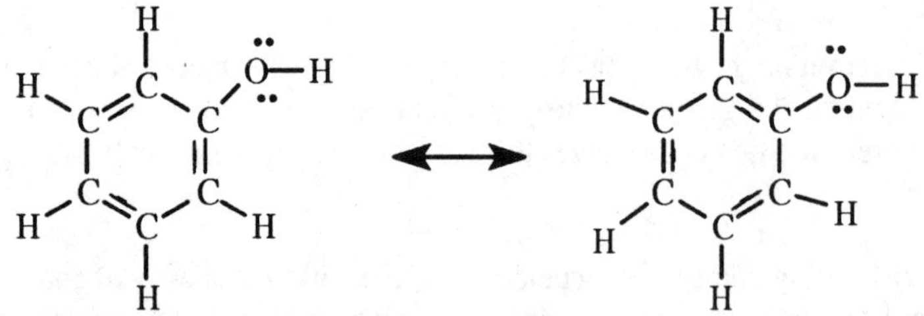

(d)

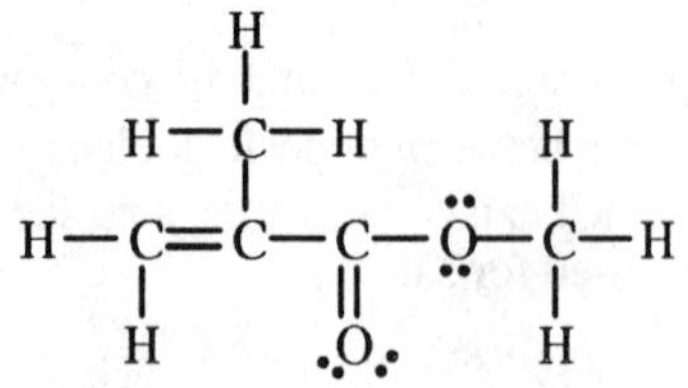

(e)

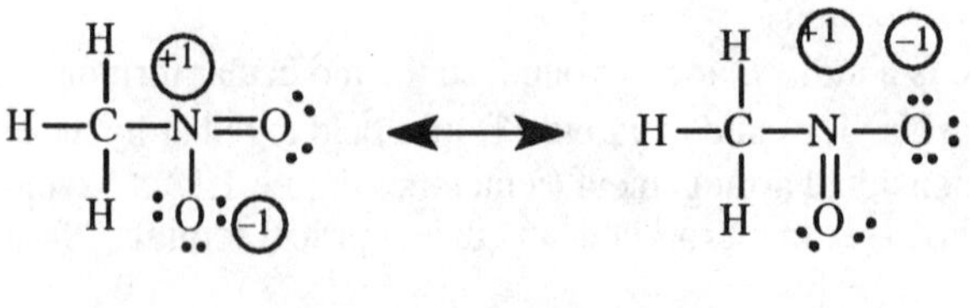

9.95

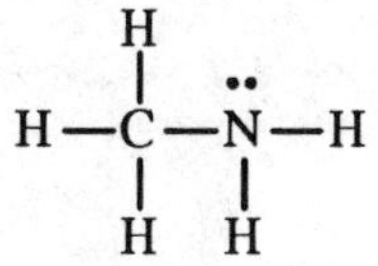

First find the empirical formula.

Assume a 100–g sample: 38.67 g C, 16.23 g H, and 45.10 g N

$$(38.67\ \cancel{g\ C})\left(\frac{1\ \text{mol C}}{12.011\ \cancel{g\ C}}\right) = 3.220\ \text{mol C}/3.220 = 1.000\ \text{mol C atoms}$$

$$(16.23\ \cancel{g\ H})\left(\frac{1\ \text{mol H}}{1.0079\ \cancel{g\ H}}\right) = 16.10\ \text{mol H}/3.220 = 5.000\ \text{mol H atoms}$$

$$(45.10\ \cancel{g\ N})\left(\frac{1\ \text{mol N}}{14.007\ \cancel{g\ N}}\right) = 3.220\ \text{mol N}/3.220 = 1.000\ \text{mol N atoms}$$

Empirical formula = CH_5N
Empirical formula mass = 31
Since the molecular mass is 32, molecular formula = empirical formula = CH_5N and the

Lewis structure is

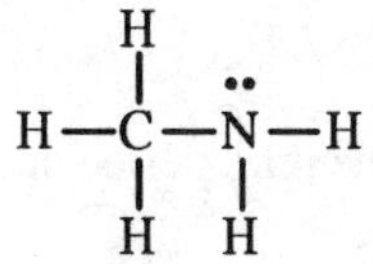

9.97

(a) $[Xe]6s^1 + [He]2s^2\,2p^5 \rightarrow [Xe] + [He]2s^2\,2p^6$ or [Ne]
Cs + F → Cs^+ + F^-
$2Cs(s) + F_2(g) \rightarrow 2CsF(s)$ cesium fluoride

(b) $[Xe]6s^2 + [He]2s^2\,2p^4 \rightarrow [Xe] + [He]2s^2\,2p^6$ or [Ne]
Ba + O → Ba^{2+} + O^{2-}
$2Ba(s) + O_2(g) \rightarrow 2BaO(s)$ barium oxide

(c) $[Xe]6s^2 + \begin{matrix}[Ne]3s^2\,3p^5 \\ [Ne]3s^2\,3p^5\end{matrix} \rightarrow [Xe] + \begin{matrix}[Ne]3s^2\,3p^6 \text{ or } [Ar] \\ [Ne]3s^2\,3p^6 \text{ or } [Ar]\end{matrix}$

Ba + 2Cl → Ba^{2+} + $2Cl^-$
$Ba(s) + Cl_2(g) \rightarrow BaCl_2(s)$ barium chloride

(d) $[Ar]4s^2$ [Ar]

$[Ar]4s^2 + \begin{matrix}[He]2s^2\,2p^3 \\ [He]2s^2\,2p^3\end{matrix} \rightarrow [Ar] + \begin{matrix}[He]2s^2\,2p^6 \text{ or } [Ne] \\ [He]2s^2\,2p^6 \text{ or } [Ne]\end{matrix}$

$[Ar]4s^2$ [Ar]
3Ca + 2N → $3Ca^{2+}$ + $2N^{3-}$
$3Ca(s) + N_2(g) \rightarrow Ca_3N_2(s)$ calcium nitride

(e) $[Ar]4s^2 + \begin{matrix} 1s^1 \\ 1s^1 \end{matrix} \rightarrow [Ar] + \begin{matrix} 1s^2 \text{ or [He]} \\ 1s^2 \text{ or [He]} \end{matrix}$

$Ca + 2H \rightarrow Ca^{2+} + 2H^-$

$Ca(s) + H_2(g) \rightarrow CaH_2(s)$ calcium hydride

9.99 (a) Molecular equation: $PCl_3(l) + 3H_2O(l) \rightarrow H_3PO_3(aq) + 3HCl(aq)$

Ionic equations (complete and net): $PCl_3(l) + 3H_2O(l) \rightarrow H_3PO_3(aq) + 3H^+ + 3Cl^-$

Molecular equation: $PCl_5(s) + 4H_2O(l) \rightarrow H_3PO_4(aq) + 5HCl(aq)$

Ionic equations (complete and net): $PCl_5(s) + 4H_2O(l) \rightarrow H_3PO_4(aq) + 5H^+ + 5Cl^-$

(b)

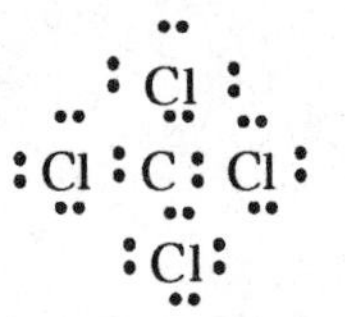

C has octet. C is in second period and cannot expand its octet. A bond must be broken before a new bond can form.

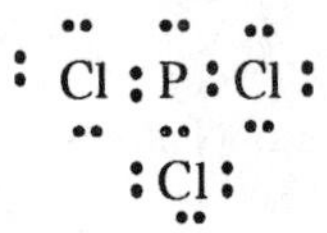

The unshared pair of electrons on P can form a bond to another atom.

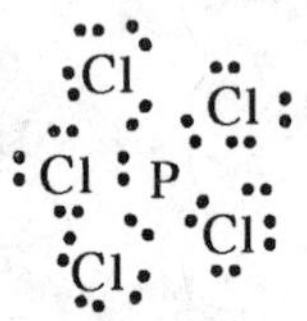

P can expand its octet further so that a bond to P can form.

(c) A reaction will take place if $SiCl_4$ is added to water because Si is in the third period and can have more than eight electrons around it. (This prediction is correct.)

(d) Density of CCl_4 is greater than the density of water, since CCl_4 sinks to the bottom when added to water.

9.101 (a) –100 kJ

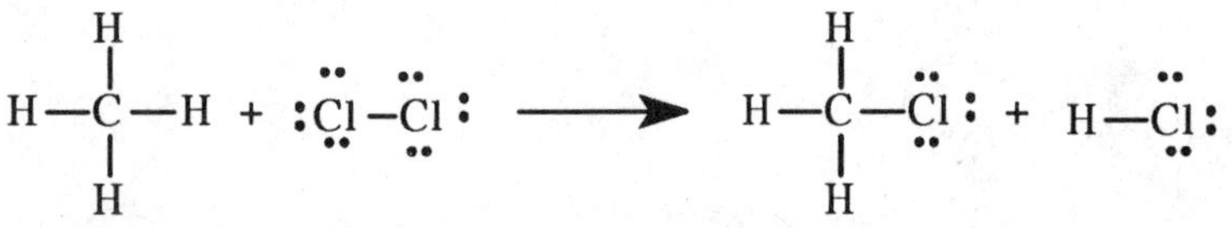

One C—H and one Cl—Cl bond must be broken. One C—Cl and one H—Cl bond are formed.

	ΔH^0, kJ/mol
Break one C—H bond:	+414
Break one Cl—Cl bond:	+244
Form one C—Cl bond:	–326
Form one H—Cl bond:	–432
	$\Delta H^0 = -100$

(b) –99 kJ

Equation	$CH_4(g)$	+	$Cl_2(g)$	$\rightarrow$	$CH_3Cl(g)$	+	$HCl(g)$
ΔH_f^0, kJ/mol	–74.85		0		–82		–92.31

$$\Delta H^0_{rxn} = \left[(1\ \cancel{\text{mol CH}_3\text{Cl}})\left(\frac{-82\ \text{kJ}}{\cancel{\text{mol CH}_3\text{Cl}}}\right) + (1\ \cancel{\text{mol HCl}})\left(\frac{-92.31\ \text{kJ}}{\cancel{\text{mol HCl}}}\right)\right]$$

$$-\left[(1\ \cancel{\text{mol CH}_4})\left(\frac{-74.85\ \text{kJ}}{\cancel{\text{mol CH}_4}}\right) + (1\ \cancel{\text{mol Cl}_2})\left(\frac{0\ \text{kJ}}{\cancel{\text{mol Cl}_2}}\right)\right]$$

$$= -99\ \text{kJ}$$

(c) Bond energies are average values. The bond energy of a given bond depends on the character of the entire molecule. For example, the bond energy of the O—H bond is not exactly the same in water, H—O—H, as it is in methanol, CH_3—O—H.

9.103 (a) C_3H_4, (c) NO_2^+, and (d) SCN^-

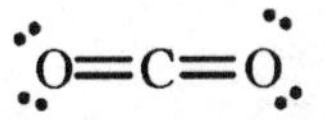

(16 valence electrons)

(a) 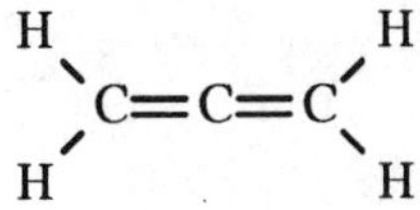

(16 valence electrons)

(b) 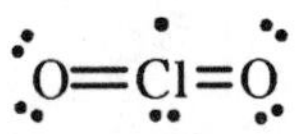

(19 valence electrons)

(c) 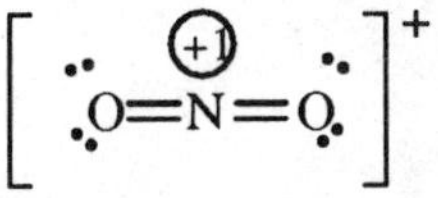

(16 valence electrons)

(d) 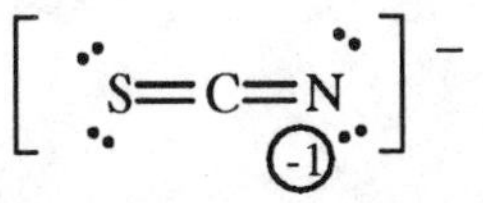

(16 valence electrons)

(e) 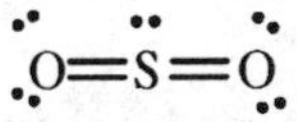

(18 valence electrons)

Applications

9.105 The best Lewis formula that can be written for the cyanate ion, OCN^-, has the minimum formal charge of −1 on the more electronegative atom, N. The best Lewis formula that can be written for the fulminate ion, CNO^-, has more formal charges and has a positive formal charge on N. Therefore, it is not surprising that fulminates are unstable.

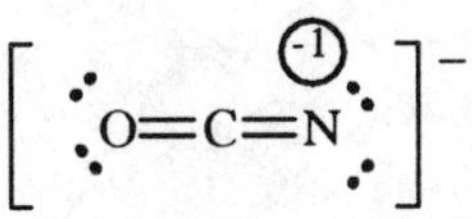

cyanate

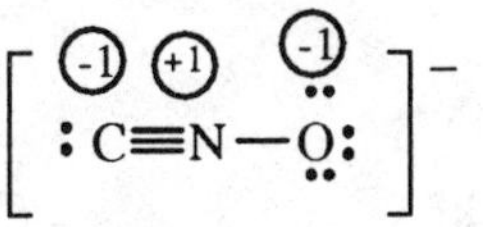

fulminate

9.107 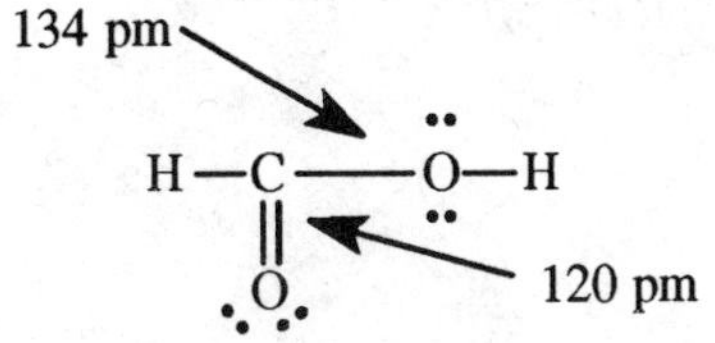

9.109 **(a, b)**

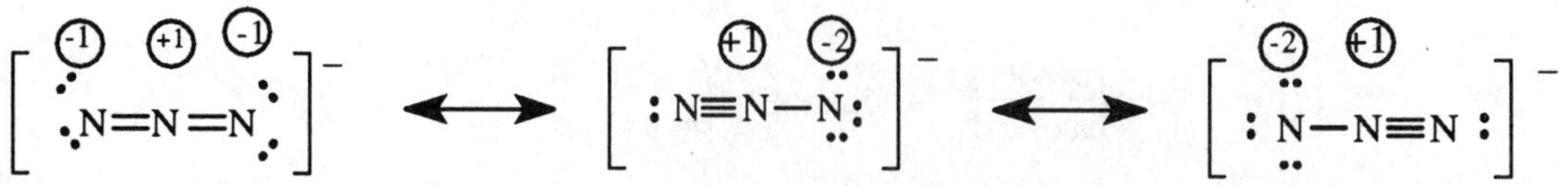

(c) The original Lewis formula contributes most to the hybrid because it has the smaller formal charges.

9.111 **(a)**

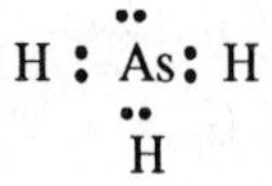

(b) $2AsH_3(g) \xrightarrow{250\text{–}300\ °C} 2As(s) + 3H_2(g)$

9.113 **(a)**

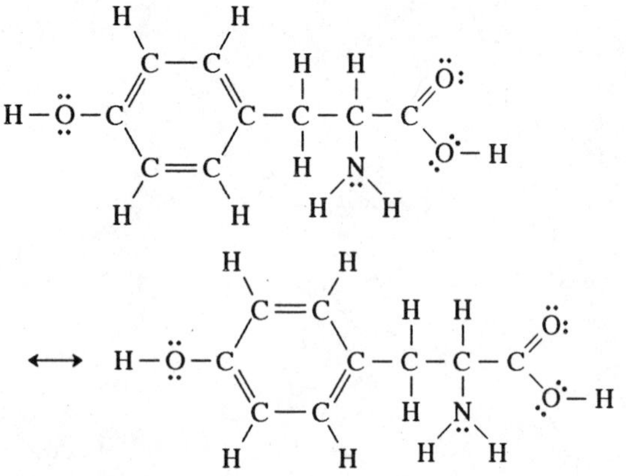

All the formal charges are zero in both structures.

(b) The bonds to H are shortest. Of the bonds between second period elements, the C═O bond is shortest. The C—C bonds in the benzene ring are intermediate between the C═C bond length of 134 pm and the C—C bond length of 154 pm. The C—N and C—O bonds are slightly shorter than C—C single bonds.

9.115 322 kJ/mol

The formation equation for $PCl_3(g)$ from $P(g)$ is

$$2P(g) + 3Cl_2(g) \rightarrow 2PCl_3(g)$$

or	$P(g)$	+	$\frac{3}{2}Cl_2(g)$	$\rightarrow$	$PCl_3(g)$
ΔH_f^0, kJ/mol	314.25 kJ		0		–287.0 kJ

$$\Delta H^0_{rxn} = \left[(1\ \cancel{mol\ PCl_3})\left(\frac{-287.0\ kJ}{\cancel{mol\ PCl_3}}\right)\right] - \left[(1\ \cancel{mol\ P})\left(\frac{314.25\ kJ}{\cancel{mol\ P}}\right) + \left(\frac{3}{2}\ \cancel{mol\ Cl_2}\right)\left(\frac{0\ kJ}{\cancel{mol\ Cl_2}}\right)\right] = -601.3\ kJ$$

Let x = bond energy P—Cl bond.

		ΔH^0, kJ/mol
Break $\frac{3}{2}$ (Cl—Cl) bond	$\frac{3}{2}$(244 kJ)	+366
Form 3 P—Cl bonds	3(x)	$-3x$
		−601.3

$366 - 3x = -601.3$

$322 = x$

The average bond energy of a P—Cl bond is 322 kJ/mol.

9.117

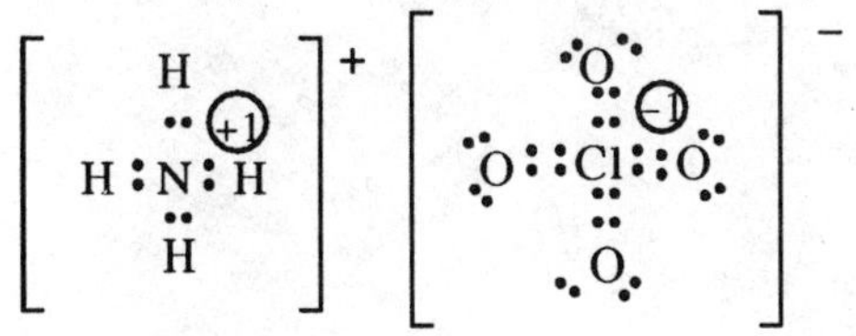

9.119 $H_2C{=}CH(OOCCH_3)$ or $CH_3COOCH{=}CH_2$

Vinyl acetate is the monomer of poly(vinyl acetate).

CHAPTER 10:
Molecular Shape and Theory of Chemical Bonding

Practice Problems

10.1 **(a)**

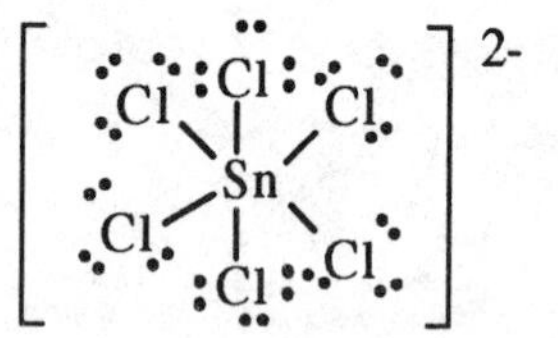

$(SnCl_6)^{2-}$	1 Sn @ 4 valence electrons	=	4 valence electrons
	6 Cl @ 7 valence electrons	=	42 valence electrons
	2 extra electrons for 2– charge	=	2 valence electrons
	Total	=	48 valence electrons

(b) octrahedral

A total coordination number of 6 for Sn and no unshared pairs gives octahedral geometry.

(c)

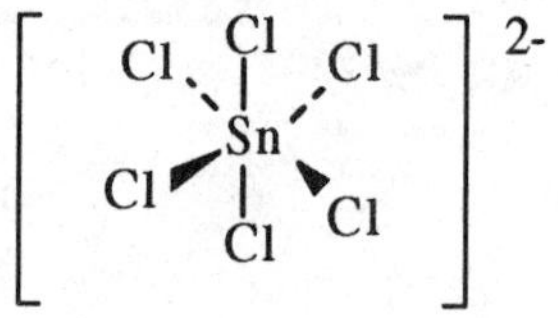

10.2 linear

1 Be @ 2 valence electrons	=	2 valence electrons
2 F @ 7 valence electrons	=	14 valence electrons
Total	=	16 valence electrons

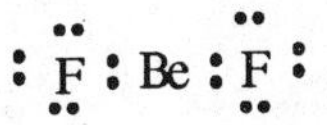

This molecule should be linear due to unshared pairs and bonded pairs being farther apart that way. The total coordination number of beryllium is 2.

10.3 **(a)**

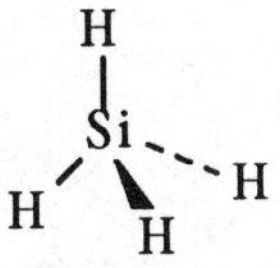

The Lewis formula for SiH_4 is

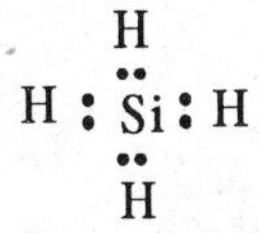

The total coordination number of Si is 4. SiH_4 has four shared pairs and no unshared pairs of electrons. Therefore, a tetrahedral shape is predicted.

(b)

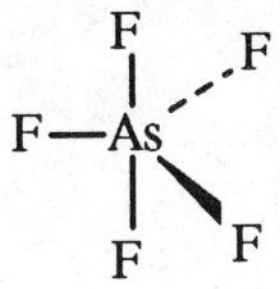

The Lewis formula for AsF_5 is

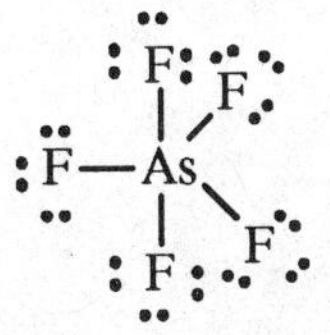

The total coordination number of As is 5. AsF_5 has five shared pairs and no unshared pairs of electrons. Therefore, a trigonal bipyramidal shape is predicted.

10.4 (i) CF_4

1 C @ 4 valence electrons	=	4 valence electrons
4 F @ 7 valence electrons	=	28 valence electrons
Total	=	32 valence electrons

The Lewis formula for CF_4 is

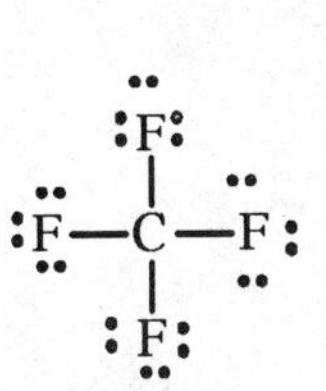

(a) The total coordination number is 4.

(b) There are four shared pairs and no unshared pairs around the central atom.

(c) The molecule is tetrahedral.

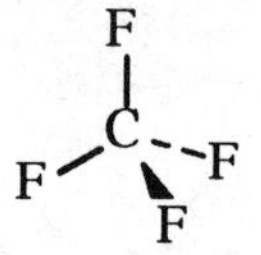

(ii) $GeBr_2$

1 Ge @ 4 valence electrons	=	4 valence electrons
2 Br @ 7 valence electrons	=	14 valence electrons
Total	=	18 valence electrons

The Lewis formula for $GeBr_2$ is

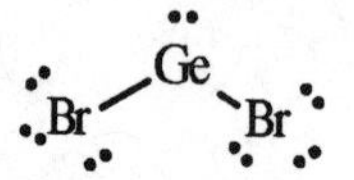

(a) The total coordination number is 3.

(b) There are two shared pairs and one unshared pair around the central atom.

(c) The molecule is bent or angular.

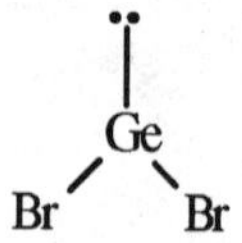

(iii) ICl_4^-

1 I @ 7 valence electrons	=	7 valence electrons
4 Cl @ 7 valence electrons	=	28 valence electrons
1 extra electron for 1− charge	=	1 valence electron
Total	=	36 valence electrons

The Lewis formula for ICl_4^- is

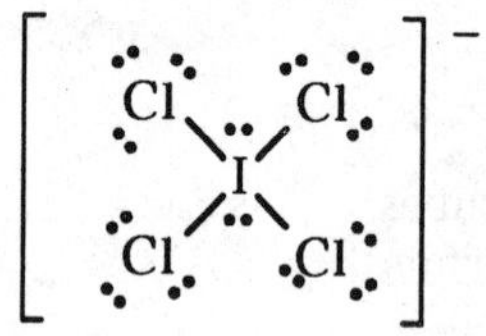

(a) The total coordination number is 6.

(b) There are four shared pairs and two unshared pairs around the central atom.

(c) The ion is square planar.

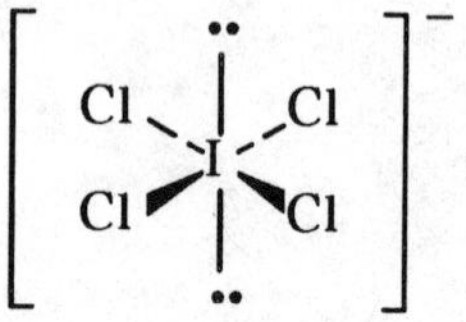

(iv) IF_4^+

1 I @ 7 valence electrons	=	7 valence electrons
4 F @ 7 valence electrons	=	28 valence electrons
Less 1 electron for 1+ charge	=	–1 valence electron
Total	=	34 valence electrons

The Lewis formula for IF_4^+ is

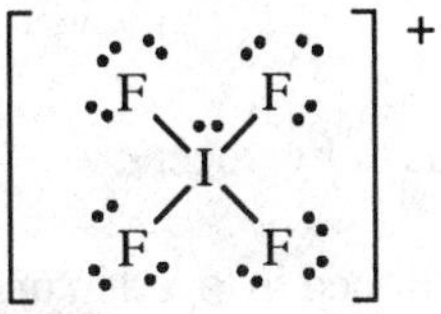

(a) The total coordination number is 5.

(b) There are four shared pairs and one unshared pair around the central atom.

(c) The ion is seesaw.

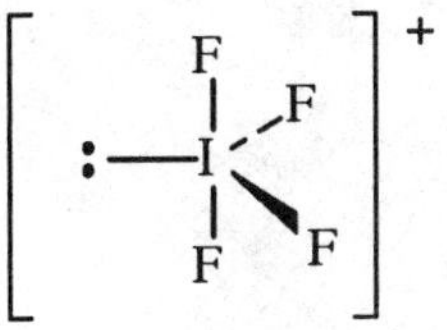

(v) NH_2^-

1 N @ 5 valence electrons	=	5 valence electrons
2 H @ 1 valence electron	=	2 valence electrons
1 extra electron for 1– charge	=	1 valence electron
Total	=	8 valence electrons

The Lewis formula for NH_2^- is

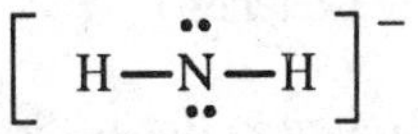

(a) The total coordination number is 4.

(b) There are two shared pairs and two unshared pairs around the central atom.

(c) The ion is bent or angular.

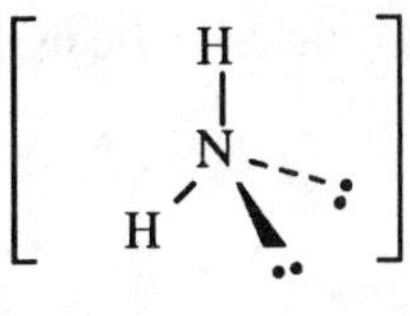

10.5 linear

H—C≡C—H

The total coordination number for each carbon atom in acetylene is 2. Each carbon has four shared pairs and no unshared pairs of electrons. Therefore, the atoms attached to each carbon lie on a straight line. Putting the two halves of the molecule together gives a linear shape for the shape of the molecule as shown in the structure above.

10.6 **(a)** $BeCl_2$, BF_3, CH_4, PCl_5, SF_6

The ions ICl_2^- and IF_4^- are also nonpolar.

(b) GeF_2, SF_4, ClF_3, IF_5

10.7 Sigma bonds. Ends of sp^3 orbitals overlap $1s$ orbitals of hydrogens.

The C—H bonds in methane are pictured as being formed by overlap of the ends of the four sp^3 orbitals of carbon with four $1s$ orbitals of four hydrogen atoms (see Figure 10.11 (a)).

10.8 **(a)** 1

Five orbitals are needed to hold the five pairs of electrons around S in SF_4. One *s*, three *p*, and one *d* orbital must be combined.

(b) sp^3d

(c) 5

See Table 10.2.

10.9 **(a)**

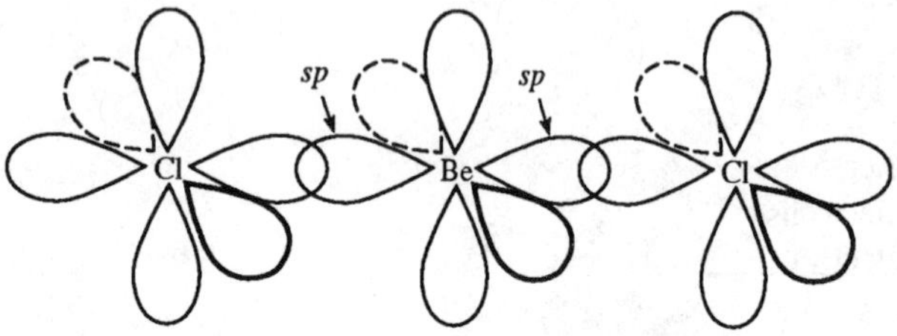

(Remember that orbitals are not drawn to scale in sketches.)

Experimental data indicate that no pi bonds are present in $BeCl_2$ (Section 9.10). Chlorine is in the third period, whereas Be is in the second period. The $3p$ orbitals of Cl are too big to overlap well sidewise with the $2p$ orbitals of Be. The Lewis formula for $BeCl_2$ is

:C̤̈l—Be—C̤̈l:

There are two shared pairs of electrons around Be, and a linear arrangement would be predicted by VSEPR. From Table 10.2, linear species are pictured as having *sp* hybridization. Two *sp* hybrid orbitals on Be overlap endwise with one *p* orbital on each Cl to form two sigma bonds.

(b)

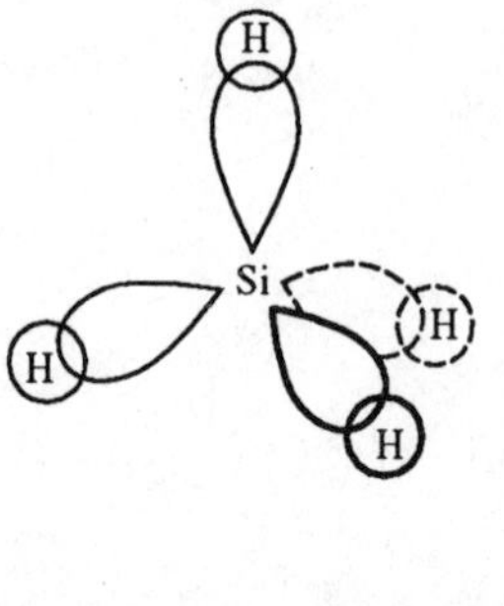

The Lewis formula for SiH_4 is

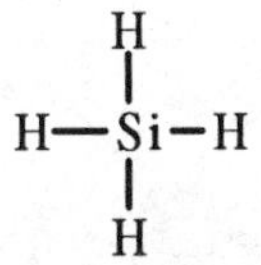

There are four shared pairs of electrons around Si, and a tetrahedral arrangement would be predicted by VSEPR. From Table 10.2, tetrahedral species are pictured as having sp^3 hybridization of the central atom. The ends of the four sp^3 orbitals on Si overlap with four 1*s* orbitals from four hydrogen atoms.

10.10 **(a)** 9 **(b)** 1 **(c)** 8

10.11 **(a)** H 1*s* overlaps with C *sp* to form a sigma bond; C *sp* overlaps with one N *sp* orbital to form a sigma bond; two C *p* orbitals overlap with two N *p* orbitals to form two pi bonds.

(b) The lefthand carbon has four sp^3 orbitals, three of which overlap with three H 1*s* orbitals to form sigma bonds and one of which overlaps with an sp^2 hybrid orbital of the middle carbon to form a sigma bond. The middle carbon also overlaps one sp^2 orbital with a hydrogen 1*s* to form a sigma bond, and another sp^2 orbital with an sp^2 from the righthand carbon to form a sigma bond. The righthand carbon forms a sigma bond with each of two hydrogen 1*s* orbitals, using its remaining two sp^2 orbitals. The remaining parallel *p* orbitals on the middle and righthand carbons overlap to form a pi bond.

10.12 **(a)** about 120° **(b)** about 120°

10.13 No. The three hydrogens on the lefthand C cannot all be in the same plane.

All the atoms except the hydrogens on the left–hand C must lie in the same plane. One of the hydrogens on the left–hand C can also lie in the plane, but it does not have to do so.

10.14 **(a)**

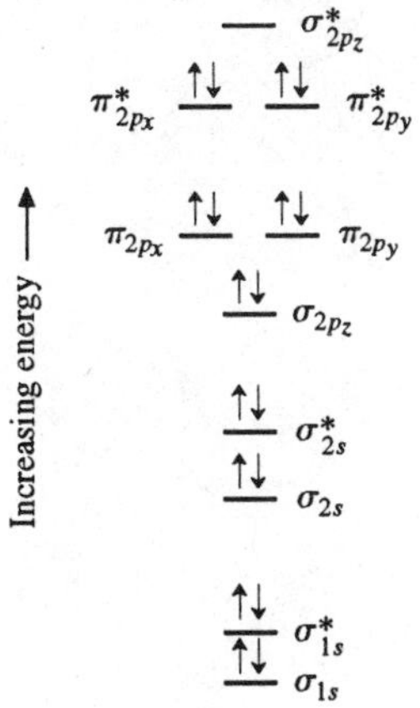

F_2 has 18 electrons. The ground-state electron configuration is:

$(\sigma_{1s})^2 (\sigma^*_{1s})^2 (\sigma_{2s})^2 (\sigma^*_{2s})^2 (\sigma_{2p_z})^2 (\pi_{2p_x})^2 (\pi_{2p_y})^2 (\pi^*_{2p_x})^2 (\pi^*_{2p_y})^2$

(b) zero

There are no π bonds in F_2 owing to the fact that both π bonding orbitals and antibonding orbitals are filled.

(c) one

There is one sigma bond in F_2, σ_{2p_z}. The $\sigma^*_{2p_z}$ orbital is unfilled.

(d) single bond, bond order = 1

10.15   **(a)**

$$\begin{array}{c}
\text{—} \;\; \sigma^*_{2p_z} \\[4pt]
\pi^*_{2p_x} \;\text{—} \;\; \text{—} \;\; \pi^*_{2p_y} \\[16pt]
\text{—} \;\; \sigma_{2p_z} \\[4pt]
\pi_{2p_x} \;\uparrow\;\; \uparrow\; \pi_{2p_y} \\[10pt]
\uparrow\downarrow \;\; \sigma^*_{2s} \\[2pt]
\uparrow\downarrow \;\; \sigma_{2s} \\[16pt]
\uparrow\downarrow \;\; \sigma^*_{1s} \\[2pt]
\uparrow\downarrow \;\; \sigma_{1s}
\end{array}$$

Increasing energy →

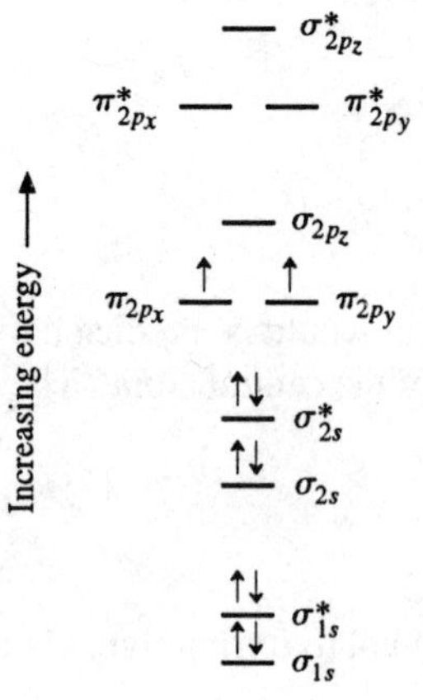

B_2 has 10 electrons. The ground–state electron configuration is:

$$(\sigma_{1s})^2 (\sigma^*_{1s})^2 (\sigma_{2s})^2 (\sigma^*_{2s})^2 (\pi_{2p_x})^1 (\pi_{2p_y})^1$$

(b) bond order = 1

$$\frac{1}{2}(6-4) = 1$$

(c) paramagnetic, two

10.16   **(a)**

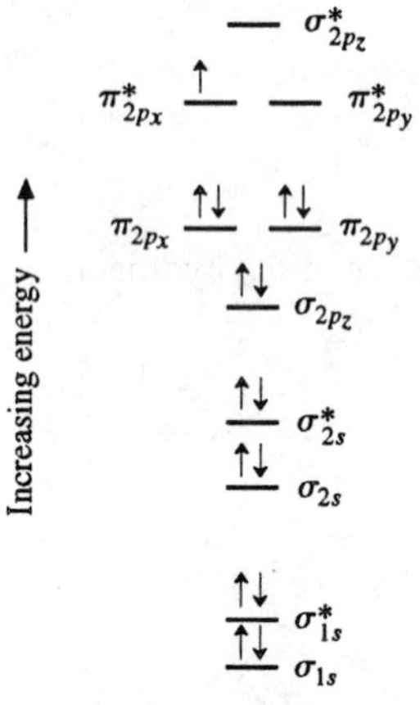

$$\begin{array}{c}
\text{—} \;\; \sigma^*_{2p_z} \\[4pt]
\pi^*_{2p_x} \;\uparrow\;\; \text{—} \;\; \pi^*_{2p_y} \\[16pt]
\pi_{2p_x} \;\uparrow\downarrow \;\; \uparrow\downarrow\; \pi_{2p_y} \\[6pt]
\uparrow\downarrow \;\; \sigma_{2p_z} \\[10pt]
\uparrow\downarrow \;\; \sigma^*_{2s} \\[2pt]
\uparrow\downarrow \;\; \sigma_{2s} \\[16pt]
\uparrow\downarrow \;\; \sigma^*_{1s} \\[2pt]
\uparrow\downarrow \;\; \sigma_{1s} \\[4pt]
\text{NO}
\end{array}$$

Increasing energy →

NO has 15 electrons.

(b) $(\sigma_{1s})^2 (\sigma^*_{1s})^2 (\sigma_{2s})^2 (\sigma^*_{2s})^2 (\sigma_{2p_z})^2 (\pi_{2p_x})^2 (\pi_{2p_y})^2 (\pi^*_{2p_x})^1$

(c) bond order = 2.5

$$\frac{1}{2}(6-1) = 2.5$$

(d) No.

The valence bond method shows a double bond (bond order = 2) for the N—O bond in NO.

$$\ddot{\cdot}\text{N} = \ddot{\text{O}}\ddot{\cdot}$$

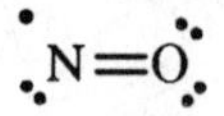

(e) bond energies and bond lengths

Bond energies and bond lengths would be used to determine which method better describes NO. If the bond is intermediate in energy and length between a double bond and a triple bond between N and O, the molecular orbital method is better. If the bond is closed in energy and length to a double bond between N and O, the valence bond method is better.

10.17 **(a)** $H_2C{=}CHCH{=}CHCH_3$ The *p* orbitals of the four carbons in the left formula overlap sidewise.

This molecule has delocalized electrons because the first four carbon atoms are sp^2 hybridized and all lie in the same plane with *p* orbitals perpendicular to the sp^2 hybrid orbitals. These *p* orbitals can overlap sidewise with each other. This situation does not exist in the second structure because there is an sp^3 hybridized C between the two double bonds.

(b) four

The *p* orbitals on four carbons overlap sidewise.

Additional Practice Problems

10.57 No. The *p* orbitals on C1 and C3 are perpendicular to each other.

The electrons in allene are not delocalized. The C_1—C_2 pi bond involves the sidewise overlap of p_y–p_y orbitals; the C_2—C_3 pi bond involves the overlap of p_z–p_z orbitals. However, since the *p* orbital systems involved in the pi bonds are not in the same plane, there is no delocalization.

10.59 Li_2 has six electrons, Be_2^+ has seven electrons, and Be_2 has eight electrons.

σ^*_{2s}	—	↑	↑↓
σ_{2s}	↑↓	↑↓	↑↓
σ^*_{1s}	↑↓	↑↓	↑↓
σ_{1s}	↑↓	↑↓	↑↓
	Li_2	Be_2^+	Be_2
Bond order:	1	0.5	0

Stability: $Li_2 > Be_2^+ > Be_2$

For Li_2, bond order $= \frac{1}{2}(4 - 2) = 1$; for Be_2^+, bond order $= \frac{1}{2}(4 - 3) = \frac{1}{2}$; for Be_2 bond order $= \frac{1}{2}(4 - 4) = 0$

Thus, the order of stability is $Li_2 > Be_2^+ > Be_2$.

10.61 **(a)** $sp \rightarrow sp^3$

(b) $sp^3 \rightarrow sp^2$

(c) no change in hybridization (sp^3 both before and after reaction)

(d) $sp^3 \rightarrow sp^3d$

(e) $sp^3d \rightarrow sp^3d^2$

10.63 (a) i, iii, and iv; ii and v

Use molecular models to determine which of these formulas are the same.

(b) i, iii, and iv; ii and v

Use molecular models to determine which of these formulas are the same.

(c) i and iv; ii, iii, and v

Use molecular models to determine which of these formulas are the same.

Stop and Test Yourself

1. (a) octahedral

The arrangement of electrons for

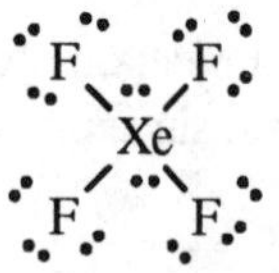

requires six different locations in space about the central Xe atom to place six pairs of valence electrons, hence, requiring an octahedral arrangement.

2. (b) square planar

The shape of a molecule of XeF_4 is square planar in which the two sets of unshared pairs of electrons are perpendicular to the plane.

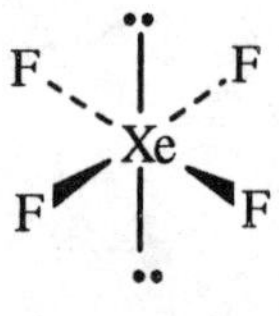

3. (e) SO_3

SO_3 has zero dipole moment because of its symmetrical Lewis formula:

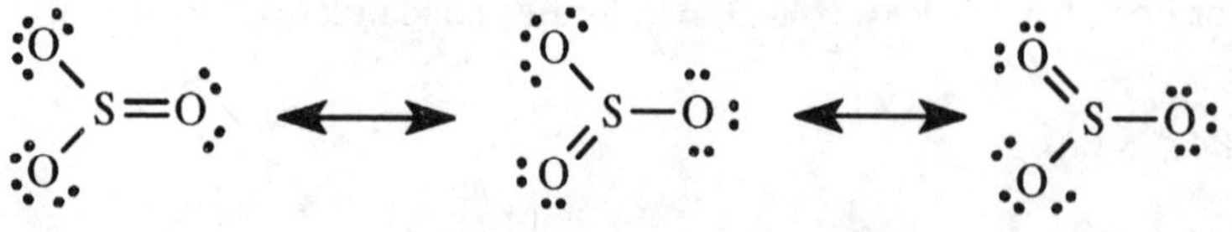

4. (e) trigonal planar

See Table 10.2.

5. (d) 5

The number of hybrid orbitals formed must be the same as the number of atomic orbitals combined.

6   (c) sp^3

The Lewis formula for $CHCl_3$ is

$$
\begin{array}{ccc}
 & :\ddot{C}l: & \\
 & | & \\
H - & C - & \ddot{C}l: \\
 & | & \\
 & :\ddot{C}l: &
\end{array}
$$

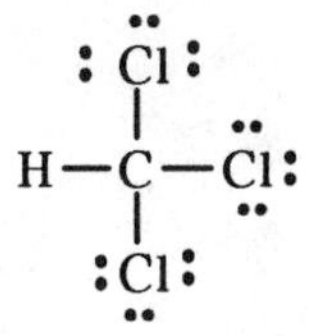

The total coordination number of C in $CHCl_3$ is four. There are four shared pairs around C, and the molecule is tetrahedral. Therefore, the C is pictured as sp^3 hybridized.

7.   (c) 3

There are two pi bonds between C–1 and C–2 and another pi bond between C–3 and C–4. (One bond of each multiple bond is a sigma bond.)

8.   (c) 10

There is one sigma bond in every bond, whether it is a single, double, or triple bond.

9.   (d) 109.5°

C–5 is sp^3 hybridized; sp^3 hybrid orbitals have the tetrahedral bond angle of about 109°.

10.   (a) C–1 and C–2

The higher the bond order, the shorter the bond.

11.   (a) Be_2

The ground–state electron configuration of Be_2 is $(\sigma_{1s})^2 (\sigma^*_{1s})^2 (\sigma_{2s})^2 (\sigma^*_{2s})^2$. The bond order is 0.

12.   (c) N_2

$(\sigma_{1s})^2 (\sigma^*_{1s})^2 (\sigma_{2s})^2 (\sigma^*_{2s})^2 (\pi_{2p_x})^2 (\pi_{2p_y})^2 (\pi_{2p_z})^2$

Bond order $= \frac{1}{2}(10 - 4) = 3$

13.   (a)

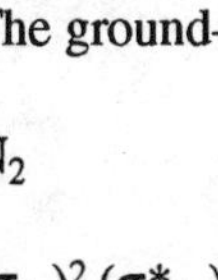

For electrons to be delocalized, three or more p orbitals must be located on adjacent atoms; the p orbitals must be parallel so that they can overlap sidewise. In structure (a), the p orbitals are not located on adjacent atoms.

14.   (a) conductors, increases

Metals conduct electricity. Since resistance decreases with decreasing temperature, the conductivity of metals increases with decreasing temperature.

Putting Things Together

10.65 (a)

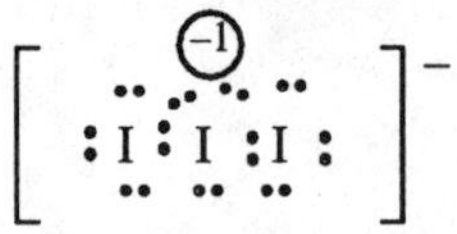

(b) Br_3^- and I_3^- are possible because Br atoms and I atoms are able to have more than eight electrons around themselves because of the availability of *d* orbitals. F_3^- does not exist under normal conditions because F atoms are unable to expand their valence shell to accommodate more than eight electrons.

(c) linear

The five pairs of electrons are distributed in a trigonal bipyramidal arrangement around the central I. Because unshared electron pairs take up more space than shared pairs, the three unshared pairs should be in the positions where they are 120° away from the other pairs. Therefore, the three iodine atoms are in a line, and the triiodide ion is linear.

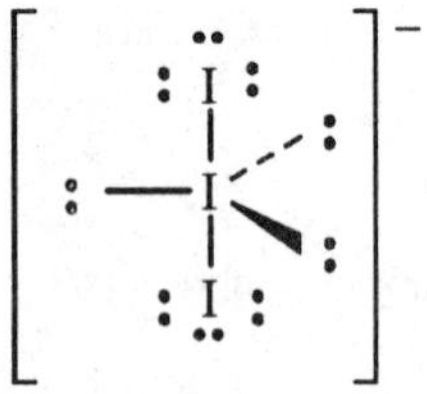

10.67 (a) superoxide ion, O_2^-; peroxide ion, O_2^{2-}; dioxygenyl ion, O_2^+

(b)

	O_2^-	O_2^{2-}	O_2^+
$\pi^*_{2p_x}$ $\pi^*_{2p_y}$	↑↓ ↑	↑↓ ↑↓	↑ __
π_{2p_x} π_{2p_y}	↑↓ ↑↓	↑↓ ↑↓	↑↓ ↑↓
σ_{2p_z}	↑↓	↑↓	↑↓
σ^*_{2s}	↑↓	↑↓	↑↓
σ_{2s}	↑↓	↑↓	↑↓
σ^*_{1s}	↑↓	↑↓	↑↓
σ_{1s}	↑↓	↑↓	↑↓

(d) Bond order: 1.5; 1; 2.5

$\frac{1}{2}(10-7) = 1.5$ $\quad \frac{1}{2}(10-8) = 1$ $\quad \frac{1}{2}(10-5) = 2.5$

	O_2^-	O_2^{2-}	O_2^+
(e) Magnetism	paramagnetic	diamagnetic	paramagnetic
# of unpaired	1	0	1

electrons

(c) O_2^-: $(\sigma_{1s})^2 (\sigma^*_{1s})^2 (\sigma_{2s})^2 (\sigma^*_{2s})^2 (\sigma_{2p_z})^2 (\pi_{2p_x})^2 (\pi_{2p_y})^2 (\pi^*_{2p_x})^2 (\pi^*_{2p_y})$

O_2^{2-}: $(\sigma_{1s})^2 (\sigma^*_{1s})^2 (\sigma_{2s})^2 (\sigma^*_{2s})^2 (\sigma_{2p_z})^2 (\pi_{2p_x})^2 (\pi_{2p_y})^2 (\pi^*_{2p_x})^2 (\pi^*_{2p_y})^2$

O_2^+: $(\sigma_{1s})^2 (\sigma^*_{1s})^2 (\sigma_{2s})^2 (\sigma^*_{2s})^2 (\sigma_{2p_z})^2 (\pi_{2p_x})^2 (\pi_{2p_y})^2 (\pi^*_{2p_x})$

10.69 Two resonance structures

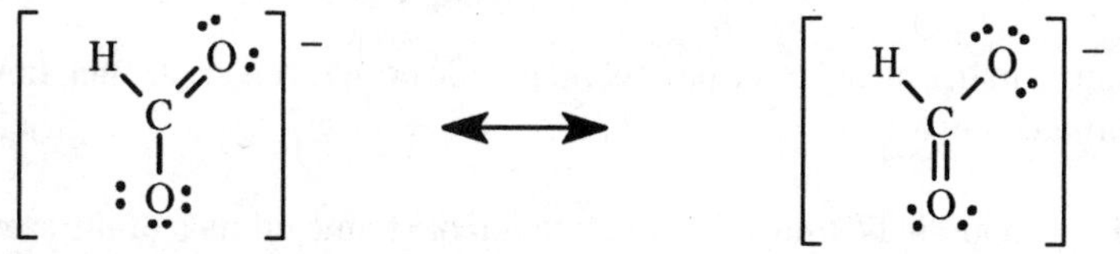

or one sketch

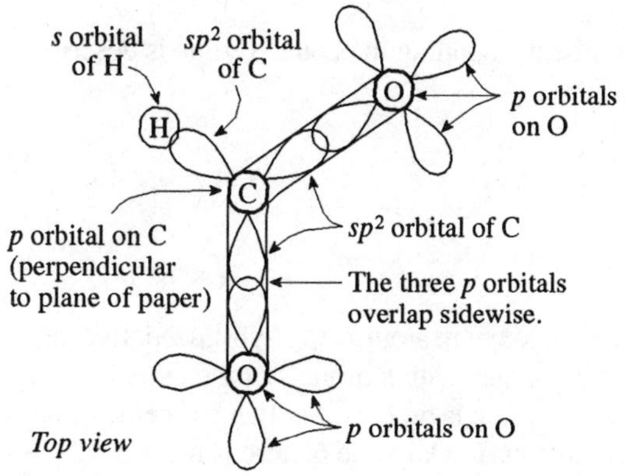

can be used. The electrons in the sigma bonds are localized between the bonded atoms. The electrons in the pi bonds are delocalized over three atoms (O, C, and O).

10.71 268 kJ/mol

According to Table 9.4, a C═C double bond has a bond energy of 615 kJ/mol and a C—C single bond has a bond energy of 347 kJ/mol; the difference between the two (615 kJ/mol – 347 kJ/mol) is 268 kJ/mol. Rotation about a C═C double bond would require breaking the second bond and would therefore require 268 kJ/mol.

10.73 A *resonance hybrid* is a molecule or polyatomic ion for which more than one Lewis formula must be written. *Hybrid orbitals* are formed by combination of two or more atomic orbitals of the same atom.

10.75 The carbide ion, C_2^{2-}, is isoelectronic with nitrogen, N_2, carbon monoxide, CO, and acetylene, C_2H_2.

Lewis formulas are

$[:C{\equiv}C:]^{2-}$ $\quad :N{\equiv}N: \quad :C{\equiv}O: \quad H{-}C{\equiv}C{-}H$

10.77 The valence bond method predicts that the oxygen molecule has a double bond and is diamagnetic. The Lewis formula is given below. The molecular orbital method predicts that oxygen has two unpaired electrons, has a double bond between the oxygen atoms, and is paramagnetic as a result of the two unpaired electrons. Oxygen is paramagnetic (Figure 10.29), in agreement with the prediction of the molecular orbital method.

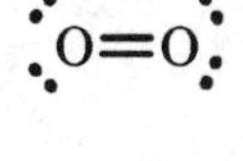

10.79 (a) three

Three atomic orbitals would form three hybrid orbitals.

(b) three

Three atomic orbitals would form three molecular orbitals.

(c) maximum number is two for both

The maximum number of electrons per hybrid atomic orbital is two; the maximum number of electrons per molecular orbital is two.

(d) Hybrid orbitals are orbitals formed by combination of atomic orbitals of the same atom; molecular orbitals are orbitals formed by combination of atomic orbitals from different atoms.

10.81 The valence bond method uses resonance to describe the bonding in ozone. Ozone is described as a hybrid of two structures.

The molecular orbital method assumes that the central oxygen atom is sp^2 hybridized, that sigma bonds are formed by endwise overlap of sp^2 hybrid orbitals of the central oxygen with *p* orbitals of the other two oxygens, and that the *p* orbital of the central oxygen overlaps sidewise with *p* orbitals on the other two oxygens, forming pi bonds. The electrons of the pi bonds are delocalized over all three oxygen atoms. Only one picture is needed.

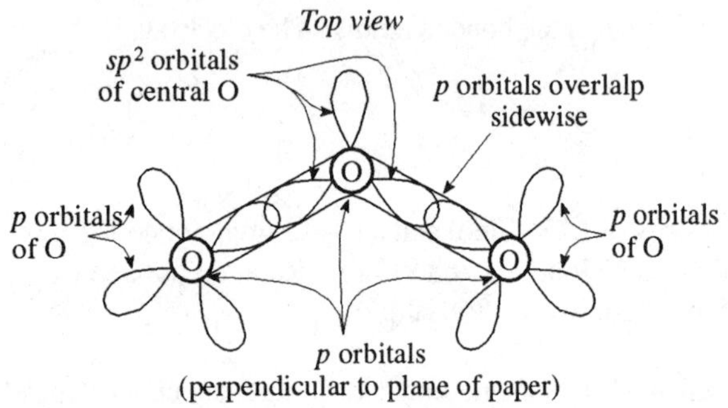

10.83 $O_2^{2-} < O_2^{-} < O_2 < O_2^{+}$

O_2 has a bond order of 2. In order of increasing bond order, the species are

	O_2^{2-}	<	O_2^{-}	<	O_2	<	O_2^{+}
Bond order	1		1.5		2		2.5

The higher the bond order, the stronger the bond. Thus, this order is also the order of increasing bond strength.

10.85 **(a)** $\Delta H^0 = 248$ kJ

The equation for the formation of benzene is:

$6C(s) + 3H_2(g) \rightarrow C_6H_6(g)$

Bond energies can only be used when all the species are in the gas phase. Thus, C(s) must be converted to the gas phase.

		ΔH^0 kJ/mol
Vaporize 6 C atoms	6(718.384)	+4310.304
Break 3 H—H bonds	3(436)	+1308
Form 3 C—C bonds	3(347)	−1041
Form 3 C═C bonds	3(615)	−1845
Form 6 C—H bonds	6(414)	−2484
	ΔH^0 =	248

(b) 165 kJ

248 kJ estimated – 83 kJ experimental = 165 kJ difference

(c)

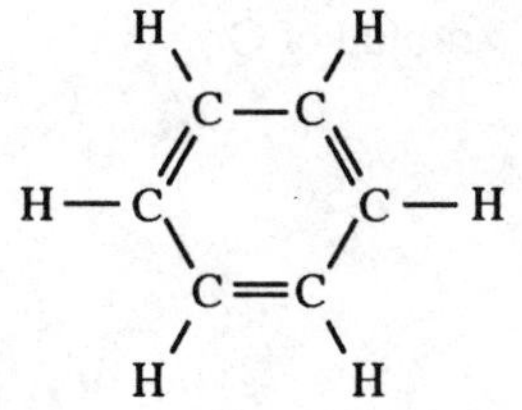

(d)

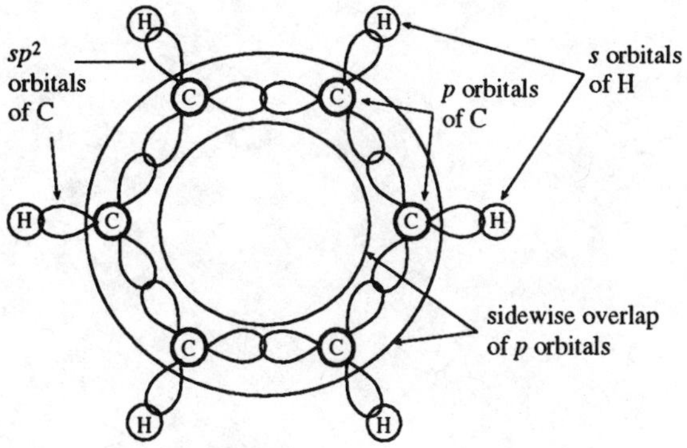

Top view

There are "doughnuts" of delocalized electrons above and below the plane of the ring. (Remember Figure 10.31.)

Applications

10.87 N_2: one sigma *sp*–*sp*, two pi *p*–*p*;
NH_3: three sigma sp^3–*s*;
HNO_3: H—O bond is sigma *s*–sp^3, O—N bond is sigma sp^3–sp^2,
N═O bond is composed of one sigma sp^2–*p* and one pi *p*–*p*, N—O bond is sp^2–*p*.

Lewis formulas are

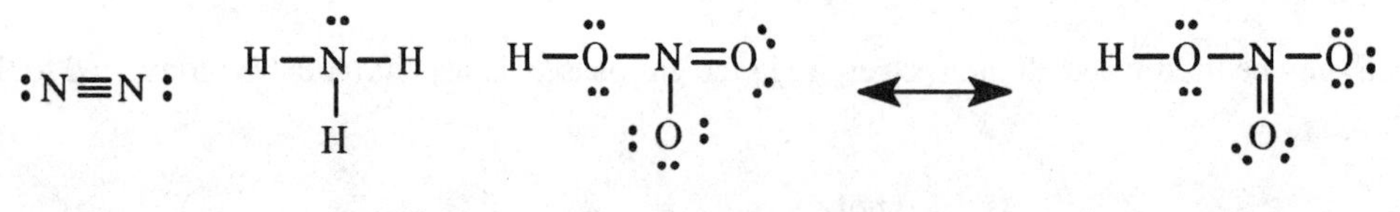

Nitrogen in N_2 is pictured as *sp* hybridized. One *sp* oribital from one nitrogen overlaps endwise with one *sp* orbital from the other nitrogen to form a sigma bond. The two *p* orbitals remaining on each nitrogen overlap sidewise to form two pi bonds. The unshared pair on each nitrogen is in the second *sp* orbital.

According to the hybrid orbital model, N in NH_3 is pictured as sp^3 hybridized. Three of the four sp^3 hybridized orbitals overlap endwise with *s* orbitals from H, forming three sigma bonds. The unshared electron pair occupies the fourth sp^3 orbital.

In HNO_3, N is pictured as sp^2 hybridized and the O to which H is bonded is pictured as sp^3 hybridized. The O—H bond is formed by endwise overlap of the *s* orbital of H with one of the sp^3 orbitals of O to form a sigma bond. Another of the sp^3 orbitals of O overlaps endwise with one of the three sp^2 orbitals of N, forming a sigma bond. The other two sp^2 orbitals of N each overlap endwise with a *p* orbital of O and sidewise with another *p* orbital of O.

10.89 **(a)**

	O_2^+	N_2^+	
$\pi^*_{2p_x}$ $\pi^*_{2p_y}$	↑ __	__ __	
π_{2p_x} π_{2p_y}	↑↓ ↑↓	↑	σ_{2p_z}
σ_{2p_z}	↑↓	↑↓ ↑↓	π_{2p_x} π_{2p_y}
σ^*_{2s}	↑↓	↑↓	
σ_{2s}	↑↓	↑↓	
σ^*_{1s}	↑↓	↑↓	
σ_{1s}	↑↓	↑↓	

	O_2^+	N_2^+
(c) Bond order	2.5	2.5
(d) Magnetism	paramagnetic	paramagnetic
# of unpaired electrons	1	1

(b) electron configuration:

O_2^+: $(\sigma_{1s})^2 (\sigma^*_{1s})^2 (\sigma_{2s})^2 (\sigma^*_{2s})^2 (\sigma_{2p_z})^2 (\pi_{2p_x})^2 (\pi_{2p_y})^2 (\pi^*_{2p_x})^1$

N_2^+: $(\sigma_{1s})^2 (\sigma^*_{1s})^2 (\sigma_{2s})^2 (\sigma^*_{2s})^2 (\pi_{2p_x})^2 (\pi_{2p_y})^2 (\sigma_{2p_z})^1$

10.91 **(a)**

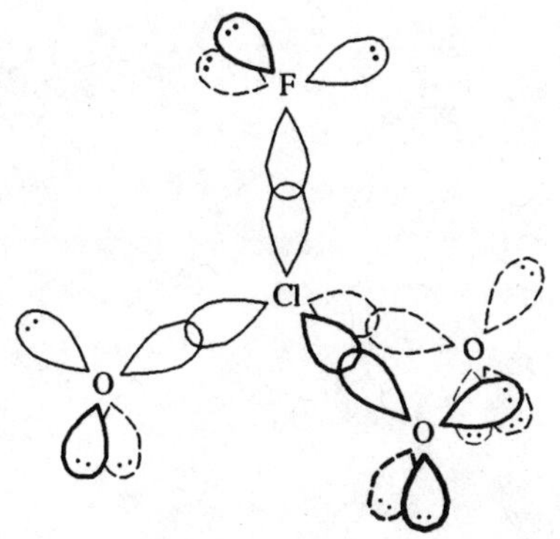

(b)

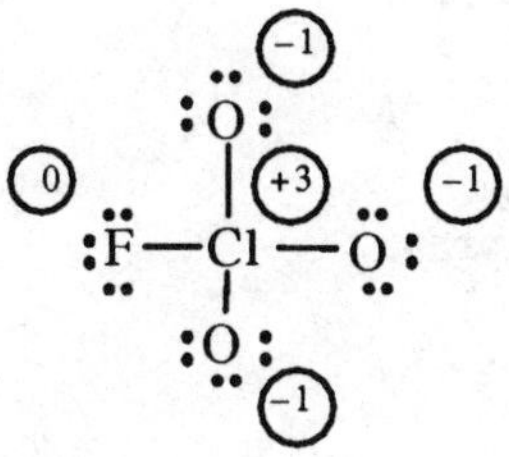

10.93 **(a)**

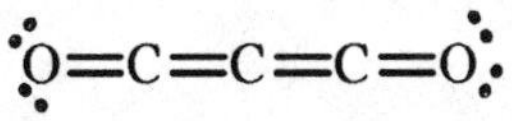

(b) *sp*

(c) 180°

(d) A polymer is a substance whose molecules are composed of thousands of atoms and which has repeating units.

(e) a monomer

10.95 **(a)**

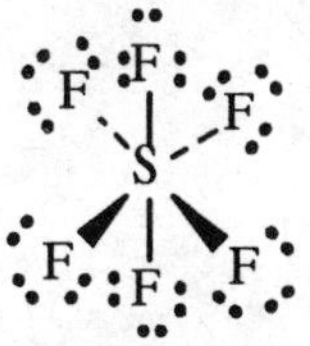

(b) 1×10^{-6} ppb, 4×10^{6} molecules SF_6

$$(1\ \text{km}^3)\left(\frac{(10^3\ \text{m})^3}{\text{km}^3}\right)\left(\frac{1\ \text{cm}^3}{(10^{-2}\ \text{m})^3}\right)\left(\frac{1.02\ \text{g}}{\text{cm}^3}\right) = 1.02 \times 10^{15}\ \text{g seawater}$$

$$\left(\frac{1\ \text{g}}{1.02 \times 10^{15}\ \text{g}}\right)(10^9) = 1 \times 10^{-6}\ \text{ppb}$$

$$(1\ \text{g}\ SF_6)\left(\frac{1\ \text{mol}\ SF_6}{146\ \text{g}\ SF_6}\right)\left(\frac{6.02 \times 10^{23}\ \text{molecules}\ SF_6}{\text{mol}\ SF_6}\right) = 4 \times 10^{6}\ \text{molecules}\ SF_6$$

(c) 108 kg F_2

Problem			? kg		139 kg
Equation	S(s)	+	$3F_2(g)$	→	$SF_6(g)$
Formula mass, u			38.00		146.1
Recipe, mol			3		1

$$(139\ \cancel{\text{kg SF}_6})\left(\frac{1\ \cancel{\text{kg-mol SF}_6}}{146.1\ \cancel{\text{kg SF}_6}}\right)\left(\frac{3\ \cancel{\text{kg-mol F}_2}}{1\ \cancel{\text{kg-mol SF}_6}}\right)\left(\frac{38.00\ \text{kg F}_2}{1\ \cancel{\text{kg-mol F}_2}}\right) = 108\ \text{kg F}_2$$

(d) sulfur hexafluoride

(e) It must not react with water or any of the solutes in seawater (such as Na^+ and Cl^-).

10.97 **(a)** H_2O, $CH_3CH_2CH_2OH$, and $CH_3(CH_2)_{14}COOCH_3$

Each of these substances is polar because of the group of atoms containing oxygen.

(b) Water has an unusually high specific heat.

(c)

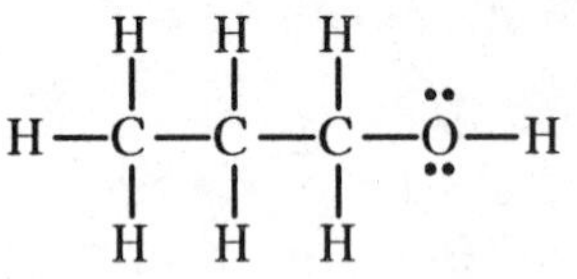

10.99 **(a)** 11–*trans*–retinal

See p. 354.

(b) vitamin A: 10 atoms; β–carotene: 22 atoms

(c) There are 22 atoms in vitamin A that must lie in a plane (see circled atoms in diagram).

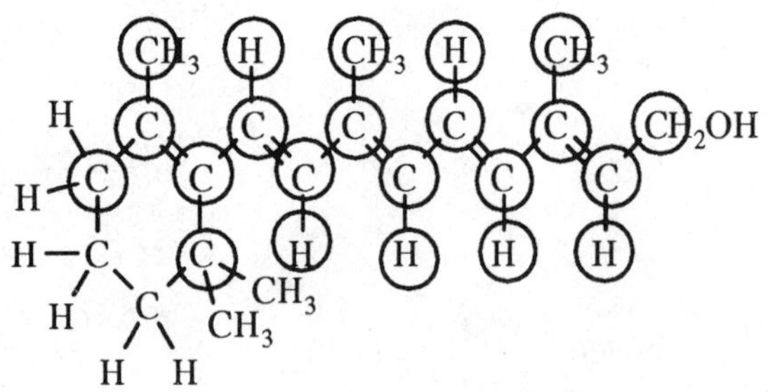

(d)

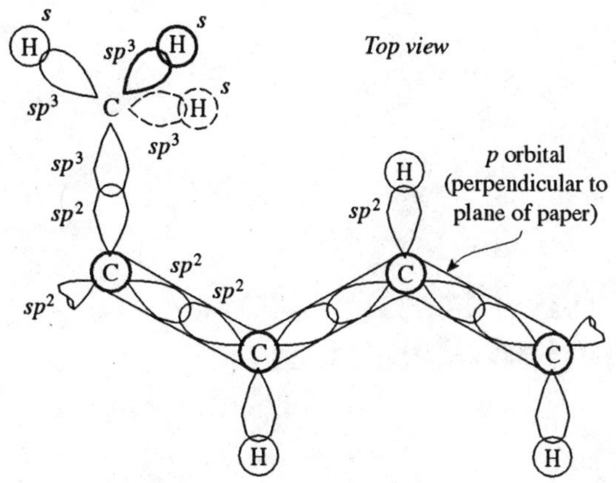

(e) 3.61×10^{-17} J

$$(550\ \cancel{\text{nm}})\left(\frac{10^{-9}\ \text{m}}{\cancel{\text{nm}}}\right) = 5.50 \times 10^{-7}\ \text{m}$$

$$E_{\text{photon}} = \frac{hc}{\lambda} = \frac{(6.626 \times 10^{-34}\ \text{J} \cdot \cancel{\text{s}})(2.998 \times 10^{8}\ \cancel{\text{m}} \cdot \cancel{\text{s}^{-1}})}{5.50 \times 10^{-7}\ \cancel{\text{m}}} = 3.61 \times 10^{-19}\ \text{J}$$

$$(100\ \cancel{\text{photons}})\left(\frac{3.61 \times 10^{-19}\ \text{J}}{\cancel{\text{photon}}}\right) = 3.61 \times 10^{-17}\ \text{J}$$

CHAPTER 11:
Oxidation–Reduction Reactions

Practice Problems

11.1 **(a)** 0

$$\overset{0}{N_2} \quad \text{(Rule 1)}$$
$$0 = 0$$

(b) N, +5; O, –2

$$\overset{+5}{N}\ \overset{-2}{O_3^-} \quad \text{(Rules 1 and 4)}$$
$$+5 - 6 = -1$$

(c) N, –3; H, +1

$$\overset{-3}{N}\ \overset{+1}{H_4^+} \quad \text{(Rules 1 and 3)}$$
$$-3 + 4 = +1$$

(d) Mg, +2; N, –3

$$\overset{+2}{Mg_3}\overset{-3}{N_2} \quad \text{(Rules 1 and 2)}$$
$$+6 - 6 = 0$$

(e) H, +1; N, –3; O, –2

$$\overset{+1}{H}\,\overset{+3}{N}\,\overset{-2}{O_2} \quad \text{(Rules 1, 3, and 4)}$$
$$1 + 3 - 4 = 0$$

(f) Fe, +2; S, +6; O, –2

The compound $FeSO_4$ contains two elements, Fe and S, that are not covered by the rules. However, you should recognize the group SO_4 as the sulfate ion, SO_4^{2-}. Because the charge on the sulfate ion is 2–, the charge on the iron ion must be 2+ in order for the compound to have zero charge. Thus, the oxidation number of iron is +2 by rule 1. The oxidation number of O is –2 by rule 4, and for the sulfate ion the total oxidation number for O is –8.

$$(S\overset{-2}{O_4})^{2-}$$
$$-8$$

For the sum of the oxidation numbers to equal –2, the charge on the sulfate ion, the oxidation number of S must be +6.

$$\overset{+6}{\text{(S}}\ \overset{-2}{\text{O}_4})^{2-}$$

$$+6 - 8 = -2$$

(g) Cr, +3; S, +6; O, –2

The compound $Cr_2(SO_4)_3$ also contains two elements, Cr and S, that are not covered by the rules. From part (f), the oxidation number of S is +6. Therefore, the oxidation number of Cr is +3.

$$\overset{+3}{\text{Cr}_2}\ (\overset{+6}{\text{S}}\ \overset{-2}{\text{O}_4})_3$$

$$+6 + 18 - 24 = 0$$

11.2 **(a)** P, +3; Cl, –1

3 Cl @ 7 valence electrons	=	21 valence electrons
1 P @ 5 valence electrons	=	5 valence electrons
Total	=	26 valence electrons

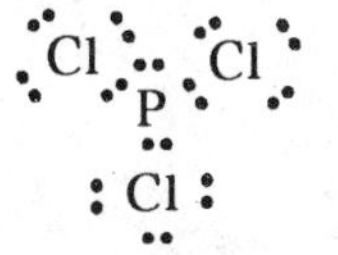

According to the definition of oxidation number, if all of the shared electrons were transferred to the more electronegative element, Cl, the charge on each chlorine would be 1– and the charge on phosphorus would be 3+.

(b) C, +2; N, –3

1 C @ 4 valence electrons	=	4 valence electrons
1 N @ 5 valence electrons	=	5 valence electrons
1 extra electron for 1– charge	=	1 valence electron
Total	=	10 valence electrons

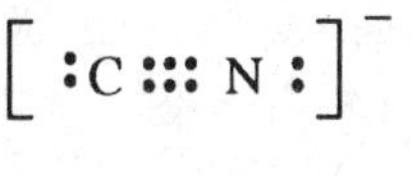

If all the shared electrons were transferred to the more electronegative element, nitrogen, the charge on nitrogen would be 3– and the charge on carbon would be 2+.

or

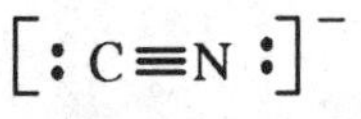

11.3 +5 and –3

The highest oxidation number observed for an element in its compounds is never higher than the group number. Phosphorus is in Group VA. Therefore, its highest oxidation number is +5, which corresponds to the transfer of all its valence electrons to other atoms. The most negative oxidation number for an element is equal to the group number minus eight, the number of electrons that must be gained to complete an octet. Therefore, for phosphorus the lowest oxidation number is (5 – 8) = –3.

11.4 HNO_3. Nitrogen is in a higher oxidation state in HNO_3 (+5) than in HNO_2 (+3), which indicates that HNO_3 is a stronger acid.

11.5 (a) chromium(VI) oxide

$$\overset{+6}{Cr}\overset{-2}{O_3}$$
$$+6 - 6 = 0$$

The oxidation number of Cr in CrO_3 is +6.

(b) calcium chlorite

The ion ClO_2^- is chlorite (Table 11.1). The oxidation number of calcium in compounds is always +2 (Rule 2). Therefore, only one calcium chlorite is known, and no Roman numeral is necessary in the name.

(c) hypochlorite ion

See Table 11.1

(d) potassium chlorate

The ion ClO_3^- is chlorate (Table 11.1). The oxidation number of potassium in compounds is always +1 (Rule 2). Therefore, only one potassium chlorate is known, and no Roman numeral is necessary in the name.

(e) periodic acid

By analogy to perchloric acid, $HClO_4$ (Table 11.1), HIO_4 is called periodic acid.

11.6 (a) $LiClO_2$

By analogy to sodium chlorite, $NaClO_2$ in Table 11.1, the formula for lithium chlorite is $LiClO_2$.

(b) $Ca(ClO)_2$

From Table 11.1, the formula for hypochlorite ion is ClO^-. Because calcium ion has a 2+ charge, the formula for calcium hypochlorite is $Ca(ClO)_2$.

(c) Na_2SO_3

The formula for sulfite ion is SO_3^{2-} (Table 1.2). Because sodium ion has a 1+ charge, the formula for sodium sulfite is Na_2SO_3.

(d) $MnCl_2$

The formula for manganese(II) ion is Mn^{2+} and for chloride ion is Cl^- (Table 1.3).

(e) Fe_2O_3

The formula for iron(III) ion is Fe^{3+} and for oxide ion is O^{2-} (Table 1.3). For iron(III) oxide to have no net charge, there must be three oxide ions for every two iron(III) ions.

11.7 **(a)** Zn(s) is oxidized, H^+ is reduced.

oxidized

oxidation number 0 +1 −1 +2 −1 0

$$Zn(s) + 2H^+ + 2Cl^- \rightarrow Zn^{2+} + 2Cl^- + H_2(g)$$

reduced

The oxidation number of Zn increases and the oxidation number of H decreases as a result of the reaction.

(b) Na(s) is oxidized, $Cl_2(g)$ is reduced.

oxidized

oxidation number 0 0 +1 −1

$$2Na(s) + Cl_2(g) \rightarrow 2NaCl(s)$$

reduced

(c) HgO is both oxidized and reduced. The O^{2-} ion in HgO is oxidized and the Hg^{2+} ion is reduced.

oxidized

oxidation number +2 −2 0 0

$$2HgO(s) \rightarrow 2Hg(l) + O_2(g)$$

reduced

11.8 **(a)** not redox

+2 +4 −2 +1 −1 +2 −1 +4 −2 +1 −2

$$CaCO_3(s) + 2HCl(aq) \rightarrow CaCl_2(aq) + CO_2(g) + H_2O(l)$$

This reaction is *not* redox since no species changed oxidation state.

(b) redox. Iron is oxidized; chlorine is reduced.

+2 −1 +1 −1 +1 +5 −2 +3 −1 +1 −1 +1 −2

$$6FeCl_2(aq) + 6HCl(aq) + NaClO_3(aq) \rightarrow 6FeCl_3(aq) + NaCl(aq) + 3H_2O(l)$$

This reaction *is* redox, with Fe being oxidized from +2 to +3 and Cl being reduced from +5 to −1. To write the complete ionic equation, write soluble strong electrolytes as ions. Write a molecular formula for the weak electrolyte water.

$$6Fe^{2+} + 12Cl^- + 6H^+ + 6Cl^- + Na^+ + ClO_3^- \rightarrow 6Fe^{3+} + 18Cl^- + Na^+ + Cl^- + 3H_2O(l)$$

Complete ionic equation:

$$6Fe^{2+} + 18Cl^- + 6H^+ + Na^+ + ClO_3^- \rightarrow 6Fe^{3+} + 19Cl^- + Na^+ + 3H_2O(l)$$

Net ionic equation:

$$6Fe^{2+} + ClO_3^- + 6H^+ \rightarrow 6Fe^{3+} + Cl^- + 3H_2O(l)$$

ClO_3^- is the oxidizing agent, and Fe^{2+} is the reducing agent.

(c) redox. Sulfur is oxidized; iodine is reduced.

This reaction can be immediately recognized as redox because the element I_2 is a reactant. S is oxidized from –2 in S^{2-} to +6 in SO_4^{2-}, and I is reduced from 0 in I_2 to –1 in I^-.

Complete ionic equation:

$2Na^+ + S^{2-} + 4I_2(s) + 4H_2O(l) \rightarrow 8H^+ + 8I^- + 2Na^+ + SO_4^{2-}$

Net ionic equation:

$S^{2-} + 4I_2(s) + 4H_2O(l) \rightarrow 8H^+ + 8I^- + SO_4^{2-}$

S^{2-} is the reducing agent, and $I_2(s)$ is the oxidizing agent.

11.9 **(a)** $2Sb(s) + 3Cl_2(g) \rightarrow 2SbCl_3(s)$

(b) $2HBr(aq) + H_2SO_4(aq) \rightarrow Br_2(l) + SO_2(g) + 2H_2O(l)$

11.10 **(a)**

0 0 +3 –1
$2Sb(s) + 3Cl_2(g) \rightarrow 2SbCl_3(s)$

The oxidation number of two Sbs changes from 0 to +3, an increase of 2(3) = 6; the oxidation number of six Cls changes from 0 to –1, a decrease of 6(1) = 6.

(b)

+1 –1 +1 +6 –2 0 +4 –2 +1 –2
$2HBr(aq) + H_2SO_4(aq) \rightarrow Br_2(l) + SO_2(g) + 2H_2O(l)$

The oxidation number of two Brs changes from –1 to 0, an increase of 2(1) = 2; the oxidation number of one S changes from +6 to +4, a decrease of 1(2) = 2.

11.11 **(a)** $Zn(s) \rightarrow Zn^{2+} + 2e^-$

(b) $2H^+ + 2e^- \rightarrow H_2(g)$

11.12 **(a)** $2I^- + Br_2(aq) \rightarrow I_2(s) + 2Br^-$

(b) $2I^- \rightarrow I_2(s) + 2e^-$

11.13 **(a)** Ca **(b)** F_2 **(c)** Fe^{2+}, SO_2, and Cu^+. Ca is in lowest oxidation state, F_2 is in highest, and Fe^{2+}, S in SO_2, and Cu^+ are in intermediate oxidation states.

Ca is in its lowest possible oxidation state and therefore can only be oxidized, not reduced. F_2 is the most electronegative element. Therefore, it can only be reduced to F^-. It cannot be in a positive oxidation state. Fe^{2+}, S in SO_2 with +4 oxidation state, and Cu^+ are in intermediate oxidation states and can be oxidized and reduced.

11.14 **(a)** $2HS_2O_4^- + H_2O(l) \rightarrow S_2O_3^{2-} + 2H_2SO_3(aq)$
or $2HS_2O_4^- + H_2O(l) \rightarrow S_2O_3^{2-} + 2H_2O(l) + 2SO_2(g)$
or $2HS_2O_4^- \rightarrow S_2O_3^{2-} + H_2O(l) + 2SO_2(g)$

oxidized (S: +3 → +4)

$$\overset{+1\ +3\ -2}{2HS_2O_4^-} + \overset{+1\ -2}{H_2O(l)} \rightarrow \overset{+2\ -2}{S_2O_3^{2-}} + \overset{+1\ +4\ -2}{2H_2SO_3(aq)}$$

reduced (S: +3 → +2)

(b) sulfur

Sulfur in $HS_2O_4^-$ is both oxidized (oxidation number increases from +3 to +4) and reduced (oxidation number decreases from +3 to +2).

11.15 Complete ionic equation: $3Br_2(aq) + 6Na^+ + 6OH^- \rightarrow 6Na^+ + BrO_3^- + 5Br^- + 3H_2O(l)$

Molecular equation: $3Br_2(aq) + 6NaOH(aq) \rightarrow NaBrO_3(aq) + 5NaBr(aq) + 3H_2O(l)$

11.16 0.446 M

Problem	25.00 mL, ?M	22.43 mL, 0.0995 M
Equation	$10FeSO_4(aq)$ +	$2KMnO_4(aq) + 8H_2SO_4(aq) \rightarrow 2MnSO_4(aq) + 5Fe_2(SO_4)_3(aq) + K_2SO_4(aq) + 8H_2O(l)$
Recipe, mol	10	2

Step 1: Calculation of moles $FeSO_4$ reacted:

$$(22.43\ \cancel{mL})\left(\frac{0.0995\ \cancel{mol\ KMnO_4}}{1000\ \cancel{mL\ soln}}\right)\left(\frac{10\ mol\ FeSO_4}{2\ \cancel{mol\ KMnO_4}}\right) = 0.0112\ mol\ FeSO_4$$

Step 2: Calculation of molarity of $FeSO_4$:

$$M = \frac{mol\ FeSO_4}{L\ soln} = \left(\frac{0.0112\ mol\ FeSO_4}{25.00\ \cancel{mL\ soln}}\right)\left(\frac{10^3\ \cancel{mL\ soln}}{L\ soln}\right) = 0.448\ M$$

(Best answer, without intermediate round off, is 0.446 M.)

11.17 (a), (c), and (d)

Substance (b) does not react with oxygen because it is an organic compound in which halogen atoms have been substituted for all of the hydrogens. Substance (e) will not react with oxygen because both Fe and S are already in their highest oxidation states (+3 for Fe and +6 for S).

11.18 **(a)** $2Ca(s) + O_2(g) \xrightarrow{heat} 2CaO(s)$

Calcium is in the same group as magnesium.

(b) $Si(s) + O_2(g) \xrightarrow{heat} SiO_2(s)$

Silicon is in the same group as carbon.

(c) $2C_6H_6(l) + 15O_2(g) \xrightarrow{heat} 12CO_2(g) + 6H_2O(l)$

Combustion of a compound containing carbon and hydrogen yields carbon dioxide and water.

(d) no reaction

(e) $4Li(s) + O_2(g) \xrightarrow{\text{heat}} 2Li_2O(s)$

See Section 1.12.

(f) $2CH_3CH=CH_2(g) + 9O_2(g) \xrightarrow{\text{heat}} 6CO_2(g) + 6H_2O(l)$

Combustion of a compound containing carbon and hydrogen yields carbon dioxide and water.

(g)

$$2CH_3CH=CH_2(g) \; + \; O_2(g) \xrightarrow{\text{catalyst}} 2CH_3CH - CH_2 \; (g)$$

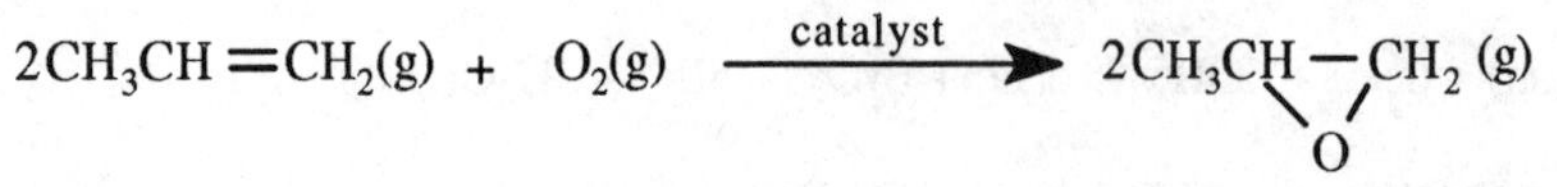

See Section 11.8. This is like the reaction of $CH_2=CH_2$.

11.19 Like gold, iridium is not very reactive.

11.20 **(a)** $MoO_3(s) + 3H_2(g) \xrightarrow{\text{heat}} Mo(s) + 3H_2O(g)$

Mo is in the same group as W.

(b) $PbO(s) + C(s) \xrightarrow{\text{heat}} Pb(s) + CO(g)$

Reduction of manganese(IV) oxide with carbon gives carbon monoxide.

Additional Practice Problems

11.49 **(a)** Al = +3, N = –3

+3 –3
Al N   (Rule 2)
+3 – 3 = 0

(b) Na = +1, O = –2, C = +4, N = –3

Sodium ion is Na^+; the anion must be OCN^-. The OCN^- ion contains two elements not covered by the rules. The definition of oxidation number must be used. The Lewis formula for OCN^- is

$$\left[\; \ddot{\overset{..}{O}} = C = \ddot{\overset{..}{N}} \; \right]^-$$

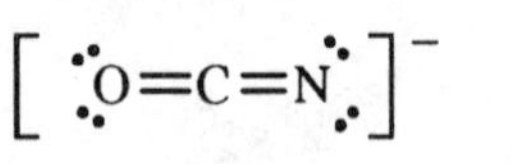

O and N are more electronegative than C. If all the shared pairs are transferred to O and N, the O would have 8 valence electrons, the C none, and the N, 8. Thus the oxidation number of O is –2, the oxidation number of C is +4, and the oxidation number of N is –3.

Check: The sum of the oxidation numbers gives the ionic charge: $-2 + 4 - 3 = -1$

(c) N = –3, H = +1, O = –2, Mo = +6

Since the anion must have a 2– charge, its formula is $Mo_2O_7^{2-}$ and

$$\begin{aligned} 2(\text{Mo}) + 7(-2) &= -2 \\ 2\text{Mo} &= +12 \\ \text{Mo} &= +6 \end{aligned}$$

(d) S = 0

(e) Ti = +4, O = –2

11.51 $MnSO_4 < Mn_2O_3 < MnO_2 < MnO_4^-$

(a) MnO_2 Mn = +4

(b) Mn_2O_3 Mn = +3

(c) MnO_4^- Mn = +7

(d) $MnSO_4$ Mn = +2

(d) $MnSO_4$ < (b) Mn_2O_3 < (a) MnO_2 < (c) MnO_4^-
(+2) < (+3) < (+4) < (+7)

11.53 Yes.

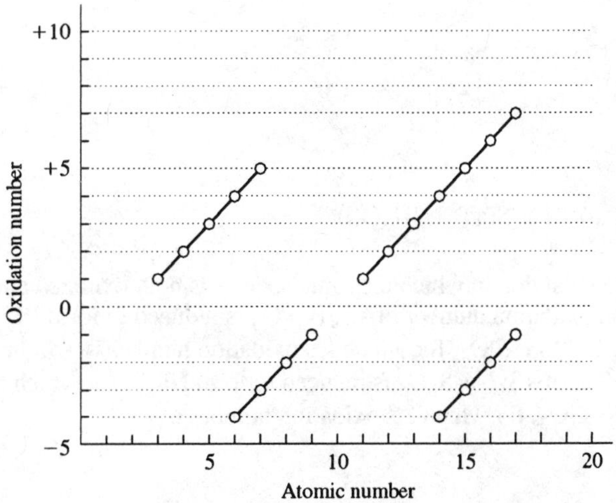

Oxidation number as a function of atomic number

11.55 **(a)** Limestone, $CaCO_3$, decomposes to form carbon dioxide gas and the basic oxide CaO. The CaO reacts with acidic oxide inpurities in the iron ore, such as SiO_2, to form a less dense liquid that floats on the surface of the iron.

(b) The incoming air reacts with carbon to form carbon monoxide and to liberate considerable amounts of thermal energy in the process. The carbon monoxide acts as a reducing agent for the iron oxides. Thus, the oxygen in the air blast generates the reducing agent from the coke and in so doing generates enough thermal energy for the rest of the purification process.

(c) The coke acts as the fuel and source of the reducing agent CO. The reaction of carbon and limited quantities of oxygen is highly exothermic; thus, the coke acts as a fuel by releasing thermal energy.

Stop and Test Yourself

1. **(d)** +4

For $NaHSO_3$,

$$\overset{+1}{Na}\overset{+1}{H}\ \overset{+4}{S}\ \overset{-2}{O_3}$$

$$+1 + 1 + 4 - 6 = 0$$

2. **(c)** +1

The Lewis formula for ICl is given below. If the shared pair were transferred to the more electronegative Cl, Cl would have eight valence electrons and I, six. Thus, the oxidation number of I is +1.

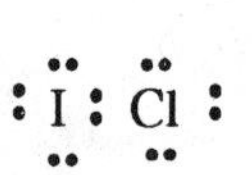

3. **(d)** +6

 has 6 valence electrons. A loss of all the valence electrons would result in an oxidation number of +6.

4. **(e)** $PF_3(g) + 3H_2O(l) \rightarrow H_3PO_3(aq) + 3HF(aq)$

It is the only reaction listed in which there are no changes in oxidation numbers.

$$\overset{+3\,-1}{PF_3}(g) + 3\overset{+1\,-2}{H_2O}(l) \rightarrow \overset{+1\,+3\,-2}{H_3PO_3}(aq) + 3\overset{+1\,-1}{HF}(aq)$$

5. **(a)** $4H_3PO_3(l) \rightarrow PH_3(g) + 3H_3PO_4(l)$

(c) $3Na_2S_2O_4(aq) + 6NaOH(aq) \rightarrow 5Na_2SO_3(aq) + Na_2S(aq) + 3H_2O(l)$

Disproportionation is an oxidation–reduction reaction in which the same species is both oxidized and reduced. In (a), the P atom in the reactant H_3PO_3 has an oxidation number of +3; H_3PO_3 is reduced to form PH_3, for which P's oxidation number is –3, and H_3PO_3 is oxidized to H_3PO_4, for which P's oxidation number is +5. In (c), the S atom in the reactant $Na_2S_2O_4$ has an oxidation number of +3; $Na_2S_2O_4$ is reduced to form Na_2S, for which S's oxidation number is –2 and $Na_2S_2O_4$ is oxidized to Na_2SO_3, for which S's oxidation number is +4.

6. **(b)** Cl_2

Cl, which has an oxidation number of zero in Cl_2, is able to be oxidized to ClO^-, ClO_2^-, ClO_3^-, and ClO_4^{4-}, in which Cl's oxidation numbers are +1, +3, +5, and +7, respectively, and is able to be reduced to Cl^-, in which the oxidation number of Cl is –1.

7. **(b)** Cl

Cl is able to have oxidation states of –1, 0, +1, +3, +5, and +7. None of the other choices has more than one oxidation state in compounds.

8. **(c)** manganese(IV) oxide

(a) is incorrect because the metal should be named first.
(b) is incorrect because the oxidation number of Mn in MnO_2 is +4, not +2.
(d) is incorrect because manganous oxide is a common name, not a systematic name.
(e) is incorrect because Roman numerals, not Greek prefixes, should be used in naming compounds of metals and nonmetals.

9. **(a)** $FeSO_4$

Iron(II) corresponds to Fe^{2+}; sulfate is SO_4^{2-}

10. **(d)** 4

$$2\overset{0}{C}(s) + 4\overset{+1}{H}\overset{+5}{N}\overset{-2}{O}_3(aq) \rightarrow 2\overset{+4}{C}\overset{-2}{O}_2(g) + 4\overset{+4}{N}\overset{-2}{O}_2(g) + 2\overset{+1}{H}_2\overset{-2}{O}(l)$$

11. **(c)** manganese

$$2\overset{+1}{K}\overset{+7}{Mn}\overset{-2}{O}_4((aq) + 5\overset{+1}{H}_2\overset{+3}{C}_2\overset{-2}{O}_4(aq) + 3\overset{+1}{H}_2\overset{+6}{S}\overset{-2}{O}_4(aq) \rightarrow 2\overset{+2}{Mn}\overset{+6}{S}\overset{-2}{O}_4(aq) + \overset{+1}{K}_2\overset{+6}{S}\overset{-2}{O}_4(aq) + 10\overset{+4}{C}\overset{-2}{O}_2(g) + 8\overset{+1}{H}_2\overset{-2}{O}(l)$$

The oxidation number of manganese decreases from +7 to +6. Therefore, manganese is reduced.

12. **(c)** $KMnO_4$

$$2\overset{+1}{K}\overset{+7}{Mn}\overset{-2}{O}_4((aq) + 5\overset{+1}{H}_2\overset{+3}{C}_2\overset{-2}{O}_4(aq) + 3\overset{+1}{H}_2\overset{+6}{S}\overset{-2}{O}_4(aq) \rightarrow 2\overset{+2}{Mn}\overset{+6}{S}\overset{-2}{O}_4(aq) + \overset{+1}{K}_2\overset{+6}{S}\overset{-2}{O}_4(aq) + 10\overset{+4}{C}\overset{-2}{O}_2(g) + 8\overset{+1}{H}_2\overset{-2}{O}(l)$$

The oxidizing agent gets reduced. Since $KMnO_4$ is the reacting species containing manganese, the element being reduced, it is the oxidizing agent.

13. **(a)** CH_3Cl

Most compounds containing carbon are easily oxidized. CCl_4 does not react because all the hydrogens that could have been bonded to carbon have been replaced with chlorine. $Mg(ClO_4)$ won't react because both Mg and Cl are in their highest oxidation states. Ne is a noble gas and Pt is unreactive.

14. **(c)** 0.0476

Problem	25.00 mL, ? M		24.12 mL, 0.0987 M		
Equation	Sn^{2+}	+	$2Ce^{4+}$	$\rightarrow$	$Sn^{4+} + 2Ce^{3+}$
Recipe, mol	1		2		

$$(24.12\,\cancel{mL})\left(\frac{0.0987\,\cancel{mol\,Ce^{4+}}}{1000\,\cancel{mL}}\right)\left(\frac{1\text{ mol }Sn^{2+}}{2\,\cancel{mol\,Ce^{4+}}}\right) = 0.001\,19\text{ mol }Sn^{2+}$$

$$M = \left(\frac{0.001\,19\text{ mol }Sn^{2+}}{25.00\,\cancel{mL}}\right)\left(\frac{1000\,\cancel{mL}}{1\text{ L}}\right) = 0.0476\text{ M}$$

or, combining steps,

$$\left(\frac{0.02412\,\cancel{L\,Ce^{4+}\,soln}}{0.02500\text{ L }Sn^{2+}\text{ soln}}\right)\left(\frac{0.0987\,\cancel{mol\,Ce^{4+}}}{1\,\cancel{L\,Ce^{4+}\,soln}}\right)\left(\frac{1\text{ mol }Sn^{2+}}{2\,\cancel{mol\,Ce^{4+}}}\right) = 0.0476\,\frac{\text{mol }Sn^{2+}}{\text{L }Sn^{2+}\text{ soln}}$$

15. **(a)** Addition of desirable impurities.

Desirable impurities are added in refining.

Putting Things Together

11.57 **(a)** $3SO_3^{2-} + 2MnO_4^- + H_2O(l) \rightarrow 3SO_4^{2-} + 2MnO_2(s) + 2OH^-$

Complete ionic: $6Na^+ + 3SO_3^{2-} + 2K^+ + 2MnO_4^- + H_2O(l) \rightarrow 6Na^+ + 3SO_4^{2-} + 2MnO_2(s) + 2K^+ + 2OH^-$
Net ionic: $3SO_3^{2-} + 2MnO_4^- + H_2O(l) \rightarrow 3SO_4^{2-} + 2MnO_2(s) + 2OH^-$

(b) Mn in MnO_4^- is reduced from +7 to +4 oxidation state.
S in SO_3^{2-} is oxidized from +4 to +6 oxidation state.

(c) MnO_4^- is the oxidizing agent; the substance reduced is always the oxidizing agent.
SO_3^{2-} is the reducing agent; the substance oxidized is always the reducing agent.

11.59 11.7 g CuS

The quantities of two reactants are given. This is a limiting reactant problem.

Problem	35.4 g		41.3 mL, 16.0 M
Equation	$3CuS(s)$	+	$8HNO_3(conc) \rightarrow 3CuSO_4(aq) + 8NO(g) + 4H_2O(l)$
Formula mass, u	95.612		
Recipe, mol	3		8

$$(35.4\ \cancel{g\ CuS})\left(\frac{1\text{ mol CuS}}{95.61\ \cancel{g\ CuS}}\right) = 0.370\text{ mol CuS} \qquad (41.3\ \cancel{mL})\left(\frac{16.0\text{ mol }HNO_3}{1000\ \cancel{mL}}\right) = 0.661\text{ mol }HNO_3$$

$$\frac{0.370\ \cancel{mol\ CuS}}{3\ \cancel{mol\ CuS}} = 0.123\text{ of a recipe} \qquad \frac{0.661\ \cancel{mol\ HNO_3}}{8\ \cancel{mol\ HNO_3}} = 0.0826\text{ of a recipe}$$

$0.0826 < 0.123$. HNO_3 is limiting.

$$(41.3\ \cancel{mL})\left(\frac{16.0\ \cancel{mol\ HNO_3}}{1000\ \cancel{mL}}\right)\left(\frac{3\ \cancel{mol\ CuS}}{8\ \cancel{mol\ HNO_3}}\right)\left(\frac{95.61\text{ g CuS}}{\cancel{mol\ CuS}}\right) = 23.7\text{ g CuS reacted.}$$

35.4 g CuS present – 23.7 g CuS reacted = 11.7 g CuS left

11.61 **(a)** Complete ionic equation:

$$14K^+ + 14MnO_4^- + 20K^+ + 20OH^- + C_3H_8O_3(aq) \rightarrow 28K^+ + 14MnO_4^{2-} + 3CO_3^{2-} + 6K^+ + 14H_2O(l)$$

(b) Molecular equation: $14KMnO_4(aq) + 20KOH(aq) + C_3H_8O_3(aq) \rightarrow 14K_2MnO_4(aq) + 3K_2CO_3(aq) + 14H_2O(l)$

11.63 **(a)** Cl_2O, dichlorine monoxide

(b) ClO_2, chlorine dioxide

(c) Cl_2O_6, dichlorine hexoxide

ClO_3 is the empirical formula.

Calculation of formula mass:

1 Cl @ 35.5 u	=	35.5 u
3 O @ 16.0 u	=	48.0 u
		83.5 u

$$\frac{\text{molecular mass}}{\text{formula mass}} = \frac{167}{83.5} = 2$$

Thus, the molecular formula is $2(ClO_3) = Cl_2O_6$

11.65 $HClO_4 > HClO_3 > HClO_2 > HClO$

Acidity increases with increasing oxidation number.

11.67 In compounds, +4 and +2. (The oxidation number of the free element would, of course, be 0).

Element 114 will be in Group IVA. From Figure 11.1, the highest oxidation number observed for Group IVA is +4. Since no negative oxidation states are listed for lead, the lowest oxidation number of element 114 in compounds is likely to be +2. The oxidation number of the free element would, of course, be 0.

11.69 **(a)** –1

$$\overset{+1\ -1}{H_2O_2}$$
$+2 - 2 = 0$

(b) Yes. As an oxidizing agent, H_2O_2 can be reduced to H_2O, in which the oxidation number of O is –2.

(c) Yes. As a reducing agent, H_2O_2 can be oxidized to O_2, in which the oxidation number of O is zero.

11.71 (b) and (c)

Polymer (a) will not burn because all hydrogens have been replaced by fluorines.

11.73 The sulfur is oxidized. Copper, which is in its highest oxidation state, and oxygen are reduced.

$$\overset{+2\ -2}{CuS}(s) + \overset{0}{O_2}(g) \rightarrow \overset{0}{Cu}(s) + \overset{+4\ -2}{SO_2}(g)$$

11.75 Two separate reactions are taking place simultaneously: $HNO_3 \rightarrow HNO_2$ and $HNO_3 \rightarrow NO$. Two separate equations must be written, not one combined equation.

Applications

11.77 **(a)** $Fe = \frac{+8}{3}$

$$\overset{+8/3\ -2}{Fe_3\ O_4}$$

+8 − 8 = 0

(b) S = +2

+1 +2 −2
$Na_2\ \underline{S}_2\ O_3$
+2 + 4 − 6 = 0

(c) Mo = +4 S = −2

$\underline{MoS}_2$

S is more electronegative than Mo (Figure 9.3). The only common negative oxidation number for S is −2 (Figure 11.1). Therefore, the oxidation number of S must be −2. If the oxidation number of S is −2, then the oxidation number of molybdenum must be +4.

+4 −2
$\underline{Mo}\ \underline{S}_2$
4 − 4 = 0

(d) Cl = +1

+1 −2
$\underline{Cl}_2\ O$
+2 − 2 = 0

(e) Sb = −3

+3 −3
$Al\ \underline{Sb}$
+3 − 3 = 0

11.79 **(a)** copper(I) iodide

I is more electronegative than Cu (Figure 9.3). The only common negative oxidation number for I is −1 (Figure 11.1). Therefore, the oxidation number of I must be −1, and the oxidation number of Cu, +1.

(b) iron(II) phosphate

The formula for phosphate ion is PO_4^{3-} (Table 1.2). For the compound $Fe_3(PO_4)_2$ to be electrically neutral, the charge on each iron ion must be 2+.

(c) iron(II) sulfate

The formula for sulfate ion is SO_4^{2-} (Table 1.2). For the compound $FeSO_4$ to be electrically neutral, the charge on each iron ion must be 2+.

(d) sulfuric acid

(e) potassium hydrogen sulfite

From Table 1.2, the ion SO_3^{2-} is called sulfite. Therefore, the ion HSO_3^- is called hydrogen sulfite. No Roman numeral is needed because the oxidation number of potassium is always +1 in compounds (Rule 2).

(f) potassium sulfate

(g) sodium sulfite

(h) sulfur dioxide

SO_2 is a compound of two nonmetals and should be named with a Greek prefix.

11.81 **(a)** $O_2(g) + 2H_2O(l) + 4e^- \rightarrow 4OH^-$

(b) $4Fe^{2+} + O_2(g) + 2H_2O(l) \rightarrow 4Fe^{3+} + 4OH^-$

$$\begin{array}{lllllll} & & 4(Fe^{2+} & \rightarrow & Fe^{3+} & + & e^-) \\ \underline{O_2(g)} & \underline{+\ 2H_2O(l)} & \underline{+\ 4e^-} & \underline{\rightarrow} & \underline{4OH^-} & & \\ 4Fe^{2+} & +\ O_2(g) & +\ 2H_2O(l) & \rightarrow & 4Fe^{3+} & + & 4OH^- \end{array}$$

11.83 61 ppm

Calculate the amount of MnO_4^- initially present:

$$(0.2500\ \cancel{\text{L soln}})\left(\frac{0.004\ 98\ \text{mol } MnO_4^-}{\cancel{\text{L soln}}}\right) = 1.25 \times 10^{-3}\ \text{mol } MnO_4^-$$

To calculate the amount of MnO_4^- that did not react with SO_2, set up a stoichiometry problem.

Problem			? mol		37.26 mL 0.0998 M				
Equation	$8H^+$	+	MnO_4^-	+	$5Fe^{2+}$	$\rightarrow$	Mn^{2+}	+ $5Fe^{3+}$ +	$4H_2O(l)$
Recipe, mol			1		5				

$$(0.037\ 26\ \cancel{\text{L soln}})\left(\frac{0.0998\ \cancel{\text{mol } Fe^{2+}}}{\cancel{\text{L soln}}}\right)\left(\frac{1\ \text{mol } MnO_4^-}{5\ \cancel{\text{mol } Fe^{2+}}}\right) = 7.43 \times 10^{-4}\ \text{mol } MnO_4^-\ \text{remaining}$$

The amount of MnO_4^- that reacted with the SO_2 is found by difference: $1.25 \times 10^{-3} - 7.43 \times 10^{-4} = 5.1 \times 10^{-4}$

Calculate the moles of SO_2 in the air by setting up another stoichiometry problem.

Problem	? mol		5.1×10^{-4} mol	
Equation	$5SO_2(g)$	+	$2MnO_4^-$ + $2H_2O(l)$	$\rightarrow 2Mn^{2+} + 5SO_4^{2-} + 4H^+$
Recipe, mol	5		2	

$$(5.1 \times 10^{-4}\ \cancel{\text{mol } MnO_4^-})\left(\frac{5\ \text{mol } SO_2}{2\ \cancel{\text{mol } MnO_4^-}}\right) = 1.3 \times 10^{-3}\ \text{mol } SO_2$$

Use the ideal gas law to calculate the volume of SO_2 (760 mmHg = 1.00 atm, 25 °C = 298 K)

$$V = \frac{nRT}{P} = \frac{(1.3 \times 10^{-3}\ \cancel{\text{mol}})(0.082\ 06\ \text{L} \cdot \cancel{\text{atm}} \cdot \cancel{\text{mol}^{-1}} \cdot \cancel{\text{K}^{-1}})(298\ \cancel{\text{K}})}{1.00\ \cancel{\text{atm}}} = 3.2 \times 10^{-2}\ \text{L } SO_2$$

$$\text{concentration } SO_2 = \left(\frac{3.2 \times 10^{-2}\ \text{L } SO_2}{5.00 \times 10^2\ \text{L air}}\right)(10^6) = 64\ \text{ppm}$$

(Best answer, without intermediate round off, is 61 ppm.)

11.85 **(a)** Molecular equation: $C_6H_{12}O_6(aq) + 6O_2(aq) \rightarrow 6CO_2(aq) + 6H_2O(l)$ $\Delta H = -2.87 \times 10^3$ kJ

(b) ΔH, the energy released or absorbed when a change takes place at constant pressure, is a thermodynamic state function whose value does not depend on the steps involved but only depends on the initial and final states.

(c) Net ionic equations for half-reactions of reduction:
$10e^- + 12H^+ + 2NO_3^- \rightarrow N_2(g) + 6H_2O(l)$
$8e^- + 8H^+ + SO_4^{2-} \rightarrow S^{2-} + 4H_2O(l)$

$NO_3^- + H^+ \rightarrow N_2 + H_2O$

Material balance: $2NO_3^- + 12H^+ \rightarrow N_2 + 6H_2O$
Charge balance: $2NO_3^- + 12H^+ + 10e^- \rightarrow N_2 + 6H_2O$

$SO_4^{2-} + H^+ \rightarrow S^{2-} + H_2O$

Material balance: $SO_4^{2-} + 8H^+ \rightarrow S^{2-} + 4H_2O$
Charge balance: $SO_4^{2-} + 8H^+ + 8e^- \rightarrow S^{2-} + 4H_2O$

11.87 (a)

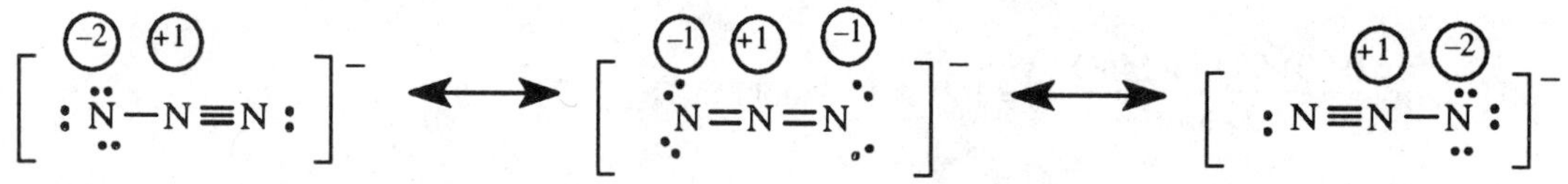

(b) $\frac{-1}{3}$

$\overset{-1/3}{\underset{-1}{N_3^-}}$

(c) No. No atoms undergo any change in oxidation number.

(d) Lead azide is probably insoluble in water.

(e) Complete ionic equation: $Pb^{2+} + 2NO_3^- + 2Na^+ + 2N_3^- \rightarrow Pb(N_3)_2(s) + 2Na^+ + 2NO_3^-$

Net ionic equation: $Pb^{2+} + 2N_3^- \rightarrow Pb(N_3)_2(s)$

(f) $Pb(N_3)_2(s) \rightarrow Pb(s) + 3N_2(g)$. Pb^{2+} is reduced. N is oxidized.

Pb^{2+} is reduced from +2 to zero.
N in N_3^- is oxidized from $\frac{-1}{3}$ to zero.

11.89 (a) All are redox.

(b) Step 3 is disproportionation.

(c) 15.6 lb HNO_3

The problem gives both reactants, so it is a limiting reactant problem. Oxygen is a reactant in the first two steps. Combine these steps to cancel the intermediate product NO.

$4NH_3(g) + 5O_2(g) \rightarrow 4NO(g) + 6\,H_2O(l)$

$\underline{4NO(g) + 2\,O_2(g) \rightarrow 4\,NO_2(g)}$
$4NH_3(g) + 7O_2(g) \rightarrow 4\,NO_2(g) + 6H_2O(l)$

23.1% of 90.0 = (0.231)(90.0) = 20.8 lb O_2

Problem	10.0 lb		20.8 lb		? lb		
Equation	$4NH_3(g)$	+	$7O_2(g)$	→	$4\,NO_2(g)$	+	$6H_2O(l)$
Formula mass, u	17.03		32.00		46.01		
Recipe, mol	4		7		4		

$$(10.0\ \cancel{\text{lb }NH_3})\left(\frac{1\text{ lb-mol }NH_3}{17.03\ \cancel{\text{lb }NH_3}}\right) = 0.587\text{ mol }NH_3 \qquad \left(\frac{0.587\ \cancel{\text{mol }NH_3}}{4\ \cancel{\text{mol }NH_3}/\text{recipe}}\right) = 0.147\text{ of a recipe}$$

$$(20.8\ \cancel{\text{lb }O_2})\left(\frac{1\text{ lb-mol }O_2}{32.00\ \cancel{\text{lb }O_2}}\right) = 0.650\text{ mol }O_2 \qquad \left(\frac{0.650\ \cancel{\text{mol }O_2}}{7\ \cancel{\text{mol }O_2}/\text{recipe}}\right) = 0.0929\text{ of a recipe}$$

Oxygen is the limiting reagent.

$$(20.8\ \cancel{\text{lb }O_2})\left(\frac{1\ \cancel{\text{lb-mol }O_2}}{32.00\ \cancel{\text{lb }O_2}}\right)\left(\frac{4\ \cancel{\text{lb-mol }NO_2}}{7\ \cancel{\text{lb-mol }O_2}}\right)\left(\frac{46.01\text{ lb }NO_2}{1\ \cancel{\text{lb-mol }NO_2}}\right) = 17.1\text{ lb }NO_2$$

Use equation 3 to calculate the theoretical yield of HNO_3.

Problem	17.1 lb				? lb		
Equation	$3NO_2(g)$	+	$H_2O(l)$	→	$2HNO_3(aq)$	+	$NO(g)$
Formula mass, u	46.01				63.01		
Recipe, mol	3				2		

$$(17.1\ \cancel{\text{lb }NO_2})\left(\frac{1\ \cancel{\text{lb-mol }NO_2}}{46.01\ \cancel{\text{lb }NO_2}}\right)\left(\frac{2\ \cancel{\text{lb-mol }HNO_3}}{3\ \cancel{\text{lb-mol }NO_2}}\right)\left(\frac{63.01\text{ lb }HNO_3}{1\ \cancel{\text{lb-mol }HNO_3}}\right) = 15.6\text{ lb }HNO_3$$

(d) 14.8 lb HNO_3

$$(15.6\text{ lb }HNO_3)\left(\frac{95}{100}\right) = 14.8\text{ lb }HNO_3$$

(e) Operating at pressures of about 100 atm means that the volumes of the gases will be smaller. Therefore, either the plant can be smaller or else the same size plant will yield a greater output.

$$(10\ \cancel{\text{MPa}})\left(\frac{10^6\ \cancel{\text{Pa}}}{\cancel{\text{MPa}}}\right)\left(\frac{1\text{ atm}}{1.01\times 10^5\ \cancel{\text{Pa}}}\right) = 99\text{ atm}$$

CHAPTER 12:
Liquids, Solids, and Changes in State

Practice Problems

12.1 (a) CH_3CH_2OH, (d) NH_3. Both have H bonded to small, electronegative atom (O, N).

Only these two molecules contain hydrogen atoms covalently bonded to the very electronegative atoms oxygen and nitrogen.

12.2 (a) CH_3CH_2OH, (b) CH_3OCH_3, (d) NH_3

All of these molecules are polar and contain small electronegative atoms with unshared electrons (O and N), whereas (c) CH_4 and (e) F_2 are nonpolar.

12.3

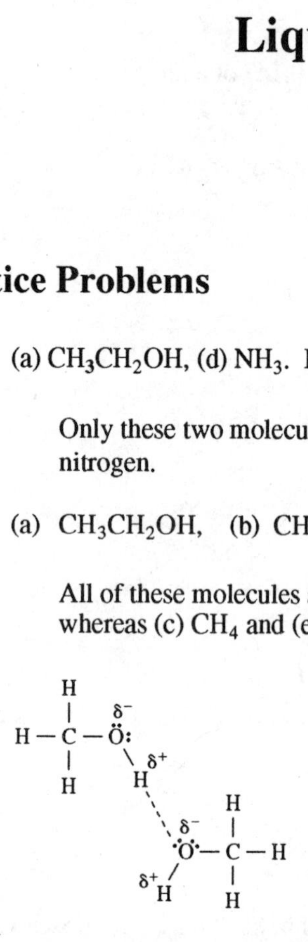

12.4 **(a)** NO. It is polar.

Because oxygen is more electronegative than N, NO is polar. The O_2 molecule is, of course, nonpolar because it is composed of two identical atoms.

(b) NH_3. It is hydrogen bonded.

NH_3 has three covalent bonds between hydrogen atoms and the very electronegative nitrogen atom. In addition, its geometry is unsymmetrical, giving it a relatively high dipole moment. These facts allow hydrogen bonding to take place between molecules. For CH_4, the electronegativity difference between carbon and hydrogen is low and the geometry of the molecule is symmetrical. Thus, no dipole–dipole interactions or hydrogen bonding are possible for CH_4.

12.5 **(a)** argon, atoms larger

Argon atoms contain more electrons and are larger than neon atoms, making argon atoms more polarizable and resulting in stronger London forces.

(b) Argon gas must only be cooled to 87.5 K; argon has stronger London forces.

Argon atoms are larger than neon atoms. Thus, argon has stronger London forces, allowing it to be liquefied at a higher temperature.

12.6 **(a)** $CH_3CH_2CH_2CH_2Cl$, molecules longer

$CH_3CH_2CH_2CH_2Cl$ is the larger of the two molecules, which causes it to have better contact with neighboring molecules. Thus, the London forces are greater.

(b) $CH_3CH_2CH_2CH_2Cl$ condenses at 78 °C because it has stronger London forces.

The stronger London forces in $CH_3CH_2CH_2CH_2Cl$ cause it to liquefy at a higher temperature than $CH_3CH_2CH_2Cl$ under the same pressure.

12.7 **(a)** $CH_3CH_2CH_2CH_2OH$, molecules linear [Molecules of $(CH_3)_3COH$ are spherical.]

$(CH_3)_3COH$ is more nearly spherical than $CH_3CH_2CH_2CH_2OH$, resulting in less opportunity for intermolecular interactions to occur. In effect, the "surface" of the molecule where all interactions occur is smaller for $(CH_3)_3COH$ than for $CH_3CH_2CH_2CH_2OH$. Also, the possible area of contact is smaller with a spherical geometry.

(b) $CH_3CH_2CH_2CH_2OH$ condenses at 117.2 °C because it has stronger London forces.

$CH_3CH_2CH_2CH_2OH$ has stronger London forces, which allows it to be liquefied at a higher temperature.

12.8 **(a)** HCl; Cl is more electronegative than H so HCl is polar

The electronegativity difference between hydrogen and chlorine in HCl causes a permanent dipole moment that results in dipole–dipole interactions between HCl molecules.

(b) HCl must be cooled to 188 K. HCl has dipole–dipole attractions in addition to London forces, which are similar for HCl and F_2.

The dipole–dipole attractions in HCl allow it to be liquefied at a higher temperature than F_2, which has no dipole–dipole attractions. London forces are about the same for HCl and F_2; both have 18 electrons and are about the same size.

12.9 **(a)** no change **(b)** no shift in equilibrium

If the volume of the liquid is decreased, the volume of the vapor will be increased. At first the concentration of molecules in the gas phase will be lower, and the rate of condensation will be reduced. However, the rate of vaporization will remain constant because the surface area of the liquid and temperature are the same. The concentration of molecules in the gas phase will eventually increase to the original value.

12.10 78 °C

The normal boiling point of ethyl alcohol is the temperature at which its vapor pressure is 760 mmHg. This occurs at 78 °C according to the figure.

12.11 83 °C

12.12 approximately 220 mmHg

12.13 $CH_3(CH_2)_6CH_3$; vapor pressure is lower

According to Figure 12.24, the vapor pressure of $CH_3(CH_2)_6CH_3$ is lower than the vapor pressure of CH_3CH_2OH. Therefore, the intermolecular attractions for $CH_3(CH_2)_6CH_3$ are stronger. While CH_3CH_2OH has the ability to form hydrogen bonds, it has far weaker London forces than $CH_3(CH_2)_6CH_3$ because molecules of CH_3CH_2OH are smaller than molecules of $CH_3(CH_2)_6CH_3$. The stronger London forces present more than make up for the lack of hydrogen bonding in $CH_3(CH_2)_6CH_3$.

12.14 decreases; solids are more orderly

The entropy of a sample is related to the disorder in the sample. When a sample changes from a state of relatively low order (as in a liquid) to a state of relatively high order (as in a crystalline solid), entropy decreases.

12.15 24.2 kJ

From Table 12.2, ΔH_{vap} for ethyl alcohol is 43.5 kJ/mol. Since the molar heat of vaporization is given in kJ/mol, we must convert the quantity of ethyl alcohol from grams to moles.

$$\text{kJ needed} = (25.64\ \cancel{\text{g } CH_3CH_2OH})\left(\frac{1\ \cancel{\text{mol } CH_3CH_2OH}}{46.07\ \cancel{\text{g } CH_3CH_2OH}}\right)\left(\frac{43.5\text{ kJ}}{1\ \cancel{\text{mol } CH_3CH_2OH}}\right) = 24.2\text{ kJ}$$

12.16 gas

The point (0 °C, 10 atm) is in the area of the phase diagram labeled "gas." Carbon dioxide is a gas under these conditions.

12.17 (a) The temperature of the solid CO_2 will increase until it reaches approximately –57 °C. Then the temperature will remain constant until all of the CO_2(s) has melted. The temperature will then again rise until it reaches approximately –35 °C. The temperature will once again remain constant until all of the CO_2(l) has vaporized. When all of the CO_2(l) has vaporized, the temperature will rise to 0 °C.

(b) Heating curve for CO_2 at 13 atm:

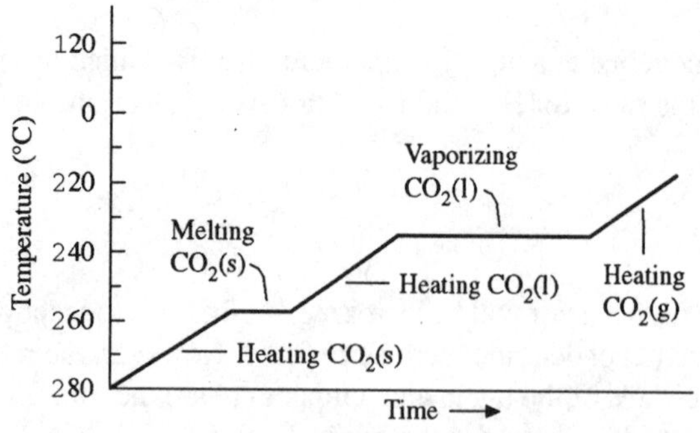

12.18 Phase diagram for O_2:

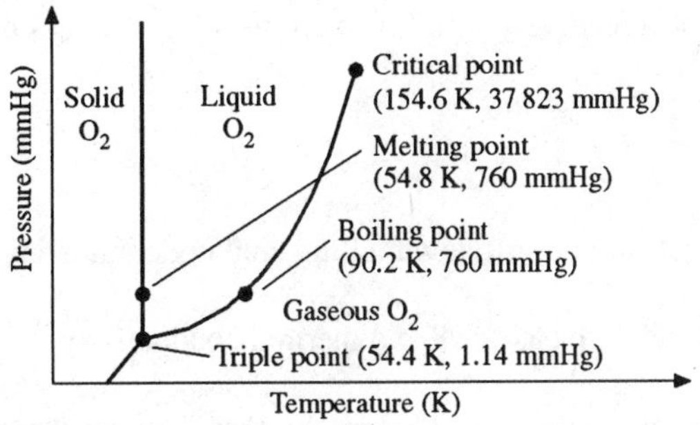

12.19 (a) molecular

The units of the crystals are P_4 molecules.

(b) molecular

Hydrogen chloride is a molecular compound because it is a compound of two nonmetals. The units of the crystals are HCl molecules.

(c) metallic

Manganese, Mn, is a metal.

(d) ionic

Potassium iodide is an ionic compound because it is a compound of a reactive metal with low ionization energy with a nonmetal that has a relatively high electron affinity. The units of the crystals are K^+ and I^- ions.

12.20 (a) London forces

The formula, P_4, shows that each phosphorus atom is bound to three other phosphorus atoms by covalent bonds. The intermolecular attractions between P_4 molecules involve only London forces because P_4 molecules are nonpolar.

(b) dipole–dipole attractions, London forces

HCl has a permanent dipole moment due to the electronegativity difference between hydrogen and chlorine. Thus, dipole–dipole attractions are important in the solid state. London forces are also present.

(c) metallic bonding

Mn is a metal that is pictured as consisting of positively charged Mn ions embedded in a "sea" of valence electrons. These electrons are "delocalized" throughout the metal. This is known as metallic bonding. The attraction between positively charged Mn ions and negatively charged electrons holds the units in place in the crystal.

(d) electrostatic attractions

Potassium iodide is formed from a Group IA metal with a relatively low ionization energy and a Group VIIA nonmetal with a relatively high electron affinity and is predominantly ionic. The electrostatic attraction between oppositely charged ions holds the units in place in the crystal.

12.21 **(a)** soft, low melting, nonconducting solid

P_4 is a molecular solid. Thus, it would be expected to be a relatively soft, low melting solid that does not conduct electricity.

(b) soft, low melting, nonconducting solid

HCl crystals are molecular solids, which would be expected to be soft, low melting, and nonconducting.

(c) shiny, hard, high melting, malleable, ductile, metallic solid with electrical and thermal conductivity

Mn is a transistion metal that is between Fe and Cr in the periodic table. One might, therefore, predict that Mn would have properties similar to Fe and Cr. Manganese should have relatively high electrical and thermal conductivity, a shiny luster, should be malleable and ductile, and should be hard and high melting.

(d) hard, brittle, high melting solid

KI is an ionic solid. Therefore, it should be hard and brittle with a high melting point. The melt should conduct electricity.

12.22 molecular

Since $SnCl_4$ is a liquid under normal conditions, one can conclude that it is a predominantly molecular substance.

12.23 two

The body–centered cubic unit cell contains one whole atom in the center and $\frac{1}{8}$ of each of the eight corner atoms.

$$\text{spheres} = 1 \text{ central sphere} + 8\left(\frac{1}{8}\text{ sphere}\right) = 2 \text{ spheres}$$

12.24 124 pm

Given: Ni unit cell edge = 352 pm

Since the unit cell is a cube, the edges are identical. That is, AB = AD = DC = BC. Also, for the face–centered cubic arrangement, the Ni atoms are in contact with one another across the diagonal (BD) of the faces as shown below:

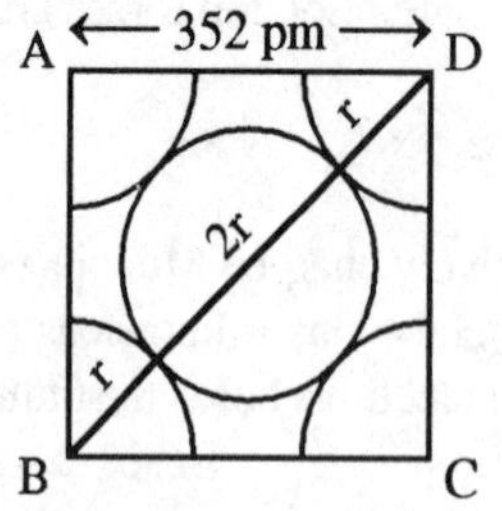

Knowing the length of the edges, we can calculate the diagonal of the face, BD, using the Pythagorean theorem as follows:

$$(BD)^2 = (352 \text{ pm})^2 + (352 \text{ pm})^2 \qquad BD = 498 \text{ pm}$$

From the figure it is obvious that BD is 4 times the radius, r, of the Ni atom. Therefore,

$$BD = 4r = 498 \text{ pm}$$
$$r = 498 \text{ pm} \div 4 = 125 \text{ pm}$$

(Best answer, without intermediate round off, is 124 pm)

12.25 **(a)** *n*-type semiconductor

Arsenic is in Group VA. Each arsenic atom contributes 5 electrons, one more than germanium. Thus, there are extra electrons to carry the current.

(b) *p*-type semiconductor

Gallium is in Group IIIA. If an atom of gallium is substituted for one of germanium, there are too few electrons, leaving electron-deficient holes in the crystal.

Additional Practice Problems

12.77 **(a)** Decrease. When a glass crystallizes, the arrangement of the units becomes ordered.

(b) Decreases. Thermal energy is released when a glass (or a liquid) crystallizes.

(c) Increase. A liquid that crystallizes rapidly when scratched with a glass stirring rod must be supercooled, that is, cooled below its normal melting point. Thermal energy is released when the liquid crystallizes, and the temperature rises.

12.79 1.18 kJ

From Table 12.2 the heat of vaporization of Br_2 is 29.5 kJ · mol^{-1}

$$(6.37\ \cancel{\text{g Br}_2})\left(\frac{1\ \cancel{\text{mol Br}_2}}{159.8\ \cancel{\text{g Br}_2}}\right)\left(\frac{29.5\ \text{kJ}}{\cancel{\text{mol Br}_2}}\right) = 1.18\ \text{kJ}$$

12.81 **(a)** Temperature must be raised until the vapor pressure of the water equals atmospheric pressure. To convert liquid to vapor at the boiling point, thermal energy is needed to overcome intermolecular attractions.

(b) Normally the partial pressure of H_2O in the atmosphere is less than the vapor pressure of H_2O at room temperature. Water molecules will escape from the liquid faster than water molecules will condense from the vapor. If the container is open, all of the water will evaporate eventually. Because evaporation of the water takes place slowly, enough thermal energy is supplied by the surroundings so that the temperature does not change.

(c) The vacuum pump reduces the atmospheric pressure to a pressure lower than the vapor pressure of the water. Thermal energy is required to vaporize the water, and this must be provided by the surroundings. If the pressure is very low, boiling will be rapid so that enough thermal energy is removed from the remaining water to cause it to freeze.

12.83 **(a)** 12.1 cm^3

$$\text{Vol. 1 mol Si} = \left(\frac{28.09\ \cancel{\text{g Si}}}{\text{mol Si}}\right)\left(\frac{1\ \text{cm}^3}{2.33\ \cancel{\text{g Si}}}\right) = \frac{12.1\ \text{cm}^3}{\text{mol Si}}$$

(b) 4.15 cm^3

$$\text{Vol. 1 mol Si atoms} = \text{Vol. 1 Si atom}\left(\frac{6.022 \times 10^{23}\ \text{atoms}}{\text{mol}}\right)$$

$$= \frac{4(3.142)(118\ \text{pm})^3}{3}\left(\frac{10^{-12}\ \text{m}}{\text{pm}}\right)^3\left(\frac{1\ \text{cm}}{10^{-2}\ \text{m}}\right)^3\left(\frac{6.022 \times 10^{23}}{\text{mol}}\right)$$

$$= \frac{4.15\ \text{cm}^3}{\text{mol}}$$

(c) 66%

empty space = total volume – occupied volume = 12.1 cm^3 – 4.15 cm^3 = 8.0 cm^3

$$\%\ \text{empty space} = \frac{8.0\ \text{cm}^3}{12.1\ \text{cm}^3} \times 100 = 66\%$$

12.85 (a) Vapor pressure of Cl_2 as a function of temperature:

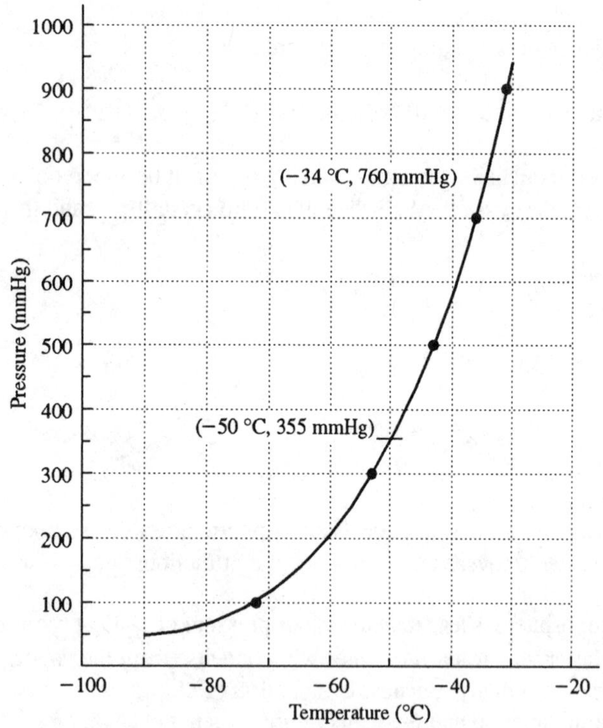

(b) Normal boiling point (vp = 760 mmHg) = –34 °C

(c) Vapor pressure at –50 °C = 355 mmHg

12.87 The quantity of ethene may be determined by gauge pressure; the quantity of ethane must be determined by weighing. The critical temperature of ethene, 9.9 °C, is below room temperature. The contents of the ethene cylinder will be gaseous no matter how high the pressure. Part of the contents of the ethane container will be liquid at pressures above the critical pressure. The gauge will give the vapor pressure of the liquid.

12.89 Phase diagram for water, with slope of ice-liquid water line exaggerated:

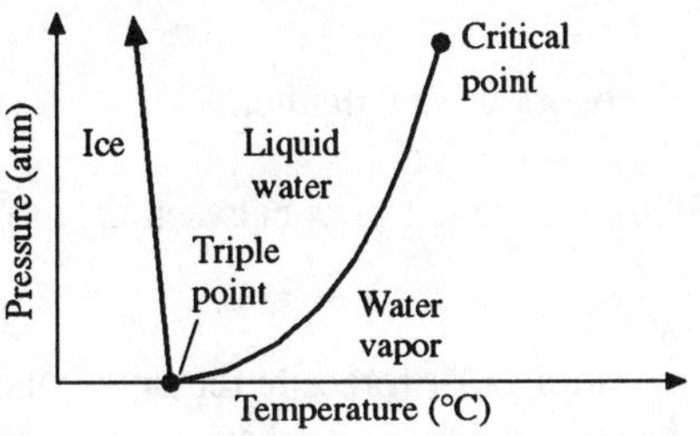

Stop and Test Yourself

1. (e) either an amorphous solid or a liquid.

The molecules touch each other but are arranged in a disorderly way. From Figure 12.2 (b), such an arrangement is found for liquids and amorphous solids.

2. (a) is the resistance of fluids to flow.

3. (d) permanent dipole forces.

Hydrogen chloride molecules are polar because chlorine is more electronegative than hydrogen. However, they do not hydrogen bond because chlorine is too large and not electronegative enough. London forces are about the same for HCl and Ar, as they have the same number of electrons in about the same volume.

4. (a) The normal boiling point of acetone is 59 °C.

The vapor pressure of acetone is 760 mmHg at 59 °C.

5. (b) Condensation rate decreases.

If some vapor escapes, fewer molecules will be available to return to the liquid phase and the condensation rate would decrease.

6. (d) Kr, boiling point –152.3 °C.

Kr has the highest boiling point.

7. (b) decrease temperature and increase pressure.

Since air is mostly O_2 and N_2, neither of which has strong intermolecular attractions, the molecules must be close together and moving slowly for liquid to form. This can be accomplished by increasing pressure and decreasing temperature.

8. (e) acetylene, C_2H_2, critical temperature 35.5 °C.

If we assume 20–25°C as room temperature, only one of these gases (C_2H_2) has a critical temperature above room temperature and, thus, is the only one that can be liquefied by pressure at room temperature.

9. (b) body–centered cubic

See Figure 12.44.

10. (c) covalent network

Only covalent network crystals are hard enough to scratch the covalent network crystal diamond.

11. (c) the attractive forces between the liquid and the surface are greater than the attractive forces between the molecules of the liquid.

If the attractive forces between liquid and surface are greater than the intermolecular forces in the liquid, the liquid molecules will be attracted to the surface more strongly than they will be attracted to each other.

12. (d) 948 J

MM $CH_3COOCH_2CH_3$ (EA) = 88.11

$$(2.57\ \cancel{g\ EA})\left(\frac{1\ \cancel{mol\ EA}}{88.11\ \cancel{g\ EA}}\right)\left(\frac{32.5\ kJ}{\cancel{mol\ EA}}\right) = 0.948\ kJ \text{ or } 948\ J$$

13. (b) methanol

For H–bonding between molecules of the same kind, there must be at least one H bonded to N, O, or F. Methanol is the only substance listed that has H bonded to N, O, or F.

14. (e) both solid and liquid

Point T is on the line that separates the liquid and solid areas of the phase diagram.

15. (b) W to V

In the graph of temperature against time, temperature is increasing with increasing time. Therefore the temperature–time diagram refers to passage from left to right on the phase diagram. This eliminates (a), (c), and (d). Only one phase change is shown, which eliminates (e). (b) involves a single phase change – a sublimation at constant pressure.

Putting Things Together

12.91 Because the sugar molecules have so many —OH groups, hydrogen bonding is strong and liquid sugars are very viscous. The molecules can't move into an orderly arrangement.

12.93 0.75 kJ (estimate)

$$(1.00\ \cancel{mL})\left(\frac{0.79\ \cancel{g}}{\cancel{mL}}\right)\left(\frac{1\ \cancel{mol\ CH_3CH_2OH}}{46.1\ \cancel{g}}\right)\left(\frac{43.5\ kJ}{\cancel{mol\ CH_3CH_2OH}}\right) = 0.75\ kJ$$

Note: Answer is an estimate because it assumes that ΔH is same at skin temperature (37 °C) as at bp (79 °C)

12.95 **(a)** 1.34 L

$$(1.00\ \cancel{\text{g H}_2\text{O}})\left(\frac{\text{mol H}_2\text{O}}{18.02\ \cancel{\text{g H}_2\text{O}}}\right) = 0.555\ \text{mol H}_2\text{O}$$

$$V = \frac{nRT}{P} = \frac{(0.0555\ \cancel{\text{mol}})\left(\frac{62.36\ \cancel{\text{mmHg}}\cdot \text{L}}{\cancel{\text{mol}}\cdot \cancel{\text{K}}}\right)(293.2\ \cancel{\text{K}})}{760.0\ \cancel{\text{mmHg}}} = 1.34\ \text{L}$$

(b) Some liquid will remain. Pressure will be the vapor pressure of water at 20.0 °C, 17.5 mmHg.

From Table 5.4, vapor pressure of water at 20.0 °C is 17.535 mmHg. Calculate the number of moles of water vapor in a 2.0 L container at this pressure.

$$n = \frac{PV}{RT} = \frac{(17.5\ \cancel{\text{mmHg}})(2.0\ \cancel{\text{L}})}{\left(\frac{62.36\ \cancel{\text{mmHg}}\cdot \cancel{\text{L}}}{\text{mol}\cdot \cancel{\text{K}}}\right)(293.2\ \cancel{\text{K}})} = 0.0019\ \text{mol}$$

0.0019 mol < 0.0555 mol. Therefore, all the water cannot evaporate. Some liquid will remain at equilibrium and the pressure will be equal to the vapor pressure of water.

(c) All water will evaporate and the pressure will be 5, 5.1, or 5.07 mmHg

$$P = \frac{(0.0555\ \cancel{\text{mol}})\left(\frac{62.36\ \text{mmHg}\cdot \cancel{\text{L}}}{\cancel{\text{mol}}\cdot \cancel{\text{K}}}\right)(293.2\ \cancel{\text{K}})}{200\ \cancel{\text{L}}} = 5.07\ \text{mmHg}$$

Depending on number of significant figures in 200 L, pressure will be 5.07 mmHg, 5.1 mmHg, or 5 mmHg.

12.97 22.1 °C

According to Table 5.4, pressure of water vapor over water at 22.0 °C is 19.827 mmHg, and at 22.2 °C it is 20.070 mmHg.

From vapor pressure curve for water (Figure 5.12), the "curve" for this small temperature range is a straight line. Thus, you can interpolate.

$$\begin{array}{r} 20.070\ \text{mmHg} \\ -19.827\ \text{mmHg} \\ \hline 0.243\ \text{mmHg} \end{array} \qquad \begin{array}{r} 20.00\ \text{mmHg} \\ -19.827\ \text{mmHg} \\ \hline 0.17\ \text{mmHg} \end{array}$$

$$\frac{0.17\ \cancel{\text{mmHg}}}{0.243\ \cancel{\text{mmHg}}} \times 0.2\ ^\circ\text{C} = 0.1\ ^\circ\text{C}$$

Vapor pressure is 20.00 mmHg at 22.1 °C. Boiling point is the temperature at which the vapor pressure is equal to the pressure on the liquid.

12.99 77.73%

Cubic close packing has one octahedral hole per sphere. The oxide ions form the cubic close packed structure, so there are 4 oxide ions per unit cell. There must be 4 iron ions per unit cell. Thus, the compound must be FeO. You could also come to this conclusion from the fact that the crystal structure is like that of NaCl.

$$\% \text{ Fe} = \frac{\text{atomic mass for Fe}}{\text{formula mass for FeO}} = \frac{55.85 \text{ g Fe}}{(55.85 + 16.00) \text{ g FeO}} \times 100 = 77.73 \% \text{ Fe}$$

12.101 **(a)** Surface tension of water at different temperatures:

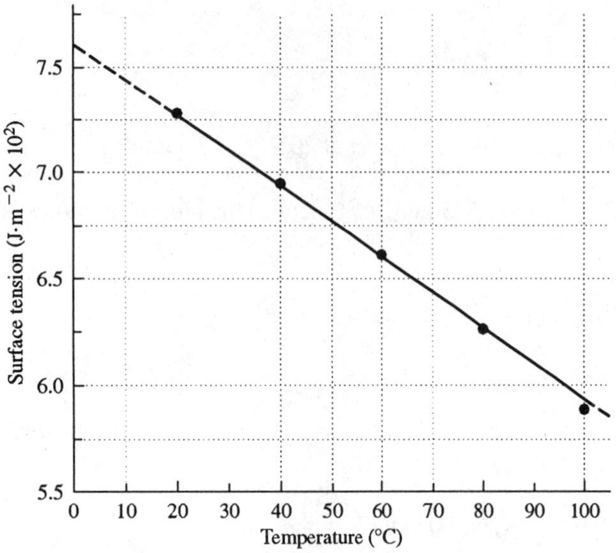

(b) This plot is approximately a straight line.

(c) surface tension $= \left(-1.74 \times 10^{-4} \frac{\text{J}}{\text{m}^2\ °\text{C}}\right) t_C + 0.0763 \frac{\text{J}}{\text{m}^2}$

surface tension at 0.0 °C = 7.63×10^{-2} J/m² from graph.

$$\text{The slope} = \frac{(5.89 \times 10^{-2} - 7.28 \times 10^{-2}) \text{ J/m}^2}{(100.0 - 20.0)\ °\text{C}} = -1.7 \times 10^{-4} \text{ J} \cdot \text{m}^{-2} \cdot °\text{C}^{-1}$$

Thus, the equation for the line is

$$\text{surface tension} = \left(-1.74 \times 10^{-4} \frac{\text{J}}{\text{m}^2 \cdot °\text{C}}\right) t_C + 0.0763 \frac{\text{J}}{\text{m}^2}$$

(See Appendix B.6 for directions for writing the equation for a straight line.)

12.103 5.32×10^{-3} g

According to Table 5.4, at 23.6 °C the vapor pressure of water is 21.845 mmHg. Use the ideal gas equation to calculate the number of moles in 250 mL at this temperature.

$$n = \frac{PV}{RT} = \frac{(21.845 \text{ mmHg})(0.250 \text{ L})}{\left(62.36 \frac{\text{mmHg} \cdot \text{L}}{\text{mol} \cdot \text{K}}\right)(296.8 \text{ K})} = 2.95 \times 10^{-4} \text{ mol}$$

$$(2.95 \times 10^{-4} \text{ mol } H_2O)\left(\frac{18.02 \text{ g } H_2O}{\text{mol } H_2O}\right) = 5.32 \times 10^{-3} \text{ g } H_2O$$

12.105 Heating curve for CH_3CH_2OH:

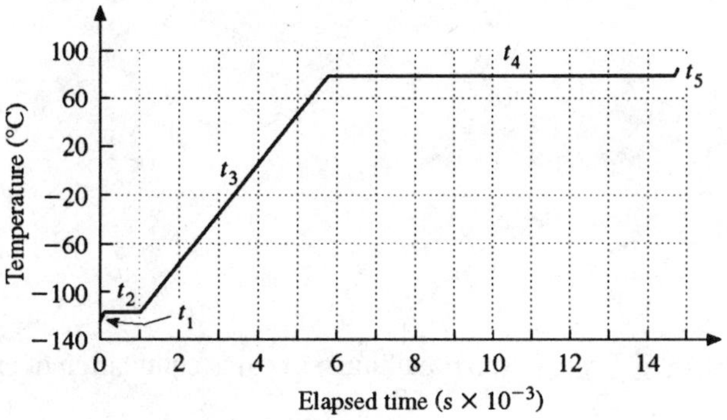

Sp.ht. $CH_3CH_2OH(s) = 0.97\ J \cdot g^{-1} \cdot {}^\circ C^{-1}$
Sp.ht. $CH_3CH_2OH(l) = 2.42\ J \cdot g^{-1} \cdot {}^\circ C^{-1}$ (Table 6.1)
Sp.ht. $CH_3CH_2OH(g) = 1.42\ J \cdot g^{-1} \cdot {}^\circ C^{-1}$

$\Delta H_{fus} = CH_3CH_2OH = 4.60 \times 10^3$ J/mol (Table12.2)
$\Delta H_{vap} = CH_3CH_2OH = 4.35 \times 10^4$ J/mol (Table 12.2)

mp = –117 °C   bp = 79 °C   4.72-g sample

Heat solid to mp:

$$(4.72\ \cancel{g})\left(\frac{0.97\ J}{\cancel{g} \cdot \cancel{^\circ C}}\right)[-117 - (-125)]\ \cancel{^\circ C} = 37\ J$$

$$(37\ \cancel{J})\left(\frac{1\ s}{0.50\ \cancel{J}}\right) = 7.4 \times 10^1\ s$$

Melt solid at mp:

$$(4.72\ \cancel{g})\left(\frac{1\ \cancel{mol}}{46.07\ \cancel{g}}\right)\left(\frac{4.60 \times 10^3\ J}{\cancel{mol}}\right) = 471\ J$$

$$(471\ \cancel{J})\left(\frac{1\ s}{0.50\ \cancel{J}}\right) = 9.4 \times 10^2\ s$$

Heat liquid to bp:

$$(4.72\ \cancel{g})\left(\frac{2.42\ J}{\cancel{g} \cdot \cancel{^\circ C}}\right)[79 - (-117)]\ \cancel{^\circ C} = 2.24 \times 10^3\ J$$

$$(2.24 \times 10^3\ \cancel{J})\left(\frac{1\ s}{0.50\ \cancel{J}}\right) = 4.5 \times 10^3\ s$$

Evaporate liquid at bp:

$$(4.72\ \cancel{g})\left(\frac{1\ \cancel{mol}}{46.07\ \cancel{g}}\right)\left(\frac{4.35 \times 10^4\ J}{\cancel{mol}}\right) = 4.46 \times 10^3\ J$$

$$(4.46 \times 10^3\ \cancel{J})\left(\frac{1\ s}{0.50\ \cancel{J}}\right) = 8.9 \times 10^3\ s$$

Heat vapor to 85 °C:

$$(4.72\ \text{g})\left(\frac{1.42\ \text{J}}{\text{g}\cdot{}^\circ\text{C}}\right)(85-79)\ {}^\circ\text{C} = 4.0\times10^{1}\ \text{J}$$

$$(4.0\times10^{1}\ \text{J})\left(\frac{1\ \text{s}}{0.50\ \text{J}}\right) = 8.0\times10^{1}\ \text{s}$$

Table of data for graph:

	Temperature, °C	Elapsed Time, s (no rounding to avoid accumulation of errors.)
Start	–125	0
Heat solid	–125 to –117	73 (t_1)
Melt	constant at –117	1016 (t_2)
Heat liquid	–117 to 79	5794 (t_3)
Evaporate	constant at 79	14 707 (t_4)
Heat vapor	79 to 85	14 787 (t_5)

12.107 (b), (c), (e)

(a) CO_2 is linear (see structures below) so that C═O dipoles cancel and molecule does not have a permanent dipole.

(b) SO_2 is bent so there is probably a small permanent dipole by the unshared pair on the sulfur. The effect of the electronegative O's and the unshared pair on S work in opposite directions.

(c) PCl_3 is trigonal pyramidal. A permanent dipole probably exists. The dipole effect resulting from the difference in electronegativity between P and Cl acts in the opposite direction to the effect of the unshared pair of electrons on the phosphorus.

(d) PCl_5 (in gas phase) is trigonal bipyramidal and should probably have no permanent dipole.

(e) ICl_3 is T–shaped. As a result of the lack of symmetry, this molecule is probably polar. It is unlikely that the dipoles of the $I^{\delta+}$—$Cl^{\delta-}$ bonds are exactly cancelled by the dipoles of the unshared electron pairs.

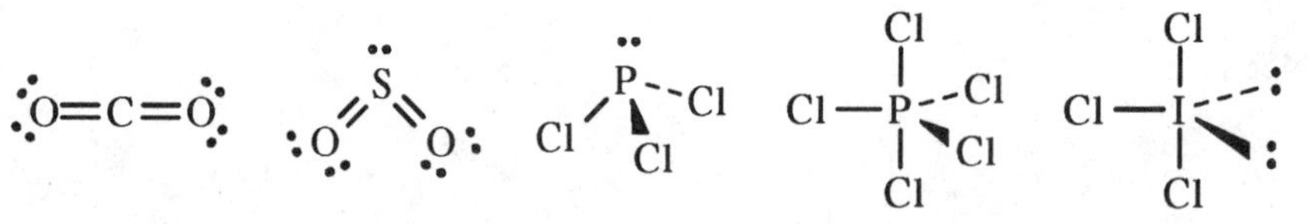

12.109 The height of the liquid will be higher on the moon than on Earth because the force of gravity pulling down on the liquid will be less. Adhesive forces between liquid and glass and surface tension will remain the same.

12.111 **(a)** (*i*) N_2, (*ii*) Cl_2, (*iii*) NH_3

(*i*) N_2 has a triple bond holding the atoms together. Triple bonds are much stronger than single bonds.

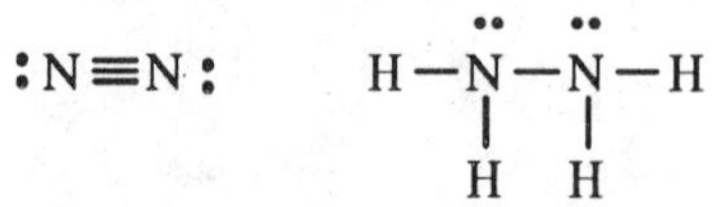

(*ii*) The Cl—Cl bond is shorter than the Br—Br bond because chlorine atoms are smaller. Shorter single bonds are stronger than longer single bonds.

(*iii*) The N—H bond is shorter and therefore stronger than the P—H bond because nitrogen atoms are smaller than phosphorus atoms. Table 9.4 lists the bond energy for N—H bonds as 389 kJ/mol and that for P—H bonds as 322 kJ/mol.

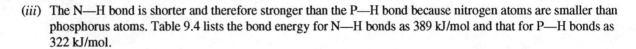

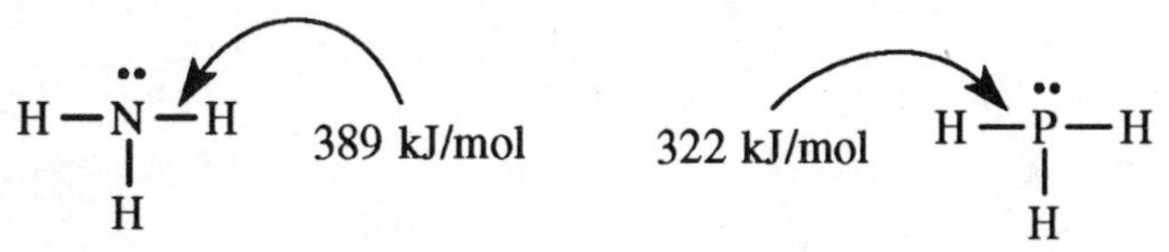

(b) (*i*) H_2NNH_2, (*ii*) Br_2, (*iii*) NH_3

(*i*) H_2NNH_2 has hydrogen bonding.

(*ii*) Br_2 is larger and more polarizable than Cl_2.

(*iii*) NH_3 has hydrogen bonding.

Applications

12.113 The surface tension of molten Pb is high enough that, while falling through air, the atoms move to the lowest energy shape, spherical. When the drops hit the water, cooling is so rapid that the drops stay round.

12.115 Wetting agents decrease the surface tension of water so that the water is attracted more to solid surfaces than normally and spreads out.

12.117 The surface tension of molten or softened glass is high, so that the surface area tends to decrease to a minimum, which causes the rounding.

12.119 **(a)** The liquid is under pressure in the container so that boiling is prevented.

(b) The critical temperature must be above room temperature or the butane in the lighters could not be liquid.

12.121 47 mmHg

Plot vapor pressure of water (in mmHg) against temperature (in °C) using data from table in Appendix C. Read vapor pressure at 37 °C from the graph.

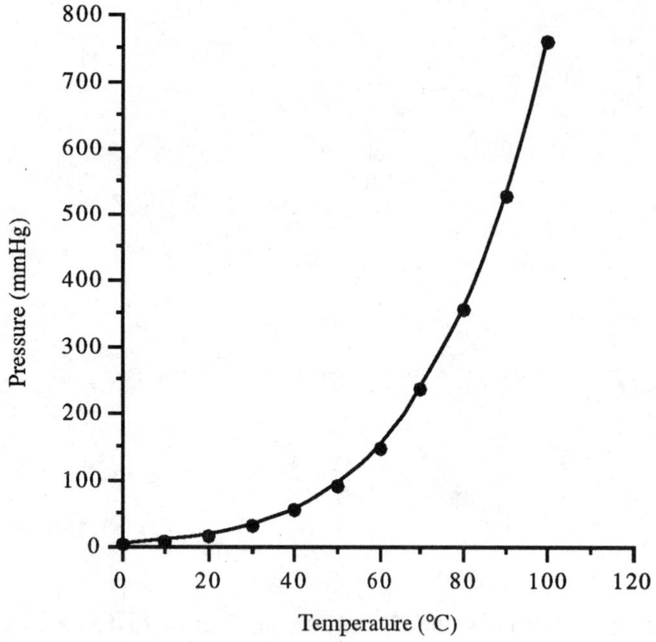

12.123 TiO_2

Ti: 8 at corners $\times \frac{1}{8} = 1$ O: 4 in faces $\times \frac{1}{2} = 2$

1 at center $\times$ 1 = 1 2 in interior $\times$ 1 = 2

2 Ti and 4 O

Ratio of Ti:O is 2:4, or 1:2

12.125 molecular crystals

The softness, low melting point, and failure to conduct electricity in the liquid state all indicate that crystals of OsO_4 are molecular.

12.127 **(a)** C_3N_4

Assume 100.00 g sample.

$$(39.13\ \text{g C})\left(\frac{1\ \text{mol C}}{12.011\ \text{g C}}\right) = 3.258\ \text{mol C}$$

$$(60.87\ \text{g N})\left(\frac{1\ \text{mol N}}{14.007\ \text{g N}}\right) = 4.345\ \text{mol N}$$

$$\frac{3.258}{3.258} = 1.000; \quad \frac{4.345}{3.258} = 1.334$$

$CN_{1.334}$ Multiply each subscript by 3 to get whole numbers.

C_3N_4 is the empirical formula.

(b) tricarbon tetranitride

(c) covalent network crystal

12.129 **(a)** Relative humidity = 70.9%

From Table 5.4, the vapor pressure of water at 22.4 °C is 20.316 mmHg.

$$\text{Relative humidity} = \left(\frac{14.4\ \cancel{\text{mmHg}}}{20.32\ \cancel{\text{mmHg}}}\right)100 = 70.9\%$$

(b) Assuming no further evaporation, the equilibrium vapor pressure will increase while partial pressure remains the same. Therefore, the relative humidity will decrease with increasing temperature.

12.131 **(a)** *o*–dichlorobenzene

Both *o*– and *m*–dichlorobenzene are liquids at room temperature and would make suitable solvents. However, the *m*– form is more expensive than the *o*– form, so the *o*–dichlorobenzene would be used.

(b) *p*–dichlorobenzene

Normally, moth repellents are solid. A gas would escape and a liquid might spot the clothes. Therefore, *p*–dichlorobenzene, the only solid at room temperature of this group of compounds, is the one sold as a moth repellent.

(c) The triple point of the moth repellent must be at a pressure greater than 1 atm, like the triple point of carbon dioxide (Figure 12.32).

12.133 183 g NH_3

First calculate how much thermal energy must be removed from the water in the ice cube tray to cool it to 0 °C and then freeze it.

From Table 6.1, the specific heat of $H_2O(l)$ is $4.18\ J \cdot g^{-1} \cdot {}^\circ C^{-1}$.

To cool 5.00×10^2 g water from 40.0 °C to 0.0 °C,

$$(5.00 \times 10^2\ \cancel{g\ H_2O})\left(\frac{4.18\ J}{\cancel{g\ H_2O} \cdot \cancel{^\circ C}}\right)(0.0 - 40.0)\ \cancel{^\circ C} = -8.36 \times 10^4\ J$$

must be removed from the water.

From Table 12.2, ΔH_{fus} for H_2O is 6.01 kJ/mol.

To freeze 5.00×10^2 g water at 0 °C,

$$(5.00 \times 10^2 \text{ g } H_2O)\left(\frac{1 \text{ mol } H_2O}{18.02 \text{ g } H_2O}\right)\left(\frac{6.01 \text{ kJ}}{\text{mol } H_2O}\right) = 167 \text{ kJ must be removed from the water.}$$

Thus a total of

$$(8.36 \times 10^4 \text{ J})\left(\frac{1 \text{ kJ}}{10^3 \text{ J}}\right) + 167 \text{ kJ} = 251 \text{ kJ must be removed from the water.}$$

Then calculate how many grams of ammonia would have to vaporize to absorb this much thermal energy.

$$(251 \text{ kJ})\left(\frac{1 \text{ mol } NH_3}{23.35 \text{ kJ}}\right)\left(\frac{17.03 \text{ g } NH_3}{\text{mol } NH_3}\right) = 183 \text{ g } NH_3$$

12.135 **(a)** nitrogen

Carbon has a radius of 77 pm (Figure 8.8). It should be doped with nitrogen, radius = 74 pm, to make an *n*-type semiconductor.

(b) boron

Carbon should be doped with boron, radius 80 pm, to make a *p*-type semiconductor.

12.137 The water in the squirrels' blood must supercool.

Blood is mostly water. If the water froze, the resulting expansion would burst the cell walls. Since the blood does not contain an "antifreeze" and it doesn't freeze, it must supercool.

CHAPTER 13:
Solutions Revisited

Practice Problems

13.1 **(a)** not miscible

Water is quite polar with the capability of hydrogen bonding to itself, whereas pentane is nonpolar.

(b) miscible

Heptane is nonpolar and similar in structure to pentane. It is predicted to be miscible.

(c) miscible

Although 1–propanol is polar and capable of hydrogen bonding to itself, the $CH_3CH_2CH_2$ part of the molecule is structually similar to pentane. CH_3CH_2OH is miscible with hexane, so $CH_3CH_2CH_2OH$ would be expected to be miscible with pentane.

(d) miscible

CCl_4 is nonpolar and would, thus, be expected to be miscible with pentane.

13.2 $C_{12}H_{22}O_{11}$ can form hydrogen bonds with water. When sugar crystals and water are mixed, fast–moving water molecules hit vigorously vibrating sugar molecules at the surface of the crystal and knock them out of the crystal. In solution, sugar molecules are quickly surrounded by spherical shells of water molecules.

13.3 A lump of $CuSO_4 \cdot 5H_2O$ could be dissolved more quickly by (a) breaking it into smaller pieces (increasing the surface area), (b) raising the temperature, and (c) stirring the mixture.

13.4 **(a)** Mg^{2+}, smaller

Mg^{2+} is smaller than Ca^{2+}. This will cause it to be more strongly hydrated.

(b) Ba^{2+}, higher charge

Ba^{2+} and K^+ are of similar size. The higher charge on Ba^{2+} will cause it to be more strongly hydrated.

(c) Na^+, smaller

Na^+ is smaller than Cl^- (Figure 8.9).

13.5 (a) KI(s), (d) $CH_3OH(l)$

(a) KI(s) – soluble

According to Table 4.4, all common potassium compounds are soluble.

(b) Ti(s) – insoluble

Ti(s) is a metal. Metals are insoluble in water.

(c) CCl_4 – insoluble

CCl_4 is nonpolar and is, thus, insoluble in water.

(d) CH_3OH – soluble

CH_3OH can hydrogen bond with water, is polar, and only has one carbon atom. Thus, it is soluble in water.

(e) C(diamond) – insoluble

C is a covalent network crystal. It is, therefore, insoluble in water.

13.6 $CaSO_4$

Ca^{2+} is smaller than Ba^{2+}. The charges are the same.

13.7 about 25 g $KClO_3$/100 g H_2O

13.8 unsaturated

The solubility of sucrose ($C_{12}H_{22}O_{11}$) is about 210 g/100 g H_2O at 25 °C.

13.9 about 25 g H_2O

The solubility of KNO_3 at 100 °C is about 247 g/100 g H_2O according to Figure 13.5.

$$(62\ \cancel{g\ KNO_3})\left(\frac{100\ g\ H_2O}{247\ \cancel{g\ KNO_3}}\right) \approx 25\ g\ H_2O$$

13.10 lower the temperature of the water.

This will allow more O_2 to dissolve, since the solvation process is exothermic.

13.11 1.4×10^{-3} g N_2/100 g H_2O

According to Henry's law, concentration of a gas is directly proportional to pressure at constant temperature. The final nitrogen pressure, 0.80 atm, is 80% of 1.00 atm. Thus, the solubility of N_2 at 0.80 atm is 80% the solubility at 1.00 atm:

$$\frac{80}{100}(0.0018 \text{ g } N_2/100 \text{ g } H_2O) = 1.4 \times 10^{-3} \text{ g } N_2/100 \text{ g } H_2O$$

Alternatively: Use Henry's law to calculate the value of k for nitrogen at 25 °C. Then use the value of k and Henry's law to calculate the concentration of nitrogen at 0.80 atm.

$$k = \frac{C_g}{P_g} = \frac{\dfrac{0.0018 \text{ g } N_2}{100 \text{ g } H_2O}}{1.00 \text{ atm}} = 0.0018 \frac{\text{g } N_2}{100 \text{ g } H_2O \cdot \text{atm}}$$

$$C_{\text{nitogen}} = 0.0018 \left(\frac{\text{g } N_2}{100 \text{ g } H_2O \cdot \cancel{\text{atm}}}\right) 0.80 \, \cancel{\text{atm}} = 0.0014 \frac{\text{g } N_2}{100 \text{ g } H_2O}$$

13.12 **(a)** 0.25 m

$$m = \frac{\text{moles solute}}{\text{mass solution in kg}} = \frac{0.53 \text{ mol NaCl}}{2.1 \text{ kg } H_2O} = 0.25\ m$$

(b) 1.03×10^{-1} m

Convert quantity of NaCl to moles:

$$(6.00 \, \cancel{\text{g NaCl}}) \left(\frac{1 \text{ mol NaCl}}{58.44 \, \cancel{\text{g NaCl}}}\right) = 0.103 \text{ mol NaCl}$$

Convert quantity of water to kilograms:

$$(996.5 \, \cancel{\text{g } H_2O}) \left(\frac{1 \text{ kg}}{10^3 \, \cancel{\text{g}}}\right) = 0.9965 \text{ kg } H_2O$$

Use the definition to calculate molality:

$$m = \frac{0.103 \text{ mol NaCl}}{0.9965 \text{ kg } H_2O} = 0.103\ m$$

or, combining steps,

$$(6.00 \, \cancel{\text{g NaCl}}) \left(\frac{1 \text{ mol NaCl}}{58.44 \, \cancel{\text{g NaCl}}}\right) \left(\frac{1}{996.5 \, \cancel{\text{g } H_2O}}\right) \left(\frac{10^3 \, \cancel{\text{g } H_2O}}{\text{kg } H_2O}\right) = 1.03 \times 10^{-1}\ m$$

13.13 methyl alcohol – 0.21, ethyl alcohol – 0.26, water – 0.53

$$X_{MA} = \frac{2.9 \, \cancel{\text{mol}}}{2.9 \, \cancel{\text{mol}} + 3.6 \, \cancel{\text{mol}} + 7.2 \, \cancel{\text{mol}}} = 0.21$$

$$X_{EA} = \frac{3.6 \, \cancel{\text{mol}}}{2.9 \, \cancel{\text{mol}} + 3.6 \, \cancel{\text{mol}} + 7.2 \, \cancel{\text{mol}}} = 0.26$$

$$X_{H_2O} = \frac{7.2\ \cancel{mol}}{2.9\ \cancel{mol} + 3.6\ \cancel{mol} + 7.2\ \cancel{mol}} = 0.53$$

Check: 0.21 + 0.26 + 0.53 = 1.00

13.14 NaCl, 1.86×10^{-3}; H_2O, 0.9982

$$(6.00\ \cancel{g\ NaCl})\left(\frac{1\ mol\ NaCl}{58.44\ \cancel{g\ NaCl}}\right) = 0.103\ mol\ NaCl$$

$$(996.5\ \cancel{g\ H_2O})\left(\frac{1\ mol\ H_2O}{18.015\ \cancel{g\ H_2O}}\right) = 55.32\ mol\ H_2O$$

$$X_{NaCl} = \frac{0.103\ mol}{0.103\ mol + 55.32\ mol} = \frac{0.103\ \cancel{mol}}{(55.32 + 0.103)\ \cancel{mol}} = 1.86 \times 10^{-3}$$

$$X_{H_2O} = \frac{55.32\ \cancel{mol}}{(55.32 + 0.103)\ \cancel{mol}} = 0.9982$$

Check: $(1.86 \times 10^{-3}) + 0.9982 = 1.0001$

13.15 $p_{N_2} = 717$ mmHg, $p_{H_2O} = 35.4$ mmHg

$$p_{N_2} = \left(\frac{4.36\ \cancel{mol}}{4.36\ \cancel{mol} + 0.215\ \cancel{mol}}\right)(752.8\ mmHg) = 717\ mmHg$$

$$p_{H_2O} = \left(\frac{0.215\ \cancel{mol}}{4.36\ \cancel{mol} + 0.215\ \cancel{mol}}\right)(752.8\ mmHg) = 35.4\ mmHg$$

Check: $P_{total} = p_{N_2} + p_{H_2O} = 717\ mmHg + 35.4\ mmHg = 752\ mmHg$

13.16 98.7 *m*

First find the mole fraction of water in the mixture.

$$X_{H_2O} = 1 - 0.640 = 0.360$$

Then consider a one–mole sample. It will consist of 0.640 mol C_2H_5OH and 0.360 mol H_2O. From the wording of the problem ". . . in an *aqueous* solution of CH_3CH_2OH . . .," water is to be considered the solvent. Find the mass of water in kilograms.

$$(0.360\ \cancel{mol\ H_2O})\left(\frac{18.02\ \cancel{g\ H_2O}}{\cancel{mol\ H_2O}}\right)\left(\frac{1\ kg\ H_2O}{10^3\ \cancel{g\ H_2O}}\right) = 0.006\ 49\ kg\ H_2O$$

$$m = \frac{n_{C_2H_5OH}}{kg\ H_2O} = \frac{0.640\ mol\ C_2H_5OH}{0.006\ 49\ kg\ H_2O} = 98.6\ molal$$

(Best answer, without intermediate round off, is 98.7 molal.)

13.17 14.9 M

From Practice Problem 13.16 we know that a one–mole sample of this solution contains 0.640 mol CH_3CH_2OH and 0.360 mol H_2O.

The mass of the one–mole sample is

$$(0.640\ \cancel{\text{mol CH}_3\text{CH}_2\text{OH}})\left(\frac{46.07\ \text{g CH}_3\text{CH}_2\text{OH}}{\cancel{\text{mol CH}_3\text{CH}_2\text{OH}}}\right) + (0.360\ \cancel{\text{mol H}_2\text{O}})\left(\frac{18.02\ \text{g H}_2\text{O}}{\cancel{\text{mol H}_2\text{O}}}\right) = 36.0\ \text{g}$$

The volume of the one–mole sample is

$$(36.0\ \cancel{\text{g soln}})\left(\frac{1\ \cancel{\text{cm}^3\ \text{soln}}}{0.8387\ \cancel{\text{g soln}}}\right)\left(\frac{1\ \text{L soln}}{10^3\ \cancel{\text{cm}^3\ \text{soln}}}\right) = 0.0429\ \text{L soln}$$

$$\text{M} = \frac{0.640\ \text{mol C}_2\text{H}_5\text{OH}}{0.0429\ \text{L soln}} = 14.9\ \text{M}$$

13.18 1.09×10^{-1}

Assume 1 kg H_2O used to make solution. Then 6.80 mol sucrose were present.

$$(1\ \cancel{\text{kg H}_2\text{O}})\left(\frac{10^3\ \cancel{\text{g H}_2\text{O}}}{\cancel{\text{kg H}_2\text{O}}}\right)\left(\frac{1\ \text{mol H}_2\text{O}}{18.015\ \cancel{\text{g H}_2\text{O}}}\right) = 55.509\ \text{mol H}_2\text{O}$$

$$x_{\text{sucrose}} = \frac{6.80\ \text{mol sucrose}}{6.80\ \text{mol sucrose} + 55.509\ \text{mol water}} = 1.09 \times 10^{-1}$$

13.19 **(a)** 335.5 mmHg

From the problem,

$X_{\text{hexane}} = 0.2674$; $\quad VP_{\text{pentane}} = 420.8$ mmHg; $\quad VP_{\text{hexane}} = 101.9$ mmHg

$X_{\text{pentane}} = 1 - X_{\text{hexane}} = 1.0000 - 0.2674 = 0.7326$

$VP_{\text{pentane in soln}} = X_{\text{pentane}}\ VP_{\text{pentane}} = (0.7326)(420.8\ \text{mmHg}) = 308.3\ \text{mmHg}$

$VP_{\text{hexane in soln}} = X_{\text{hexane}}\ VP_{\text{hexane}} = (0.2674)(101.9\ \text{mmHg}) = 27.25\ \text{mmHg}$

$VP_{\text{solution}} = (308.3\ \text{mmHg} + 27.25\ \text{mmHg}) = 335.6\ \text{mmHg}$

(Best answer, without intermediate round off, is 335.5 mmHg.)

(b) It increases. Compare the mixture in this problem with the mixture in Sample Problem 13.6.

Comparing the pentane/hexane mixture in Sample Problem 13.6 with the pentane/hexane mixture in this problem, the more volatile component, pentane, increased in mole fraction from 0.5443 to 0.7326. This resulted in an increase in the vapor pressure of the mixture from 275.4 mmHg to 335.5 mmHg. By definition, the more volatile component has a higher vapor pressure in its pure state than the less volatile component. Thus, in a "mixture" containing 1 mole fraction of the more volatile component, the vapor pressure would be higher than that containing any mole fraction of the more volatile component less than 1.

13.20 **(a)** 23.114 mmHg

From the problem,

$X_{\text{urea}} = 0.015\,425 \quad X_{H_2O} = 1.000\,000 - 0.015\,425 = 0.984\,575$

$VP_{H_2O \text{ at } 24.8\ °C} = 23.476$ mmHg according to Table 5.4

$VP_{soln} = (0.984\ 575)(23.476 \text{ mmHg}) = 23.114 \text{ mmHg}$

(b) 0.362 mmHg

Vapor pressure lowering = (23.476 mmHg – 23.114 mmHg) = 0.362 mmHg

13.21 **(a)** 79.9 °C

$$\Delta bp = K_{bp}m = \left(\frac{1.22\ °C}{\cancel{m}}\right)(1.344\ \cancel{m}) = 1.64\ °C$$

$bp_{soln} = (78.3\ °C + 1.64\ °C) = 79.9\ °C$

(b) –120.0 °C

$$\Delta fp = K_{fp}m = \left(\frac{-1.99\ °C}{\cancel{m}}\right)(1.344\ \cancel{m}) = -2.67\ °C$$

$fp_{soln} = (-117.3\ °C - 2.67\ °C) = -120.0\ °C$

13.22 1.9 *m*

$$\Delta fp = K_{fp}m \quad m = \frac{\Delta fp}{K_{fp}} = \frac{[-121.0\ \cancel{°C} - (-117.3\ \cancel{°C})]}{[-1.99\ \cancel{°C} \cdot m^{-1}]} = 1.9\ m$$

13.23 **(a)** 0.1 *m*

$C_{12}H_{22}O_{11}$ is a nonelectrolyte.

(b) 0.2 *m*

NaCl is a strong electrolyte; each formula unit yields two particles, a Na^+ ion and a Cl^- ion, in aqueous solution.

(c) 0.3 *m*

$CaCl_2$ is a strong electrolyte; each formula unit yields three particles, a Ca^{2+} ion and 2 Cl^- ions, in aqueous solution.

(d) 0.1 *m*

HCl is a strong electrolyte; each formula unit yields two particles, an H^+ ion and a Cl^- ion, in aqueous solution.

13.24 **(a)** $fp_{0.1\ m\ CaCl_2} < fp_{0.1\ m\ NaCl} < fp_{0.1\ m\ C_{12}H_{22}O_{11}}$

0.1 *m* $CaCl_2$ yields 0.3 *m* particles in aqueous solution and results in the largest freezing point depression. Therefore, this solution has the lowest freezing point. 0.1 *m* $C_{12}H_{22}O_{11}$ yields only 0.1 *m* particles in aqueous solution. This solution has the highest freezing point.

(b) $fp_{0.1\ m\ HCl} < fp_{0.1\ m\ HC_2H_3O_2}$ (weak acid) $< fp_{0.05\ m\ HCl}$

HCl is a strong acid. 0.05 *m* HCl → 0.1 *m* particles. If $HC_2H_3O_2$ were a nonelectrolyte, 0.1 *m* $HC_2H_3O_2$ would yield 0.1 *m* particles. However, $HC_2H_3O_2$ is a weak acid and yields more than 0.1 *m* particles (but less than 0.2 *m* particles).

13.25 –0.0558 °C

Total number of particles per mole in K_2SO_4 = 3.
Thus, ionic molality in aqueous solution = 3(0.0100) = 0.0300 molal

$$\Delta fp = K_{fp}m = \left(\frac{-1.86\ °C}{\cancel{m}}\right)(0.0300\ \cancel{m}) = -0.0558\ °C$$

$$fp_{soln} = [0.000\ °C + (-0.0558\ °C)] = -0.0558\ °C$$

13.26 9.69 mmHg

First, calculate molarity, M,

$$M = \frac{\text{mol solute}}{\text{vol soln L}} = (0.887\ \cancel{\text{g thrombin}})\left(\frac{1\ \text{mol thrombin}}{33\ 580\ \cancel{\text{g thrombin}}}\right)\left(\frac{1}{50.0\ \cancel{\text{mL soln}}}\right)\left(\frac{10^3\ \cancel{\text{mL soln}}}{\text{L soln}}\right) = 5.28 \times 10^{-4}\ M$$

and convert temperature to K:

$$T = (273.2\ \cancel{°C} + 20.8\ \cancel{°C})\left(\frac{1\ K}{1\ \cancel{°C}}\right) = 294.0\ K$$

Then use the osmotic pressure law to calculate the osmotic pressure.

$$\pi = MRT = \left(\frac{5.28 \times 10^{-4}\ \cancel{\text{mol}}}{\cancel{\text{L soln}}}\right)\left(\frac{62.36\ \text{mmHg} \cdot \cancel{L}}{\cancel{\text{mol}} \cdot \cancel{K}}\right)(294.0\ \cancel{K}) = 9.68\ \text{mmHg}$$

(Best answer, without intermediate round off, is 9.69.)

13.27 3.45×10^4 u

$$T = (19.8\ \cancel{°C} + 273.2\ \cancel{°C})\left(\frac{1\ K}{1\ \cancel{°C}}\right) = 293.0\ K$$

Solving the osmotic pressure law for M gives

$$M = \frac{\pi}{RT} = \frac{(7.46\ \cancel{\text{mmHg}})}{\left(62.36\ \frac{\cancel{\text{mmHg}} \cdot L}{\text{mol} \cdot \cancel{K}}\right)(293.0\ \cancel{K})} = 4.08 \times 10^{-4}\ M$$

Use M and the concentration given, 0.3521 g pepsin/25.0 mL solution, to find g/mol.

$$\left(\frac{0.3521\ \text{g pepsin}}{25.0\ \cancel{\text{mL soln}}}\right)\left(\frac{10^3\ \cancel{\text{mL soln}}}{\cancel{\text{L soln}}}\right)\left(\frac{1\ \cancel{\text{L soln}}}{4.08 \times 10^{-4}\ \text{mol pepsin}}\right) = 3.45 \times 10^4\ \text{g/mol}$$

The molecular mass of pepsin is 3.45×10^4 u.

Additional Practice Problems

13.77 32 g pentane, 78 g hexane, 115 g heptane

mole fraction heptane = 1 – (0.18 + 0.36) = 0.46

$$(2.50\ \cancel{\text{mol}})(0.18)\left(\frac{72.2\ \text{g pentane}}{1\ \cancel{\text{mol pentane}}}\right) = 32\ \text{g pentane}$$

$$(2.50\ \cancel{\text{mol}})(0.36)\left(\frac{86.2\ \text{g hexane}}{1\ \cancel{\text{mol hexane}}}\right) = 78\ \text{g hexane}$$

$$(2.50\ \cancel{\text{mol}})(0.46)\left(\frac{100\ \text{g heptane}}{1\ \cancel{\text{mol heptane}}}\right) = 115\ \text{g heptane} \quad \text{(rounded to nearest gram)}$$

13.79 **(a)** For dilute solutions (0.1 M and lower), molality and molarity are equal for all practical purposes.

(b) No. By definition, $m = \frac{\text{mol solute}}{\text{kg solvent}}$ and $\text{M} = \frac{\text{mol solute}}{\text{L solution}}$, and m only equals M if kg solvent = L solution. This is true for dilute aqueous solutions because the density of water = 1.00 g/mL under ordinary conditions. However, few other common solvents have densities of 1.00 g/mL.

13.81 **(a)** A lyophobic colloid is a solvent–hating colloid.
A lyophilic colloid is a solvent–loving colloid.

(b) A hydrophobic colloid is a water–hating colloid.
A hydrophilic colloid is a water–loving colloid.

(c) A sol is a solid colloid suspended in a liquid medium.
A gel is a semisolid sol.

(d) A foam is a gas suspended in a liquid or solid.
An emulsion is a liquid suspended in a liquid or solid.

(e) An aerosol is a liquid or a solid suspended in a gas.
A foam is a gas suspended in a liquid or solid.

13.83 **(a)** 1.3×10^2 u

$$\Delta\text{fp} = 4.4\ °\text{C} - 5.5\ °\text{C} = -1.1\ °\text{C}$$

$$\Delta\text{fp} = K_{\text{fp}}m, \qquad m = \frac{\Delta\text{fp}}{K_{\text{fp}}} = \frac{-1.1\ \cancel{°\text{C}}}{-5.12\ \cancel{°\text{C}} \cdot m^{-1}} = 0.21\ m$$

$$m = \frac{n_{\text{solute}}}{\text{kg solvent}} \text{ and } n_{\text{solute}} = m \cdot \text{kg solvent}$$

$$n_{\text{solute}} = \left(\frac{0.21\ \text{mol solute}}{\cancel{\text{kg solvent}}}\right)\left(\frac{1\ \cancel{\text{kg solvent}}}{10^3\ \cancel{\text{g solvent}}}\right)(0.9476\ \cancel{\text{g solvent}}) = 2.0 \times 10^{-4}\ \text{mol solute}$$

$$\frac{0.0263\ \text{g solute}}{2.0 \times 10^{-4}\ \text{mol solute}} = 1.3 \times 10^2\ \text{g/mol}$$

The molecular mass of the unknown compound is 1.3×10^2 u.

(b) The melting point of camphor is above room temperature, making it easier to measure precisely and accurately. The K_{fp} for camphor is almost eight times that of benzene, which means for a given concentration of solute, a much larger Δfp is observed. This, too, should improve accuracy and precision. In addition, camphor is not a carcinogen (substance that tends to produce cancer) as benzene is.

13.85

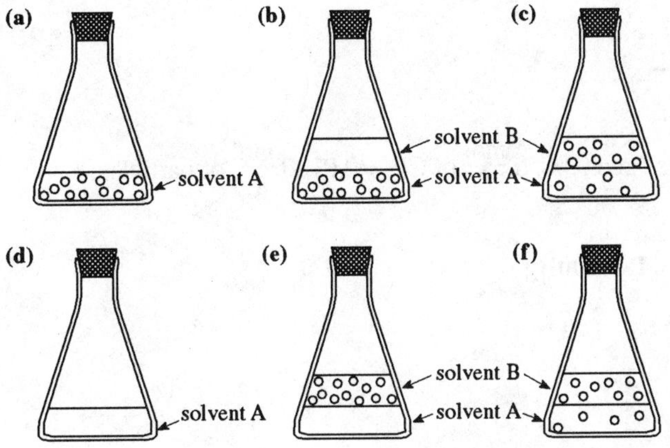

Stop and Test Yourself

1. (d) CH_3OH

   CH_3OH is most like water with the alcohol —OH able to hydrogen bond with water and very little oil–like character with only a one-carbon CH_3— group.

2. (b) higher temperature and lower pressure

   All gases are less soluble at higher temperatures and lower pressures, which favor an increase in entropy.

3. (b) $CH_3(CH_2)_3CH_3$ and $CH_3(CH_2)_4CH_3$

   These molecules differ by only one —CH_2— group and, thus, are more similar than any other listed pair.

4. (c) 0.10 *m* $CaCl_2$

   The freezing point depression is dependent on the molality of the dissolved particles. For one mole of $CaCl_2$ there are three moles of ions produced, thus yielding an effective molality of about 0.30 *m*.

5. (e) osmotic pressure

   Osmotic pressure is the colligative property most sensitive to changes in concentration. It is difficult to obtain large concentrations of high molecular mass substances, so osmotic pressure is the only useful colligative property in these cases.

6. (b) 0.030 M aqueous KI

   Because the density of water is 1.00 g/mL, a liter of this dilute aqueous solution contains close to 1000 g of solvent.

7. (c) 36

   An aqueous solution that is saturated with $KClO_3$ at 86 °C contains about 44 g $KClO_3$/100 g water. An aqueous solution that is saturated with $KClO_3$ at 22 °C contains about 8 g $KClO_3$. The difference equals the number of grams that will precipitate.

   44 g – 8 g = 36 g

8. **(e)** 183 mmHg

$\pi = MRT$

Note that solution will contain $0.0050 \frac{\text{mol}}{\text{L}}$ of Na^+ and $0.0050 \frac{\text{mol}}{\text{L}}$ of $Cl^- = 0.0100 \frac{\text{mol}}{\text{L}}$ of particles.

$$\pi = \left(\frac{0.0100\ \cancel{\text{mol}}}{\cancel{\text{L}}}\right)\left(\frac{62.36\ \text{mmHg}\cdot\cancel{\text{L}}}{\cancel{\text{mol}}\cdot\cancel{\text{K}}}\right)(292.9\ \cancel{\text{K}}) = 183\ \text{mmHg}$$

9. **(d)** 1060

$\pi = MRT$

$$12.60\ \cancel{\text{mmHg}} = \text{M}\left(\frac{62.36\ \cancel{\text{mmHg}}\cdot\text{L}}{\text{mol}\cdot\cancel{\text{K}}}\right)(294.8\ \cancel{\text{K}})$$

$$\text{M} = 6.854\times10^{-4}\frac{\text{mol}}{\text{L}}$$

$$\left(\frac{1\ \cancel{\text{L}}}{6.854\times10^{-4}\ \text{mol}}\right)\left(\frac{0.07265\ \text{g}}{100.00\ \cancel{\text{mL}}}\right)\left(\frac{10^3\ \cancel{\text{mL}}}{1\ \cancel{\text{L}}}\right) = 1.060\times10^3\ \frac{\text{g}}{\text{mol}}$$

10. **(b)** 12.1 mmHg

$$\text{VP}_{H_2O} = X_{H_2O}\text{VP}_{\text{pure }H_2O}$$

$$X_{H_2O} = 1 - 0.247 = 0.753$$

From Table 5.4, the vapor pressure of water at 18.6 °C = 16.071 mmHg.

$$p_{H_2O} = 0.753(16.071\ \text{mmHg}) = 12.1\ \text{mmHg}$$

11. **(d)** 100.12 °C

$$m_{\text{ions}} = 2m = 0.244\ m$$

$$\Delta\text{bp} = K_{\text{bp}}m = \left(\frac{0.512\ °\text{C}}{\cancel{m}}\right)(0.244\ \cancel{m}) = 0.125$$

Therefore, bp solution = 100.00 °C + 0.125 °C = 100.13 °C

(Best answer, without intermediate round off, is 100.12 °C.)

12. **(d)** 2.29

$$m = \frac{\text{mol solute}}{\text{kg solvent}}$$

$$(35.2\ \cancel{\text{g KI}})\left(\frac{\text{mol KI}}{166.0\ \cancel{\text{g KI}}}\right)\left(\frac{1}{92.4\ \cancel{\text{g } H_2O}}\right)\left(\frac{10^3\ \cancel{\text{g } H_2O}}{1\ \text{kg } H_2O}\right) = 2.29\ \frac{\text{mol KI}}{\text{kg } H_2O}$$

13. **(a)** 0.0397

$$(35.2\,\cancel{\text{g KI}})\left(\frac{1\text{ mol K}}{166.0\,\cancel{\text{g KI}}}\right) = 0.212\text{ mol KI}$$

$$(92.4\,\cancel{\text{g H}_2\text{O}})\left(\frac{1\text{ mol H}_2\text{O}}{18.02\,\cancel{\text{g H}_2\text{O}}}\right) = 5.13\text{ mol H}_2\text{O}$$

$$X_{\text{KI}} = \frac{\cancel{\text{mol}}\text{ KI}}{\cancel{\text{mol}}\text{ KI} + \cancel{\text{mol}}\text{ H}_2\text{O}} = \frac{0.212}{0.212 + 5.13} = 0.0397$$

14. **(b)** emulsion

See Table 13.4

15. **(a)** $CaCl_2(s)$

Ca^{2+} is the more positively charged ion.

Putting Things Together

13.87 **(a)** energy required to separate solute particles and to separate solvent molecules, energy released by solvation, entropy

Solubility depends on the energy required to separate solute particles from each other, the energy required to separate solvent molecules from each other, the energy released by solvation of solute particles by solvent molecules, and the entropy change that accompanies solution.

(b) Small size and high charge both increase interionic attractions and result in strongly hydrated ions; entropy may increase or decrease.

The same factors that result in strongly hydrated ions – small size and high charge – also make interionic attractions in the solid strong. In additon, prediction of the entropy change that accompanies solution is difficult when water is the solvent. Entropy usually increases when a solution forms. However, formation of hydrated ions orders water, and the entropy of the system may decrease when an aqueous solution forms.

13.89 **(a)** 13.8 mmHg

$$\text{VP}_{\text{H}_2\text{O}} = X_{\text{H}_2\text{O}}\,\text{VP}_{\text{pure H}_2\text{O}}$$

$$(22.5\,\cancel{\text{g NaCl}})\left(\frac{1\text{ mol NaCl}}{58.44\,\cancel{\text{g NaCl}}}\right) = 0.385\text{ mol NaCl} = 0.770\text{ mol particles}$$

$$(83.9\,\cancel{\text{g H}_2\text{O}})\left(\frac{1\text{ mol H}_2\text{O}}{18.02\,\cancel{\text{g H}_2\text{O}}}\right) = 4.66\text{ mol H}_2\text{O}$$

$$X_{\text{H}_2\text{O}} = \frac{4.66\,\cancel{\text{mol}}\text{ H}_2\text{O}}{4.66\,\cancel{\text{mol}}\text{ H}_2\text{O} + 0.770\,\cancel{\text{mol}}\text{ particles}} = 0.858$$

$$\text{VP}_{\text{H}_2\text{O}} = 0.858(16.071\text{ mmHg}) = 13.8\text{ mmHg}$$

(Note that the solution in this problem is quite concentrated and probably does not behave ideally.)

(b) 2.3 mmHg

From Table 5.4, the vapor pressure of water at 18.6 °C is 16.071 mmHg.

(16.1 – 13.8) mmHg = 2.3 mmHg

(c) –17.1 °C

$$\Delta fp = -1.86\frac{°C}{m} \cdot m\ (\text{ions})$$

$$m\ (\text{ions}) = \frac{0.770\ \text{mol ions}}{0.0839\ \text{kg}\ H_2O} = 9.18\ m$$

$$\Delta fp = \left(-1.86\frac{°C}{\cancel{m}}\right)(9.18\ \cancel{m}) = -17.1\ °C$$

(d) 104.70 °C

$$\Delta bp = \left(0.512\frac{°C}{\cancel{m}}\right)(9.18\ \cancel{m}) = 4.70\ °C$$

bp = 100.00 °C + 4.70 °C = 104.70 °C

13.91 $Ba(OH)_2$

Ba^{2+} has a 2+ charge; K^+ only has a 1+ charge. The two ions are about the same size (Figure 8.9).

13.93 **(a)** right – increasing pressure favors solution of gases

Increasing the pressure of a gas increases the solubility of a gas (Henry's law).

(b) right – the solubility of most gases in water increases as temperature decreases

Because solution of a gas is an exothermic process, lowering temperature will shift the equilibrium to the right according to Le Chatelier's principle.

13.95 **(a)** $C_6H_{12}O_6(s) \rightleftarrows C_6H_{12}O_6(aq)$

(b) Some of the $C_6H_{12}O_6(s)$ will disappear.

13.97 **(a)** 239 u

Use freezing point depression to calculate molality.

$$\Delta fp = K_{fp} \cdot m \text{ and } m = \frac{\Delta fp}{K_{fp}}$$

From Table 13.3, K_{fp} for benzene is –5.12 °C/m. Thus,

$$m = \frac{(5.317 - 5.513)\ \cancel{°C}}{-5.12\ \cancel{°C}\ m^{-1}} = 0.0383\ m$$

From the molality and the data given in the problem, calculate the number of moles of benzoic acid (BA) used.

$$m = \frac{\text{mols BA}}{\text{kg benzene}}$$

$$\left(\frac{0.0383 \text{ mol BA}}{\cancel{\text{kg benzene}}}\right)\left(\frac{1 \cancel{\text{kg benzene}}}{10^3 \cancel{\text{g benzene}}}\right)(23.65 \cancel{\text{g benzene}}) = 9.06 \times 10^{-4} \text{ mol BA}$$

Then calculate the number of grams of benzoic acid per mole of benzoic acid.

$$(1 \cancel{\text{mol BA}})\left(\frac{0.2168 \text{ g BA}}{9.06 \times 10^{-4} \cancel{\text{mol BA}}}\right) = 239 \text{ g BA/mol BA}$$

The molecular mass of benzoic acid is 239 u.

(b) The experimentally determined molecular mass of 239 u is about double the value of 122.123 u calculated for C_6H_5COOH.

(c) The molecule must form a dimer:

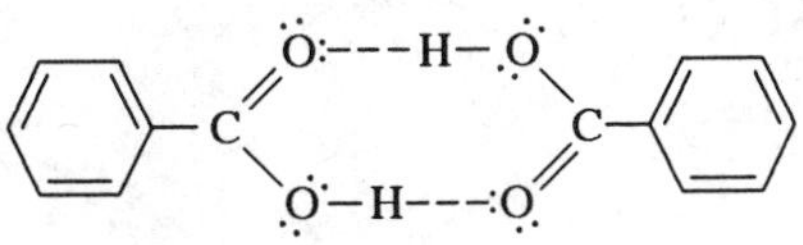

13.99 **(a)** $\frac{45 \text{ g } NH_3}{100 \text{ g } H_2O}$

100 g solution contains 31 g NH_3 and 69 g H_2O

$$(100 \cancel{\text{g } H_2O})\left(\frac{31 \text{ g } NH_3}{69 \cancel{\text{g } H_2O}}\right) = \frac{45 \text{ g } NH_3}{100 \text{ g } H_2O}$$

(b) Since gases are usually not very soluble in water, the $NH_3(g)$ must have reacted with the water.

(c) < 25 °C. Gases are more soluble in water at lower temperatures.

13.101 **(a)** 1.96 m

$$m = \frac{\text{mol solute}}{\text{mass solvent in kg}} = \frac{(3.16 \cancel{\text{g } NH_4Cl})\left(\frac{1 \text{ mol } NH_4Cl}{53.49 \cancel{\text{g } NH_4Cl}}\right)}{0.030\ 14 \text{ kg } H_2O} = 1.96\ m$$

(b) 1.82 M

$$M = \frac{\text{mol solute}}{\text{L soln}}$$

From the mass of solution and density, calculate the volume of the solution.

mass of solution = 3.16 g NH_4Cl + 30.14 g H_2O = 33.30 g soln

$$(33.30 \cancel{\text{g soln}})\left(\frac{1 \cancel{\text{cm}^3}}{1.0272 \cancel{\text{g soln}}}\right)\left(\frac{1 \text{ L}}{10^3 \cancel{\text{cm}^3}}\right) = 0.03242 \text{ L}$$

Then calculate the molarity from the definition of molarity.

$$M = \frac{\text{mol solute}}{\text{vol soln L}} = \frac{(3.16\ \cancel{\text{g NH}_4\text{Cl}})\left(\dfrac{1\ \text{mol NH}_4\text{Cl}}{53.49\ \cancel{\text{g NH}_4\text{Cl}}}\right)}{0.032\,42\ \text{L}} = 1.82\ \text{M}$$

(c) $X_{NH_4Cl} = 0.0341$, $X_{H_2O} = 0.9659$

$$(30.14\ \cancel{\text{g H}_2\text{O}})\left(\frac{1\ \text{mol H}_2\text{O}}{18.015\ \cancel{\text{g H}_2\text{O}}}\right) = 1.673\ \text{mol H}_2\text{O}$$

$$(3.16\ \cancel{\text{g NH}_4\text{Cl}})\left(\frac{1\ \text{mol NH}_4\text{Cl}}{53.49\ \cancel{\text{g NH}_4\text{Cl}}}\right) = 0.0591\ \text{mol NH}_4\text{Cl}$$

$$\frac{1.673\ \cancel{\text{mol}}\ \text{H}_2\text{O}}{1.673\ \cancel{\text{mol}}\ \text{H}_2\text{O} + 0.0591\ \cancel{\text{mol}}\ \text{NH}_4\text{Cl}} = 0.9659 \quad (\text{mol fraction H}_2\text{O})$$

$$\frac{0.0591\ \cancel{\text{mol}}\ \text{NH}_4\text{Cl}}{1.673\ \cancel{\text{mol}}\ \text{H}_2\text{O} + 0.0591\ \cancel{\text{mol}}\ \text{NH}_4\text{Cl}} = 0.0341 \quad (\text{mol fraction NH}_4\text{Cl})$$

Check: $0.0341 + 0.9659 = 1.0000$

(d) 9.49% NH_4Cl, 90.51% H_2O

$$\%\ \text{NH}_4\text{Cl} = \frac{3.16\ \text{g NH}_4\text{Cl}}{(3.16 + 30.14)\ \text{g total}} \times 100 = 9.49\%$$

$$\%\ \text{H}_2\text{O} = \frac{30.14\ \text{g H}_2\text{O}}{(3.16 + 30.14)\ \text{g total}} \times 100 = 90.51\%$$

Check: $9.49\% + 90.51\% = 100.00\%$

13.103 **(a)** benzene

The structure of benzene is similar to the six-membered rings in buckyballs. (See Figure 12.36.) Like dissolves like.

(b) $—NH_2$

This group is capable of hydrogen bonding with water.

(c) sodium benzoate

Sodium benzoate is an ionic compound. All sodium salts are water soluble (Table 4.4).

Applications

13.105 The pressure of gas above the liquid must be maintained to keep the CO_2 dissolved in the desired concentration.

13.107 Slowly freeze the seawater. While some liquid remains, separate the ice, which is free of dissolved materials.

Slowly cool the seawater until part of it has frozen. The solid will be pure ice; dissolved materials will remain in solution. Separate the ice by filtering, centrifuging, or decanting. Wash the ice with a little cold pure water obtained from a previous preparation. Let the ice melt to obtain water free from dissolved materials. (Sterilization, as well as removal of dissolved materials, may be necessary to obtain water suitable for drinking.)

13.109 Osmosis occurs, removing water from the plant cells.

Water passes through the cell walls into the salad dressing by osmosis because the concentration of solute particles in the salt–containing dressing is greater than the concentration of solute particles inside the cells. The cells shrink and the greens wilt.

13.111 The water in the plant cells contains dissolved material, which lowers the freezing point.

Solutes lower the freezing points of the aqueous solution in the plants below that of pure water. Some kinds of plants have higher concentrations of solute particles in their fluids than others and are more resistant to freezing.

13.113 The dissolved electrolytes in seawater precipitate the colloidal dispersion of mud in river water.

The mouths of rivers are where rivers run into the sea. The concentration of electrolytes is much higher in seawater than in fresh water and causes colloidal particles of mud in the river water to precipitate, forming a delta.

13.115 $\dfrac{0.580 \text{ g CO}_2}{100 \text{ g H}_2\text{O}}$

$$c_{CO_2} = kp_{CO_2}$$

$$\frac{0.145 \text{ g CO}_2}{100 \text{ g H}_2\text{O} \cdot 1 \text{ atm}} = k$$

$$c_{CO_2} = \left(\frac{0.145 \text{ g CO}_2}{100 \text{ g H}_2\text{O} \cdot 1 \cancel{\text{atm}}}\right)(4 \cancel{\text{atm}}) = \frac{0.580 \text{ g CO}_2}{100 \text{ g H}_2\text{O}}$$

13.117 582 u

From the osmotic pressure law, $\pi = MRT$,

$$\text{M} = \frac{\pi}{RT} = \frac{8.72 \cancel{\text{mmHg}}}{\left(\dfrac{62.36 \cancel{\text{mmHg}} \cdot \text{L}}{\text{mol} \cdot \cancel{\text{K}}}\right)(296.8 \cancel{\text{K}})} = 4.71 \times 10^{-4} \frac{\text{mol}}{\text{L soln}}$$

The M and the concentration given in the problem, 0.013 70 g/50.0 mL solution, can be used to find molar mass:

$$\left(\frac{0.013\ 70 \text{ g}}{50.0 \cancel{\text{mL soln}}}\right)\left(\frac{1000 \cancel{\text{mL soln}}}{1 \cancel{\text{L soln}}}\right)\left(\frac{1 \cancel{\text{L soln}}}{4.71 \times 10^{-4} \text{ mol}}\right) = 582 \frac{\text{g}}{\text{mol}}$$

The molecular mass of streptomycin is 582 u.

13.119 **(a)** 29.5 atm

Note that there are 2 mol of ions for every mol of NaCl.

$$\pi = 2\left(\frac{0.604\ \cancel{\text{mol}}}{\cancel{\text{L}}}\right)\left(\frac{0.082\ 06\ \text{atm}\cdot\cancel{\text{L}}}{\cancel{\text{mol}}\cdot\cancel{\text{K}}}\right)(298\ \cancel{\text{K}}) = 29.5\ \text{atm}$$

(b) A pressure greater than 29.5 atm is needed.

(c) As salt–free water is pushed out, M increases. This, in turn, increases the amount of pressure needed to continue the reverse osmosis process.

(d) 5.68 L

First use the osmotic pressure law to find the molarity of the solution at 100.0 atm.

$$100.0\ \cancel{\text{atm}} = \text{M}\left(\frac{0.082\ 06\ \text{L}\cdot\cancel{\text{atm}}}{\text{mol}\cdot\cancel{\text{K}}}\right)(298\ \cancel{\text{K}})$$

$$4.09 = \text{M},\quad \frac{4.09}{2} = 2.05 = \text{M}_{\text{NaCl}}$$

Start with 0.604 M – end with 2.05 M.

Let x = L at start. Then $x - 4$ = L at finish. Because all the salt present at the beginning is left in the solution at the end,

$$\left(\frac{0.604\ \text{mol}}{\text{L}}\right)x\ \text{L} = \left(\frac{2.05\ \text{mmol}}{\text{L}}\right)(x-4)\ \text{L}$$

and $x = 5.66$ L

(Best answer, without intermediate round off, is 5.68 L.)

13.121 **(a)** 1.6×10^2 g urea, 1.0×10^2 g $CaCl_2$, 7.8×10^1 g NaCl

$$\Delta\text{fp} = K_{\text{fp}} \cdot m \qquad m = \frac{\text{mol solute}}{\text{kg solvent}}$$

$$(-5.0\ ^\circ\text{C})\left(\frac{m}{-1.86\ ^\circ\text{C}}\right) = \frac{2.7\ \text{mol}}{1\ \text{kg H}_2\text{O}}$$

$$(2.7\ \cancel{\text{mol urea}})\left(\frac{60.1\ \text{g urea}}{\cancel{\text{mol urea}}}\right) = 1.6 \times 10^2\ \text{g urea}$$

$$(2.7\ \cancel{\text{mol ions}})\left(\frac{111\ \text{g CaCl}_2}{\cancel{\text{mol CaCl}_2}}\right)\left(\frac{1\ \cancel{\text{mol CaCl}_2}}{3\ \cancel{\text{mol ions}}}\right) = 1.0 \times 10^2\ \text{g CaCl}_2$$

$$(2.7\ \text{mol}\ \cancel{\text{ions}})\left(\frac{58.4\ \text{g NaCl}}{\cancel{\text{mol NaCl}}}\right)\left(\frac{1\ \cancel{\text{mol NaCl}}}{2\ \cancel{\text{mol ions}}}\right) = 7.9 \times 10^1\ \text{g NaCl}$$

(Best answer, without intermediate round off, is 7.8×10^1 g NaCl.)

(b) \$0.70 (urea), \$0.73 (NaCl), \$0.75 ($CaCl_2$)

$$\left(\frac{\$4.32}{1\text{ kg urea}}\right)\left(\frac{1\text{ kg}}{10^3\text{ g}}\right)(1.6\times10^2\text{ g urea}) = \$0.69$$ (Best answer, without intermediate round off from part (a), is \$0.70.)

$$\left(\frac{\$7.59}{1\text{ kg CaCl}_2}\right)\left(\frac{1\text{ kg}}{10^3\text{ g}}\right)(1.0\times10^2\text{ g CaCl}_2) = \$0.76$$ (Best answer, without intermediate round off from part (a), is \$0.75.)

$$\left(\frac{\$9.25}{1\text{ kg NaCl}}\right)\left(\frac{1\text{ kg}}{10^3\text{ g}}\right)(7.8\times10^1\text{ g NaCl}) = \$0.72$$ (Best answer, without intermediate round off from part (a), is \$0.73.)

(c) A larger mass of urea is required to obtain the same amount of protection from freezing. Also, too much N in runoff is not good for the water supply.

13.123 **(a)** Freezing–point depression calculations are based on colligative properties of dilute, ideal solutions in which the solute is assumed to be nonvolatile.

(b) 0.802 L methyl alcohol

To make 100 g of 38.8% by mass methyl alcohol in water.

$$(38.8\text{ g CH}_3\text{OH})\left(\frac{1\text{ cm}^3}{0.791\text{ g CH}_3\text{OH}}\right)\left(\frac{1\text{ L}}{10^3\text{ cm}^3}\right) = 4.91\times10^{-2}\text{ L CH}_3\text{OH}$$

and

$$[(100-38.8)\text{ g H}_2\text{O}]\left(\frac{1\text{ cm}^3}{1.00\text{ g H}_2\text{O}}\right)\left(\frac{1\text{ L}}{10^3\text{ cm}^3}\right) = 6.12\times10^{-2}\text{ L H}_2\text{O}$$

are required. The liters of methyl alcohol per liter of water are

$$\frac{4.91\times10^{-2}\text{ L CH}_3\text{OH}}{6.12\times10^{-2}\text{ L H}_2\text{O}} = \frac{8.02\times10^{-1}\text{ L CH}_3\text{OH}}{1\text{ L H}_2\text{O}}$$

(c) \$1.30 per liter (methyl alcohol–water), \$2.71 per liter (ethylene glycol–water)

For the methyl alcohol–water mixture, the cost is

$$\left(\frac{0.935\text{ g solution}}{\text{cm}^3}\right)\left(\frac{10^3\text{ cm}^3}{1\text{ L}}\right)\left(\frac{38.8\text{ g CH}_3\text{OH}}{100\text{ g solution}}\right)\left(\frac{1\text{ kg CH}_3\text{OH}}{10^3\text{ g CH}_3\text{OH}}\right)\left(\frac{\$3.58}{\text{kg CH}_3\text{OH}}\right) = \frac{\$1.30}{\text{L}}$$

For the ethylene glycol–water mixture, the cost is

$$\left(\frac{1.071\text{ g solution}}{\text{cm}^3}\right)\left(\frac{10^3\text{ cm}^3}{1\text{ L}}\right)\left(\frac{55.1\text{ g ethylene glycol}}{100\text{ g solution}}\right)\left(\frac{1\text{ kg ethylene glycol}}{10^3\text{ g ethylene glycol}}\right)\left(\frac{\$4.60}{\text{kg ethylene glycol}}\right) = \frac{\$2.71}{\text{L}}$$

(d) The boiling points of methyl alcohol–water mixtures are probably between the boiling points of CH_3OH (65.0 °C) and water (100 °C). Thus, the methyl alcohol solutions will evaporate much faster than ethylene glycol–water solutions and may even boil over.

13.125 **(a)** water: carbon tetrachloride; air: heptane

Carbon tetrachloride has a low value for Henry's law constant and will therefore accumulate in the water. Heptane, which has a high value of *H*, will accumulate in the air.

(b) smaller

The constants will be smaller at 5 °C because the solubility of gases in water increases with decreasing temperature.

13.127 **(a)** The ability to use a wide variety of solvents for freezing point depression is desirable because the method only works for soluble solutes. The wider the variety of solvents, the larger the number of substances whose molecular masses can be obtained.

(b) –37.7 °C/*m*

$$m = \frac{\text{mol solute}}{\text{kg solvent}} = \frac{(0.6381\ \cancel{\text{g } C_{12}H_{26}})\left(\dfrac{1\ \text{mol } C_{12}H_{26}}{170.34\ \cancel{\text{g } C_{12}H_{26}}}\right)}{0.030\,4112\ \text{kg solvent}} = 0.1232\ m$$

$$K_{fp} = \frac{\Delta fp}{m} = \frac{33.06\ °C - 37.70\ °C}{0.1232\ m} = \frac{-4.64\ °C}{0.1232\ m} = -37.7\ °C/m$$

13.129 **(a)** 0.8

$$\frac{5\ \text{mol He}}{5\ \text{mol He} + 1\ \text{mol Ne}} = 0.8$$

(b) orange-red

$$(633\ \cancel{\text{nm}})\left(\frac{10^{-9}\ \text{m}}{1\ \cancel{\text{nm}}}\right) = 6.33 \times 10^{-7}\ \text{m}$$

See Figure 7.15.

(c) $4.74 \times 10^{14}\ s^{-1}$

$$\nu = \frac{c}{\lambda} = \frac{2.998 \times 10^{8}\ \cancel{\text{m}} \cdot s^{-1}}{6.33 \times 10^{-7}\ \cancel{\text{m}}} = 4.74 \times 10^{14}\ s^{-1}$$

(d) 3.14×10^{-19} J

$$E = h\nu = (6.626 \times 10^{-34}\ \text{J} \cdot \cancel{s})(4.74 \times 10^{14}\ \cancel{s^{-1}}) = 3.14 \times 10^{-19}\ \text{J}$$

CHAPTER 14: Chemical Equilibrium

Practice Problems

14.1 **(a)** [HI] = 0.0200 M, $[H_2] = [I_2] = 0.0000$ M

(b) 0.0175 M

(c) 0.0025 mol

[(0.0200 M)(1L) – (0.0175 M)(1 L)] = 0.0025 mol

(d) 0.0013 mol H_2, 0.0013 mol I_2

$$(0.0025 \, \cancel{\text{mol HI}}) \left(\frac{1 \text{ mol } H_2}{2 \, \cancel{\text{mol HI}}}\right) = 0.0013 \text{ mol } H_2$$

$$(0.0025 \, \cancel{\text{mol HI}}) \left(\frac{1 \text{ mol } I_2}{2 \, \cancel{\text{mol HI}}}\right) = 0.0013 \text{ mol } I_2$$

(e) $[H_2] = 0.0013$ M, $[I_2] = 0.0013$ M

14.2 **(a)** $K_c = \dfrac{[SO_2]^2[O_2]}{[SO_3]^2}$

(b) $K_c = \dfrac{[Sn^{2+}][Fe^{3+}]^2}{[Sn^{4+}][Fe^{2+}]^2}$

(c) $K_c = \dfrac{[H^+][NO_2^-]}{[HNO_2]}$

(d) $K_c = \dfrac{[NH_3]^2}{[N_2][H_2]^3}$

(e) $K_c = \dfrac{[O_3]^2}{[O_2]^3}$

14.3 (a) homogeneous

All reactants and products are in the same phase (gas).

(b) heterogeneous

The AgCl is a solid, whereas Ag^+ and Cl^- are in aqueous solution.

(c) heterogeneous

Mg and MgO are solids while O_2 is a gas.

(d) homogeneous

All reactants and products are in the same phase (aqueous solution).

14.4 (a) $K_c = [NH_3]^2[CO_2][H_2O]$

This is a heterogeneous equilibrium. The concentration of the solid reactant, $(NH_4)_2CO_3$, is constant and is not included in the equilibrium constant expression. (It is included in the equilibrium constant.)

(b) $K_c = [Ag^+][Cl^-]$

This is also a heterogeneous equilibrium involving a solid reactant, AgCl.

(c) $K_c = \dfrac{1}{[O_2]}$

(d) $K_c = \dfrac{[OH^-]^2}{[O_2][SH^-]^2}$

A solid, a gas, and an aqueous solution are all present in this heterogeneous equilibrium.

14.5 No. In order for equilibrium to be reached, the products must be in contact with the reactants. In an open container, the $CO_2(g)$ is lost continuously from the reaction mixture, making it unavailable for the reverse reaction.

14.6 (a) (iv)

The equilibrium constant, K_c, is very large, which indicates that at equilibrium the concentration of the product is very much higher than the concentrations of the reactants.

(b) (i)

K_c is very small, indicating that at equilibrium the concentrations of the reactants are very high compared to the concentration of the product.

(c) (ii) and (iii)

The values of the equilibrium constants for these reactions are neither very large nor very small. They are between 10^{-2} and 10^{2}.

14.7 2.11×10^{-1}

For the equilibrium,

$$N_2O_4(g) \rightleftarrows 2NO_2(g), \quad K_c = \frac{[NO_2]^2}{[N_2O_4]} = \frac{(1.72 \times 10^{-2})^2}{(1.40 \times 10^{-3})} = 2.11 \times 10^{-1}$$

14.8 **(a)** 0.030 96 M

Set up the summary table for the equilibrium reaction:

	$N_2O_4(g)$	$\rightleftarrows$	$2NO_2(g)$
Initial concentration, M	0.020 00		0.000 00
Change in concentration, ΔM	–0.015 48		0.030 96
Equilibrium concentration, M	0.004 52		0.030 96

In step 1 calculate the change in concentration of N_2O_4, $\Delta M_{N_2O_4}$:

$$\Delta M_{N_2O_4} = (0.020\,00 - 0.004\,52)\text{ M} = 0.015\,48\text{ M}$$

Assume the reaction takes place in a 1–L container.
The ΔM (1 L) = Δmol.

$$\Delta mol_{N_2O_4} = -0.015\,48\text{ mol}$$

$$\Delta mol_{NO_2} = (0.015\,48\ \text{mol } N_2O_4)\left(\frac{2\text{ mol } NO_2}{1\text{ mol } N_2O_4}\right) = 0.030\,96\text{ mol } NO_2$$

$$[NO_2]_{eq} = (0.030\,96\text{ M} + 0.000\,00\text{ M}) = 0.030\,96\text{ M}$$

(b) 0.212 or 2.12×10^{-1}

$$K_c = \frac{[NO_2]^2}{[N_2O_4]} = \frac{(0.030\,96)^2}{(0.004\,52)} = 0.212 \text{ or } 2.12 \times 10^{-1}$$

14.9 1.38×10^{5}

From the reaction, $2H_2(g) + S_2(g) \rightleftarrows 2H_2S(g)$, write the equilibrium constant expression:

$$K_c = \frac{[H_2S]^2}{[H_2]^2[S_2]} = 1.10 \times 10^{7}$$

Equation 14.7 gives the relationship between K_p and K_c.

$$K_p = K_c(RT)^{\Delta n_g}$$

Δn_g = moles of gaseous products – moles of gaseous reactants = 2 – 3 = –1

Express the temperature in K:

$$T = (273\ \cancel{°C} + 700\ \cancel{°C})\left(\frac{1\ K}{1\ \cancel{°C}}\right) = 973\ K$$

For correct cancellation of units, use $R = 0.082\ 06\ atm \cdot L \cdot mol^{-1} \cdot K^{-1}$

$$K_p = [1.10 \times 10^7\ L \cdot mol^{-1}][(0.082\ 06\ atm \cdot L \cdot mol^{-1} \cdot K^{-1})(973\ K)]^{-1}$$

$$= \left(\frac{1.10 \times 10^7\ \cancel{L}}{\cancel{mol}}\right)\left(\frac{\cancel{mol} \cdot \cancel{K}}{0.082\ 06\ atm \cdot \cancel{L}}\right)\left(\frac{1}{973\ \cancel{K}}\right) = 1.38 \times 10^5\ atm^{-1}$$

14.10 **(a)** $K_p = \dfrac{p_{H_2S}{}^2}{p_{H_2}{}^2\ p_{S_2}}$

If partial pressures are in atmospheres, the equilibrium expression, K_p, is similar to K_c, where partial pressures of gaseous reactants and products have replaced concentrations.

(b) $K_p = p_{CO_2}$

CO_2 is the only gaseous substance in this reaction.

14.11 **(a)** 2.43

Since all products and reactants are gases, K_p can be calculated in terms of partial pressures:

$$K_p = \frac{p_{SO_2}{}^2\ p_{O_2}}{p_{SO_3}{}^2} = \frac{(0.722\ \cancel{atm})^{\cancel{2}}\ (0.361\ atm)}{(0.278\ \cancel{atm})^{\cancel{2}}} = 2.43\ atm$$

(b) 2.70×10^{-2}

Equation 14.7 gives the relationship between K_p and K_c:

$$K_p = K_c(RT)^{\Delta n_g}, \quad T = (827\ \cancel{°C} + 273\ \cancel{°C})\left(\frac{1\ K}{1\ \cancel{°C}}\right) = 1.100 \times 10^3\ K,$$

$$\Delta n_g = 3 - 2 = 1$$

$$K_c = \frac{K_p}{(RT)^{\Delta n_g}} = \frac{(2.43\ \cancel{atm})}{\left[\left(0.082\ 06\ \frac{\cancel{atm} \cdot L}{mol \cdot \cancel{K}}\right)(1.100 \times 10^3\ \cancel{K})\right]^1} = 2.69 \times 10^{-2}\ L \cdot mol^{-1}$$

(Best answer, without intermediate round off in part (a), is 2.70×10^{-2}.)

14.12 **(a)** 9.050×10^{-8}

The equilibrium constant expression, K'_c, for this reaction is the reciprocal of the equilibrium constant expression given, K_c.

$$K_c = \frac{[H_2S]^2}{[H_2]^2[S_2]} = 1.105 \times 10^7$$

$$K'_c = \frac{[H_2]^2[S_2]}{[H_2S]^2} = \frac{1}{1.105 \times 10^7} = 9.050 \times 10^{-8}$$

(b) 3.324×10^3

$$K''_c = \frac{[H_2S]}{[H_2][S_2]^{1/2}} = \sqrt{\frac{[H_2S]^2}{[H_2]^2[S_2]}} = \sqrt{1.105 \times 10^7} = 3.324 \times 10^3$$

14.13 **(a)** not at equilibrium, reverse direction

The reaction for this problem is given in Sample Problem 14.7.

$H_2(g) + I_2(g) \rightleftarrows 2HI(g)$

The reaction quotient, Q, is

$$Q = \frac{[HI]^2}{[H_2][I_2]} = \frac{(2.57 \times 10^{-2})^2}{(8.48 \times 10^{-3})(9.91 \times 10^{-4})} = 78.6$$

The equilibrium constant for this reaction at 425.4 °C is 54.5. Therefore, $Q > K_c$ (78.6 > 54.5). This means that the reaction is not at equilibrium, and the reaction will take place in the reverse direction until $Q = K_c$.

(b) at equilibrium

$$Q = \frac{[HI]^2}{[H_2][I_2]} = \frac{(1.369 \times 10^{-2})^2}{(6.42 \times 10^{-5})(5.36 \times 10^{-2})} = 54.5$$

Since 54.5 = 54.5, the reaction is at equilibrium.

14.14 $[COCl_2]_{eq} = 0.074$, $[CO]_{eq} = [Cl_2]_{eq} = 4.0 \times 10^{-6}$

Set up a table of information, letting $x = [CO]_{eq} = [Cl_2]_{eq}$ and assuming initial concentration of products to be 0.000 M.

	$COCl_2(g)$ $\rightleftarrows$	$CO(g)$	+	$Cl_2(g)$
Initial concentration, M	0.074	0.000		0.000
Change in concentration, ΔM	$-x$	$+x$		$+x$
Equilibrium concentration, M	$0.074 - x$	x		x

Write the equilibrium constant expression for the reaction and plug in known and unknown quantities:

$$K_c = \frac{[CO]_{eq}[Cl_2]_{eq}}{[COCl_2]_{eq}} = 2.2 \times 10^{-10}$$

$$\frac{x^2}{(0.074 - x)} = 2.2 \times 10^{-10}$$

Assume x is negligible compared to 0.074:

$$\frac{x^2}{0.074} = 2.2 \times 10^{-10}$$

$x^2 = (0.074)(2.2 \times 10^{-10}) = 1.6 \times 10^{-11}$

$x = 4.0 \times 10^{-6}$

$[COCl_2]_{eq} = 0.074\text{ M} - (4.0 \times 10^{-6}\text{ M}) = 0.074\text{ M}$. Therefore,

x is negligible compared to 0.074 and $[CO]_{eq} = [Cl_2]_{eq} = 4.0 \times 10^{-6}$;

$[COCl_2]_{eq} = 0.074$

14.15 $[PCl_5]_{eq} = 0.0236$, $[PCl_3]_{eq} = [Cl_2]_{eq} = 0.0169$

$$K_C = \frac{[PCl_3][Cl_2]}{[PCl_5]} = 1.21 \times 10^{-2}$$

Let $x = [PCl_3] = [Cl_2]$. Set up table of information:

	$PCl_5(g) \rightleftarrows$	$PCl_3(g)$	+	$Cl_2(g)$
Initial concentration, M	0.0405	0.000		0.000
Change in concentration, ΔM	$-x$	$+x$		$+x$
Equilibrium concentration, M	$0.0405 - x$	x		x

Plug the appropriate values into the equilibrium constant expression:

$$\frac{(x)(x)}{(0.0405 - x)} = 1.21 \times 10^{-2}$$

Using successive approximations, first assume x is neglibible compared to 0.0405:

$$\frac{x^2}{0.0405} = 1.21 \times 10^{-2}$$

$$x^2 = 4.90 \times 10^{-4}$$

$$x = 0.0221$$

This value of x is not negligible compared to 0.0405. For the second approximation, we use $x = 0.0221$ to calculate a better value for $[PCl_3]_{eq} = 0.0405 - x$:

$$\frac{x^2}{0.0405 - 0.0221} = \frac{x^2}{0.0184} = 1.21 \times 10^{-2}, x = 0.0149$$

Repeating, using $x = 0.0149$ for our new approximation of $0.0405 - x$, we have

$$\frac{x^2}{0.0405 - 0.0149} = \frac{x^2}{0.0256} = 1.21 \times 10^{-2}, x = 0.0176$$

A fourth approximation gives

$$\frac{x^2}{0.0229} = 1.21 \times 10^{-2}, x = 0.0166$$

A fifth approximation gives

$$\frac{x^2}{0.0239} = 1.21 \times 10^{-2}, x = 0.0170$$

and a sixth approximation gives

$$\frac{x^2}{0.0235} = 1.21 \times 10^{-2}, x = 0.0169$$

Therefore, $[PCl_5]_{eq} = 0.0405 - 0.0169 = 0.0236$ and $[PCl_3]_{eq} = [Cl_2]_{eq} = 0.0169$

To check the answers, substitute them in the equilibrium constant expression:

$$K_c = \frac{(0.0169)(0.0169)}{(0.0236)} = 1.21 \times 10^{-2}, \text{ which is correct.}$$

14.16 $p_{N_2O_4} = 2.5$ atm, $p_{NO_2} = 0.53$ atm

Let $x = p_{NO_2}$ and $p_{N_2O_4} = 3.0 - x$

We then have

$$K_p = \frac{x^2}{3.0 - x} = 0.113 \text{ or } x^2 + 0.113x - 0.339 = 0$$

Using successive approximations, first assume that x is negligible compared to 3.0:

$$\frac{x^2}{3.0} = 0.113$$

$x^2 = 0.339, x = 0.582$

This value of x is not negligible compared to 3.0. For the second approximation, we use $x = 0.582$ to calculate a better value for $p_{N_2O_4} = 3.0 - x$:

$$\frac{x^2}{2.418} = 0.113, x = 0.523$$

Repeating, using $x = 0.523$ for our new approximation of $3.0 - x$, we have

$$\frac{x^2}{2.477} = 0.113, x = 0.529$$

A fourth approximation gives

$$\frac{x^2}{2.472} = 0.113, x = 0.528$$

and a fifth approximation gives

$$\frac{x^2}{2.472} = 0.113, x = 0.529$$

To two significant digits, $p_{NO_2} = x = 0.53$
$p_{N_2O_4} = 3.0 - x = 2.5$

Check: $\frac{(0.53)^2}{3.0 - 0.53} = 0.11$

14.17 $[H_2] = 4.199 \times 10^{-3}$, $[I_2] = 3.76 \times 10^{-4}$, $[HI] = 9.28 \times 10^{-3}$

	$H_2(g)$ +	$I_2(g)$ $\rightleftarrows$	$2HI(g)$
Original equilibrium conc., M	4.562×10^{-3}	7.384×10^{-4}	1.355×10^{-2}
Initial equilibrium conc., M	4.562×10^{-3}	7.384×10^{-4}	$(1.355 \times 10^{-2}) - (0.5000 \times 10^{-2}) = 8.55 \times 10^{-3}$
Change in equilibrium conc., ΔM	$-x$	$-x$	$+2x$
Equilibrium conc. at new equilibrium, M	$[(4.562 \times 10^{-3}) - x]$	$[(7.384 \times 10^{-4}) - x]$	$[(8.55 \times 10^{-3}) + 2x]$

$$K_c = \frac{[HI]^2}{[H_2][I_2]} = 54.5$$

The value of K_c is relatively large, which indicates that x is not insignificant with respect to the initial equilibrium concentrations of H_2 and I_2 nor is $2x$ small compared to the initial equilibrium concentration of HI. The quadratic formula or successive approximations must be used to solve for x.

$$K_c = \frac{[(8.55 \times 10^{-3}) + 2x]^2}{[(4.562 \times 10^{-3}) - x][(7.384 \times 10^{-4}) - x]} = 54.5$$

To use the quadratic formula, the equation must first be rearranged into the standard quadratic form.

$$\frac{(8.55 \times 10^{-3})^2 + 2(8.55 \times 10^{-3})(2x) + 4x^2}{(3.369 \times 10^{-6}) - (5.3004 \times 10^{-3})x + x^2} = 54.5$$

$$(7.310 \times 10^{-5}) + (3.420 \times 10^{-2})x + 4x^2 = (1.8359 \times 10^{-4}) - (2.8887 \times 10^{-1})x + 54.5x^2$$

$$0 = 50.5x^2 - (3.2307 \times 10^{-1})x + 1.1049 \times 10^{-4} \text{ or } 50.5x^2 - 0.323\,07x + 1.1049 \times 10^{-4} = 0$$

Substitution of the coefficients in the quadratic formula

$$x = \frac{-b \pm \sqrt{b^2 - 4ac}}{2a} \text{ gives } x = \frac{-(-0.323\,07) \pm \sqrt{(-0.323\,07)^2 - 4(50.5)(1.1049 \times 10^{-4})}}{2(50.5)}$$

$$= \frac{(0.323\,07) \pm \sqrt{(0.104\,37) - (0.022\,319)}}{101.0} = \frac{(0.323\,07) \pm \sqrt{0.082\,05}}{101.0}$$

$$= \frac{(0.323\,07) \pm (0.286\,45)}{101.0} = 6.03 \times 10^{-3} \text{ or } 3.63 \times 10^{-4}$$

(Note that in this problem and others involving the solution of a quadratic equation, extra digits have been retained in intermediate steps to avoid significant rounding errors.)

Since 6.03×10^{-3} is greater than 4.562×10^{-3}, the initial concentration of H_2, it is not a possible value. Therefore,

$$[H_2]_{eq} = (4.562 \times 10^{-3}) - (3.63 \times 10^{-4}) = 4.199 \times 10^{-3}$$

$$[I_2]_{eq} = (7.384 \times 10^{-4}) - (3.63 \times 10^{-4}) = 3.75 \times 10^{-4}$$

$$[HI]_{eq} = (8.55 \times 10^{-3}) + 2(3.63 \times 10^{-4}) = 9.28 \times 10^{-3}$$

(Best answer, without intermediate round off, is $[H_2] = 4.199 \times 10^{-3}$, $[I_2] = 3.76 \times 10^{-4}$, and $[HI\} = 9.28 \times 10^{-3}$.)

Check: $K_c = \dfrac{(9.28 \times 10^{-3})^2}{(4.199 \times 10^{-3})(3.76 \times 10^{-4})} = 54.5$

14.18 (a) equilibrium shifts to right, $[H_2O(g)]$ decreases, $[CO_2(g)]$ and $[H_2(g)]$ increase.

Increasing the concentration of CO(g) will cause the equilibrium

$$CO(g) + H_2O(g) \rightleftarrows CO_2(g) + H_2(g)$$

to shift to the right. Both the concentration and the quantity of $H_2O(g)$ will decrease and the concentrations and quantities of $CO_2(g)$ and $H_2(g)$ will increase.

(b) equilibrium shifts to right, [CO(g)] and $[H_2O(g)]$ decrease, $[CO_2(g)]$ increases.

Decreasing the concentration of $H_2(g)$ will also cause the equilibrium to shift to the right. The concentrations and quantities of CO(g) and $H_2O(g)$ will decrease and the concentration and quantity of $CO_2(g)$ will increase.

(c) equilibrium shifts to left, [CO(g)] increases, $[CO_2(g)]$ and $[H_2(g)]$ decrease.

Decreasing the concentration of $H_2O(g)$ will cause the equilibrium to shift to the left. The concentration and quantity of CO(g) will increase and the quantities and concentrations of $CO_2(g)$ and $H_2(g)$ will decrease.

(d) equilibrium shifts to left, $[H_2(g)]$ decreases, [CO(g)] and $[H_2O(g)]$ increase.

Increasing the concentration of $CO_2(g)$ will also cause the equilibrium to shift to the left. The concentrations and quantities of CO(g) and $H_2O(g)$ will increase and the concentration and quantity of $H_2(g)$ will decrease.

14.19 (a) equilibrium shifts to left.

Increasing pressure by decreasing the volume shifts the equilibrium toward the side with fewer moles of gas.

(b) equilibrium will not shift.

Under ideal conditions, addition of inert gases does not shift equilibria.

(c) equilibrium shifts to right.

The forward reaction is endothermic, which means that an increase in temperature will be "relieved" by a shift of the equilibrium to the right.

14.20 (a) equilibrium shifts to left.

The left side of the reaction has fewer moles of gas than the right side.

(b) equilibrium will not shift.

(c) equilibrium shifts to right.

The forward reaction is endothermic.

(d) equilibrium will not shift.

Catalysts make reactions reach equilibrium faster but do not shift equilibria.

14.21 **(a)** (c) Increasing temperature at constant pressure will change the value of the equilibrium constant.

(b) increase

Increasing temperature at constant pressure shifts the equilibrium to the right. The concentrations of the products, which are in the numerator of the equilibrium constant expression, increase and the concentrations of the reactants, which are in the denominator of the equilibrium constant expression, decrease. Thus, the value of the equilibrium constant increases.

Additional Practice Problems

14.59 1.24×10^{-5}

$$K_p = \frac{(p_{HI})^2}{p_{H_2S}} = \frac{(3.50 \times 10^{-3})^2}{9.90 \times 10^{-1}} = 1.24 \times 10^{-5}$$

14.61 **(a)** $F_2(g) \rightleftarrows 2F(g)$

(b) $4NH_3(g) + 5O_2(g) \rightleftarrows 4NO(g) + 6H_2O(g)$

(c) $C_6H_5COOH(aq) \rightleftarrows C_6H_5COO^- + H^+$

(d) $2H_2O(l) \rightleftarrows 2H_2(g) + O_2(g)$

14.63 **(a)** $[CO]_{eq} = 3.8 \times 10^{-6}$, $[COCl_2]_{eq} = 6.6 \times 10^{-2}$

$$K_c = 2.2 \times 10^{-10} = \frac{[CO][Cl_2]}{[COCl_2]}$$

$$[CO]_{eq} = [Cl_2]_{eq} = 3.8 \times 10^{-6}$$

$$[COCl_2]_{eq} = \frac{(3.8 \times 10^{-6})^2}{2.2 \times 10^{-10}} = 6.6 \times 10^{-2}$$

(b) 6.6×10^{-2} M

$COCl_2 \rightleftarrows CO + Cl_2$

From (a), $[COCl_2]_{eq} = 6.6 \times 10^{-2}$ and 3.8×10^{-6} mol $COCl_2$ was required to produce the 6.6×10^{-2} mol of CO and Cl_2. Thus, the initial concentration was $(6.6 \times 10^{-2} + 3.8 \times 10^{-6}) = 6.6 \times 10^{-2}$.

14.65 **(a)** low temperature

The forward reaction is exothermic (ΔH^0 negative). Thus, according to Le Chatelier's principle, the forward reaction is favored at *low* temperature.

(b) low pressure

There are two moles of gaseous product and only one mole of gaseous reactant. Therefore, the forward reaction would be favored by *low* pressure.

(c) Add a catalyst.

This would increase the rates of both the forward and reverse reactions and would make the reaction reach equilibrium faster without shifting the equilibrium. Increasing the temperature would also make the reaction reach equilibrium faster but would shift the equilibrium to the left so that reaction would be less complete at equilibrium.

14.67 (a) 1.6×10^{-21}

$K_p = K_c(RT)^{\Delta n_g}$ $\Delta n_g = n_{\text{gas products}} - n_{\text{gas reactants}} = 1$

$$K_p = (6.5 \times 10^{-23})\left[\left(\frac{0.082\,06\ \text{L}\cdot\text{atm}}{\cancel{\text{K}}\cdot\text{mol}}\right)(298\,\cancel{\text{K}})\right]^{+1} = 1.6 \times 10^{-21}$$

(b) 3.0×10^{-2}

$$2.2 = K_c\left[\left(\frac{0.082\,06\ \text{L}\cdot\text{atm}}{\cancel{\text{K}}\cdot\text{mol}}\right)(900\,\cancel{\text{K}})\right]^{+1}, \quad K_c = \frac{2.2}{(0.082\,06\cdot 900)} = 3.0 \times 10^{-2}$$

14.69 (a) $[N_2O_4]_{eq} = 4.9 \times 10^{-2}$, $[NO_2]_{eq} = 1.02 \times 10^{-1}$

The summary table is

	$N_2O_4(g)$ $\rightleftarrows$	$2NO_2(g)$
Init. conc., M	0.100	0.000
Change, ΔM	$-x$	$2x$
Eq. conc., M	$0.100 - x$	$2x$

The equilibrium constant expression is

$$K_c = \frac{[NO_2]^2}{[N_2O_4]} = 0.21$$

Substitution of the equilibrium concentrations in terms of x in the equilibrium constant expression gives

$$\frac{(2x)^2}{0.100 - x} = 0.21$$

Because the value of K_C for this equilibrium is *not* very small, either successive approximations or the quadratic formula must be used to solve for x.

The quadratic equation in standard form is

$$4x^2 + 0.21x - 0.021 = 0$$

and $x = 0.051$ or 5.1×10^{-2}

$[N_2O_4]_{eq} = 0.100 - 5.1 \times 10^{-2} = 4.9 \times 10^{-2}$

$[NO_2]_{eq} = 2(5.1 \times 10^{-2}) = 1.02 \times 10^{-1}$ (i.e., $0.051 + 0.051 = 0.102$)

(b) $[N_2O_4] = 0.075$

According to the equation for the equilibrium, to form 0.050 mol NO_2,

$$(0.050\ \cancel{\text{mol } NO_2})\left(\frac{1\text{ mol } N_2O_4}{2\ \cancel{\text{mol } NO_2}}\right) = 0.025\text{ mol } N_2O_4$$

are required.

0.100 mol	N_2O_4 present at start
– 0.025 mol	N_2O_4 reacted
0.075 mol	N_2O_4 left

(c) $[NO_2] = 0.080$

When $[N_2O_4] = 0.60$, $0.100 - 0.060 = 0.040$ mol N_2O_4 have reacted and

$$(0.040\ \cancel{\text{mol } N_2O_4})\left(\frac{2\text{ mol } NO_2}{1\ \cancel{\text{mol } N_2O_4}}\right) = 0.080\text{ mol } NO_2$$

has formed.

(d) Change in $[N_2O_4]$ and $[NO_2]$ as a function of time:

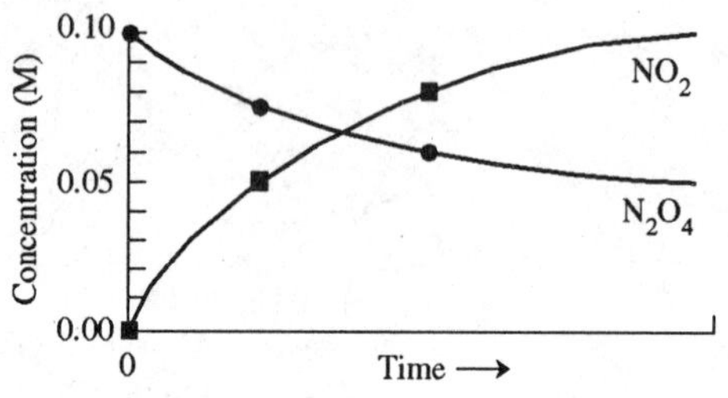

Summarize the information for the graph in a table.

$[N_2O_4]_0 = 0.100$	$[NO_2] = 0.000$
$[N_2O_4]_{t_1} = 0.075$	$[NO_2]_{t_1} = 0.050$
$[N_2O_4]_{t_2} = 0.060$	$[NO_2]_{t_2} = 0.080$
$[N_2O_4]_{eq} = 0.049$	$[NO_2]_{eq} = 0.102$

14.71 $[N_2O_4]_{eq} = 0.009\,48$, $[NO_2]_{eq} = 0.006\,64$

$$K_c = \frac{[NO_2]^2_{eq}}{[N_2O_4]_{eq}} = 4.64 \times 10^{-3}$$

Let $2x$ = mol NO_2 reacted at equilibrium. Then x = mol N_2O_4 formed at equilibrium. Set up table of information:

	$N_2O_4(g)$	$\rightleftarrows$	$2NO_2(g)$
Initial concentration, M	0.0000		0.0256
Change in concentration, ΔM	$+x$		$-2x$
Equilibrium concentration, M	x		$(0.0256 - 2x)$

Plug in the appropriate values into the equilibrium constant expression:

$$\frac{(0.0256 - 2x)^2}{x} = 4.64 \times 10^{-3}$$

Assuming that $2x$ is negligible compared to 0.0256 gives $x = 0.141$ and $2x = 0.242$. This is impossible; 0.242 mol NO_2 cannot react if only 0.0256 mol NO_2 are present at the start of the reaction. The quadratic formula must be used to solve this problem. To use the quadratic formula, the equation must first be rearranged into the standard quadratic form.

$$(0.0256 - 2x)^2 = (4.64 \times 10^{-3})x$$

$$(0.0256)^2 + 2(0.0256)(-2)x + (-2x)^2 = (4.64 \times 10^{-3})x$$

$$(6.553 \times 10^{-4}) + (-0.1024)x + 4x^2 - (4.64 \times 10^{-3})x = 0$$

$$4x^2 - (0.1024)x - (4.64 \times 10^{-3})x + 6.553 \times 10^{-4} = 0$$

$$4x^2 - 0.107\,04x + 6.553 \times 10^{-4} = 0$$

Substitution of the coefficients in the quadratic formula

$$x = \frac{-b \pm \sqrt{b^2 - 4ac}}{2a} \text{ gives } x = \frac{-(-0.107\,04) \pm \sqrt{(-0.107\,04)^2 - 4(4)(6.553 \times 10^{-4})}}{2(4)}$$

$$= \frac{(0.107\,04) \pm \sqrt{(0.011\,4575) - (0.010\,48)}}{8} = \frac{(0.107\,04) \pm \sqrt{9.727 \times 10^{-4}}}{8}$$

$$= \frac{(0.107\,04) \pm (0.031\,188)}{8} = 0.0173 \text{ or } 0.009\,48$$

(Note that in this problem and others involving the solution of a quadratic equation, extra digits have been retained in intermediate steps to avoid significant rounding errors.)

Since 2(0.0173) = 0.0346 is greater than 0.0256, the initial concentration of NO_2, the $[N_2O_4]_{eq}$ must be 0.009 48 M and $[NO_2]_{eq}$ = 0.0256 M – 2(0.009 48 M) = 0.006 64 M

To check the answers, substitute them into the equilibrium constant expression:

$$K_c = \frac{[NO_2]^2_{eq}}{[N_2O_4]_{eq}} = \frac{(0.006\,64 \text{ M})^2}{0.009\,48 \text{ M}} = 4.65 \times 10^{-3}\text{, which is correct to 3 significant figures}$$

14.73 $p_{PCl_3} = p_{Cl_2} = 0.691$ atm, $p_{PCl_5} = 0.96$ atm

The summary table is

	$PCl_5(g)$ ⇄	$PCl_3(g)$	+	$Cl_2(g)$
Initial partial pressure, atm	1.65	0.000		0.000
Change in pressure, atm	$-x$	$+x$		$+x$
Equilibrium pressure, atm	$1.65 - x$	x		x

The equilibrium constant expression is

$$K_p = \frac{(p_{PCl_3})(p_{Cl_2})}{p_{PCl_5}} = 0.497$$

Substitution of the equilibrium partial pressures in terms of x in the equilibrium constant expression gives

$$\frac{(x)(x)}{1.65 - x} = 0.497$$

Because the value of K_p for this equilibrium is not very small, either successive approximations or the quadratic formula must be used to solve for x.

The quadratic equation in standard form is

$$x^2 + 0.497x - 0.820 = 0$$

and $x = 0.691$ or -1.19 (the latter being impossible)

$p_{PCl_3} = p_{Cl_2} = x = 0.691$ atm

$p_{PCl_5} = 1.65 - 0.691 = 0.96$ atm

14.75 5.3×10^{-5}

The second equilibrium is the reverse of the first. Therefore $K'_c = \frac{1}{K_c} = \frac{1}{3.6 \times 10^8} = 2.8 \times 10^{-9}$.

The quantities shown in the second equation are half those shown in the first equation.
Therefore, $K''_c = \sqrt{K'_c} = 5.3 \times 10^{-5}$

Stop and Test Yourself

1. (c) $2C(s) + O_2(g) \rightarrow 2CO(g) \quad K_c = 1 \times 10^{16}$

The greater the value of K_c, the greater the concentrations of the products.

2. (a) N_2, 0.075 M and H_2, 0.225 M

To form 0.350 mol NH_3

$$(0.350\ \cancel{\text{mol NH}_3})\left(\frac{1\ \text{mol N}_2}{2\ \cancel{\text{mol NH}_3}}\right) = 0.175\ \text{mol N}_2$$

and

$$(0.350\ \cancel{\text{mol NH}_3})\left(\frac{3\ \text{mol H}_2}{2\ \cancel{\text{mol NH}_3}}\right) = 0.525\ \text{mol H}_2 \text{ must react}$$

The summary table is

	$N_2(g)$	+	$3H_2(g)$	$\rightleftarrows$	$2NH_3(g)$
Init. conc., M	0.250		0.750		0.000
Change, ΔM	–0.175		–0.525		+0.350
Eq. conc., M	0.250 – 0.175		0.750 – 0.525		0.350

$[N_2]_{eq} = 0.250 - 0.175 = 0.075$ and $[H_2]_{eq} = 0.750 - 0.525 = 0.225$

3. (b) $K_c = \dfrac{[N_2O][H_2O]^3}{[NH_3]^2[O_2]^2}$

4. (d) $K_c = \dfrac{[CO]^2}{[CO_2]}$

The [C(s)] is included in K_c and does not appear in the equilibrium constant expression.

5. (b) 4.56×10^{-2}

$[CO] = [Cl_2] = 2.32 \times 10^{-2}$ $[COCl_2] = 0.0350 - 0.0232 = 1.18 \times 10^{-2}$

$$K_c = \frac{[CO][Cl_2]}{[COCl_2]} = \frac{(2.32 \times 10^{-2})^2}{1.18 \times 10^{-2}} = 4.56 \times 10^{-2}$$

6. (e) 3.5×10^{-5}

	$N_2(g)$	+	$O_2(g)$	$\rightleftarrows$	$2NO(g)$
Init. conc., M	0.036		0.075		0
Change, ΔM	$-x$		$-x$		$+2x$
Eq. conc., M	$0.036 - x$		$0.075 - x$		$2x$

The equilibrium constant expression is

$$K_c = \frac{[NO]^2}{[N_2][O_2]} = 4.59 \times 10^{-7}$$

Substitution of the equilibrium concentrations in terms of x in the equilibrium constant expression gives

$$\frac{(2x)^2}{(0.036 - x)(0.075 - x)} = 4.59 \times 10^{-7}$$

Since K_c is small, assume x is negligible compared to 0.036 and 0.075. Then

$$K_c = 4.59 \times 10^{-7} \approx \frac{4x^2}{(0.036)(0.075)}$$

and

$$x = 1.8 \times 10^{-5}$$

The assumption that x is negligible compared to 0.036 and 0.075 is correct.

$[NO]_{eq} = 2x = 3.6 \times 10^{-5}$

(Best answer, without intermediate round off, is 3.5×10^{-5}.)

7. (c) 0.034

	$Cl_2(g)$	$\rightleftarrows$	$2Cl(g)$
Init. conc., M	0.049		0.000
Change, ΔM	$-x$		$+2x$
Eq. conc., M	$0.049 - x$		$2x$

$$K_c = 3.6 \times 10^{-2} = \frac{[Cl]^2}{[Cl_2]} = \frac{(2x)^2}{(0.049 - x)}$$

$$4x^2 + (3.6 \times 10^{-2})x - 1.8 \times 10^{-3} = 0$$

Solving for x with the quadratic formula gives

$x = 1.7 \times 10^{-2}$ $[Cl] = 2x = 3.4 \times 10^{-2} = 0.034$

8. (e) $[CH_4] = 0.025$, $[H_2O] = 0.036$, $[CO] = 0.041$, $[H_2] = 0.178$

For the equilibrium

$$CO(g) + 3H_2(g) \rightleftarrows CH_4(g) + H_2O(g)$$

the reaction quotient $Q = \dfrac{[CH_4][H_2O]}{[CO][H_2]^3}$

To determine which mixture is at equilibrium, substitute the concentrations given and compare the value of Q obtained with the value given for K_c, 3.89. If $Q = K_c$, the system is at equilibrium.

(a) $Q = \dfrac{(0.047)(0.058)}{(0.033)(0.156)^3} = 22 > K_c$

(b) $Q = \dfrac{(0.068)(0.072)}{(0.029)(0.251)^3} = 11 > K_c$

(c) $Q = \dfrac{(0.052)(0.034)}{(0.296)(0.145)^3} = 2.0 < K_c$

(d) $Q = \dfrac{(0.236)(0.029)}{(0.368)(0.124)^3} = 9.8 > K_c$

(e) $Q = \dfrac{(0.025)(0.036)}{(0.041)(0.178)^3} = 3.9$ (3.89 before rounding) $Q = K_c$

The system is at equilibrium.

9. (c) $[CH_4] = 0.052$, $[H_2O] = 0.034$, $[CO] = 0.296$, $[H_2] = 0.145$

If $Q < 3.89$, then reaction will take place in the forward direction to increase the concentrations of the products and decrease the concentrations of the reactants until $Q = K_c$ and the system is at equilibrium.

Only in (c) is $Q < K_c$.

10. **(e)** $\frac{(0.0275 - x)(0.0525 - x)}{(0.0195 + x)} = 3.9 \times 10^{-2}$

	$PCl_5(g)$ $\rightleftarrows$	$PCl_3(g)$ +	$Cl_2(g)$
Eq. conc., M	0.0195	0.0275	0.0275
Init. conc., M	0.0195	0.0275	0.0275 + 0.0250 = 0.0525
Change, ΔM	$+x$	$-x$	$-x$
Eq.' conc., M	$0.0195 + x$	$0.0275 - x$	$0.0525 - x$

$$K_c = 3.9 \times 10^{-2} = \frac{[PCl_3][Cl_2]}{[PCl_5]} = \frac{(0.0275 - x)(0.0525 - x)}{(0.0195 + x)}$$

11. **(e)** removing water

(a) Adding more SiO_2 would not shift the equilibrium because SiO_2 is a solid. Its concentration is constant. (b) Increasing the temperature would shift the equilibrium to the right because the forward reaction is endothermic; thermal energy is a "reactant." (c) Adding a catalyst does not shift equilibria. (d) Decreasing the pressure by increasing the volume would shift this equilibrium to the right because, in the forward reaction, two moles of gas (H_2O) yield four moles of gas (HF). (e) Water is a reactant so removing water would shift the equilibrium to the left.

12. **(b)** increasing the temperature

Only increasing the temperature will result in a change in the value of the equilibrium constant.

13. **(e)** $N_2(g) + O_2(g) \rightleftarrows 2NO(g)$

Only in equilibrium (e) is the number of moles of gaseous products the same as the number of moles of gaseous reactants.

14. **(b)** 8.31×10^{-6}

$$K_p = K_c(RT)^{\Delta n_g}$$

$$K_p = (0.176)\left(\left[\frac{0.082\ 06\ \text{atm} \cdot \text{L}}{\text{mol} \cdot \text{K}}\right](1773\ \text{K})\right)^{-2}$$

$$K_p = \frac{0.176}{(0.082\ 06 \cdot 1773)^2} = 8.31 \times 10^{-6}$$

15. **(d)** 0.873 atm

$$K_p = \frac{p^2_{Cl}}{p_{Cl_2}} = 6.71$$

$P_{total} = 0.987$ atm $= p_{Cl} + p_{Cl_2}$

Let $x = p_{Cl_2}$. Then $p_{Cl} = 0.987 - x$ and

$$6.71 = \frac{(0.987 - x)^2}{x}$$

or

$$x^2 - 8.684x + 9.74 \times 10^{-1} = 0$$

Solving for x by successive approximations or with the quadratic formula gives

$x = 0.114$. Therefore, $p_{Cl} = (0.987 - 0.114)$ atm $= 0.873$ atm

Putting Things Together

14.77 **(a)** 0.403

$$(1.00 \text{ mol } CO_2)\left(\frac{54.8 \text{ parts decomposed}}{100 \text{ parts total}}\right) = 0.548 \text{ mol } CO_2 \text{ decomposed}$$

From the equation for the equilibrium

$$(0.548 \text{ mol } CO_2)\left(\frac{2 \text{ mol CO}}{2 \text{ mol } CO_2}\right) = 0.548 \text{ mol CO}$$

and $(0.548 \text{ mol } CO_2)\left(\frac{1 \text{ mol } O_2}{2 \text{ mol } CO_2}\right) = 0.274 \text{ mol } O_2$

were formed. Because the volume of the container is 1.00 L,

$[CO]_{eq} = 0.548$ and $[O_2]_{eq} = 0.274$

At equilibrium, 1.00 mol CO_2 – 0.548 mol CO_2 = 0.452 mol O_2 are left and $[CO]_{eq} = 0.452$.

Thus, $K_c = \frac{[CO]^2[O_2]}{[CO_2]^2} = \frac{(0.548)^2(0.274)}{(0.452)^2} = 0.403$

(b) Either the reaction must be endothermic so that K_c is much smaller at room temperature or the rate of decomposition must be very low at room temperature.

14.79 $[SO_3] = 6.87 \times 10^{-3}$, $[SO_2] = 2.44 \times 10^{-3}$, $[O_2] = 1.22 \times 10^{-3}$

Because concentrations, not pressures, are given, we first need to find K_c.

$$90.8 = K_c\left[\left(\frac{0.082\,06 \text{ atm} \cdot \text{L}}{\text{mol} \cdot \text{K}}\right)(873 \text{ K})\right]^{-1}$$

$$6.50 \times 10^3 = K_c$$

Then we need to make the "ICE" table:

	$2SO_2(g)$	+ $O_2(g)$ $\rightleftarrows$	$2SO_3(g)$
Init. conc., M	9.31×10^{-3}	4.65×10^{-3}	0.000
Change, ΔM	$-2x$	$-x$	$+2x$
Eq. conc., M	$9.31 \times 10^{-3} - 2x$	$4.65 \times 10^{-3} - x$	$2x$

Substituting the equilibrium concentrations in terms of x in the equilibrium constant expression:

$$K_c = \frac{[SO_3]^2}{[SO_2]^2[O_2]} = \frac{(2x)^2}{(9.31 \times 10^{-3} - 2x)^2(4.65 \times 10^{-3} - x)} = 6.50 \times 10^3$$

Solving by successive approximations gives

$2x = 0.006\ 87$
$x = 0.003\ 43$

$[SO_3] = 6.87 \times 10^{-3}$
$[SO_2] = 2.44 \times 10^{-3}$
$[O_2] = 1.22 \times 10^{-3}$

Check: $\frac{(6.87 \times 10^{-3})^2}{(2.44 \times 10^{-3})^2(1.22 \times 10^{-3})} = 6.50 \times 10^3$

14.81 **(a)** 0.20

$$K_c = \frac{[NO_2]^2}{[N_2O_4]} = \frac{(0.134)^2}{0.090} = 0.20$$

(b) Decreasing the volume by a factor of two by increasing the pressure will result in a doubling of concentrations for both NO_2 and N_2O_4, that is $[NO_2] = 0.268$ and $[N_2O_4] = 0.180$.

(c) No. $Q = \frac{(0.268)^2}{0.180} = 0.40$. Thus $Q > K_c$ and the reaction will proceed from right to left to reestablish equilibrium.

(d) $[N_2O_4] = 0.211$, $[NO_2] = 0.206$

	$N_2O_4(g)$ $\rightleftarrows$	$2NO_2(g)$
Init. conc., M	0.180	0.268
Change, ΔM	x	$-2x$
Eq. conc., M	$0.180 + x$	$0.268 - 2x$

$$K_c = \frac{[NO_2]^2}{[N_2O_4]} = \frac{(0.268 - 2x)^2}{(0.180 + x)} = 2.0 \times 10^{-1}$$

Solution by the quadratic formula gives $x = 0.031$ and $[N_2O_4] = 0.180 + 0.031 = 0.211$ and $[NO_2] = 0.268 - 2(0.031) = 0.206$.

Check: $\frac{(0.206)^2}{0.211} = 0.201$

14.83 **(a)** $H_2O(l) \rightleftarrows H_2O(g)$

(b) $K_p = p_{H_2O(g)}$

(c) 29.354 mmHg (Table 5.4)

(d) 1.560×10^{-3}

$K_p = K_c(RT)^{\Delta n}$ $\quad \Delta n = 1 - 0$

$$(29.354\ \cancel{\text{mmHg}})\left(\frac{1\ \text{atm}}{760\ \cancel{\text{mmHg}}}\right) = 3.8624 \times 10^{-2}\ \text{atm}$$

$$K_c = \frac{K_p}{(RT)^{\Delta n}} = \frac{3.8624 \times 10^{-2}}{(0.082\ 06[273.2 + 28.6])}$$

$$K_c = 1.560 \times 10^{-3}$$

(e) 1.560×10^{-3} M

$$K_c = [H_2O] = 1.560 \times 10^{-3}\ \text{M}$$

14.85 1.6×10^{-15}

$$2Cl_2(g) + 2H_2O(g) \rightleftarrows 4HCl(g) + O_2(g)$$

$$K_c = \frac{[HCl]^4[O_2]}{[Cl_2]^2[H_2O]^2}$$

$$[Cl_2]_0 = (8.4\ \cancel{\text{g Cl}_2})\left(\frac{1\ \text{mol Cl}_2}{70.9\ \cancel{\text{g Cl}_2}}\right)\left(\frac{1}{4.2\ \text{L}}\right) = 2.8 \times 10^{-2}\ \frac{\text{mol}}{\text{L}}\ (\text{M})\ Cl_2$$

$$[H_2O]_0 = (2.7\ \cancel{\text{g H}_2\text{O}})\left(\frac{1\ \text{mol}}{18.0\ \cancel{\text{g H}_2\text{O}}}\right)\left(\frac{1}{4.2\ \text{L}}\right) = 3.6 \times 10^{-2}\ \frac{\text{mol}}{\text{L}}\ (\text{M})\ H_2O$$

Because $[O_2]_0 = 0.00$ and $[O_2]_{eq} = 2.3 \times 10^{-5}$, 2.3×10^{-5} mol $O_2(g)$ (per liter) were formed.

From the equation for the reaction,

$$(2.3 \times 10^{-5}\ \cancel{\text{mol O}_2})\left(\frac{4\ \text{mol HCl}}{1\ \cancel{\text{mol O}_2}}\right) = 9.2 \times 10^{-5}\ \text{mol HCl}$$

were also formed and

$$(2.3 \times 10^{-5}\ \cancel{\text{mol O}_2})\left(\frac{2\ \text{mol H}_2\text{O}}{1\ \cancel{\text{mol O}_2}}\right) = 4.6 \times 10^{-5}\ \text{mol H}_2\text{O reacted.}$$

Because the coefficient of Cl_2 in the equation is also two, 4.6×10^{-5} mol Cl_2 also reacted.

The "ICE" table is

	$2Cl_2(g)$ +	$2H_2O(g)$ ⇄	$4HCl(g)$ +	$O_2(g)$
Init. conc., M	2.8×10^{-2}	3.6×10^{-2}	0.000	0.000
Change, ΔM	-4.6×10^{-5}	-4.6×10^{-5}	$+9.2 \times 10^{-5}$	$+2.3 \times 10^{-5}$
Eq. conc., M	2.8×10^{-2}	3.6×10^{-2}	9.2×10^{-5}	2.3×10^{-5}

$$K_c = \frac{(9.2 \times 10^{-5})^4(2.3 \times 10^{-5})}{(2.8 \times 10^{-2})^2(3.6 \times 10^{-2})^2} = 1.6 \times 10^{-15}$$

14.87 **(a)** 0.75

$$CO_2(g) + H_2(g) \rightleftarrows CO(g) + H_2O(g)$$

$$K_c = \frac{[CO][H_2O]}{[CO_2][H_2]} = \frac{(0.022)(0.022)}{(0.019)(0.034)} = 0.75$$

(b) $X_{CO} = X_{H_2O} = 0.23$, $X_{CO_2} = 0.20$, $X_{H_2} = 0.35$

$0.022 + 0.022 + 0.019 + 0.034 = 9.7 \times 10^{-2}$ total mol

mol fraction: $CO = H_2O = \frac{0.022}{9.7 \times 10^{-2}} = 0.23$

$CO_2 = \frac{0.019}{9.7 \times 10^{-2}} = 0.20$ $\qquad H_2 = \frac{0.034}{9.7 \times 10^{-2}} = 0.35$

Check: $0.23 + 0.23 + 0.20 + 0.35 = 1.01$

(c) Values are the same.

$$K_c = \frac{(0.23)^2}{(0.20)(0.35)} = 0.76 \approx 0.75$$

(d) No. In the first equilibrium, there are the same number of molecules of products as there are molecules of reactants. This is not true of the second equilibrium.

$$K_x = \frac{\left(\frac{\text{mol CO}}{\text{tot}}\right)\left(\frac{\text{mol H}_2\text{O}}{\text{tot}}\right)}{\left(\frac{\text{mol CO}_2}{\text{tot}}\right)\left(\frac{\text{mol H}_2}{\text{tot}}\right)} = K_c \text{ for } CO_2(g) + H_2(g) \rightleftarrows CO(g) + H_2O(g)$$

$$\text{But } K_x = \frac{\left(\frac{\text{mol CH}_4}{\text{tot}}\right)\left(\frac{\text{mol H}_2\text{O}}{\text{tot}}\right)}{\left(\frac{\text{mol CO}}{\text{tot}}\right)\left(\frac{\text{mol H}_2}{\text{tot}}\right)^3} \neq K_c \text{ for } CO(g) + 3H_2(g) \rightleftarrows CH_4(g) + H_2O(g)$$

14.89 **(a)** 0.450 g

Problem	0.003 55 g		0.556 g		? g
Equation	$H_2(g)$	+	$I_2(g)$	→	$2HI(g)$
Formula mass, u	2.016		253.8		127.9
Recipe, mol	1		1		2

The quantities of two reactants are given. This is a limiting reactant problem.

$$(0.00355\ \cancel{g\ H_2})\left(\frac{1\ \text{mol } H_2}{2.016\ \cancel{g\ H_2}}\right) = 0.001\ 76 \text{ or } 1.76 \times 10^{-3} \text{ mol } H_2$$

$$(0.556\ \cancel{g\ I_2})\left(\frac{1\ \text{mol } I_2}{253.8\ \cancel{g\ I_2}}\right) = 0.002\ 19 \text{ or } 2.19 \times 10^{-3} \text{ mol } I_2$$

$\frac{0.0176\ \cancel{\text{mol } H_2}}{1\ \cancel{\text{mol } H_2}} = 0.001\ 76$ of a recipe $\qquad \frac{0.002\ 19\ \cancel{\text{mol } I_2}}{1\ \cancel{\text{mol } I_2}} = 0.002\ 19$ of a recipe

0.001 76 < 0.002 19. Therefore H_2 is limiting.

The theoretical yield of HI is

$$(0.00355\ \cancel{g\ H_2})\left(\frac{1\ \cancel{mol\ H_2}}{2.016\ \cancel{g\ H_2}}\right)\left(\frac{2\ \cancel{mol\ HI}}{1\ \cancel{mol\ H_2}}\right)\left(\frac{127.9\ g\ HI}{\cancel{mol\ HI}}\right) = 4.50\ g\ HI$$

(b) 0.389 g HI

	$H_2(g)$	+	$I_2(g)$	$\rightleftarrows$	$2HI(g)$
Init. conc., M	0.001 76		0.002 19		0.000 00
Change, ΔM	$-x$		$-x$		$+2x$
Eq. conc., M	$0.001\ 76 - x$		$0.002\ 19 - x$		$2x$

Substitution of the equilibrium concentrations in terms of x in the equilibrium constant expression gives

$$K_c = \frac{[HI]^2}{[H_2][I_2]} = \frac{(2x)^2}{(1.76\times10^{-3} - x)(2.19\times10^{-3} - x)} = 55.2$$

or

$$51.2x^2 - 0.218x + 2.13\times10^{-4} = 0$$

Solving the equation for x gives

$$x = 1.52\times10^{-3}$$

Therefore

$$[2(1.52\times10^{-3})\ \cancel{mol\ HI}]\left(\frac{127.9\ g\ HI}{1\ \cancel{mol\ HI}}\right) = 0.389\ g\ HI$$

are present

(c) Increase the concentration of H_2 or remove HI as it is formed.

14.91 27.1 M

$$\left(\frac{2.71\ \cancel{g\ CaCO_3}}{\cancel{cm^3}}\right)\left(\frac{1000\ \cancel{cm^3}}{L}\right)\left(\frac{1\ mol\ CaCO_3}{100.1\ \cancel{g\ CaCO_3}}\right) = 27.1\ \frac{mol\ CaCO_3}{L} = 27.1\ M$$

14.93 **(a)** 5.64×10^{-2}

	$N_2O_4(g)$	$\rightleftarrows$	$2NO_2(g)$
Init. conc., M	0.0645		0.0000
Change, ΔM	$-x$		$+2x$
Eq. conc., M	$0.0645 - x$		$+2x$

$$K_c = \frac{[NO_2]^2}{[N_2O_4]} = 4.61\times10^{-3} = \frac{(2x)^2}{(0.0645 - x)}$$

The quadratic equation is $4x^2 + 4.61\times10^{-3}x - 2.97\times10^{-4} = 0$.

Solution by the quadratic formula gives $x = 0.008\ 06$ and $[N_2O_4] = 0.0645 - 0.008\ 06 = 5.64\times10^{-2}$.

(b) 12.5%

$$\% \text{ decomposition} = \frac{x}{[N_2O_4]_0} \times 100 = \frac{0.008\,06}{0.0645} \times 100 = 12.5\%$$

14.95 Ammonia will decompose.

$$\text{molarity } NH_3\text{:} \left(\frac{0.400 \text{ mol}}{40.0 \text{ L}}\right) = 0.0100 \text{ M}$$

$$\text{molarity } N_2\text{:} \left(\frac{0.800 \text{ mol}}{40.0 \text{ L}}\right) = 0.0200 \text{ M}$$

$$\text{molarity } H_2\text{:} \left(\frac{2.40 \text{ mol}}{40.0 \text{ L}}\right) = 0.0600 \text{ M}$$

$$Q = \frac{[NH_3]^2}{[N_2][H_2]^3} = \frac{(0.0100)^2}{(0.0200)(0.0600)^3} = 23.1$$

$K_c = 0.500$; $Q > K_c$. Thus, the reaction will move to the left or reverse and ammonia will decompose.

14.97 5.1×10^{-4}

$$[SO_3]_0 = \left(\frac{0.0220 \text{ mol}}{1.50 \text{ L}}\right) = 0.0147 \text{ M}$$

$$[SO_3]_{eq} = \left(\frac{0.0150 \text{ mol}}{1.50 \text{ L}}\right) = 0.0100 \text{ M}$$

	$2SO_3(g)$ $\rightleftarrows$	$2SO_2(g)$ +	$O_2(g)$
Init. conc., M	0.0147	0.0000	0.0000
Change, ΔM	−0.0047	+0.0047	+0.0024
Eq. conc., M	0.0100	0.0047	+0.0024

$$K_c = \frac{[SO_2]^2[O_2]}{[SO_3]^2} = \frac{(0.0047)^2(0.0024)}{(0.0100)^2} = 5.3 \times 10^{-4}$$

(Best answer, without intermediate round off, is 5.1×10^{-4}.)

Applications

14.99 **(a)** The presence of water and carbon dioxide allows equilibrium to be established. When the water moves, removal of the ions causes the equilibrium to shift to the right, and $CaCO_3(s)$ dissolves..

(b) Because the forward reaction is exothermic, heating shifts the equilibrium to the left, causing $CaCO_3(s)$ to precipitate.

(c) As water evaporates, the equilibrium is shifted to the left. This results in precipitation of $CaCO_3(s)$.

14.101 Don't print the story. Since catalysts speed up the rates of *both* the forward and reverse reactions, they do not change the position of the equilibrium.

14.103 **(a)** 203.25 kJ

Equation	$CH_4(g)$	+	$H_2O(g)$	$\rightarrow$	$CO(g)$	+	$3H_2(g)$
ΔH_f^0, kJ/mol	–74.85		–238.92		–110.52		0

$$\Delta H^0_{rxn} = \left[(1\ \cancel{\text{mol CO}})\left(\frac{-110.52\text{ kJ}}{\cancel{\text{mol CO}}}\right) + (3\ \cancel{\text{mol H}_2})\left(\frac{0\text{ kJ}}{\cancel{\text{mol H}_2}}\right)\right]$$

$$-\left[(1\ \cancel{\text{mol CH}_4})\left(\frac{-74.85\text{ kJ}}{\cancel{\text{mol CH}_4}}\right) + (1\ \cancel{\text{mol H}_2\text{O}})\left(\frac{-238.92\text{ kJ}}{\cancel{\text{mol H}_2\text{O}}}\right)\right]$$

$= 203.25$ kJ

(b) High temperature, low pressure, and a high concentration of $H_2O(g)$ would favor the forward reaction. Low temperature, high pressure, and removal of $H_2O(g)$ [or $CH_4(g)$] would favor the reverse reaction.

14.105 **(a)** $p_{H_2O} = 2.51 \times 10^{-2}$ atm for $Na_2SO_4 \cdot 10H_2O$, $p_{H_2O} = 6.088 \times 10^{-8}$ atm for $CaCl_2 \cdot 6H_2O$

$K_p = (p_{H_2O})^{10} = 9.99 \times 10^{-17}$ for Na_2SO_4

$p_{H_2O} = (9.99 \times 10^{-17})^{1/10} = 2.51 \times 10^{-2}$ atm

$K_p = (p_{H_2O})^6 = 5.090 \times 10^{-44}$ for $CaCl_2$

$p_{H_2O} = (5.090 \times 10^{-44})^{1/6} = 6.088 \times 10^{-8}$ atm

(b) $CaCl_2 \cdot 6H_2O$

The vapor pressure of water over $CaCl_2 \cdot 6H_2O$ (6.1×10^{-8} atm) is much lower than the vapor pressure of water over $Na_2SO_4 \cdot 10H_2O$ (2.51×10^{-2} atm). To dry a sample, anhydrous (without water) $CaCl_2(s)$ should be added.

14.107 **(a)** $[CO] = [H_2] = 0.3$, $[H_2O] = 0.5$

	$C(s)$	+	$H_2O(g)$	$\rightleftarrows$	$CO(g)$	+	$H_2(g)$
Init. conc., M			0.8		0.0		0.0
Change, ΔM			$-x$		$+x$		$+x$
Eq. conc., M			$0.8 - x$		x		x

$$K_c = \frac{[CO][H_2]}{[H_2O]} = \frac{(x)(x)}{(0.8 - x)} = 0.2$$

Solving by successive approximations or by the quadratic formula gives $x = 0.3$.

$[CO] = [H_2] = 0.3$ and $[H_2O] = (0.8 - 0.3) = 0.5$

(b) endothermic

Heating shifts the equilibrium to the right. The process is endothermic.

(c) The pressure should be low. One mole of gas is converted to two moles of gas by the reaction.

Lower pressure favors products because there are two moles of gas on the product side and only one mole of gas on the reactant side.

(d) At *low pressures*, gases take up a lot of room. The plant would have to be enormous. In addition, at low pressures the concentrations are low and reactions take place slowly compared to their rates at high pressures. *High pressures* require special strong pressure–containment vessels.

14.109 **(a)** 490

$CoO(s) + CO(g) \rightleftarrows Co(s) + CO_2(g)$

Because there is no change in the number of moles of gas,

$K_p = K_c = 490$

(b) $p_{CO_2} = 12.4$ atm, $p_{CO} = 2.53 \times 10^{-2}$ atm

Let $x = p_{CO}$

$$K_p = 490 = \frac{p_{CO_2}}{p_{CO}} = \frac{p_{CO_2}}{x}, \quad p_{CO_2} = 490x$$

$$P_{total} = 12.4 \text{ atm} = p_{CO} + p_{CO_2}, \qquad 12.4 \text{ atm} = x + 490x = 491x$$

$$x = \frac{12.4}{491} = 0.0253 \text{ atm}$$

$$x = p_{CO} = 0.0253 \text{ atm}, \quad p_{CO_2} = 490x = 490(0.0253) = 12.4 \text{ atm}$$

(c) redox reactions

14.111 **(a)** 0.470

	$N_2(g)$	+	$3H_2(g)$	$\rightleftarrows$	$2NH_3(g)$
Init. conc., M	0.653		1.958		0.000
Change, ΔM	$-x$		$-3x$		$+2x$
Eq. conc., M	$0.653 - x$		$1.958 - 3x$		$2x$

$$K_c = \frac{[NH_3]^2}{[N_2][H_2]^3} = \frac{(2x)^2}{(0.653 - x)(1.958 - 3x)^3} = 2.68 \times 10^{-1}$$

$$\frac{(2x)^2}{(0.653 - x)[3(0.653 - x)]^3} = 2.68 \times 10^{-1}$$

$$\frac{(2x)^2}{27(0.653 - x)^4} = 2.68 \times 10^{-1}$$

$$\frac{(2x)^2}{(0.653 - x)^4} = 7.24$$

Taking the square root of both sides of the equation,

$$\frac{2x}{(0.653 - x)^2} = 2.69$$

or, in standard form,

$2.69x^2 - 5.51x + 1.15 = 0$
$x = 0.236$

(Without intermediate round off, $x = 0.235$)

$[NH_3] = 2x = 0.470$
$[N_2] = (0.653 - x) = 0.653 - 0.235 = 0.418$
$[H_2] = (1.958 - 3x) = 1.958 - 0.705 = 1.253$

Check: $\dfrac{(0.470)^2}{(0.418)(1.253)^3} = 2.69 \times 10^{-1}$

(b) 1.31

The reactants are present in the proportions shown by the equation—3 mol H_2 to 1 mol N_2. Either H_2 or N_2 can be used to calculate the theoretical yield.

If N_2 is the limiting reactant, then

$$(0.653 \text{ mol } N_2)\left(\frac{2 \text{ mol } NH_3}{1 \text{ mol } N_2}\right) = 1.31 \text{ mol } NH_3$$

or

$$(1.958 \text{ mol } H_2)\left(\frac{2 \text{ mol } NH_3}{3 \text{ mol } H_2}\right) = 1.31 \text{ mol } NH_3$$

Thus, the theoretical yield of NH_3 is 1.31 mol.

(c) 35.9%

$$\text{percent yield} = \frac{\text{yield from equilibrium calculations}}{\text{theoretical yield from stoichiometry}} \times 100$$

$$= \frac{0.470 \text{ mol } NH_3}{1.31 \text{ mol } NH_3} \times 100 = 35.9\% \text{ yield}$$

14.113 **(a)** 14

$K_p = K_c(RT)^{\Delta n_g} \quad \Delta n_g = 0 - 3$

$$K_c = \frac{K_p}{(RT)^{\Delta n_g}} = \frac{1.3 \times 10^{-4}}{[(0.082\,06)(273 + 300)]^{-3}}$$

$K_c = 14$

(b) to the left

Equation	$CO(g)$	+	$2H_2(g)$	→	$CH_3OH(l)$
ΔH_f^0, kJ/mol	–110.52		0		–238.64

$$\Delta H^0_{rxn} = \left[(1\ \text{mol}\ CH_3OH)\left(\frac{-238.64\ \text{kJ}}{\text{mol}\ CH_3OH}\right)\right] - \left[(1\ \text{mol}\ CO)\left(\frac{-110.52\ \text{kJ}}{\text{mol}\ CO}\right) + (2\ \text{mol}\ H_2)\left(\frac{0\ \text{kJ}}{\text{mol}\ H_2}\right)\right]$$

$= -128.12$ kJ (exothermic)

Reaction will shift to the left if the temperature is increased.

(c) $p_{CO} = 10$ atm, $p_{H_2} = 20$ atm

The stoichiometric ratio is 1 mol CO to 2 mol H_2. Therefore, the ratio of partial pressures is 1:2 or 10:20.

CHAPTER 15: Acids and Bases

Practice Problems

15.1

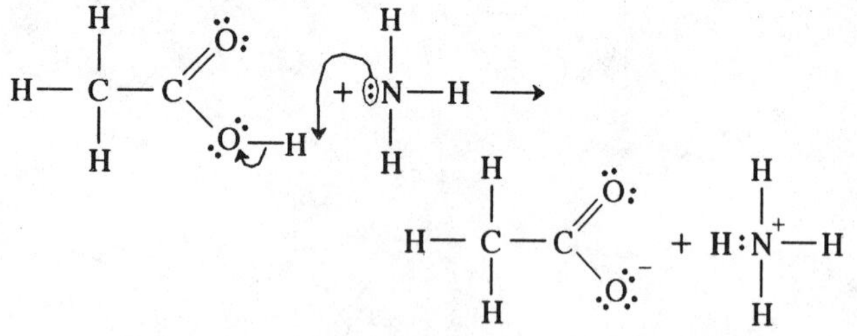

15.2 (a)

$H_2PO_4^-(aq)$	+	$H_2O(l)$	$\rightleftharpoons$	$HPO_4^{2-}(aq)$	+	$H_3O^+(aq)$
$acid_1$		$base_2$		$base_1$		$acid_2$

Species that donate (lose) H^+ are Brønsted–Lowry acids. Species that accept (gain) H^+ are Brønsted–Lowry bases. $Acid_1$ and $base_1$ are a conjugate acid–base pair, as are $acid_2$ and $base_2$.

(b) $H_2PO_4^-$

The fact that the arrow pointing to the left is longer than the arrow pointing to the right shows that equilibrium lies to the left. Thus, there are more $H_2PO_4^-$ ions than HPO_4^{2-} ions.

15.3 (a) PO_4^{3-} (b) NH_2^- (c) $Cu(H_2O)_3(OH)^+$

A conjugate base differs from its conjugate acid by having one less H and a charge one unit more negative (one unit less positive).

15.4 (a) HNO_2 (b) $H_3C\ddot{\underset{\cdot\cdot}{O}}{:}H^+$ (with H above O) (c) $Al(H_2O)_6^{3+}$

A conjugate acid differs from its conjugate base by having one more H and a charge one unit less negative (one unit more positive).

15.5 (a) autoionization

Autoionization is the reaction of one molecule of a substance with another molecule of the same substance to form ions.

(b) amphoteric substances

Substances that can act both as acids and as bases are called amphoteric.

15.6 (a) 1.7×10^{-6}

$[H^+][OH^-] = 1.0 \times 10^{-14}$

$$[OH^-] = \frac{1.0 \times 10^{-14}}{6.0 \times 10^{-9}} = 1.7 \times 10^{-6}$$

(b) basic

$[OH^-] > 1.0 \times 10^{-7}$. Therefore, the solution is basic.

15.7 (a) 1.3×10^{-11}

$$[H^+] = \frac{1.0 \times 10^{-14}}{8.0 \times 10^{-4}} = 1.3 \times 10^{-11}$$

(b) basic

$[H^+] < 1.0 \times 10^{-7}$. Therefore, the solution is basic.

15.8 (a) 5.00

$pH = -\log[H^+] = -\log[1.0 \times 10^{-5}] = 5.00$

(b) 3.08

$pH = -\log[8.3 \times 10^{-4}] = 3.08$

15.9 (a) 2×10^{-9}

$pH = -\log[H^+] = 8.7$

$\log [H^+] = -8.7$, $[H^+] = 2 \times 10^{-9}$

(b) 2.3×10^{-6}

$pH = -\log [H^+] = 5.64$

$\log [H^+] = -5.64$, $[H^+] = 2.3 \times 10^{-6}$

15.10 **(a)** 4.00

$pOH = -\log [OH^-] = -\log [1.0 \times 10^{-4}] = 4.00$

(b) 8.21

$pOH = -\log [6.2 \times 10^{-9}] = 8.21$

(c) 1.0×10^{-11}

$pOH = -\log [OH^-] = 11.00$, $\log [OH^-] = -11.00$, $[OH^-] = 1.0 \times 10^{-11}$

(d) 2.8×10^{-9}

$pOH = -\log [OH^-] = 8.55$, $\log [OH^-] = -8.55$, $[OH^-] = 2.8 \times 10^{-9}$

(e) 6.01

$pH + pOH = 14.00$, $pOH = 14.00 - 7.99 = 6.01$

15.11 3.1×10^{-4}

HCl is a strong acid that is 100% ionized in aqueous solution. Therefore, the H^+ concentration is the same as the concentration of HCl.

$HCl(aq) \rightarrow H^+ + Cl^-$

15.12 **(a)** $HCN(aq) \rightleftharpoons H^+(aq) + CN^-(aq)$ or $HCN(aq) \rightleftharpoons H^+ + CN^-$

(b) $K_a = \dfrac{[H^+][CN^-]}{[HCN]}$

15.13 1.8×10^{-4}

$$\% \text{ ionization}_{HCOOH} = \frac{[HCOO^-]_{eq}}{[HCOOH]_0} \times 100 \text{ and}$$

$$\frac{(\% \text{ ionization}_{HCOOH})([HCOOH]_0)}{100} = [HCOO^-]_{eq}.$$

Substituting the information given in the problem gives

$$[HCOO^-]_{eq} = \frac{(5.8)(0.050)}{100} = 2.9 \times 10^{-3}$$

The completed summary table is

	$HCOOH(aq)$	$\rightleftarrows$	H^+	+	$HCOO^-$
Init. conc., M	0.050		0.000		0.000
Change, ΔM	$-(2.9 \times 10^{-3})$		$+(2.9 \times 10^{-3})$		$+(2.9 \times 10^{-3})$
Eq. conc., M	4.7×10^{-2}		2.9×10^{-3}		2.9×10^{-3}

Substitution of the equilibrium concentrations in the expression for K_a

$$K_a = \frac{[H^+][HCOO^-]}{[HCOOH]} \text{ gives}$$

$$K_a = \frac{(2.9 \times 10^{-3})^2}{4.7 \times 10^{-2}} = 1.8 \times 10^{-4}$$

15.14 4.2×10^{-3} M

For the ionization of formic acid, the equation is

$$HCOOH(aq) \rightleftarrows H^+ + HCOO^-$$

and $K_a = \dfrac{[H^+][HCOO^-]}{[HCOOH]} = 1.8 \times 10^{-4}$ (Table 15.3)

Let $x = \text{mol} \cdot L^{-1}\ H^+$ formed = $\text{mol} \cdot L^{-1}\ HCOO^-$ formed.
$[HCOOH] = 0.10 - x$

The ICE table is

	$HCOOH(aq)$	$\rightleftarrows$	$H^+(aq)$	+	$HCOO^-(aq)$
Init. conc., M	0.10		0.00		0.00
Change, ΔM	$-x$		$+x$		$+x$
Eq. conc., M	$0.10 - x$		x		x

Substitution of the concentrations in terms of x in the expression for K_a gives

$$\frac{x^2}{(0.10 - x)} = 1.8 \times 10^{-4}$$

Assume x is negligible compared to 0.10:

$$\frac{x^2}{0.10} = 1.8 \times 10^{-4} \qquad x^2 = (1.8 \times 10^{-4})(1.0 \times 10^{-1}) = 1.8 \times 10^{-5}$$

$$x = 4.2 \times 10^{-3} = [H^+] = 4.2 \times 10^{-3}\ M$$

Check to see if the assumption that x is negligible compared to 0.10 is correct:

$[HCOOH] = 0.10 - x = 0.10 - 0.0042 = 0.10$

Therefore, x is negligible compared to 0.10.

Note that the number of significant figures given in the problems in this chapter is limited by the fact that concentrations are used rather than activities.

15.15 so that the graph fits on the page

Because pK_a is a logarithmic function of K_a, the graph takes up far less space. The K_a's corresponding to one pK_a unit differ by a factor of ten.

15.16 0.050 M

KOH is a strong base that is 100% ionized in aqueous solution. Therefore, the OH^- concentration is the same as the concentration of KOH.

$KOH(aq) \rightarrow K^+ + OH^-$

15.17 **(a)** $CH_3NH_2(aq) + H_2O(l) \rightleftarrows CH_3NH_3^+ + OH^-$

(b) $K_b = \dfrac{[CH_3NH_3^+][OH^-]}{[CH_3NH_2]}$

15.18 **(a)** 6.2×10^{-6} M

$$\% \text{ ionization} = \frac{[OH^-]_{eq}}{[C_6H_5NH_2]_0} \times 100 \text{ and}$$

$$\frac{[C_6H_5NH_2]\ \% \text{ ionization}}{100} = [OH^-]_{eq}$$

Substituting the information given in the problem,

$$[OH^-] = \frac{(0.090)(0.0069)}{100} = 6.2 \times 10^{-6} \text{ M}$$

(b) Yes. Just as the H^+ from water can be neglected if $[H^+] > 1.0 \times 10^{-6}$, the OH^- from water can be neglected if $[OH^-] > 1.0 \times 10^{-6}$.

(c) 4.3×10^{-10}

The ICE table is

	$C_6H_5NH_2(aq) + H_2O(l)$	$\rightleftarrows$	$C_6H_5NH_3^+(aq)$	+	$OH^-(aq)$
Init. conc., M	0.090		0.000		0.000
Change, ΔM	-6.2×10^{-6}		$+6.2 \times 10^{-6}$		$+6.2 \times 10^{-6}$
Eq. conc., M	0.090		6.2×10^{-6}		6.2×10^{-6}

$$K_b = \frac{[C_6H_5NH_3^+][OH^-]}{[C_6H_5NH_2]} = \frac{(6.2 \times 10^{-6})^2}{9.0 \times 10^{-2}} = 4.3 \times 10^{-10}$$

15.19 **(a)** 5.7×10^{-6} M

For the ionization of aniline, the equation is

$C_6H_5NH_2 + H_2O \rightleftarrows C_6H_5NH_3^+ + OH^-$

The expression for K_b is

$$K_b = \frac{[C_6H_5NH_3^+][OH^-]}{[C_6H_5NH_2]} = 4.3 \times 10^{-10} \text{ (Table 15.4)}$$

Let x = mol/L $C_6H_5NH_3^+$ formed = mol/L OH^- formed

The summary table is

	$C_6H_5NH_2 + H_2O$	$\rightleftarrows$	$C_6H_5NH_3^+$	+	OH^-
Init. conc., M	0.075		0.000		0.000
Change, ΔM	$-x$		$+x$		$+x$
Eq. conc., M	$0.075 - x$		x		x

Substitution of the concentrations in terms of x in the expression for K_b gives

$$K_b = \frac{x^2}{0.075 - x} = 4.3 \times 10^{-10}$$

Assuming that x is negligible compared to 0.075, this equation simplifies to

$$\frac{x^2}{0.075} = 4.3 \times 10^{-10}$$

$$x^2 = 3.2 \times 10^{-11}$$

and $x = 5.7 \times 10^{-6}$

The assumption that x is negligible compared to 0.075 is correct and $[C_6H_5NH_3^+] = [OH^-] = 5.7 \times 10^{-6}$, $[C_6H_5NH_2] = 0.075$.

(b) 5.7×10^{-6} M

$[OH^-]_{eq} = [C_6H_5NH_3^+]_{eq} = 5.7 \times 10^{-6}$ M

(c) 0.075 M

$[C_6H_5NH_2]_{eq} = 0.075$ M

15.20 **(a)** chloroacetic acid; K_a for chloroacetic acid is larger.

Table 15.3 lists the common weak acids in order of decreasing strength. The value of K_a for chloroacetic acid, 1.4×10^{-3}, is larger than the value of K_a for acetic acid, 1.8×10^{-5}.

(b) acetate ion; the conjugate of the weaker acid is the stronger base.

The acetate ion is the stronger base because it is the conjugate base of the weaker acid. K_b for the acetate ion is greater than K_b for the chloroacetate ion.

For acetate ion, $K_b = \frac{K_w}{K_a} = \frac{1.0 \times 10^{-14}}{1.8 \times 10^{-5}} = 5.6 \times 10^{-10}$

For chloroactate ion, $K_b = \frac{1.0 \times 10^{-14}}{1.4 \times 10^{-3}} = 7.7 \times 10^{-12}$

15.21 **(a)** 7.7×10^{-5}

K_a for phenol = 1.3×10^{-10}

$$K_b = \frac{K_w}{K_a} = \frac{1.0 \times 10^{-14}}{1.3 \times 10^{-10}} = 7.7 \times 10^{-5}$$

(b) stronger; K_b for the phenoxide ion is larger.

K_b for the ammonium ion = 1.8×10^{-5} (Table 15.4)

Phenoxide is a stronger base than ammonia because it is the anion of a weaker acid and its K_b value is larger.

15.22 **(a)** 2.4×10^{-3} M

The equation for the equilibrium that exists in an aqueous solution of sodium phenoxide is

$$C_6H_5O^-(aq) + H_2O(l) \rightleftarrows C_6H_5OH(aq) + OH^-(aq)$$

The expression for K_b is

$$K_b = \frac{[C_6H_5OH][OH^-]}{[C_6H_5O^-]} \quad \text{and} \quad K_b = \frac{K_w}{K_a} = \frac{1.0 \times 10^{-14}}{1.3 \times 10^{-10}} = 7.7 \times 10^{-5}$$

The ICE table is

	$C_6H_5O^- + H_2O(l)$	$\rightleftarrows$	$C_6H_5OH(aq)$	+	OH^-
Init. conc., M	0.075		0.000		0.000
Change, ΔM	$-x$		$+x$		$+x$
Eq. conc., M	$0.075 - x$		x		x

Substitute the equilibrium concentrations in terms of x in the K_b expression.

$$K_b = \frac{x^2}{(0.075 - x)} = 7.7 \times 10^{-5}$$

Assuming that x is negligible compared to 0.075

$$\frac{x^2}{0.075} = 7.7 \times 10^{-5}$$

$$x^2 = 5.8 \times 10^{-6}$$

and $$x = 2.4 \times 10^{-3}$$

$0.075 - (2.4 \times 10^{-3}) = 0.073$

The assumption that x is negligible compared to 0.075 is borderline. Making a second approximation or using the quadratic formula gives $x = 2.4 \times 10^{-3}$ and $[OH^-] = 2.4 \times 10^{-3}$.

(b) more basic; phenol is a weaker acid than acetic acid.

Phenol is a weaker acid than acetic acid. Therefore, the phenoxide ion is a stronger base than the acetate ion. The $[OH^-]$ in the sodium phenoxide solution (2.4×10^{-3}) is more than 300 times as high as the $[OH^-]$ in sodium acetate solution (6.5×10^{-6} from Sample Problem 15.9).

15.23 **(a)** acidic

NH_4Br is a salt of a strong acid, HBr, and a weak base, NH_3. Therefore, Br^- will not hydrolyze but NH_4^+ will:

$NH_4^+(aq) + H_2O(l) \rightleftarrows NH_3(aq) + H_3O^+(aq)$

$NH_4Br(aq)$ is *acidic.*

(b) basic

NaClO is a salt of a strong base, NaOH, and a weak acid, HClO. Therefore, the Na^+ will not hydrolyze but ClO^- will:

$ClO^-(aq) + H_2O(l) \rightleftarrows HClO(aq) + OH^-(aq)$

NaClO(aq) is *basic.*

(c) acidic

$AlCl_3 \cdot 6H_2O$ is a salt of a strong acid, HCl, and a highly charged, relatively small cation, Al^{3+}. Therefore, Cl^- will not hydrolyze but $Al(H_2O)_6^{3+}$ will:

$Al(H_2O)_6^{3+}(aq) + H_2O(l) \rightarrow Al(H_2O)_5(OH)^{2+}(aq) + H_3O^+(aq)$

$AlCl_3 \cdot 6H_2O$ is *acidic.*

(d) neutral

KI is a salt of a strong base, KOH, and a strong acid, HI. Therefore, neither K^+ nor I^- will hydrolyze and KI(aq) is *neutral.*

15.24 1.8×10^{-4}

The equation for the equilibrium between molecules of formic acid and hydrogen ions and formate ions is

$HCOOH(aq) \rightleftarrows HCOO^- + H^+$

The expression for K_a is

$$K_a = \frac{[H^+][HCOO^-]}{[HCOOH]} = 1.8 \times 10^{-4} \text{ (from Table 15.3)}$$

The ICE table is

	$HCOOH(aq)$	$\rightleftarrows$	$HCOO^-(aq)$ +	$H^+(aq)$
Init. conc., M	0.083		0.085	0.000
Change, ΔM	$-x$		$+x$	$+x$
Eq. conc., M	$0.083 - x$		$0.085 + x$	x

Solution of the equilibrium constant expression for $[H^+]$ and substitution of the equilibrium concentrations gives

$$x = \frac{K_a\,[HCOOH]}{[HCOO^-]} = \frac{(1.8 \times 10^{-4})(0.083 - x)}{(0.085 + x)}$$

Assuming that x is negligible compared to 0.083 and 0.085,

$$x = \frac{(1.8 \times 10^{-4})(0.083)}{(0.085)} = 1.8 \times 10^{-4}$$

The assumption that x is negligible compared to 0.083 and 0.085 is correct and $[H^+] = 1.8 \times 10^{-4}$

15.25 3.9×10^{-4}

	$CH_3NH_2(aq) + H_2O(l)$	$\rightleftarrows$	$CH_3NH_3^+$	+	OH^-
Init. conc., M	0.074		0.082		0.000
Change , ΔM	$-x$		$+x$		$+x$
Eq. conc., M	$0.074 - x$		$0.082 + x$		x

$$K_b = \frac{[CH_3NH_3^+][OH^-]}{[CH_3NH_2]} = 4.3 \times 10^{-4} \text{ (Table 15.4)} = \frac{(0.082 + x)(x)}{(0.074 - x)}$$

$$x = \frac{(4.3 \times 10^{-4})(0.074 - x)}{(0.082 + x)}$$

Assume x is negligible compared to 0.074 and 0.082.

$$x = \frac{(4.3 \times 10^{-4})(7.4 \times 10^{-2})}{(8.2 \times 10^{-2})} = 3.9 \times 10^{-4}$$

The assumption that x is negligible compared to 0.074 and 0.082 is correct and $[OH^-] = 3.9 \times 10^{-4}$.

15.26 (a) and (b)

These have equal concentrations of a weak base or acid and the corresponding salt and they have sufficient capacity, which (c) does not have.

15.27 **(a)** $[H^+] = 1.9 \times 10^{-4}$, pH = 3.72

$$[H^+]_{eq} = K_a \frac{[\text{Acid molecules}]_0}{[\text{Anion}]_0} = \frac{0.080}{0.075}(1.8 \times 10^{-4}) = 1.9 \times 10^{-4}$$

$$pH = -\log [H^+] = -\log (1.9 \times 10^{-4}) = -(-3.72) = 3.72$$

(b) $[H^+] = 2.2 \times 10^{-4}$, pH = 3.66

The equilibrium is

$$HCOOH(aq) \rightleftarrows HCOO^- + H^+$$

and $K_a = \frac{[HCOO^-][H^+]}{[HCOOH]} = 1.8 \times 10^{-4}$

For the reverse reaction,

$$HCOO^-(aq) + H^+(aq) \rightleftarrows HCOOH(aq) + H_2O(l),$$

$$K_c = \frac{[HCOOH]}{[HCOO^-][H^+]} = \frac{1}{K_a} = 5.6 \times 10^3$$

K_c is fairly large. Assume that this reaction is complete. 0.0050 mol HCl will react with 0.0050 mol $HCOO^-$. After the reaction has taken place:

$[HCOO^-] = [(0.075) - (0.0050)] = 0.070$ mol $HCOO^-$ left in one liter

$[HCOOH] = [(0.080) + (0.0050)] = 0.085$ mol HCOOH left in one liter

Then consider the reverse reaction

$$HCOOH(aq) \rightleftharpoons H^+ + HCOO^-$$

(because K_c is not very large).

From the K_a expression,

$$[H^+] = \frac{(1.8 \times 10^{-4})[HCOOH]}{[HCOO^-]} = \frac{(1.8 \times 10^{-4})(0.085)}{(0.070)} = 2.2 \times 10^{-4}$$

$pH = -\log (2.2 \times 10^{-4}) = 3.66$

15.28 **(a)** methylamine – methylammonium ion

(b) 0.70

For CH_3NH_2, $K_b = 4.3 \times 10^{-4}$, which is close in value to the desired $[OH^-]$.

$$CH_3NH_2(aq) + H_2O(l) \rightleftharpoons CH_3NH_3^+ + OH^-$$

$$K_b = \frac{[CH_3NH_3^+][OH^-]}{[CH_3NH_2]} = 4.3 \times 10^{-4}$$

$$\frac{[CH_3NH_2]}{[CH_3NH_3^+]} = \frac{[OH^-]}{4.3 \times 10^{-4}} = \frac{3.0 \times 10^{-4}}{4.3 \times 10^{-4}} = 0.70$$

15.29 3.37

$$pH = pK_a + \log\frac{[F^-]}{[HF]} = -\log (3.5 \times 10^{-4}) + \log\frac{[0.090]}{[0.110]}$$

$pH = 3.46 + \log (0.82) = 3.46 + (-0.086) = 3.37$

15.30 **(a)** yellow (pH >> 2.8 and < 8.0)

(b) 2

For thymol blue, the acid color is red and the base color is yellow. The pH range of the color change is 1.2–2.8. Thymol blue will be orange, a mixture of red and yellow, when the pH is in the middle of the color change range.

15.31 1.954

The equation for the neutralization reaction is

$HCl(aq) + NaOH(aq) \rightarrow NaCl(aq) + H_2O(l)$

$$(25.00\ \cancel{\text{mL soln}})\left(\frac{1\ \cancel{\text{L soln}}}{10^3\ \cancel{\text{mL soln}}}\right)\left(\frac{0.1000\ \text{mol HCl}}{1\ \cancel{\text{L soln}}}\right) = 2.500 \times 10^{-3}\ \text{mol HCl at start}$$

$$(20.00\ \cancel{\text{mL soln}})\left(\frac{1\ \cancel{\text{L soln}}}{10^3\ \cancel{\text{mL soln}}}\right)\left(\frac{0.1000\ \cancel{\text{mol NaOH}}}{1\ \cancel{\text{L soln}}}\right)\left(\frac{1\ \text{mol HCl}}{1\ \cancel{\text{mol NaOH}}}\right) = 2.000 \times 10^{-3}\ \text{mol HCl neutralized}$$

$(2.500 \times 10^{-3}$ mol HCl present) – $(2.000 \times 10^{-3}$ mol HCl neutralized) = 5.00×10^{-4} mol HCl left

Total volume = 25.00 mL + 20.00 mL = 45.00 mL = 0.045 00 L or 4.500×10^{-2} L

$$\text{M HCl} = \frac{5.00 \times 10^{-4}\ \text{mol HCl}}{4.500 \times 10^{-2}\ \text{L soln}} = 1.11 \times 10^{-2}\ \text{M}$$

$[H^+] = 1.11 \times 10^{-2}$ and pH = 1.955

(Best answer, without intermediate round off, is 1.954.)

15.32 11.959

The equation for the neutralization reaction is

$HCl(aq) + NaOH(aq) \rightarrow NaCl(aq) + H_2O(l)$

$$(25.00\ \cancel{\text{mL soln}})\left(\frac{0.1000\ \text{mol HCl}}{\cancel{\text{L soln}}}\right)\left(\frac{1\ \cancel{\text{L soln}}}{10^3\ \cancel{\text{mL soln}}}\right) = 2.500 \times 10^{-3}\ \text{mol HCl at start}$$

This quantity of HCl will neutralize 2.500×10^{-3} mol NaOH

$$(30.00\ \cancel{\text{mL soln}})\left(\frac{0.1000\ \text{mol NaOH}}{\cancel{\text{L soln}}}\right)\left(\frac{1\ \cancel{\text{L soln}}}{10^3\ \cancel{\text{mL soln}}}\right) = 3.000 \times 10^{-3}\ \text{mol NaOH added}$$

$(3.000 \times 10^{-3}$ mol added) – $(2.500 \times 10^{-3}$ mol neutralized) = 5.00×10^{-4} mol NaOH excess

Total volume = 25.00 mL + 30.00 mL = 55.00 mL or 0.055 00 L

$$[OH^-] = \frac{5.00 \times 10^{-4}\ \text{mol OH}^-}{0.055\ 00\ \text{L soln}} = 9.09 \times 10^{-3}\ \text{M}$$

$p[OH^-] = -\log (9.09 \times 10^{-3}) = 2.041$

pH = 14.00 – pOH = 11.959

15.33 **(a)** 3.68

The net ionic equation for the neutralization reaction is

$CH_3COOH(aq) + OH^-(aq) \rightarrow CH_3COO^-(aq) + H_2O(l)$

$$(25.00\ \cancel{\text{mL soln}})\left(\frac{1\ \cancel{\text{L soln}}}{10^3\ \cancel{\text{mL soln}}}\right)\left(\frac{0.1000\ \text{mol}\ CH_3COOH}{\cancel{\text{L soln}}}\right) = 2.500 \times 10^3\ \text{mol}\ CH_3COOH\ \text{at start}$$

$$(2.00\ \cancel{\text{mL soln}})\left(\frac{1\ \cancel{\text{L soln}}}{10^3\ \cancel{\text{mL soln}}}\right)\left(\frac{0.1000\ \text{mol NaOH}}{\cancel{\text{L soln}}}\right) = 2.00 \times 10^{-4}\ \text{mol NaOH added}$$

After addition of NaOH, the moles of CH_3COO^- formed = 2.00×10^{-4} and the moles of CH_3COOH left is $(2.500 \times 10^{-3}) - (2.00 \times 10^{-4}) = 2.300 \times 10^{-3}$ mol.

The initial concentrations are:

$$[CH_3COO^-]_0 = \left(\frac{2.00 \times 10^{-4}\ \text{mol}\ CH_3COO^-}{27.00\ \cancel{\text{mL soln}}}\right)\left(\frac{10^3\ \cancel{\text{mL soln}}}{\text{L soln}}\right) = 7.41 \times 10^{-3}\ \text{M}$$

$$[CH_3COOH]_0 = \left(\frac{2.300 \times 10^{-3}\ \text{mol}\ CH_3COOH}{27.00\ \cancel{\text{mL soln}}}\right)\left(\frac{10^3\ \cancel{\text{mL soln}}}{\text{L soln}}\right) = 8.519 \times 10^{-2}\ \text{M}$$

	$CH_3COOH(aq)$	$\rightleftarrows$	CH_3COO^-	+	H^+
Init. conc., M	0.085 19		0.007 41		0.0000
Change, ΔM	$-x$		$+x$		$+x$
Eq. conc., M	$0.085\ 19 - x$		$0.007\ 41 + x$		$0.0000 + x$

Assume x is negligible compared to 0.085 19 and 0.007 41.

$$x = [H^+] = \frac{K_a\,[CH_3COOH]}{[CH_3COO^-]} = \frac{(1.8 \times 10^{-5})(0.085\ 19)}{(0.007\ 41)} = 2.1 \times 10^{-4}\ \text{M}$$

$$\text{pH} = -\log\,[H^+] = -\log\,(2.1 \times 10^{-4}) = 3.68$$

(b) 8.72

$$(25.00\ \cancel{\text{mL soln}})\left(\frac{1\ \cancel{\text{L soln}}}{10^3\ \cancel{\text{mL soln}}}\right)\left(\frac{0.1000\ \text{mol NaOH}}{\cancel{\text{L soln}}}\right) = 2.500 \times 10^{-3}\ \text{mol NaOH added}$$

After addition of NaOH, the moles CH_3COO^- formed = 2.500×10^{-3} and the moles CH_3COOH left is $(2.500 \times 10^3) - (2.500 \times 10^{-3}) = 0$.

For this equilibrium, $K_b = \frac{K_w}{K_a} = \frac{1.0 \times 10^{-14}}{1.8 \times 10^{-5}} = 5.6 \times 10^{-10}$

For the equilibrium involving CH_3COO^- hydrolysis,

$$CH_3COO^- + H_2O(l) \rightleftarrows CH_3COOH(aq) + OH^-,$$

the initial concentrations are:

$$[CH_3COO^-]_0 = \left(\frac{2.500 \times 10^{-3}\ \text{mol}\ CH_3COO^-}{50.00\ \cancel{\text{mL soln}}}\right)\left(\frac{10^3\ \cancel{\text{mL soln}}}{\text{L soln}}\right) = 5.000 \times 10^{-2}\ \text{M}$$

$$[CH_3COOH]_0 = [OH^-]_0 = 0.000$$

The ICE table is

	$CH_3COO^- + H_2O(l)$	$\rightleftarrows$	$CH_3COOH(aq)$	+	OH^-
Init. conc., M	5.000×10^{-2}		0.0000		0.0000
Change, ΔM	$-x$		$+x$		$+x$
Eq. conc., M	$(5.000 \times 10^{-2}) - x$		x		x

Substituting the equilibrium concentrations in terms of x in the K_b expression,

$$5.6 \times 10^{-10} = \frac{x^2}{(5.000 \times 10^{-2}) - x}$$ Assume x is negligible compared to 5.000×10^{-2}

$$x^2 = (5.6 \times 10^{-10})(5.000 \times 10^{-2}) = 2.8 \times 10^{-11}$$

$$x = 5.3 \times 10^{-6} = [OH^-]$$

pOH = 5.28 pH = 14.00 – 5.28 = 8.72

(c) 12.3632

$$(40.00 \text{ mL soln})\left(\frac{1 \text{ L soln}}{10^3 \text{ mL soln}}\right)\left(\frac{0.1000 \text{ mol NaOH}}{\text{L soln}}\right) = 4.000 \times 10^{-3} \text{ mol NaOH added}$$

Since only 2.500×10^{-3} mol CH_3COOH is available at the start, $(4.000 \times 10^{-3}) - (2.500 \times 10^{-3}) = 1.500 \times 10^{-3}$ mol NaOH is in excess. 1.500×10^{-3} mol NaOH gives 1.500×10^{-3} mol OH^-. Thus,

$$[OH^-] = \left(\frac{1.500 \times 10^{-3} \text{ mol OH}^-}{65.00 \text{ mL soln}}\right)\left(\frac{10^3 \text{ mL soln}}{\text{L soln}}\right) = 2.308 \times 10^{-2} \text{ M}$$

$$\text{pOH} = -\log(2.308 \times 10^{-2}) = 1.6368$$

$$\text{pH} = 14.0000 - 1.6368 = 12.3632$$

15.34 thymol blue

The pH range of the color change for thymol blue is 8.0–9.6 (Table 15.6). The pH at the equivalence point in the titration of 0.1000 M CH_3COOH with 0.1000 M NaOH is 8.72. Phenolphthalein could also be used, as the pH range of its color change is 8.2–10.0.

15.35 **(a)** $[H^+] = [HC_6H_6O_6^-] = 2 \times 10^{-3}$

$$H_2C_6H_6O_6(aq) \rightleftarrows HC_6H_6O_6^- + H^+ \qquad K_{a_1} = 7 \times 10^{-5}$$

$$HC_6H_6O_6^-(aq) \rightleftarrows C_6H_6O_6^{2-} + H^+ \qquad K_{a_2} = 3 \times 10^{-12}$$

$$K_{a_1} = \frac{[HC_6H_6O_6^-][H^+]}{[H_2C_6H_6O_6]} \qquad K_{a_2} = \frac{[C_6H_6O_6^{2-}][H^+]}{[HC_6H_6O_6^-]}$$

Since $K_{a_1} >> K_{a_2}$, we can assume that all H^+ comes from the first ionization, and that the autoionization of water is negligible.

	$H_2C_6H_6O_6(aq)$	$\rightleftarrows$	$HC_6H_6O_6^-$	+ H^+
Init. conc., M	0.05		0.00	0.00
Change, ΔM	$-x$		$+x$	$+x$
Eq. conc., M	$(0.05 - x)$		x	x

$7 \times 10^{-5} = \dfrac{x^2}{0.05 - x}$ Assume x is negligible compared to 0.05.

$x^2 = (7 \times 10^{-5})(0.05) = 4 \times 10^{-6}$

$x = 2 \times 10^{-3} = [H^+] = [HC_6H_6O_6^-]$

(b) 3×10^{-12}

	$HC_6H_6O_6^-(aq)$	$\rightleftarrows$	$C_6H_6O_6^{2-}$	+ H^+
Init. conc., M	2×10^{-3}		0.0	2×10^{-3}
Change, ΔM	$-y$		$+y$	$+y$
Eq. conc., M	$(2 \times 10^{-3}) - y$		y	$(2 \times 10^{-3}) + y$

$K_a = \dfrac{(y)[y + (2 \times 10^{-3})]}{(2 \times 10^{-3} - y)} = 3 \times 10^{-12}$

Assume y is negligible compared to 2×10^{-3}.

$(3 \times 10^{-12})(2 \times 10^{-3}) = (2 \times 10^{-3})y$

$y = \dfrac{(3 \times 10^{-12})(2 \times 10^{-3})}{(2 \times 10^{-3})} = 3 \times 10^{-12} = [C_6H_6O_6^{2-}]$

Additional Practice Problems

15.91 **(a)** 1.62×10^{-7} M

Let $x = [OH^-]$ from water $= [H^+]$

Then total $[OH^-] = [(1.00 \times 10^{-7}) + x]$

Substituting in the ion product for water $[H^+][OH^-] = 1.0 \times 10^{-14}$,

$[(1.00 \times 10^{-7}) + x]x = 1.00 \times 10^{-14}$ and from the quadratic formula

$x = 6.20 \times 10^{-8}$

total $[OH^-] = (1.00 \times 10^{-7}) + (6.20 \times 10^{-8}) = 1.62 \times 10^{-7}$

(b) 38.3%

Since in *pure water*, $[OH^-] = 1.00 \times 10^{-7}$, water makes a significant contribution to $[OH^-]$. In this solution, $[OH^-]$ from water $= 6.20 \times 10^{-8}$ M.

$\% = \dfrac{6.20 \times 10^{-8}}{1.62 \times 10^{-7}} \times 100 = 38.3\%$

15.93 $[H^+] > 1$

By definition, $pH = -\log [H^+]$

If $[H^+] = 1$, pH = 0. If $[H^+] > 1$, pH is negative.

For example, if $[H^+] = 1.1$

$pH = -\log 1.1 = -0.041$

15.95 **(a)** 2.30×10^{-7} M

$K_w = 5.31 \times 10^{-14} = [H^+][OH^-]$

In pure water, $[H^+] = [OH^-]$ and $5.31 \times 10^{-14} = [H^+]^2$

$\sqrt{5.31 \times 10^{-14}} = [H^+]$

$2.30 \times 10^{-7} = [H^+] = [OH^-]$

(b) basic

The solution is basic. The $[H^+]$ is less than $[H^+]$ for pure water at 50 °C.

1.7×10^{-7}	$<$	2.30×10^{-7}
$[H^+]$ in solution		$[H^+]$ in water

15.97 6.5×10^{-2} M

The formate ion from potassium formate will shift the equilibrium between formate ions and molecules of formic acid

$$HCOOH(aq) \rightleftarrows HCOO^-(aq) + H^+(aq)$$

to the left. In the presence of potassium formate, all formate ions will come from potassium formate. Ionization of formic acid will be negligible.

Substituting the data given in the problem in the expression for K_a

$$K_a = \frac{[HCOO^-][H^+]}{[HCOOH]} = 1.8 \times 10^{-4} = \frac{[HCOO^-](2.0 \times 10^{-4})}{0.072}$$

and solving for $[HCOO^-]$,

$6.5 \times 10^{-2} = [HCOO^-] \approx$ concentration of potassium formate

15.99 0.1059 M

The empirical formula for potassium hydrogen phthalate is $C_8H_5KO_4$, or, writing the acidic H at the left, $HC_8H_4KO_4$.

The equation for the neutralization of potassium hydrogen phthalate with sodium hydroxide is

$$HC_8H_4KO_4(aq) + NaOH(aq) \rightarrow H_2O(l) + C_8H_4KNaO_4(aq)$$

The definition of molarity is molarity, $M = \frac{\text{moles solute}}{\text{volume of solution in liters}}$

From the data given, calculate the number of moles of sodium hydroxide used.

$$(0.5690\ \cancel{g\ HC_8H_4KO_4})\left(\frac{1\ \cancel{mol\ HC_8H_4KO_4}}{204.22\ \cancel{g\ HC_8H_4KO_4}}\right)\left(\frac{1\ mol\ NaOH}{1\ \cancel{mol\ HC_8H_4KO_4}}\right) = 2.786 \times 10^{-3}\ mol\ NaOH$$

The volume of sodium hydroxide solution required is given in the problem.

$$M_{NaOH} = \frac{2.786 \times 10^{-3}\ mol\ NaOH}{26.32\ \cancel{mL}\left(\frac{1\ L}{10^3\ \cancel{mL}}\right)} = 0.1059\ \frac{mol\ NaOH}{L}$$

$M = 0.1059$

or, combining steps,

$$(0.5690\ \cancel{g\ HC_8H_4KO_4})\left(\frac{1\ \cancel{mol\ HC_8H_4KO_4}}{204.22\ \cancel{g\ HC_8H_4KO_4}}\right)\left(\frac{1}{26.32\ \cancel{mL\ NaOH}}\right)\left(\frac{10^3\ \cancel{mL}}{L}\right)\left(\frac{1\ mol\ NaOH}{1\ \cancel{mol\ HC_8H_4KO_4}}\right) = 0.1059\ M$$

15.101 Differential plot of titration of 0.1 M HOCl with 0.1 M NaOH.

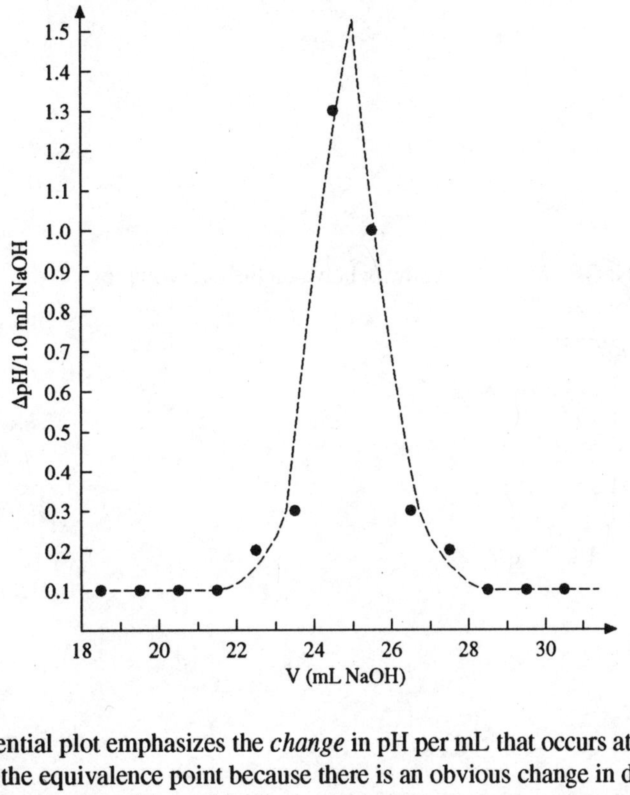

The differential plot emphasizes the *change* in pH per mL that occurs at the equivalence point in a titration. It is easier to determine the equivalence point because there is an obvious change in direction on the graph. In the ordinary titration curve for this titration (Figure 15.13), the equivalence point is not obvious because the steep portion of the curve is short and far from vertical.

Data for differential plot:

pH	mL NaOH	ΔpH/1.0 mL NaOH
8.0	18.0	
		0.1
8.1	19.0	
		0.1
8.2	20.0	
		0.1
8.3	21.0	
		0.1
8.4	22.0	
		0.2
8.6	23.0	
		0.3
8.9	24.0	
		0.3
10.2	25.0	
		1.0
11.2	26.0	
		0.3
11.5	27.0	
		0.2
11.7	28.0	
		0.1
11.8	29.0	
		0.1
11.9	30.0	
		0.1
12.0	31.0	

In the graph, the points for ΔpH/1.0 mL NaOH are plotted midway between the two volumes.

15.103 **(a)**

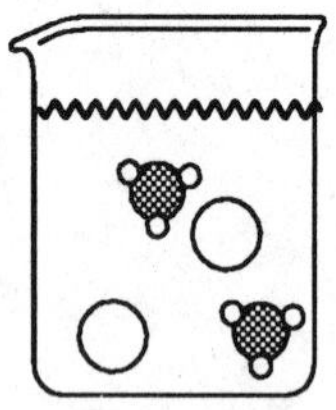

HCl(aq)

(b)

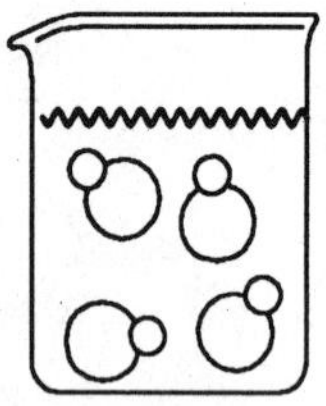

HOCl(aq)

15.105

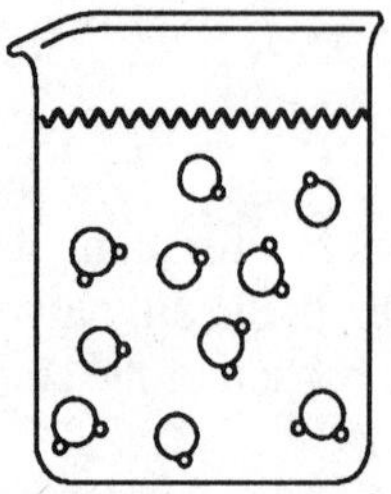

$H_2PO_4^-$ — HPO_4^{2-} buffer

Stop and Test Yourself

1. **(c)** $(CH_3)_3COH + H_3O^+ \rightarrow (CH_3)_3COH_2^+ + H_2O$

Transfer of a proton from an acid to a base occurs.

2. **(d)** $2HPO_4^{2-} \rightarrow H_2PO_4^- + PO_4^{3-}$

The HPO_4^{2-} ion reacts with itself to form ions.

3. **(e)** H_3PO_4

H_3PO_4 has one more H and one unit less of negative charge than $H_2PO_4^-$

4. **(a)** 2.6×10^{-10}, acidic

$1.00 \times 10^{-14} = 3.8 \times 10^{-5}\ [OH^-]$

$2.6 \times 10^{-10} = [OH^-]$

The solution is acidic because $[H^+] > 1.0 \times 10^{-7}$

5. **(e)** 2×10^{-10}

$9.7 = -\log\ [H^+]$

$2 \times 10^{-10} = [H^+]$

6. **(c)** 0.0053

	$ClCH_2COOH$	$\rightleftarrows$	$ClCH_2COO^-$	+	H^+
Init. conc., M	0.025		0.000		0.000
Change, ΔM	$-x$		$+x$		$+x$
Eq. conc., M	$0.025 - x$		x		x

$$1.4 \times 10^{-3} = \frac{x^2}{0.025 - x}, \quad x^2 + 0.0014x - 0.000\,035 = 0$$

$x = [H^+] = 5.3 \times 10^{-3} = 0.0053$

7. **(a)** 6.8×10^{-6}

	HB	$\rightleftarrows$	H^+	+	B^-
Init. conc., M	0.040		0		0

$$\left(\frac{1.3}{100}\right) \times 0.040 = 5.2 \times 10^{-4} = [H^+] = [B^-]$$

$0.040 - 5.2 \times 10^{-4} = [HB] = 3.9 \times 10^{-2}$

$$K_a = \frac{[H^+][B^-]}{[HB]} = \frac{(5.2 \times 10^{-4})^2}{3.9 \times 10^{-2}} = 6.9 \times 10^{-6}$$

(Best answer, without intermediate round off, is 6.8×10^{-6}.)

8. **(e)** $K_a = \dfrac{[CH_3CH_2NH_2][H^+]}{[CH_3CH_2NH_3^+]}$

$CH_3CH_2NH_3^+ \rightleftharpoons H^+ + CH_3CH_2NH_2$

$$K_a = \frac{[H^+][CH_3CH_2NH_2]}{[CH_3CH_2NH_3^+]}$$

9. **(a)** 0.1 M solution of weak acid with $K_a = 4 \times 10^{-5}$

For the same concentration of weak acid, percent ionization is lower the smaller the value of K_a (Figure 15.4). Therefore, (a) < (b) < (c). For weak acids with the same value of K_a, percent ionization increases with increasing dilution (Figure 15.5). Therefore, (a) < (d).

10. **(b)** $NaNO_2$

An aqueous solution of a salt of a strong base (NaOH) and a weak acid (HNO_2) is basic.

11. **(e)** 1.9×10^{-3}

$$K_b = \frac{K_w}{K_a} = \frac{[OH^-][C_6H_5OH]}{[C_6H_5O^-]}$$

$$K_b = \frac{1.0 \times 10^{-14}}{1.3 \times 10^{-10}} = 7.7 \times 10^{-5} = \frac{[OH^-][C_6H_5OH]}{[C_6H_5O^-]}$$

Let $x = [OH^-] = [C_6H_5OH]$.

$$7.7 \times 10^{-5} = \frac{x^2}{0.050 - x}$$

If you assume x is negligible compared to 0.050, then $x = \sqrt{3.9 \times 10^{-6}} = 2.0 \times 10^{-3}$

(Best answer, without intermediate round off and using quadratic formula, is 1.9×10^{-3}.)

12. **(b)** $[H^+] = 3.6 \times 10^{-4}$

$HCOOH \rightleftharpoons H^+ + HCOO^-$

$$K_a = 1.8 \times 10^{-4} = \frac{[H^+][HCOO^-]}{[HCOOH]}$$

Assume that in the presence of HCOOK, ionization of HCOOH is negligible and

$[HCOOH] = 0.040$ M and $[HCOO^-] = 0.020$ M.

Then $1.8 \times 10^{-4} = \dfrac{[H^+]\ 0.020}{0.040}$

$[H^+] = 3.6 \times 10^{-4}$

13. (b) 0.025 M CH_3NH_2 and 0.100 M CH_3NH_3Cl

For the equilibrium

$$CH_3NH_2 + H_2O \rightleftarrows CH_3NH_3^+ + OH^-$$

$$K_b = \frac{[CH_3NH_3^+][OH^-]}{[CH_3NH_2]} = 4.6 \times 10^{-4}$$

A buffer with pH = 10.06 has $[H^+] = 8.7 \times 10^{-11}$ and $[OH^-] = 1.2 \times 10^{-4}$.

In this buffer $\frac{[CH_3NH_3^+]}{[CH_3NH_2]}$ must equal

$$\frac{[CH_3NH_3^+](1.2 \times 10^{-4})}{[CH_3NH_2]} = 4.6 \times 10^{-4}$$

$$\frac{[CH_3NH_3^+]}{[CH_3NH_2]} = \frac{4.6 \times 10^{-4}}{1.2 \times 10^{-4}} = 3.8$$

(Best answer, without intermediate round off, is 4.0.)

Both (b) and (d) have a 4:1 ratio of $CH_3NH_3^+$ to CH_3NH_2 and both (b) and (d) yield buffers of pH = 10.06, but (b) has larger concentrations of both acid and base, thus giving it a greater capacity to neutralize added acid or base than the mixture in (d).

14. (e) 0.1 M HNO_3 with 0.1 M KOH

Titrations of strong acids with strong bases give the sharpest end points; 0.1 M solutions give sharper end points than 0.01 M solutions. (See Figures 15.12 and 15.13.)

15. (d) $H_2C_2O_4$

The acid must contain more than one acidic H. Acidic H's are generally written at the left of the formula.

Putting Things Together

15.107 (a) 1.8×10^{-7} %

$[H_2O] = 55.3$ M (see Section 15.2)
$[H^+] = [OH^-] =$ moles per liter ionized $= 1.0 \times 10^{-7}$

$$\% \text{ ionized} = \frac{1.0 \times 10^{-7}}{55.3} \times 100 = 1.8 \times 10^{-7}$$

(b) 6.0×10^{16}

$$[H^+] = 1.0 \times 10^{-7} \frac{\text{mol}}{\text{L}}$$

$$\left(\frac{1.0 \times 10^{-7} \cancel{\text{mol}}}{\text{L}}\right)\left(\frac{6.02 \times 10^{23} \text{ ions}}{\cancel{\text{mol}}}\right) = 6.0 \times 10^{16} \frac{\text{ions}}{\text{L}}$$

15.109 7.0

In pure water, $[H^+] = 1.0 \times 10^{-7}$. The $[H^+]$ from 1.0×10^{-12} is negligible compared to 1.0×10^{-7} and pH = 7.00, the pH of pure water. Note that a solution of an acid can never be basic (pH > 7)

15.111 **(a)** 1.60

Because HCl is a strong acid, $[H^+]$ from HCl = 0.025 M. Let $x = [H^+]$ from CH_3COOH.

	$CH_3COOH(aq)$ $\rightleftarrows$	H^+ +	CH_3COO^-
Init. conc., M	0.032	0.025	
Change, ΔM	$-x$	$+x$	$+x$
Eq. conc., M	$0.032 - x$	$0.025 + x$	x

$$K_a = \frac{[H^+][CH_3COO^-]}{[CH_3COOH]} = \frac{(0.025 + x)\,x}{(0.032 - x)} = 1.8 \times 10^{-5}$$

Assuming that x is negligible compared to 0.025 and 0.032,

$$\frac{0.025\,x}{0.032} = 1.8 \times 10^{-5}$$

and $x = 2.3 \times 10^{-5}$

The assumption that x is negligible compared to 0.025 and 0.032 is correct. $[H^+] = 0.025$ and pH = 1.60.

(b) 1.90

When equal volumes of the two solutions are mixed, the concentrations are halved and the solution is 0.0125 M in HCl and 0.016 M in CH_3COOH. A calculation similar to that in part (a) shows that the contribution of the CH_3COOH to $[H^+]$ is still negligible. Thus, $[H^+] = 0.0125$ and pH = 1.90.

15.113 **(a)** 1.46

$HClO_4$ is a strong acid and is completely ionized in aqueous solution. Therefore, $[H^+] = 0.035$ M and pH = $-\log [H^+]$ = 1.46.

(b) 11.10

For the equilibrium $(CH_3)_3N + H_2O \rightleftarrows (CH_3)_3NH^+ + OH^-$,

$$K_b = \frac{[(CH_3)_3NH^+][OH^-]}{[(CH_3)_3N]} = 6.3 \times 10^{-5}$$

Let $x = [OH^-] = [(CH_3)_3NH^+]$; then $[(CH_3)_3N] = 0.027 - x$

$$6.3 \times 10^{-5} = \frac{x^2}{0.027 - x}$$

$x = 1.3 \times 10^{-3} = [OH^-]$

pOH = 2.89

pH + pOH = 14.00; pH = 14.00 – pOH = 14.00 – 2.89 = 11.11

(Best answer, without intermediate round off, is 11.10.)

(c) 3.01

For the equilibrium $C_6H_5NH_3^+ \rightleftarrows C_6H_5NH_2 + H^+$

$$K_a = \frac{[H^+][C_6H_5NH_2]}{[C_6H_5NH_3^+]} = \frac{K_w}{K_b} = \frac{1.0 \times 10^{-14}}{4.3 \times 10^{-10}} = 2.3 \times 10^{-5}$$

Let $x = [H^+] = [C_6H_5NH_2]$; then $[C_6H_5NH_3^+] = (0.041 - x)$

$$2.3 \times 10^{-5} = \frac{x^2}{(0.041 - x)} \qquad x = 9.7 \times 10^{-4} = [H^+]$$

pH = –log [H⁺] = 3.01

(d) 5.1

For the equilibrium

$[Zn(H_2O)_6]^{2+} \rightleftarrows Zn(H_2O)_5(OH)^+ + H^+$

$$K_a = \frac{[Zn(H_2O)_5(OH)^+][H^+]}{[Zn(H_2O)_6^{2+}]} = 1 \times 10^{-9} = \frac{x^2}{0.052 - x}$$

where $x = [H^+] = [Zn(H_2O)_5(OH)^+]$ and $[Zn(H_2O)_6^{2+}] = 0.052 - x$

$x = 7 \times 10^{-6} = [H^+]$ and pH = –log $[H^+]$ = 5.2

(Best answer, without intermediate round off, is 5.1.)

(e) 2.57

The equation for the equilibrium is $HF(aq) \rightleftarrows H^+ + F^-$

$$K_a = \frac{[H^+][F^-]}{[HF]} = 3.5 \times 10^{-4}$$

Let $x = [H^+] = [F^-]$; $[HF] = 0.024 - x$

$$3.5 \times 10^{-4} = \frac{x^2}{(0.024 - x)}$$

If x is assumed negligible compared with 0.024, $x = [H^+] = 2.9 \times 10^{-3}$ M = 12% of 0.024. Thus, successive approximations or the quadratic formula must be used to solve for x. The result is $x = 2.7 \times 10^{-3} = [H^+]$.

pH = –log $[H^+]$ = 2.57

15.115 **(a)** SO_3^{2-}, sulfite ion

$HSO_3^- \rightleftarrows H^+ + SO_3^{2-}$

(b) I^-, iodide ion

$HI \rightleftarrows H^+ + I^-$

(c) NH_3, ammonia

$NH_4^+ \rightleftarrows H^+ + NH_3$

(d) O^{2-}, oxide ion

$OH^- \rightleftarrows H^+ + O^{2-}$

(e) CH_3COO^-, acetate ion

$CH_3COOH \rightleftarrows CH_3COO^- + H^+$

15.117 1.22

First, find the molarity of the sulfuric acid solution:

$$\left(\frac{5.32\ \text{g conc}}{1.00\ \text{L diluted soln}}\right)\left(\frac{95\ \text{g}\ H_2SO_4}{100\ \text{g conc}}\right)\left(\frac{1\ \text{mol}\ H_2SO_4}{98.1\ \text{g}\ H_2SO_4}\right) = 0.052\ \text{M}$$

Sulfuric acid is a diprotic acid for which the first ionization goes to completion:

$$H_2SO_4 \longrightarrow H^+ + HSO_4^-$$

Therefore, 0.052 mol $H_2SO_4 \rightarrow$ 0.052 mol H^+ and 0.052 mol HSO_4^-.

HSO_4^- is a "weak" acid for which $K_a = 1.2 \times 10^{-2}$

	HSO_4^- $\rightleftarrows$	H^+ +	SO_4^{2-}
Init. conc., M	0.052	0.052	0.000
Change, ΔM	$-x$	$+x$	$+x$
Eq. conc., M	$0.052 - x$	$0.052 + x$	x

$$K_a = \frac{[H^+][SO_4^{2-}]}{[HSO_4^-]} = \frac{(0.052 + x)(x)}{(0.052 - x)} = 1.2 \times 10^{-2}$$

and $x^2 + 0.064x - 0.000\,624 = 0$

By successive approximations or the quadratic formula, $x = 0.0086$

$[H^+] = 0.052 + 0.0086 = 0.061$ M, pH = 1.21

(Best answer, without intermediate round off, is 1.22.)

15.119 **(a)** Hydrochloric acid is an aqueous solution of a strong electrolyte and contains H^+ ions and Cl^- ions. Hydrogen chloride gas is a molecular compound.

(b) The charge is carried by the movement of ions through the solution between the electrodes.

(c) Hydrogen chloride is not ionized in benzene solution. Benzene is a nonpolar solvent and cannot solvate ions. Thus, the H—Cl bond is not broken.

15.121 Ammonia and acetic acid are weak electrolytes. However, when their solutions are mixed, a Brønsted–Lowry acid–base reaction takes place and ions are formed. The ionic equation for this reaction is

$$NH_3(aq) + CH_3COOH(aq) \rightarrow NH_4^+ + CH_3COO^-$$

15.123 **(a)** Molecular equation: $NaH_2PO_4(aq) + HCl(aq) \rightarrow H_3PO_4(aq) + NaCl(aq)$

Complete ionic equation: $Na^+ + H_2PO_4^- + H^+ + Cl^- \rightarrow H_3PO_4(aq) + Na^+ + Cl^-$

Net ionic equation: $H_2PO_4^- + H^+ \rightarrow H_3PO_4(aq)$

(b) Molecular equation: $NaH_2PO_4(aq) + NaOH(aq) \rightarrow Na_2HPO_4(aq) + H_2O(l)$

Complete ionic equation: $Na^+ + H_2PO_4^- + Na^+ + OH^- \rightarrow 2Na^+ + HPO_4^{2-} + H_2O(l)$

Net ionic equation: $H_2PO_4^- + OH^- \rightarrow HPO_4^{2-} + H_2O(l)$

15.125 2.58

First calculate the molarity of the dilute solution. Then calculate [H⁺] and pH in the usual way.

$$d = \frac{g}{cm^3} \quad (22.3\ \cancel{mL})\left(\frac{1\ \cancel{cm^3}}{1\ \cancel{mL}}\right)\left(\frac{1.0454\ g\ CH_3COOH}{\cancel{cm^3}}\right) = 23.3\ g\ CH_3COOH$$

$$(23.3\ \cancel{g\ CH_3COOH})\left(\frac{1\ mol\ CH_3COOH}{60.05\ \cancel{g\ CH_3COOH}}\right)\left(\frac{1}{L}\right) = 3.88 \times 10^{-1}\ M = [CH_3COOH]$$

$$K_a = 1.8 \times 10^{-5} = \frac{[H^+][CH_3COO^-]}{[CH_3COOH]}$$

Let $x = [H^+] = [CH_3COO^-]$; then $[CH_3COOH] = 3.88 \times 10^{-1} - x$

$$1.8 \times 10^{-5} = \frac{x^2}{(3.88 \times 10^{-1} - x)}$$

$x = 2.6 \times 10^{-3} = [H^+]$

$pH = -\log [H^+] = 2.59$

(Best answer, without intermediate round off, is 2.58.)

15.127 Either the sample contained a mixture of two monoprotic acids with K_a's differing by 10^{-5} or so or else the sample was a diprotic acid with $K_{a2} << K_{a1}$.

15.129 10.12

For HCO_3^-, K_a (Table 15.3) $= 5.6 \times 10^{-11}$.

For $HCO_3^- + OH^- \rightleftharpoons H_2O + CO_3^{2-}$

$$K = \frac{[CO_3^{2-}]}{[HCO_3^-][OH^-]} = \frac{K_a}{K_w} = \frac{5.6 \times 10^{-11}}{1.0 \times 10^{-14}} = 5.6 \times 10^3$$

Since K is large, assume that reaction takes place to completion.

$$(0.0107 \cancel{L})\left(\frac{0.10 \text{ mol NaOH}}{\cancel{L}}\right) = 1.1 \times 10^{-3} \text{ mol NaOH}$$

$$(0.0500 \cancel{L})\left(\frac{0.50 \text{ mol NaHCO}_3}{\cancel{L}}\right) = 2.5 \times 10^{-3} \text{ mol NaHCO}_3$$

Thus, 1.1×10^{-3} mol of CO_3^{2-} is formed and (2.5×10^{-3} mol – 1.1×10^{-3} mol) = 1.4×10^{-3} mol HCO_3^- remains.

$$[CO_3^{2-}] = \left(\frac{1.1 \times 10^{-3} \text{ mol}}{0.0607 \text{ L}}\right) = 0.018 \text{ M}$$

$$[HCO_3^-] = \left(\frac{1.4 \times 10^{-3} \text{ mol}}{0.0607 \text{ L}}\right) = 0.023 \text{ M}$$

Now calculate K for the reverse reaction:

$$H_2O + CO_3^{2-} \rightleftharpoons HCO_3^- + OH^- \quad K = \left(\frac{1}{5.6 \times 10^3}\right) = 1.8 \times 10^{-4}$$

Set up table of information

	H_2O	+	CO_3^{2-}	$\rightleftharpoons$	HCO_3^-	+	OH^-
Init. conc., M			0.018		0.023		0
Change, ΔM			$-x$		$+x$		$+x$
Eq. conc., M			$0.018 - x$		$0.023 + x$		x

Inserting the above values into the equilibrium constant expression yields

$$\frac{(0.023 + x)(x)}{(0.018 - x)} = 1.8 \times 10^{-4}$$

$$x^2 + 0.023x - 3.2 \times 10^{-6} = 0$$

$$x = 1.4 \times 10^{-4} = [OH^-]$$

pOH = 3.85

pH = 10.15

(Best answer, without intermediate round off, is pH = 10.12.)

Alternatively, the Henderson-Hasselbalch equation may be used. The best answer, without intermediate round off, using the Henderson-Hasselbalch equation, is 10.13.

15.131 **(a)** C_6H_5

(b) $CH_3CH_2NH_2$, ethylamine
$(CH_3CH_2)_2NH$, diethylamine
$(CH_3CH_2)_3N$, triethylamine

(c)

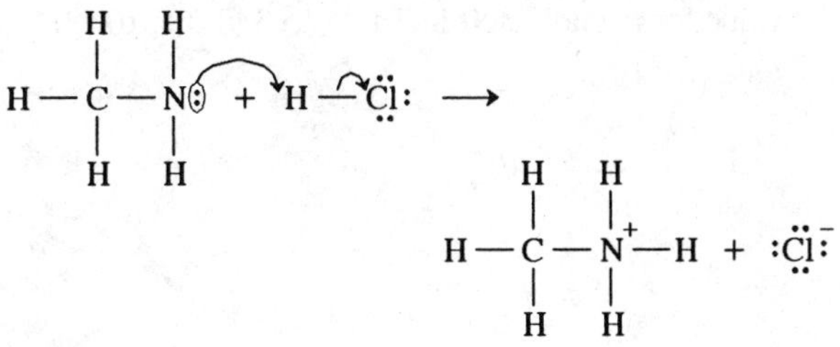

(d) $CH_3NH_2 + H^+ \rightarrow CH_3NH_3^+$

Applications

15.133 $Al(H_2O)_6^{3+} \rightleftarrows Al(H_2O)_5(OH)^{2+} + H^+$

$Fe(H_2O)_6^{3+} \rightleftarrows Fe(H_2O)_5(OH)^{2+} + H^+$

15.135 acidotic, basic

15.137 The basic group —NH_2 and the acidic group —COOH must react to give a saltlike structure $H_3\overset{+}{N}CH_2COO^-$.

15.139 3×10^{-1} or 0.3

For the equilibrium

$H_2PO_4^- \rightleftarrows H^+ + HPO_4^{2-}$

$$K_a = \frac{[H^+][HPO_4^{2-}]}{[H_2PO_4^-]} = 8.0 \times 10^{-8}$$

$[H^+]$ = antilog $(-6.6) = 3 \times 10^{-7}$

$$\frac{[HPO_4^{2-}]}{[H_2PO_4^-]} = \frac{8.0 \times 10^{-8}}{3 \times 10^{-7}} = 3 \times 10^{-1} = 0.3$$

15.141 (a)

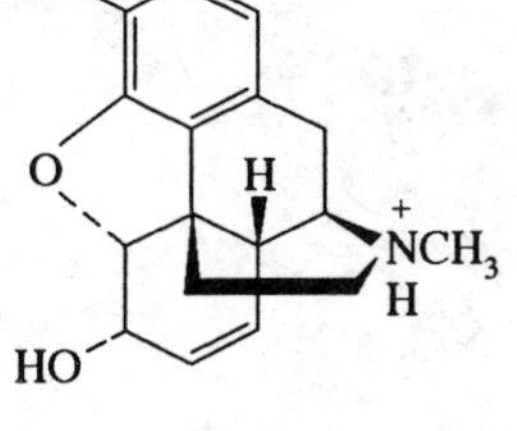

(b) No. The K_a value is for the production of H^+ from the compound as given.

(c) The acidic H is the one from the —OH attached to the benzene ring

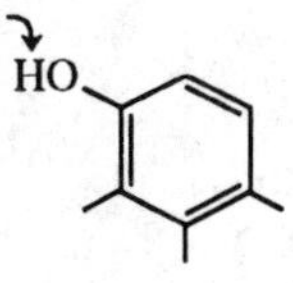

The K_a value given (1.4×10^{-10}) is similar to the value for phenol itself in Table 15.3 (1.3×10^{-10}).

15.143 **(a)** 1.7×10^{-5}

From the problem, $pK_a = 9.22$

$pK_a + pK_b = 14.00$

$pK_b = 14.00 - 9.22 = 4.78$

$K_b = \text{antilog}\ (-4.78) = 1.7 \times 10^{-5}$

(b) 5.37

$K_a = \text{antilog}\ (-9.22) = 6.0 \times 10^{-10}$

Let $x = [H^+] = [\text{free base}]$; then $[\text{base}] = 0.030 - x$

$$6.0 \times 10^{-10} = \frac{x^2}{0.030 - x}$$

$x = 4.2 \times 10^{-6} = [H^+]$

$pH = -\log\ (4.2 \times 10^{-6}) = 5.38$

(Best answer, without intermediate round off, is 5.37.)

15.145 $K_b = 2.5 \times 10^{-6}$

$\pi = MRT$

$$46.32\ \cancel{\text{mmHg}} = M\left(\frac{62.36\ \cancel{\text{mmHg}} \cdot L}{\text{mol} \cdot \cancel{K}}\right)(288.2\ \cancel{K})$$

$M = 2.577 \times 10^{-3}$

$\text{cocaine} + H_2O \rightleftharpoons (\text{cocaine})^+ + OH^-$

$$K_b = \frac{[(\text{cocaine})^+][OH^-]}{[\text{cocaine}]}$$

Let $x = [(\text{cocaine})^+] = [OH^-]$; then $(2.5000 \times 10^{-3} - x) = [\text{cocaine}]$

The total concentration of particles is $2x + (2.5000 \times 10^{-3} - x) = 2.577 \times 10^{-3}$

and $x = 7.7 \times 10^{-5}$

$$K_b = \frac{(7.7 \times 10^{-5})(7.7 \times 10^{-5})}{2.5000 \times 10^{-3}} = 2.4 \times 10^{-6}$$

(Best answer, without intermediate round off, is 2.5×10^{-6}.)

(From the Merck Index, the value of K_b for cocaine is 2.6×10^{-6} at 15 °C.)

15.147 (a) 0.01 M

$[H^+]$ = antilog (–pH) = antilog (–2.0) = 1×10^{-2}

(b) 0.4 g

$[H^+] = [HCl]_0 = 0.01$ M

$$(1\ \cancel{L})\left(\frac{0.01\ \cancel{mol\ HCl}}{\cancel{L}}\right)\left(\frac{36.5\ g\ HCl}{\cancel{mol\ HCl}}\right) = 0.4\ g$$

(c) increases

As plasma $[H^+]$ decreases, the value of plasma pH increases.

(d) Classic Coke has a pH of 2.5 (Table 15.2). Drinking colas replaces the acid and water lost by vomiting.

15.149 (a) 9×10^{-6} M

For carbonic acid, $K_{a1} = 4.3 \times 10^{-7}$ and $K_{a2} = 5.6 \times 10^{-11}$ at 25 °C (Table 15.7). Because the value of K_{a2} is very much lower than the value of K_{a1}, we can assume that all the H^+ in the solution is formed by the first ionization.

$$CO_2(aq) + H_2O(l) \rightleftharpoons HCO_3^- + H^+$$

Water in equilibrium with air has a pH of 5.7 as a result of dissolved CO_2. Therefore, the $[H^+] = 2 \times 10^{-6}$. The $[HCO_3^-]$ is also 2×10^{-6}. For the first ionization

$$K_{a1} = \frac{[HCO_3^-][H^+]}{[CO_2]} = 4.3 \times 10^{-7}$$

$$\text{and } [CO_2] = \frac{[HCO_3^-][H^+]}{4.3 \times 10^{-7}} = \frac{(2 \times 10^{-6})^2}{4.3 \times 10^{-7}} = 9 \times 10^{-6}$$

(b) CO_2 reacts with water producing HCO_3^-, which buffers the solution so that the end point is not sharp. CO_2–free water can be prepared by boiling the water and cooling it in a CO_2–free atmosphere.

15.151 Ammonium nitrate is the salt of a strong acid (HNO_3) with a weak base (NH_3) and is acidic. It lowers the pH of soil. (Soil must be moist or plants will not grow.)

CHAPTER 16:
More About Equilibria

Practice Problems

16.1 **(a)** to the right

According to Table 15.3, CH_3COOH is a stronger acid than HOCl and OCl^- is a stronger base than CH_3COO^-. Strong acids and strong bases react to form weak acids and bases. Therefore, equilibrium lies to the right.

(b) to the left

Table 15.3 shows that HF is a stronger acid than CH_3COOH and F^- is a weaker base than CH_3COO^-. Therefore, the equilibrium lies to the left.

(c) to the left

According to Table 15.3, CH_3COOH is a much stronger acid than NH_4^+ and NH_3 is a much stronger base than CH_3COO^-. Therefore, the equilibrium lies to the left.

16.2 **(a)** 5.6×10^2

The equilibrium

$$CH_3COOH(aq) + OCl^-(aq) \rightleftharpoons CH_3COO^-(aq) + HOCl(aq) \quad (16.1\ a)$$

is the sum of the equilibria

$$\begin{array}{rl} & CH_3COOH(aq) \rightleftharpoons CH_3COO^- + H^+ \\ + & [H^+ + OCl^- \rightleftharpoons HOCl(aq)] \\ \hline & CH_3COOH(aq) + OCl^- \rightleftharpoons CH_3COO^- + HOCl(aq) \end{array}$$

Therefore, the equilibrium constant is the product of the equilibrium constants. The equilibrium $H^+ + OCl^- \rightleftarrows HOCl(aq)$ is the reverse of the equilibrium $HOCl(aq) \rightleftarrows H^+ + OCl^-$ and its equilibrium constant $= \dfrac{1}{K_{a\,(HOCl)}}$.

The equilibrium constant for reaction 16.1 (a) is

$$K_{(16.1\text{ a})} = \frac{K_{a\,(CH_3COOH)}}{K_{a\,(HOCl)}} = \frac{1.8 \times 10^{-5}}{3.2 \times 10^{-8}} = 5.6 \times 10^2$$

(b) 5.1×10^{-2}

The equilibrium

$$CH_3COOH(aq) + F^- \rightleftarrows CH_3COO^- + HF(aq) \qquad (16.1\text{ b})$$

is the sum of the equilibria

$$\begin{array}{rl} & CH_3COOH(aq) \rightleftarrows CH_3COO^- + H^+ \\ + & [F^- + H^+ \rightleftarrows HF(aq)] \\ \hline & CH_3COOH(aq) + F^- \rightleftarrows CH_3COO^- + HF(aq) \end{array}$$

The equilibrium constant is the product of the equilibrium constants. The equilibrium

$$F^- + H^+ \rightleftarrows HF(aq)$$

is the reverse of the equilibrium

$$HF(aq) \rightleftarrows H^+ + F^-$$

and its equilibrium constant $= \dfrac{1}{K_{a\,(HF)}}$

Thus, the equilibrium constant for the reaction 16.1 (b) is

$$K_{(16.1\text{ b})} = \frac{K_{a\,(CH_3COOH)}}{K_{a\,(HF)}} = \frac{1.8 \times 10^{-5}}{3.5 \times 10^{-4}} = 5.1 \times 10^{-2}$$

16.3 structure (a) has one extra oxygen like phosphoric acid; monoprotic acid

Like phosphoric acid, hypophosphorous acid has one extra oxygen (oxygen that is not part of an —OH group) bound to the phosphorus atom. Thus, structure (a) shows hypophosphorous acid. Hypophosphorous acid is a monoprotic acid; only one of the three hydrogens is bonded to oxygen and is acidic.

16.4 **(a)** CH_3SH; S—H bond is longer and weaker than O—H bond

The S—H bond is longer and weaker than the O—H bond because sulfur is larger than oxygen (CH_3OH is neutral while CH_3SH is a weak acid).

(b) H_2S; bonds same length but S more electronegative than P

Sulfur and phosphorus are both in the third period and are similar in size. Sulfur is more electronegative than phosphorus. Therefore, H_2S is more acidic than PH_3 (H_2S is a weak acid while PH_3 is neutral).

(c) H_2SO_3; same number of extra oxygens, S more electronegative than P

The Lewis structure for H_2SO_3 is

$$
\begin{array}{c}
\ddot{O} \\
\parallel \\
H-\ddot{O}-S-\ddot{O}-H
\end{array}
$$

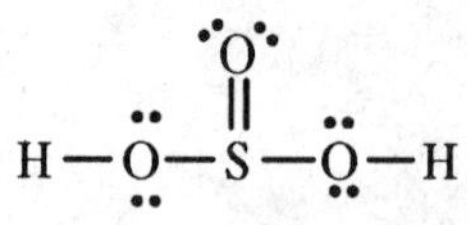

Like H_3PO_4, H_2SO_3 has one extra oxygen. Because sulfur is more electronegative than phosphorus, H_2SO_3 is the stronger acid. [K_a for H_2SO_3 is 1.5×10^{-2}, while K_a for H_3PO_4 is 7.1×10^{-3} (Table 15.7).]

(d) H_2SO_4; S has one more extra O in H_2SO_4

H_2SO_4 has two extra oxygens not bonded to hydrogen while H_2SO_3 has only one. This makes H_2SO_4 a stronger acid. Another way to arrive at the conclusion that H_2SO_4 is stronger than H_2SO_3 that does not involve writing Lewis stuctures is to note that the oxidation number of S in H_2SO_4 (+6) is higher than the oxidation number of S in H_2SO_3 (+4). An increase in acidity with increasing oxidation number is common (Section 11.2). (H_2SO_4 is a strong acid, H_2SO_3 is a weak acid.)

(e) CH_3COOH; C in CH_3COOH has an extra O

Lewis structures for CH_3OH and CH_3COOH are

$$
\begin{array}{cc}
\begin{array}{c}
H \\
\mid \\
H-C-\ddot{O}-H \\
\mid \\
H
\end{array}
&
\begin{array}{c}
H \quad \ddot{O} \\
\mid \quad \parallel \\
H-C-C-\ddot{O}-H \\
\mid \\
H
\end{array}
\end{array}
$$

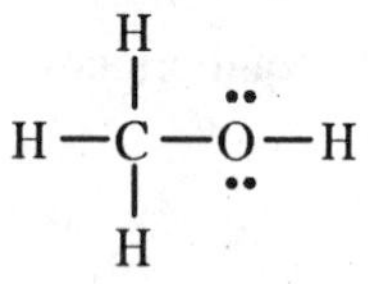

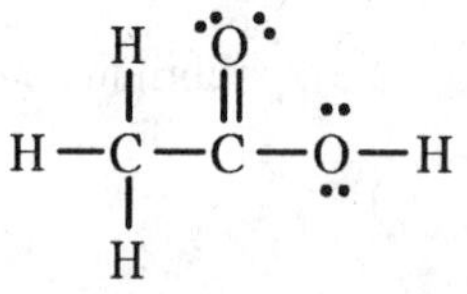

Carbon is bonded to one extra oxygen not bonded to hydrogen in CH_3COOH, making CH_3COOH a stronger acid than CH_3OH. (CH_3OH is neutral; CH_3COOH is a weak acid.)

16.5   hypobromous acid

The oxo acids of the halogens have analogous names. Since HOCl is called hypochlorous acid, HOBr is called hypobromous acid.

16.6   **(c)** 13.5; Ba^{2+} is larger than Sr^{2+}

Ba^{2+} has the same charge as Sr^{2+} but a larger radius. The electronegativities of both metals are 1.0. Therefore, the hydrated Ba^{2+} will be less acidic than the hydrated Sr^{2+} ion and its pK_a will be larger than that of Sr^{2+}. Thus, 13.5 is the correct answer.

16.7 CH_3COO^-; conjugate of weaker acid

CH_3COO^- is the conjugate base of the weaker acid. Thus, CH_3COO^- is a stronger base.

16.8   **(a)** basic; oxide of active metal

Barium is an alkaline earth metal (Group IIA). As such, its oxide is basic.

(b) acidic; nonmetal oxide

Phosphorus is a nonmetal. Therefore, its oxides are acidic.

(c) amphoteric; zinc is near metal-nonmetal border

ZnO is on the list of amphoteric oxides. Zinc is near the metal–nonmetal border in the periodic table.

16.9 (a) Net ionic equation: $O^{2-} + H_2O(l) \rightarrow 2OH^-$

Complete ionic equation: $2K^+ + O^{2-} + H_2O(l) \rightarrow 2K^+ + 2OH^-$

Molecular equation: $K_2O(s) + H_2O(l) \rightarrow 2KOH(aq)$

(Note that the symbol "(aq)" is not necessary when writing symbols for ions in aqueous solutions. Ions are assumed to be in aqueous solution unless otherwise marked, for example, $Na^+(g)$.)

(b) Net ionic equation: $Fe_2O_3(s) + 6H^+ \rightarrow 2Fe^{3+} + 3H_2O(l)$

Complete ionic equation: $Fe_2O_3(s) + 6H^+ + 6Cl^- \rightarrow 2Fe^{3+} + 6Cl^- + 3H_2O(l)$

Molecular equation: $Fe_2O_3(s) + 6HCl(aq) \rightarrow 2FeCl_3(aq) + 3H_2O(l)$

(c) Net and complete ionic equations: $N_2O_5(s) + H_2O(l) \rightarrow 2H^+ + 2NO_3^-$

Molecular equation: $N_2O_5(s) + H_2O(l) \rightarrow 2HNO_3(aq)$

16.10 (a) Cr^{3+}, $BeCl_2$

Cr^{3+} is a small ion with a high positive charge. The Lewis structure for $BeCl_2$ is

$$:\!\overset{\displaystyle ..}{\underset{\displaystyle ..}{Cl}} - Be - \overset{\displaystyle ..}{\underset{\displaystyle ..}{Cl}}\!:$$

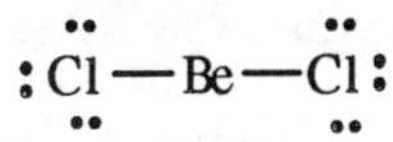

This structure indicates the Be is "electron deficient." Therefore, $BeCl_2$ acts as a Lewis acid.

(b) CH_3NH_2

The Lewis formula for CH_3NH_2 is

$$H - \overset{\displaystyle \overset{H}{|}}{\underset{\displaystyle \underset{H}{|}}{C}} - \overset{\displaystyle \overset{H}{|}}{\underset{\displaystyle \underset{H}{|}}{N}}:$$

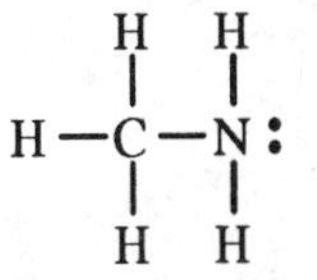

It acts as a Lewis base because it has an unshared pair of electrons on N. Although halogen atoms, such as Cl, have three unshared pairs of electrons, they do not usually act as bases.

(c) CH_4, Na^+

The Lewis structure for CH_4 is

$$H - \overset{\displaystyle \overset{H}{|}}{\underset{\displaystyle \underset{H}{|}}{C}} - H$$

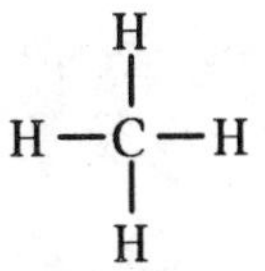

CH_4 cannot act as a Lewis acid because neither C nor H can accept an electron pair. It cannot act as a Lewis base because there is no unshared electron pair. The Na^+ ion does not act as a Lewis acid because it is relatively large and has only a 1+ charge. It does not act as a Lewis base because it does not have an unshared pair of valence electrons.

16.11 **(a)** Lewis acid – $Al(OH)_3$, Lewis base – OH^-

(b) Lewis acid – Ag^+, Lewis base – NH_3

(c) Lewis acid – BF_3, Lewis base – $(CH_3CH_2)_2O$

Lewis acids accept an electron pair; Lewis bases donate an electron pair.

16.12

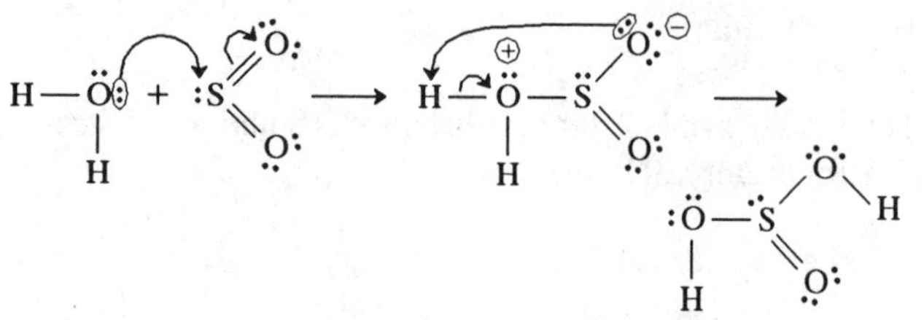

The electronegativity of oxygen, 3.5, is greater than the electronegativity of sulfur, 2.4. Therefore the S—O bonds in SO_2 are polar with a partial positive charge on sulfur. Sulfur dioxide acts as a Lewis acid and accepts an unshared pair of electrons from the oxygen in water forming a covalent bond. A proton shifts from the positively charged oxygen to the negatively charged oxygen to give an H_2SO_3 molecule.

16.13 1.1×10^{-12} M

The equilibrium constant for the equilibrium

$$Cu^{2+} + 4NH_3(aq) \rightleftarrows Cu(NH_3)_4^{2+}$$

is 5.6×10^{11} (Table 16.4), a very large number, and reaction is essentially complete at equilibrium. So that the change in concentration will be small compared to the initial concentration, assume that formation of the complex is complete and consider the reverse reaction, the dissociation of the complex.

$$Cu(NH_3)_4^{2+} \rightleftarrows Cu^{2+} + 4NH_3(aq)$$

For the dissociation of the complex

$$K_d = \frac{[Cu^{2+}][NH_3]^4}{[Cu(NH_3)_4^{2+}]} = \frac{1}{K_f} = \frac{1}{5.6 \times 10^{11}} = 1.8 \times 10^{-12}$$

The equation for the formation of the complex shows us that 4 moles of NH_3 react with 1 mole of Cu^{2+}. Therefore, 0.080 mol NH_3 will react with 0.020 mol Cu^{2+}. Thus, there will be 0.50 mol NH_3 – 0.080 mol NH_3 = 0.42 mol NH_3 left after reaction and $[NH_3]$ = 0.42 M

The "ICE" table is

	$Cu(NH_3)_4^{2+}$	$\rightleftarrows$	Cu^{2+}	+	$4NH_3(aq)$
Init. conc., M	0.020		0.00		0.42
Change, ΔM	$-x$		$+x$		$+4x$
Eq. conc., M	$0.020 - x$		x		$0.42 + 4x$

Substitution of the equilibrium concentrations in terms of x in the equilibrium constant expression gives

$$K_d = \frac{(x)(0.42 + 4x)^4}{(0.020 - x)} = 1.8 \times 10^{-12}$$

Assume $4x$ is negligible compared to 0.42 and that x is negligible compared to 0.020.

$$1.8 \times 10^{-12} = \frac{(x)(0.42)^4}{0.020}$$

$$x = \frac{(0.020)(1.8 \times 10^{-12})}{(0.42)^4} = 1.2 \times 10^{-12} = [Cu^{2+}]_{eq}$$

Thus, the assumption that x is negligible compared to 0.020 and that $4x$ is negligible compared to 0.42 is correct.

(Best answer, without intermediate round off, is 1.1×10^{-12} M.)

16.14 (a) $AgCl(s) \rightleftarrows Ag^+ + Cl^-$ $K_{sp} = [Ag^+][Cl^-]$

(b) $Fe(OH)_2(s) \rightleftarrows Fe^{2+} + 2OH^-$ $K_{sp} = [Fe^{2+}][OH^-]^2$

(c) $PbBr_2(s) \rightleftarrows Pb^{2+} + 2Br^-$ $K_{sp} = [Pb^{2+}][Br^-]^2$

(d) $Li_2CO_3(s) \rightleftarrows 2Li^+ + CO_3^{2-}$ $K_{sp} = [Li^+]^2[CO_3^{2-}]$

(e) $Ag_2SO_4(s) \rightleftarrows 2Ag^+ + SO_4^{2-}$ $K_{sp} = [Ag^+]^2[SO_4^{2-}]$

16.15 1.8×10^{-4} g AgCl/100 cm^3 solution

	$AgCl(s)$	$\rightleftarrows$	Ag^+	+	Cl^-	$K_{sp} = [Ag^+][Cl^-] = 1.6 \times 10^{-10}$
Init. conc., M			0.0		0.0	
Change, ΔM			$+x$		$+x$	
Eq. conc., M			x		x	

$K_{sp} = x^2 = 1.6 \times 10^{-10}$

$x = 1.3 \times 10^{-5}$ M

$$\left(\frac{1.3 \times 10^{-5}\ \cancel{\text{mol AgCl}}}{\cancel{\text{L soln}}}\right)\left(\frac{1\ \cancel{\text{L soln}}}{10^3\ \cancel{\text{cm}^3\ \text{soln}}}\right)(100\ \cancel{\text{cm}^3\ \text{soln}})\left(\frac{143\ \text{g AgCl}}{\cancel{\text{mol AgCl}}}\right) = 1.9 \times 10^{-4}\ \text{g AgCl}$$

(Best answer, without intermediate round off, is 1.8×10^{-4} g AgCl.)

16.16 3×10^{-8}

$TlI(s) \rightleftarrows Tl^+ + I^-$

$$\left(\frac{0.006\ \cancel{\text{g TlI}}}{100\ \cancel{\text{cm}^3\ \text{soln}}}\right)\left(\frac{10^3\ \cancel{\text{cm}^3\ \text{soln}}}{\text{L soln}}\right)\left(\frac{1\ \text{mol TlI}}{331\ \cancel{\text{g TlI}}}\right) = 2 \times 10^{-4}\ \text{M}$$

$K_{sp} = [Tl^+][I^-] = (2 \times 10^{-4})^2 = 4 \times 10^{-8}$

(Best answer, without intermediate round off, is 3×10^{-8}.)

16.17 (a) $MgCO_3$, (b) $Ca(OH)_2$, (d) ZnS

CO_3^{2-}, OH^-, and S^{2-} are all conjugate bases of weak acids while Br^- is the conjugate base of a strong acid. The three soluble materials dissolve by the following reactions in which weak electrolytes are formed:

(a) $MgCO_3(s) + 2H^+ \rightarrow Mg^{2+} + H_2O(l) + CO_2(g)$

(b) $Ca(OH)_2(s) + 2H^+ \rightarrow Ca^{2+} + 2H_2O(l)$

(d) $ZnS(s) + 2H^+ \rightarrow Zn^{2+} + H_2S(g)$

AgBr does not dissolve in dilute nitric acid because both HBr and $AgNO_3$ are soluble strong electrolytes.

16.18 (a) Net ionic equation: $Mg(OH)_2(s) + 2H^+ \rightarrow Mg^{2+} + 2H_2O(l)$

Complete ionic equation: $Mg(OH)_2(s) + 2H^+ + 2Cl^- \rightarrow Mg^{2+} + 2Cl^- + 2H_2O(l)$

Molecular equation: $Mg(OH)_2(s) + 2HCl(aq) \rightarrow MgCl_2(aq) + 2H_2O(l)$

(b) Net and complete ionic equations: $AgCl(s) + 2NH_3(aq) \rightarrow Ag(NH_3)_2^+ + Cl^-$

Molecular equation: $AgCl(s) + 2NH_3(aq) \rightarrow Ag(NH_3)_2Cl(aq)$

(c) Net ionic equation: $AgCl(s) + Cl^- \rightarrow AgCl_2^-$

Complete ionic equation: $AgCl(s) + H^+ + Cl^- \rightarrow AgCl_2^- + H^+$

Molecular equation: $AgCl(s) + HCl(aq) \rightarrow HAgCl_2(aq)$

16.19 The solubility product constant, K_{sp}, for sulfides assumes equilibrium between undissolved solid and metal and sulfide ions (for example, $ZnS \rightleftarrows Zn^{2+} + S^{2-}$). The solubilities of sulfides are much higher than predicted because S^{2-} is the conjugate base of a very weak acid, HS^-. This causes hydrolysis to take place when S^{2-} ions are present in aqueous solution (for example, $ZnS(s) + H_2O(l) \rightleftarrows Zn^{2+} + SH^- + OH^-$). This type of equilibrium increases the solubility of sulfides in water.

16.20 8×10^{-4}

$AgBr(s) \rightleftarrows Ag^+ + Br^- \quad K_{sp} = [Ag^+][Br^-] = 5 \times 10^{-13}$

$$[Br^-] = \frac{K_{sp}}{[Ag^+]} = \frac{5 \times 10^{-13}}{6 \times 10^{-10}} = 8 \times 10^{-4}$$

16.21 No.

$Ag_2C_2O_4(s) \rightleftarrows 2Ag^+ + C_2O_4^{2-} \quad K_{sp} = [Ag^+]^2[C_2O_4^{2-}] = 5 \times 10^{-12}$

$$(25.0\ \text{mL soln})\left(\frac{2 \times 10^{-4}\ \text{mol}\ (NH_4)_2C_2O_4}{\text{L soln}}\right)\left(\frac{1\ \text{L soln}}{10^3\ \text{mL soln}}\right) = 5 \times 10^{-6}\ \text{mol}\ (NH_4)_2C_2O_4$$

$$(75.0\ \text{mL soln})\left(\frac{3 \times 10^{-5}\ \text{mol}\ AgNO_3}{\text{L soln}}\right)\left(\frac{1\ \text{L soln}}{10^3\ \text{mL soln}}\right) = 2 \times 10^{-6}\ \text{mol}\ AgNO_3$$

25.0 mL + 75.0 mL = 100.0 mL soln

$$[C_2O_4^{2-}] = \left(\frac{5 \times 10^{-6}\ \text{mol}\ C_2O_4^{2-}}{100.0\ \text{mL soln}}\right)\left(\frac{10^3\ \text{mL soln}}{\text{L soln}}\right) = 5 \times 10^{-5}\ \text{M}\ C_2O_4^{2-}$$

$$[Ag^+] = \left(\frac{2 \times 10^{-6}\ \text{mol}\ Ag^+}{100.0\ \text{mL soln}}\right)\left(\frac{10^3\ \text{mL soln}}{\text{L soln}}\right) = 2 \times 10^{-5}\ \text{M}\ Ag^+$$

$[Ag^+]^2[C_2O_4^{2-}] = (2 \times 10^{-5})^2(5 \times 10^{-5}) = 2 \times 10^{-14}$

$2 \times 10^{-14} < 5 \times 10^{-12}$ Therefore, $Ag_2C_2O_4$ will not precipitate.

16.22 **(a)** 9×10^{-9} M

	$AgI(s)$	$\rightleftarrows$	Ag^+	+	I^-
Init. conc., M			0.00		0.00
Change, ΔM			$+x$		$+x$
Eq. conc., M			x		x

$K_{sp} = [Ag^+][I^-] = 9 \times 10^{-17} = x^2$

$x = 9 \times 19^{-9}$ M

(b) 9×10^{-9} M

HNO_3 has no effect on the solubility of AgI because both $AgNO_3$ and HI are soluble strong electrolytes. Therefore, AgI does not react with HNO_3 and AgI is not soluble in HNO_3.

(c) 2×10^{-5} M

Two equilibria are important to the solution of this problem.

$AgI(s) \rightleftarrows Ag^+ + I^- \quad K_{sp} = [Ag^+][I^-] = 9 \times 10^{-17}$

$Ag^+ + 2NH_3(aq) \rightleftarrows Ag(NH_3)_2^+ \quad K_f = 1.7 \times 10^7$

Addition of these two equilibria gives the overall equation for the solution of AgI(s) in $NH_3(aq)$.

$AgI(s) + 2NH_3(aq) \rightleftarrows Ag(NH_3)_2^+ + I^-$

Therefore, the equilibrium constant for the overall reaction is equal to the product of K_{sp} and K_f.

$$K_{eq} = \frac{[Ag(NH_3)_2^+][I^-]}{[NH_3]^2} = K_{sp} \cdot K_f = (9 \times 10^{-17})(1.7 \times 10^7) = 2 \times 10^{-9}$$

Now proceed as usual and let x equal mol/L $Ag(NH_3)_2^+$ and I^- formed. Then $2x$ equals the mol/L of NH_3 reacted.

The summary table is

	$AgI(s)$	+	$2NH_3(aq)$	$\rightleftarrows$	$Ag(NH_3)_2^+$	+	I^-
Init. conc., M			0.50		0.00		0.00
Change, ΔM			$-2x$		$+x$		$+x$
Eq. conc., M			$0.50 - 2x$		x		x

Substitution of the equilibrium concentrations in terms of x in the equilibrium constant expression gives

$$2 \times 10^{-9} = \frac{x^2}{(0.50 - 2x)^2}$$

Both numerator and denominator on the right are perfect squares. Take the square root of both sides.

$$\sqrt{2 \times 10^{-9}} = \frac{x}{0.050 - 2x}$$

and $x = 2 \times 10^{-5}$

Note how the solubility of AgI is increased by complex ion formation.

(d) 4×10^{-15} M

The summary table is

	AgI(s)	$\rightleftarrows$	Ag^+	+	I^-
Init. conc., M			0.025		0.000
Change, ΔM			$+x$		$+x$
Eq. conc., M			$0.025 + x$		x

Substitution of the equilibrium concentrations in terms of x in the solubility product expression gives

$$K_{sp} = [Ag^+][I^-] = 9 \times 10^{-17} = (0.025 + x)x$$

which simplifies to

$$9 \times 10^{-17} = 0.025x$$

if x is asumed to be negligible compared to 0.025. Solution for x then gives

$$x = 4 \times 10^{-15}$$

The assumption that x is negligible compared to 0.025 is correct. The molar solubility of AgI in 0.025 M $AgNO_3$ is 4×10^{-15}. Note how the solubility of AgI is decreased by the presence of the common ion (Ag^+).

Additional Practice Problems

16.73 **(a)**

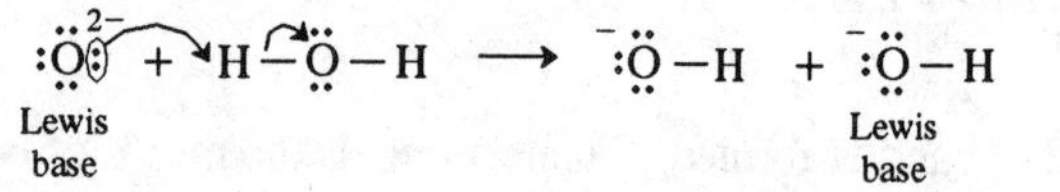

(b)

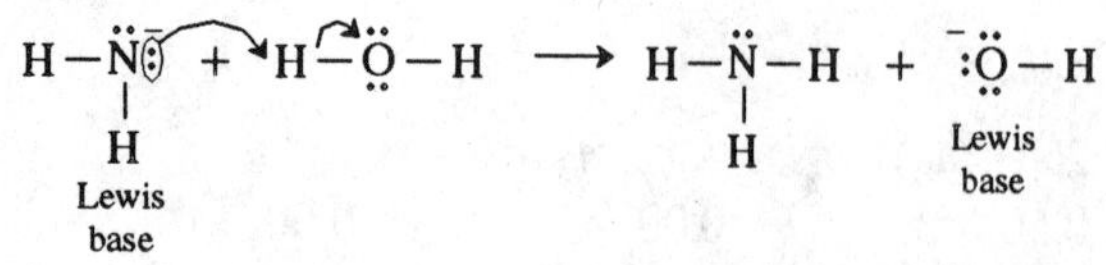

(c)

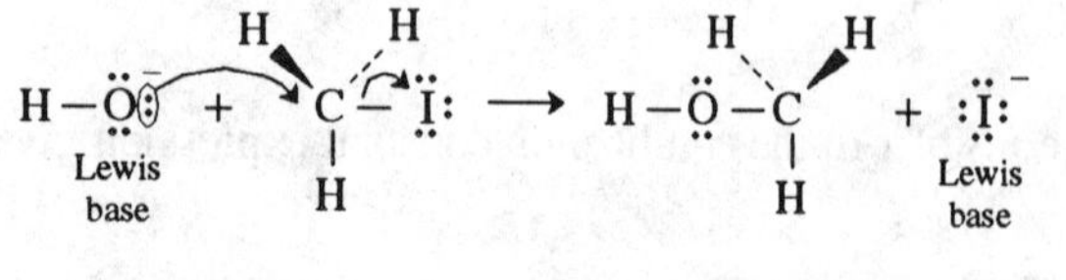

(d)

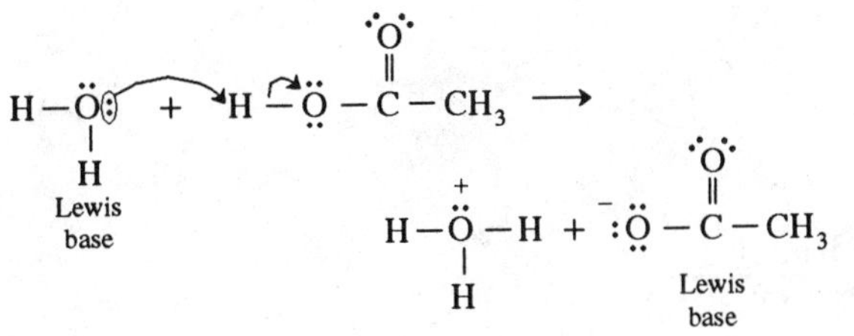

16.75 (a) AgI

As $AgNO_3(s)$ is added to the KI–NaCl solution, $[Ag^+]$ will increase. Whether AgCl or AgI will precipitate first depends on which will precipitate at the lower $[Ag^+]$. Calculate the concentrations of Ag^+ necessary to precipitate AgCl from 0.015 M NaCl and AgI from 0.010 M KI and see which is lower.

For AgCl the solubility equilibrium is

$$AgCl(s) \rightleftharpoons Ag^+ + Cl^-$$

and $K_{sp} = [Ag^+][Cl^-] = 1.6 \times 10^{-10}$ (Table 16.6). If $[Cl^-] = 0.015$, the $[Ag^+]$ necessary to precipitate AgCl is

$$\frac{1.6 \times 10^{-10}}{0.015} = [Ag^+] = 1.1 \times 10^{-8}$$

For AgI the solubility equilibrium is

$$AgI(s) \rightleftharpoons Ag^+ + I^-$$

and $K_{sp} = [Ag^+][I^-] = 9 \times 10^{-17}$. If $[I^-] = 0.010$, the $[Ag^+]$ necessary to precipitate AgI is

$$\frac{9 \times 10^{-17}}{0.010} = [Ag^+] = 9 \times 10^{-15}$$

Because $9 \times 10^{-15} < 1.1 \times 10^{-8}$, AgI will precipitate first.

(b) 8×10^{-9}, $8 \times 10^{-5}\%$

For AgCl to begin to precipitate, $[Ag^+]$ must be 1.1×10^{-8}. When $[Ag^+] = 1.1 \times 10^{-8}$, $[I^-]$ must be

$$\frac{9 \times 10^{-17}}{1.1 \times 10^{-8}} = [I^-] = 8 \times 10^{-9}$$

The percent of the original $[I^-]$ that is still in solution is

$$\frac{8 \times 10^{-9}}{1.0 \times 10^{-2}} \times 100 = 8 \times 10^{-5}\ \%$$

Stop and Test Yourself

1. (b) $CH_3CH_2COOH(aq) + SO_4^{2-} \rightleftharpoons CH_3CH_2COO^-(aq) + HSO_4^-(aq)$

According to their positions in Table 15.3, $HSO_4^-(aq)$ is a stronger acid than CH_3CH_2COOH.

2. **(d)** 3.5×10^1

The equilibrium is the sum of the equilibria.

$$\begin{array}{rl} & HNO_2(aq) \rightleftarrows H^+ + NO_2^- \\ \text{and} & CH_3CH_2COO^- + H^+ \rightleftarrows CH_3CH_2COOH(aq) \\ \hline & HNO_2(aq) + CH_3CH_2COO^- \rightleftarrows NO_2^- + CH_3CH_2COOH(aq) \end{array}$$

Therefore, $K = K_{HNO_2} \cdot \dfrac{1}{K_{CH_3CH_2COOH}} = (4.6 \times 10^{-4})\left(\dfrac{1}{1.3 \times 10^{-5}}\right) = 3.5 \times 10^1$

3. **(e)** $HCl > HF > H_2O$

HCl is a strong acid, HF is a weak acid, and H_2O is neutral.

4. **(e)** $H_2SO_4 > H_2SO_3 > HSO_3^-$

H_2SO_4 is a strong acid. H_2SO_3 is a weak acid. It has one less extra oxygen than H_2SO_4; the oxidation number of S is lower in H_2SO_3 (+4 as compared to +6 in H_2SO_4). HSO_3^- is a weaker acid than H_2SO_3 as a result of its negative charge.

$K_{a(H_2SO_3)} = 1.7 \times 10^{-2}$

$K_{a(HSO_3^-)} = 6.0 \times 10^{-8}$

$K_{a(H_2SO_4)} =$ very large

5. **(c)** $FCH_2COO^- < ClCH_2COO^- < CH_3COO^-$

The stronger an acid, the weaker its conjugate base. To arrange anions in order of basicity, begin by arranging their conjugate acids in order of acidity. $ClCH_2COOH$ is stronger than CH_3COOH as a result of the substitution of the more electronegative Cl for H. FCH_2COOH is stronger than $ClCH_2COOH$ because F is more electronegative than Cl. Thus, in order of decreasing acid strength the acids are $FCH_2COOH > ClCH_2COOH > CH_3COOH$, and the bases in order of increasing strength are $FCH_2COO^- < ClCH_2COO^- < CH_3COO^-$.

6. **(b)** $Cr(NO_3)_3$

The Cr^{3+} ion has the highest charge and the smallest radius. Therefore, the hydrated Cr^{3+} ion is the most acidic.

7. **(a)** BaO

Ba is a metal. Most metal oxides are basic; nonmetal oxides are usually acidic. (A few nonmetal oxides such as CO are neutral.)

8. **(a)** O^{2-}

A Lewis acid is an electron pair acceptor. O in O^{2-} has an octet and cannot accept any more electrons.

9. **(c)** 1.4×10^{-34}

Hydrolysis of CN^- is negligible (Sample Problem 16.3).

The equilibrium constant for the equilibrium

$$Fe^{2+} + 6CN^- \rightleftarrows Fe(CN)_6^{4-}$$

is 2.5×10^{35} (Table 16.4), a very large number. Reaction is essentially complete at equilibrium. Assume that formation of the complex is complete and consider the reverse reaction

$$Fe(CN)_6^{4-} \rightleftarrows Fe^{2+} + 6CN^-$$

for which $K = \dfrac{[Fe^{2+}][CN^-]^6}{[Fe(CN)_6^{4-}]} = \dfrac{1}{2.5 \times 10^{35}} = 4.0 \times 10^{-36}$

According to the equation for the formation of $Fe(CN)_6^{4-}$, 6 mol of CN^- react with 1 mol of Fe^{2+} to form 1 mol of complex ion. Therefore, 0.025 mol Fe^{2+} will react with 6(0.025) = 0.15 mol CN^- and the number of moles of CN^- left in a liter of solution will be

0.45 mol originally present – 0.15 mol reacted = 0.30 mol left

If we consider one liter of solution, $[CN^-] = 0.30$. According to the equation for the dissociation of $Fe(CN)_6^{4-}$, 1 mol $Fe(CN)_6^{4-}$ yields 1 mol Fe^{2+} and 6 mol CN^-. If we let x equal mol/L Fe^{2+} formed, then $6x$ equals mol/L CN^- formed and $-x$ equals mol/L $Fe(CN)_6^{4-}$ dissociated. The completed summary table is:

	$Fe(CN)_6^{4-}$	$\rightleftarrows$	Fe^{2+}	+	$6CN^-$
Init. conc., M	0.025		0.000		0.30
Change, ΔM	$-x$		$+x$		$+6x$
Eq. conc., M	$0.025 - x$		x		$0.30 + 6x$

Substitution of the equilibrium concentrations in terms of x in the equilibrium constant expression gives

$$\frac{x(0.30 + 6x)^6}{(0.025 - x)} = 4.0 \times 10^{-36}$$

and $x = 1.4 \times 10^{-34} = [Fe^{2+}]$

10. **(b)** 3×10^{-6}

For the equilibrium

$$Zn(OH)_2 \rightleftarrows Zn^{2+} + 2OH^-$$

$K_{sp} = [Zn^{2+}][OH^-]^2 = 8 \times 10^{-17}$
Let $x = [Zn^{2+}]$; then $[OH^-] = 2x$

$8 \times 10^{-17} = x(2x)^2$

$x = 3 \times 10^{-6} = [Zn^{2+}]$ = molar solubility

11. **(d)** 3.2×10^{-9}

For the equilibrium

$$AgCl \rightleftarrows Ag^+ + Cl^-$$

$K_{sp} = [Ag^+][Cl^-] = 1.6 \times 10^{-10}$

$1.6 \times 10^{-10} = (0.050)\ [Cl^-]$

$3.2 \times 10^{-9} = [Cl^-] = \text{molar solubility}$

12. (e) 4.3×10^{-6}

Two equilibria are involved:

$AgCl(s) \rightleftarrows Ag^+ + Cl^- \quad K_{sp} = [Ag^+][Cl^-] = 1.6 \times 10^{-10}$

$Ag^+ + 2Cl^- \rightleftarrows AgCl_2^- \quad K_f = 1.8 \times 10^5$

The overall equation, $AgCl(s) + Cl^- \rightleftarrows AgCl_2^-$, is the sum of the other two and therefore, for this equilibrium,

$$K = \frac{[AgCl_2^-]}{[Cl^-]} = K_{sp} \cdot K_f = (1.6 \times 10^{-10})(1.8 \times 10^5) = 2.9 \times 10^{-5}$$

	AgCl(s)	+	Cl^-	$\rightleftarrows$	$AgCl_2^-$
Init. conc., M			0.15		0.00
Change, ΔM			$-x$		$+x$
Eq. conc., M			$0.15 - x$		x

Substitution in the equilibrium constant expression gives

$$\frac{x}{0.15 - x} = 2.9 \times 10^{-5} \text{ and } x = 4.4 \times 10^{-6}$$

(Best answer, without intermediate round off, is 4.3×10^{-6}.)

13. (b) $PbCl_2$

Both $Pb(NO_3)_2$ and HCl are soluble strong electrolytes. No reaction will take place. Thus, $PbCl_2$ is not soluble in dilute nitric acid. All of the other compounds react with H^+ to form weak electrolytes, for example,

$$PbCO_3(s) + 2H^+ \rightarrow H_2O(l) + CO_2(g) + Pb^{2+}$$

14. (c) or (d) (because Q_{sp} must be $1000 \times K_{sp}$ for the precipitate to be visible)

Because equal volumes of the two solutions are mixed, the concentrations must be halved before calculating Q_{sp}. For AgCl, $K_{sp} = 1.6 \times 10^{-10}$.

(a) $Q_{sp} = (0.8 \times 10^{-5})(2.0 \times 10^{-5}) = 1.6 \times 10^{-10} = K_{sp}$ no precipitate will form

(b) $Q_{sp} = (4.0 \times 10^{-6})(2.0 \times 10^{-5}) = 8.0 \times 10^{-11} < K_{sp}$ no precipitate will form

(c) $Q_{sp} = (0.5 \times 10^{-4})(0.7 \times 10^{-4}) = 3.5 \times 10^{-9} > K_{sp}$ a precipitate will form

However, because the value of Q_{sp} must be about 1000 times K_{sp} for the precipitate to be visible, (d) is also an acceptable answer. If you chose (e), you forgot the effect of the increase in volume on concentration.

Putting Things Together

16.77 (a), (c); $Zn(OH)_2$ and $Al(OH)_3$ are amphoteric

The hydroxides of elements near the borderline between metals and nonmetals in the periodic table are amphoteric.

16.79 2.6×10^{-6} M

First calculate $[Cl^-]$ in the calcium chloride solution.

$$\left(\frac{5.0\ \cancel{g\ CaCl_2}}{1\ L}\right)\left(\frac{1\ \cancel{mol\ CaCl_2}}{111\ \cancel{g\ CaCl_2}}\right)\left(\frac{2\ mol\ Cl^-}{1\ \cancel{mol\ CaCl_2}}\right) = 0.090\ \frac{mol}{L} = 0.090\ M = [Cl^-]$$

Two equilibria are involved:

$$AgCl(s) \rightleftharpoons Ag^+ + Cl^- \qquad K_{sp} = [Ag^+][Cl^-] = 1.6 \times 10^{-10}$$

$$Ag^+ + 2Cl^- \rightleftharpoons AgCl_2^- \qquad K_f = \frac{[AgCl_2^-]}{[Cl^-]^2} = 1.8 \times 10^5$$

The overall equation, $AgCl(s) + Cl^- \rightleftharpoons AgCl_2^-$, is the sum of the other two and therefore, for this equilibrium,

$$K = \frac{[AgCl_2^-]}{[Cl^-]} = K_{sp} \cdot K_f = (1.6 \times 10^{-10})(1.8 \times 10^5) = 2.9 \times 10^{-5}$$

	$AgCl(s)$	+	Cl^-	$\rightleftharpoons$	$AgCl_2^-$
Init. conc., M			0.090		0.000
Change, ΔM			$-x$		$+x$
Eq. conc., M			$0.090 - x$		x

$$2.9 \times 10^{-5} = \frac{x}{0.090 - x}$$

and $x = 2.6 \times 10^{-6}$

16.81 **(a)** $NaH(s) + H_2O(l) \rightarrow NaOH(aq) + H_2(g)$

(b) $CaH_2(s) + 2H_2O(l) \rightarrow Ca(OH)_2(s) + 2H_2(g)$

(c) $NaNH_2(s) + H_2O(l) \rightarrow NaOH(aq) + NH_3(aq)$

(d) $Na_2O(s) + H_2O(l) \rightarrow 2NaOH(aq)$

(e) $SO_2(g) + H_2O(l) \rightleftharpoons H^+ + HSO_3^-$

(f) no reaction

16.83 **(a)** $Al(OH)_3(s) + 3HCl(aq) \rightarrow AlCl_3(aq) + 3H_2O(l)$

(b) $Al(OH)_3(s) + NaOH(aq) \rightarrow NaAl(OH)_4(aq)$

(c) $NaHCO_3(s) + H_2O(l) \rightleftharpoons NaOH(aq) + H_2O(l) + CO_2(aq)$

(d) no reaction

(e) $NH_4Cl(s) + H_2O(l) \rightleftharpoons H_3O^+ + NH_3(aq) + Cl^-$

(f) $CaO(s) + SO_3(g) \rightarrow CaSO_4(s)$

16.85 (a)

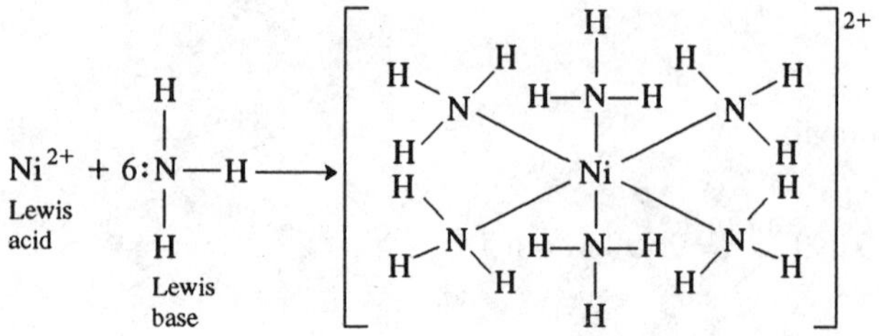

(b)

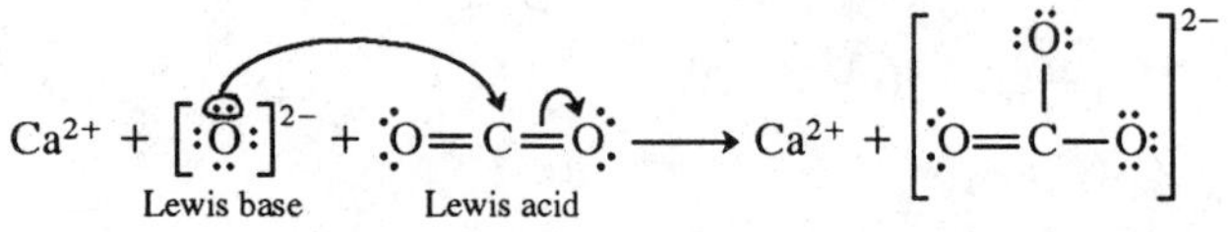

(c)

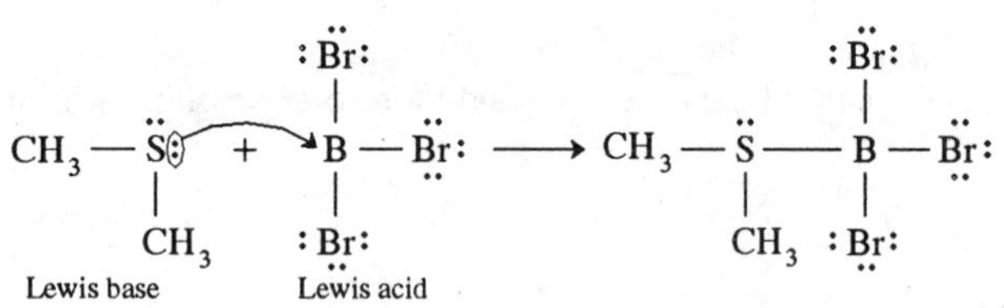

(d)

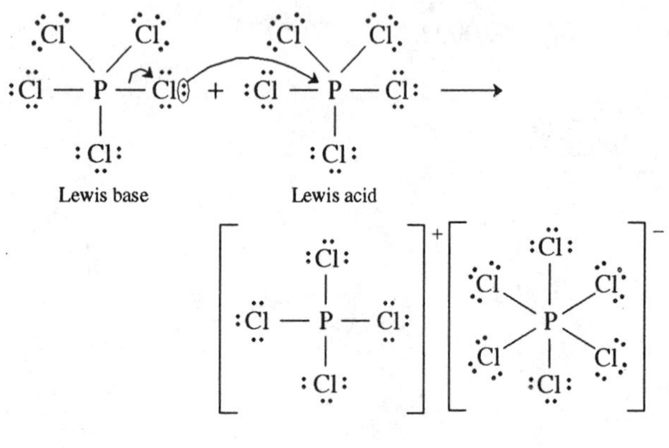

(e)

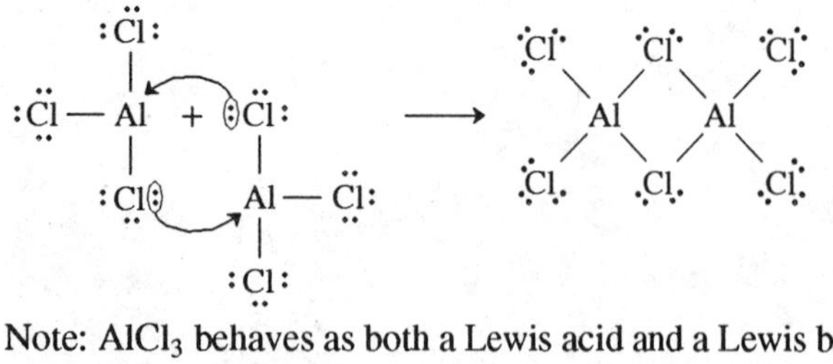

Note: $AlCl_3$ behaves as both a Lewis acid and a Lewis base.

16.87 Both Lewis and Brønsted–Lowry acid–base definitions can be applied to reactions in the gas phase and in solvents other than water as well as to reactions in aqueous solutions. The same substances are viewed as bases according to both definitions. However, Brønsted–Lowry acids must contain hydrogen, whereas Lewis acids do not need to contain hydrogen. Reactions such as the reaction between CaO and CO_2 to form $CaCO_3$ are acid–base reactions according to the Lewis definitions but not according to the Brønsted–Lowry definitions. On the other hand, typical hydrogen-containing acids such as HCl and CH_3COOH are more easily classified as acids using the Brønsted–Lowry definitions. Brønsted–Lowry acid–base reactions can easily be treated quantitatively; Lewis acid–base reactions cannot.

16.89 Al^{3+} + $6H_2O \rightleftarrows Al(H_2O)_6^{3+}$
Lewis acid Lewis base

$Al(H_2O)_6^{3+} + H_2O \rightleftarrows Al(H_2O)_5(OH)^{2+} + H_3O^+$
B–L acid B–L base

$Al(NO_3)_3$ is an ionic compound composed of Al^{3+} ions and NO_3^- ions. Aluminum ions are highly charged and small and are hydrated in aqueous solution. Hydrated aluminum ions undergo a Brønsted–Lowry acid–base reaction with water.

16.91 No.

For AgBr,

$K_{sp} = 5 \times 10^{-13} = [Ag^+][Br^-]$

Since equal volumes are mixed, concentrations will be halved.

$[Ag^+]_0 = \dfrac{4 \times 10^{-6}}{2} = 2 \times 10^{-6}$ $[Br^-]_0 = \dfrac{8 \times 10^{-8}}{2} = 4 \times 10^{-8}$

$Q_{sp} = (2 \times 10^{-6})(4 \times 10^{-8}) = 8 \times 10^{-14}$

$8 \times 10^{-14} < 5 \times 10^{-13}$ so no precipitate will form.

16.93 $[Ag^+] = 3.0 \times 10^{-6}$, $[Cl^-] = 5.4 \times 10^{-5}$

When a precipitate of AgCl forms, silver ions and chloride ions react in a 1:1 ratio.

$Ag^+ + Cl^- \rightleftarrows AgCl(s)$
Let x = mol/L Ag^+ = mol/L Cl^- that react. Then $[Ag^+] = 4.2 \times 10^{-6} - x$ and $[Cl^-] = 5.5 \times 10^{-5} - x$ after reaction.

After precipitation is complete, the solution is saturated and

$[Ag^+][Cl^-] = K_{sp} = 1.6 \times 10^{-10}$

Substituting concentrations in terms of x

$(4.2 \times 10^{-6} - x)(5.5 \times 10^{-5} - x) = 1.6 \times 10^{-10}$

x will *not* be negligible. In standard form the equation is

$x^2 - 5.9 \times 10^{-5}\, x + 0.7 \times 10^{-10} = 0$

and, by the quadratic formula,

$x = 1.2 \times 10^{-6}$

$[Ag^+] = 4.2 \times 10^{-6} - (1.2 \times 10^{-6}) = 3.0 \times 10^{-6}$
$[Cl^-] = 5.5 \times 10^{-5} - (1.2 \times 10^{-6}) = 5.4 \times 10^{-5}$

16.95 (a) 27 mL

For AgCl to begin to precipitate

$$Q_{sp} = [Ag^+][Cl^-] \text{ must exceed } 1.6 \times 10^{-10}$$

For this to be true, $[Cl^-]$ must exceed 4.4×10^{-3} M since $[Ag^+] = 3.6 \times 10^{-8}$:

$K_{sp} = 1.6 \times 10^{-10} = 3.6 \times 10^{-8}\,[Cl^-]$

$[Cl^-] = 4.4 \times 10^{-3}$ M

$$\left(\frac{4.4 \times 10^{-3}\text{ mol}}{\cancel{L}}\right)(254\,\cancel{mL})\left(\frac{1\,\cancel{L}}{10^3\,\cancel{mL}}\right) = 1.1 \times 10^{-3}\text{ mol } Cl^- \text{ needed from HCl}$$

$$V_{HCl} = (1.1 \times 10^{-3}\,\cancel{mol})\left(\frac{62.36\,\cancel{L} \cdot \cancel{mmHg}}{\cancel{mol} \cdot \cancel{K}}\right)\left(\frac{1}{764\,\cancel{mmHg}}\right)(295.0\,\cancel{K})\left(\frac{10^3\text{ mL}}{\cancel{L}}\right) = 27\text{ mL}$$

NOTE: Calculations in this problem give the point when AgCl(s) should theoretically start to precipitate. However, a precipitate will not be *observed* until Q_{sp} is about 1000 times K_{sp}.

(b) 23 mL

From part (a), 1.1×10^{-3} mol of HCl are needed

$$(1.1 \times 10^{-3}\,\cancel{mol})\left(\frac{1\,\cancel{L}}{0.050\,\cancel{mol}}\right)\left(\frac{10^3\text{ mL}}{\cancel{L}}\right) = 22\text{ mL}$$

(Best answer, without intermediate round off, is 23 mL.)

(c) 66 mg

$$(1.1 \times 10^{-3}\,\cancel{mol})\left(\frac{58.4\,\cancel{g\ NaCl}}{\cancel{mol}}\right)\left(\frac{10^3\text{ mg}}{\cancel{g}}\right) = 64\text{ mg}$$

(Best answer, without intermediate round off, is 66 mg.)

16.97 In the buffer solution, the $[CO_3^{2-}]$ is high enough to make $Q_{sp} > K_{sp}$ for $CaCO_3$ but not for $MgCO_3$.

Because K_{sp} for $MgCO_3$ is 2300 times as large as K_{sp} for $CaCO_3$, a concentration of CO_3^{2-} that is high enough to precipitate $CaCO_3$ is not high enough to precipitate $MgCO_3$. This procedure is used to separate Mg^{2+} from Ca^{2+} in qualitative analysis. The buffer keeps $[OH^-]$ low enough that $Mg(OH)_2$ does not precipitate and $[CO_3^{2-}]$ high enough that $CaCO_3$ precipitates completely (by shifting the equilibrium $CO_3^{2-} + H_2O(l) \rightleftarrows HCO_3^- + OH^-$ to the left).

16.99 (a) $KClO_2$ (b) $Ca(ClO)_2$ (c) NH_4ClO_4 (d) $NaClO_3$ (e) $Al(BrO_3)_3 \cdot 9H_2O$

16.101 (a) periodic acid, H_5IO_6 iodic acid, HIO_3

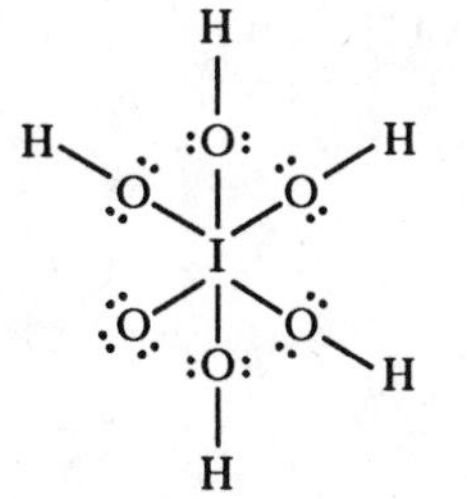

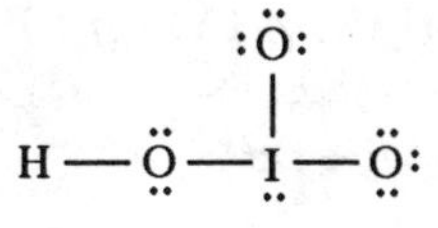

(b) Iodic acid is stronger because it has more extra oxygens (2) than does periodic acid (1). (This prediction is correct. For HIO_3, $K_a = 1.57 \times 10^{-1}$ while for H_5IO_6, $K_a = 5.1 \times 10^{-4}$. Note that the oxidation number of I in H_5IO_6 is +7 and the oxidation number of I in HIO_3 is +5. Thus, the acidities of these acids do not follow the rule that greater acidity is associated with higher oxidation number.)

(c) Periodic acid – octahedral;
Iodic acid – trigonal pyramidal.

The I in H_5IO_6 has six electron pairs (all shared) around it. VSEPR predicts octahedral geometry.

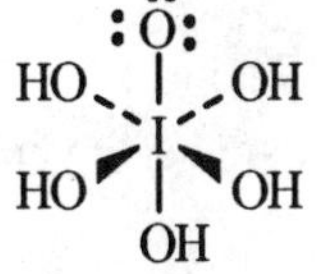

The I in HIO_3 has four electron pairs – three shared and one unshared – around it. VSEPR predicts a trigonal pyramidal shape for this molecule.

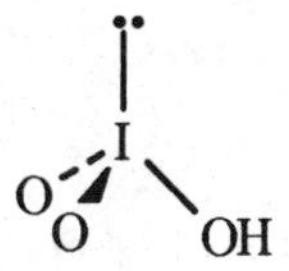

16.103 The ionization of weak acids at infinite dilution is limited by the ionization of water. The H^+ from the water shifts the equilibrium between the weak acid and its ions in solution to the left, thus limiting the percent ionization. Water does not produce any ions that participate in the equilibrium involving the solubility of silver chloride. Thus, silver chloride can be completely ionized in dilute solution (and a saturated solution is dilute), and the calculated solubility closely matches the observed solubility in spite of the fact that silver chloride is a weak electrolyte.

16.105 (a) Molecular equation: $Ni(OH)_2(s) + 2HCl(aq) \rightarrow NiCl_2(aq) + 2H_2O(l)$

Complete ionic equation: $Ni(OH)_2(s) + 2H^+ + 2Cl^- \rightarrow Ni^{2+} + 2Cl^- + 2H_2O(l)$

Net ionic equation: $Ni(OH)_2(s) + 2H^+ \rightarrow Ni^{2+} + 2H_2O(l)$

(b) Molecular equation: $FeCO_3(s) + 2HCl(aq) \rightarrow H_2O(l) + CO_2(g) + FeCl_2(aq)$

Complete ionic equation: $FeCO_3(s) + 2H^+ + 2Cl^- \rightarrow H_2O(l) + CO_2(g) + Fe^{2+} + 2Cl^-$

Net ionic equation: $FeCO_3(s) + 2H^+ \rightarrow H_2O(l) + CO_2(g) + Fe^{2+}$

(c) Molecular equation: $ZnS(s) + 2HCl(aq) \rightarrow ZnCl_2(aq) + H_2S(g)$

Complete ionic equation: $ZnS(s) + 2H^+ + 2Cl^- \rightarrow Zn^{2+} + 2Cl^- + H_2S(g)$

Net ionic equation: $ZnS(s) + 2H^+ \rightarrow Zn^{2+} + H_2S(g)$

(d) Molecular equation: $CoCl_3(aq) + 6NH_3(aq) \rightarrow Co(NH_3)_6Cl_3(aq)$

Complete ionic equation: $Co^{3+} + 3Cl^- + 6NH_3(aq) \rightarrow Co(NH_3)_6{}^{3+} + 3Cl^-$

Net ionic equation: $Co^{3+} + 6NH_3(aq) \rightarrow Co(NH_3)_6{}^{3+}$

(e) Molecular equation: $Al(OH)_3(s) + NaOH(aq) \rightarrow NaAl(OH)_4(aq)$

Complete ionic equation: $Al(OH)_3(s) + Na^+ + OH^- \rightarrow Na^+ + Al(OH)_4{}^-$

Net ionic equation: $Al(OH)_3(s) + OH^- \rightarrow Al(OH)_4{}^-$

(f) Molecular equation: $2AgNO_3(aq) + K_2CrO_4(aq) \rightarrow Ag_2CrO_4(s) + 2KNO_3(aq)$

Complete ionic equation: $2Ag^+ + 2NO_3{}^- + 2K^+ + CrO_4{}^{2-} \rightarrow Ag_2CrO_4(s) + 2K^+ + 2NO_3{}^-$

Net ionic equation: $2Ag^+ + CrO_4{}^{2-} \rightarrow Ag_2CrO_4(s)$

16.107 **(a)** 1.9×10^{12} ions

The complete ionic equation for the equilibrium is

$Bi_2S_3(s) + 3H_2O(l) \rightleftharpoons 2Bi^{3+} + 3HS^- + 3OH^-$

$K_{sp} = [Bi^{3+}]^2[HS^-]^3[OH^-]^3 = 1.6 \times 10^{-72}$

First calculate the molar solubility of Bi_2S_3. From it, calculate the number of ions in 1.0 mL.

Let x = molar solubility. Then $[Bi^{3+}] = 2x$, $[HS^-] = 3x$, and $[OH^-] = 3x$.

Substitute the concentrations in terms of x in the solubility product expression and solve for x.

$(2x)^2(3x)^3(3x)^3 = 1.6 \times 10^{-72}$

$x^8 = 5.5 \times 10^{-76}$

and $x = 3.9 \times 10^{-10}$

Each mole of Bi_2S_3 yields 8 mol ions.

$$\left(\frac{3.9 \times 10^{-10}\ \cancel{\text{mol Bi}_2\text{S}_3}}{\cancel{\text{L}}}\right)\left(\frac{8\ \cancel{\text{mol ions}}}{\cancel{\text{mol Bi}_2\text{S}_3}}\right)\left(\frac{1\ \cancel{\text{L}}}{10^3\ \text{mL}}\right)\left(\frac{6.02 \times 10^{23}\ \text{ions}}{\cancel{\text{mol ions}}}\right) = 1.9 \times 10^{12}\ \frac{\text{ions}}{\text{mL}}$$

(b) 3×10^3 ions

The complete ionic equation for the equilibrium is

$$HgS(s) + H_2O(l) \rightleftarrows Hg^{2+} + HS^- + OH^-$$

$$K_{sp} = [Hg^{2+}][HS^-][OH^-] = 4 \times 10^{-54}$$

Let y = molar solubility. Then $[Hg^{2+}] = [HS^-] = [OH^-] = y$

and $y^3 = 4 \times 10^{-54}$

$y = 1.6 \times 10^{-18}$ (carrying one extra digit; value is 2×10^{-18} to one significant figure)

Each mole of HgS yields 3 mol ions.

$$\left(\frac{1.6 \times 10^{-18}\ \cancel{\text{mol HgS}}}{\cancel{\text{L}}}\right)\left(\frac{3\ \cancel{\text{mol ions}}}{\cancel{\text{mol HgS}}}\right)\left(\frac{1\ \cancel{\text{L}}}{10^3\ \text{mL}}\right)\left(\frac{6.02 \times 10^{23}\ \text{ions}}{\cancel{\text{mol ions}}}\right) = 3 \times 10^{3}\ \frac{\text{ions}}{\text{mL}}$$

(c) No. Bi_2S_3, which has smaller K_{sp}, is more soluble.

$$3.9 \times 10^{-10}\ \frac{\text{mol}}{\text{L}} > 1.6 \times 10^{-18}\ \frac{\text{mol}}{\text{L}}; \quad 1.9 \times 10^{12}\ \frac{\text{ions}}{\text{mL}} > 3 \times 10^{3}\ \frac{\text{ions}}{\text{mL}}$$

Thus Bi_2S_3, which has a much smaller K_{sp} than HgS, is much more soluble than HgS both in terms of molar solubility and in terms of number of ions in solution. Calculations for the concentrations of ions in solutions of slightly soluble salts are complicated by the stoichiometry of the solution equilibria. Since K_{sp} is the product of the various ion concentrations *raised to the appropriate power* as determined by their coefficients in the equations for the solubility equilibria, salts that produce different numbers of ions cannot easily be compared with one another.

16.109 **(a)** The *solubility product* is K_c for a solubility equilibrium; the concentrations in a solubility product are equilibrium concentrations. The *ion product* is Q for a solubility equilibrium; the concentrations may be any concentrations.

(b) *Activity* is the "effective" concentration of species in solution; it includes effects due to interactions of ionic species. For dilute solutions, activity and *concentration* in molarity are approximately equal.

(c) A *saturated solution* is in equilibrium with undissolved solute (or has the same concentration as a solution that is in equilibrium with undissolved solute.) If a crystal of solute is added to a saturated solution, the crystal neither dissolves nor grows. A *supersaturated solution* contains more solute than is permitted by equilibrium considerations. If a crystal of solute is added to a supersaturated solution, the crystal grows.

(d) The *common ion effect* is a decrease in solubility of an ionic solute in a solution containing one of the ions of the solute. The *salt effect* is an increase in the solubility of an ionic solute in a solution containing ions that are not common to the added salt.

(e) A *Brønsted–Lowry acid* is a proton (H^+) donor. A *Lewis acid* is an electron pair acceptor.

Applications

16.111 **(a)** Lewis acid-base reaction

$$CaO(s) + SiO_2(s) \xrightarrow{heat} CaSiO_3(l)$$

(b) –65.17 kJ/mol

Equation	$CaO(g)$	+	$H_2O(l)$	→	$Ca(OH)_2(s)$
ΔH_f^0, kJ/mol	–635.09		–285.83		–986.09

$$\Delta H^0_{rxn} = \left[(1 \text{ mol } Ca(OH)_2)\left(\frac{-986.09 \text{ kJ}}{\text{mol } Ca(OH)_2}\right)\right] - \left[(1 \text{ mol } CaO)\left(\frac{-635.09 \text{ kJ}}{\text{mol } CaO}\right) + (1 \text{ mol } H_2O)\left(\frac{-285.83 \text{ kJ}}{\text{mol } H_2O}\right)\right]$$

$$= (-986.09 \text{ kJ}) - [(-635.09 \text{ kJ}) + (-285.83 \text{ kJ})] = -65.17 \text{ kJ}$$

(c) $Ca(OH)_2(s) + SO_2(g) \rightarrow CaSO_3(s) + H_2O(l)$

16.113 $Ca(OH)_2(s) + CO_2(g) \rightarrow CaCO_3(s) + H_2O(l)$

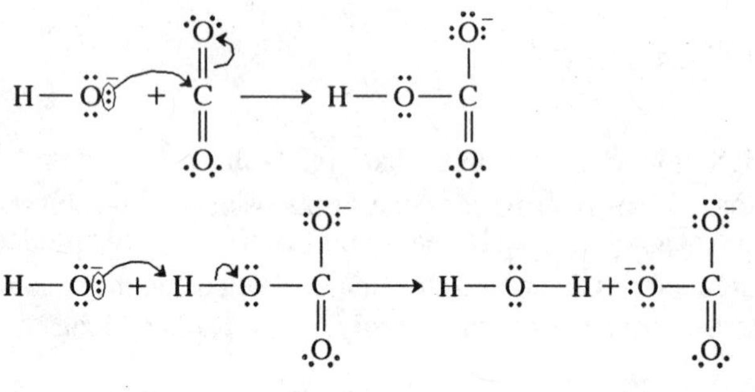

Ca^{2+} is a spectator ion.

16.115 9×10^5

The equilibrium is the sum of the equilibria

$$\begin{array}{l} Ca(OH)_2(s) \rightleftharpoons Ca^{2+} + 2OH^- \\ + \quad Mg^{2+} + 2OH^- \rightleftharpoons Mg(OH)_2(s) \\ \hline Ca(OH)_2(s) + Mg^{2+} \rightleftharpoons Ca^{2+} + Mg(OH)_2(s) \end{array}$$

Therefore, $K = K_{sp\,Ca(OH)_2}\left(\frac{1}{K_{sp\,Mg(OH)_2}}\right) = (5 \times 10^{-6})\left(\frac{1}{5.66 \times 10^{-12}}\right) = 9 \times 10^5$

16.117 **(a)** probably acidic; oxidation number of Cr is higher

Since Cr in CrO_3 has an oxidation number of +6 and Cr in Cr_2O_3 has an oxidation number of +3, CrO_3 will be more acidic than Cr_2O_3. Since Cr_2O_3 is amphoteric, CrO_3 is probably acidic.

(b) Cr_2O_3, chromium(III) oxide

CrO_3, chromium(VI) oxide

(c) CrO_2

16.119 **(a)** They are equal.

$Ag^+ + SCN^- \rightarrow AgSCN(s)$

(b) 1.0×10^{-6} M

$AgSCN(s) \rightleftharpoons Ag^+ + SCN^-$

$K_{sp} = [Ag^+][SCN^-] = 1.0 \times 10^{-12} = x^2$

$x = 1.0 \times 10^{-6}\text{ M} = [SCN^-]$

(c) 6.5×10^{-3} M

$Fe^{3+} + SCN^- \rightleftharpoons FeSCN^{2+}$

$$K_f = \frac{[FeSCN^{2+}]}{[Fe^{3+}][SCN^-]} = 1.0 \times 10^3 = \frac{(6.5 \times 10^{-6})}{(x)(1.0 \times 10^{-6})}$$

$x = 6.5 \times 10^{-3} = [Fe^{3+}]$

16.121 1.6×10^{-6}

$$[Ag^+]_0 = \left(\frac{17\ \cancel{\text{g AgNO}_3}}{1\text{ L soln}}\right)\left(\frac{1\ \cancel{\text{mol AgNO}_3}}{170\ \cancel{\text{g AgNO}_3}}\right)\left(\frac{1\text{ mol Ag}^+}{1\ \cancel{\text{mol AgNO}_3}}\right) = 0.10\text{ M}$$

$K_{sp} = [Ag^+]\,[Cl^-] = 1.6 \times 10^{-10}$

Q must be 1000 (K_{sp}) = $1000(1.6 \times 10^{-10}) = 1.6 \times 10^{-7}$

$0.10\,[Cl^-] = 1.6 \times 10^{-7}$, $[Cl^-] = 1.6 \times 10^{-6}$

16.123 **(a)**

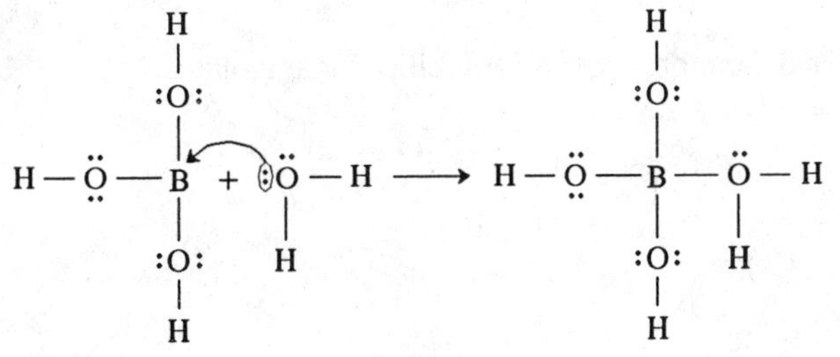

Thus, the ionic equation for the reaction of boric acid with water is

$B(OH)_3(s) + 2H_2O(l) \rightarrow B(OH)_4^-(aq) + H_3O^+(aq)$ or

$B(OH)_3(s) + H_2O(l) \rightarrow B(OH)_4^-(aq) + H^+(aq)$

(b) 6×10^{-10}

$pK_a = -\log K_a$

$K_a = \text{antilog}(-pK_a) = \text{antilog}(-9.2)$

$K_a = 6 \times 10^{-10}$

(c) 5.5

	$B(OH)_3(aq) + H_2O(l)$	$\rightleftarrows$	H^+ +	$B(OH)_4^-$
Init. conc., M	0.020		0.000	0.000
Change, ΔM	$-x$		$+x$	$+x$
Eq. conc., M	$0.020 - x$		x	x

$$K_a = \frac{[H^+][B(OH)_4^-]}{[B(OH)_3]} = 6 \times 10^{-10} = \frac{x^2}{0.020 - x}$$

$x = 3 \times 10^{-6} = [H^+]$

$pH = -\log(3 \times 10^{-6}) = 5.5$

(d) 9.2

$$6 \times 10^{-10} = \frac{[H^+][B(OH)_4^-]}{[B(OH)_3]}$$

$$6 \times 10^{-10} = \frac{[H^+](0.02)}{(0.02)}, \quad [H^+] = 6 \times 10^{-10}$$

$pH = -\log(6 \times 10^{-10}) = 9.2$

16.125 **(a)** aragonite; K_{sp} for aragonite is larger.

For calcite, $K_{sp} = 5 \times 10^{-9} = [Ca^{2+}][CO_3^{2-}]$

For aragonite, $K_{sp} = 7 \times 10^{-9} = [Ca^{2+}][CO_3^{2-}]$

7×10^{-9} shows a greater ion concentration and therefore a greater solubility for aragonite.

(b) 0.7

The equilibrium

$$CaCO_3(\text{calcite}) \rightleftarrows CaCO_3(\text{aragonite})$$

is the sum of the two equilibria

$$\begin{array}{rl} & CaCO_3(\text{calcite}) \rightleftarrows Ca^{2+} + CO_3^{2-} \\ + & Ca^{2+} + CO_3^{2-} \rightleftarrows CaCO_3(\text{aragonite}) \\ \hline & CaCO_3(\text{calcite}) \rightleftarrows CaCO_3(\text{aragonite}) \end{array}$$

Therefore the value of $K = K_{sp\ CaCO_3(calcite)}\left(\frac{1}{K_{sp\ CaCO_3(aragonite)}}\right) = (5 \times 10^{-9})\left(\frac{1}{7 \times 10^{-9}}\right) = 0.7$

(c) Hydrochloric acid will make the real carbonate coral fizz. Bubbles of $CO_2(g)$ will be given off.

16.127 **(a)** 0.30 g $Na_2S_2O_3$

$$\begin{array}{ll} Ag^+ + 2S_2O_3^{2-} \rightleftarrows Ag(S_2O_3)_2^{3-} & K_f = 5.2 \times 10^{12} \\ +\ \ AgBr(s) \rightleftarrows Ag^+ + Br^- & K_{sp} = 5 \times 10^{-13} \\ \hline AgBr(s) + 2S_2O_3^- \rightleftarrows Ag(S_2O_3)_2^{3-} + Br^- & \end{array}$$

$K = K_f \cdot K_{sp} = (5.2 \times 10^{12})(5 \times 10^{-13}) = 3$

$(0.14\ \cancel{g\ AgBr})\left(\frac{1\ mol\ AgBr}{188\ \cancel{g\ AgBr}}\right) = 7.4 \times 10^{-4}$ mol AgBr

All of the AgBr must be dissolved (reacted). Therefore, $[Ag(S_2O_3)_2^{3-}] = [Br^-] = 7.4 \times 10^{-4}$ is needed.

$K = \frac{[Ag(S_2O_3)_2^{3-}][Br^-]}{[S_2O_3^-]^2} = 3$ Let $x = [S_2O_3^{2-}]$

$\frac{(7.4 \times 10^{-4})^2}{x^2} = 3$

and $x = 4 \times 10^{-4} = [S_2O_3^{2-}]$ left

$2(7.4 \times 10^{-4}) = 1.48 \times 10^{-3} = [S_2O_3^{2-}]$ reacted

$$\begin{array}{l} 14.8 \times 10^{-4}\ [S_2O_3^{2-}] \text{ reacted} \\ +\ 4 \times 10^{-4}\ [S_2O_3^{2-}] \text{ left} \\ \hline 19 \times 10^{-4} \text{ total mol } S_2O_3^{2-} \text{ needed} \end{array}$$

$(19 \times 10^{-4}\ \cancel{mol\ Na_2S_2O_3})\left(\frac{158\ g\ Na_2S_2O_3}{\cancel{mol\ Na_2S_2O_3}}\right) = 0.30\ g\ Na_2S_2O_3$

(b) 0.47 g $Na_2S_2O_3 \cdot 5H_2O$

$0.30\ \cancel{g\ Na_2S_2O_3} \times \frac{248\ g\ Na_2S_2O_3 \cdot 5H_2O}{158\ \cancel{g\ Na_2S_2O_3}} = 0.47\ g\ Na_2S_2O_3 \cdot 5H_2O$

16.129 **(a)** An *aerosol* is a colloidal dispersion of a liquid or a solid in a gas (Table 13.4).

(b) carbon dioxide (CO_2), sulfuric acid (H_2SO_4), nitric acid (HNO_3), nitrous acid (HNO_2), and sulfur dioxide (SO_2)

(c) Only nitric and sulfuric acids are strong acids. $[H^+]$ from these two acids shifts the weak acid equilibria toward molecules.

(d) oxidation-reduction (specifically, disproportionation)

$$\overset{+4\ -2}{2NO_2(g)} + \overset{+1\ -2}{H_2O(l)} \rightarrow \overset{+1\ +5\ -2}{HNO_3(aq)} + \overset{+1\ +3\ -2}{HNO_2(aq)}$$

16.131 **(a)** The solubility of gases increases with increasing pressure. Since the pressure at deep vents is greater than the pressure at shallow vents, gases are more soluble at deep vents and there are no bubbles.

(b) The fluids are acidic. Calcium carbonate is soluble in acidic solutions:

$$CaCO_3(s) + 2H^+ \rightarrow Ca^{2+} + H_2O(l) + CO_2(g)$$

CHAPTER 17: Chemical Thermodynamics Revisited: A Closer Look at Enthalpy, Entropy, and Equilibrium

Practice Problems

17.1 **(a)** small and unpredictable; no change in moles gas

When $CH_4(g)$ reacts with $O_2(g)$ to form $CO_2(g)$ and $H_2O(g)$, no change occurs in the number of moles of gas. Thus, the entropy change is expected to be small and the sign is unpredictable.

(b) negative; moles gas decrease, volume decreases

When $CH_4(g)$reacts with $O_2(g)$ to form $CO_2(g)$ and $H_2O(l)$, the volume decreases and ΔS is –.

(c) negative; temperature decreases, volume decreases

When the temperature of $N_2(g)$ decreases at constant pressure, the volume of the sample decreases. Therefore, entropy decreases and the sign of ΔS is –.

(d) negative; pressure increases, volume decreases

When the pressure of $N_2(g)$ increases at constant temperature, the volume of the sample must decrease. Therefore, entropy decreases and the sign of ΔS is –.

(e) negative; moles gas decrease, volume decreases

When $H_2(g)$ reacts with $O_2(g)$ to form $H_2O(g)$, the number of moles of gas decreases, the volume decreases, and the sign of ΔS is –.

17.2 **(a)** decreases; solid more ordered

When liquid benzene freezes to solid benzene, the molecules become fixed in a highly ordered arrangement compared to the arrangement in the liquid state. Therefore, entropy decreases.

(b) increases; volume increases

When liquid benzene evaporates, the volume of the system increases and entropy increases.

(c) increases; volume increases

When miscible liquids are mixed, the volume of the system increases and the entropy increases.

17.3 **(a)** Ca(s); Ca below Mg in periodic table

Mg(s) and Ca(s) are both in Group IIA in the periodic table and are both in the solid state. Therefore, since Ca is below Mg in the periodic table, Ca(s) has a higher standard entropy than Mg(s).

(b) $H_2S(g)$; S below O in periodic table

Since sulfur is below oxygen in Group VIA of the periodic table, $H_2S(g)$ has a higher standard entropy than $H_2O(g)$.

(c) $PCl_5(c)$; more atoms

$PCl_5(g)$ and $PCl_3(g)$ have similar structures but $PCl_5(g)$ has more atoms in the molecule. Therefore, it has a higher standard entropy.

(d) $Cl_2(g)$; Cl below F in periodic table

Cl and F are both in Group VIIA in the periodic table and are both in the gaseous state. Cl is below F. Therefore, $Cl_2(g)$, has the higher standard entropy.

(e) unpredictable; solid more ordered but I below Br in periodic table

I is below Br in Group VIIA but I_2 is a solid, whereas Br_2 is a liquid. Solid I_2 is more ordered than liquid Br_2. Therefore, $I_2(s)$ would be predicted to have a lower standard entropy. However, I is below Br in the periodic table and the entropy of I atoms is greater than the entropy of Br atoms. Which of these two opposing trends is more important is hard to predict.

17.4 **(a)** –5.2 J/K

Equation	$CH_4(g)$	+	$2O_2(g)$	→	$CO_2(g)$	+	$2H_2O(g)$
S^0, J/K · mol	186.2		205.03		213.64		188.72

$$\Delta S^0_{rxn} = \left[(1\ \cancel{\text{mol } CO_2})\left(\frac{213.64\ \text{J}}{\text{K}\cdot\cancel{\text{mol } CO_2}}\right) + (2\ \cancel{\text{mol } H_2O})\left(\frac{188.72\ \text{J}}{\text{K}\cdot\cancel{\text{mol } H_2O}}\right)\right]$$

$$- \left[(1\ \cancel{\text{mol } CH_4})\left(\frac{186.2\ \text{J}}{\text{K}\cdot\cancel{\text{mol } CH_4}}\right) + (2\ \cancel{\text{mol } O_2})\left(\frac{205.03\ \text{J}}{\text{K}\cdot\cancel{\text{mol } O_2}}\right)\right]$$

$$= \left[213.64\frac{\text{J}}{\text{K}} + 377.44\frac{\text{J}}{\text{K}}\right] - \left[186.2\frac{\text{J}}{\text{K}} + 410.06\frac{\text{J}}{\text{K}}\right]$$

$$= -5.2\,\frac{J}{K}$$

Thus, ΔS^0_{rxn} is small. In Practice Problem 17.1, it was predicted that the entropy change would be small but that the sign was unpredictable.

(b) –242.8 J/K

Equation	$CH_4(g)$	+	$2O_2(g)$	→	$CO_2(g)$	+	$2H_2O(l)$
S^0, J/K · mol	186.2		205.03		213.64		69.91

$$\Delta S^0_{rxn} = \left[(1\ \cancel{mol\ CO_2})\left(\frac{213.64\ J}{K\cdot \cancel{mol\ CO_2}}\right) + (2\ \cancel{mol\ H_2O})\left(\frac{69.91\ J}{K\cdot \cancel{mol\ H_2O}}\right)\right]$$

$$- \left[(1\ \cancel{mol\ CH_4})\left(\frac{186.2\ J}{K\cdot \cancel{mol\ CH_4}}\right) + (2\ \cancel{mol\ O_2})\left(\frac{205.03\ J}{K\cdot \cancel{mol\ CO_2}}\right)\right]$$

$$= \left[213.64\,\frac{J}{K} + 139.82\,\frac{J}{K}\right] - \left[186.2\,\frac{J}{K} + 410.06\,\frac{J}{K}\right]$$

$$= -242.8\,\frac{J}{K}$$

In Practice Problem 17.1, it was predicted that the entropy change would be negative.

(c) –88.95 J/K

Equation	$2H_2(g)$	+	$O_2(g)$	→	$2H_2O(g)$
S^0, J/K · mol	130.68		205.03		188.72

$$\Delta S^0_{rxn} = \left[(2\ \cancel{mol\ H_2O})\left(\frac{188.72\ J}{K\cdot \cancel{mol\ H_2O}}\right)\right] - \left[(2\ \cancel{mol\ H_2})\left(\frac{130.68\ J}{K\cdot \cancel{mol\ H_2}}\right) + (1\ \cancel{mol\ O_2})\left(\frac{205.03\ J}{K\cdot \cancel{mol\ O_2}}\right)\right]$$

$$= 377.44\,\frac{J}{K} - \left[261.36\,\frac{J}{K} + 205.03\,\frac{J}{K}\right]$$

$$= -88.95\,\frac{J}{K}$$

In Practice Problem 17.1, it was predicted that the entropy change would be negative.

17.5 –281.07 J/K

Equation	$2Li(s)$	+	$C(s)$	+	$\frac{3}{2}O_2(g)$	→	$Li_2CO_3(s)$
S^0, J/K · mol	29.10		5.69		205.03		90.37

$$\Delta S^0_{rxn} = \left[(1\ \cancel{mol\ Li_2CO_3})\left(\frac{90.37\ J}{K\cdot \cancel{mol\ Li_2CO_3}}\right)\right]$$

$$- \left[(2\ \cancel{mol\ Li})\left(\frac{29.10\ J}{K\cdot \cancel{mol\ Li}}\right) + (1\ \cancel{mol\ C})\left(\frac{5.69\ J}{K\cdot \cancel{mol\ C}}\right) + \left(\frac{3}{2}\ \cancel{mol\ O_2}\right)\left(\frac{205.03\ J}{K\cdot \cancel{mol\ O_2}}\right)\right]$$

$$= \left(\frac{90.37\ \text{J}}{\text{K}}\right) - \left[\left(\frac{58.20\ \text{J}}{\text{K}}\right) + \left(\frac{5.69\ \text{J}}{\text{K}}\right) + \left(\frac{307.55\ \text{J}}{\text{K}}\right)\right]$$

$$= -281.07\ \frac{\text{J}}{\text{K}}$$

17.6 (c) spontaneous at low temperatures but nonspontaneous at high temperatures.

At low temperatures ΔH predominates. Since its sign is –, ΔG is – and the reaction is spontaneous. At high temperatures, the $T\Delta S$ term predominates. Since its sign is +, ΔG is + and the reaction is nonspontaneous.

17.7 –794.9 kJ

Equation	$CH_4(g)$	+	$2O_2(g)$	→	$CO_2(g)$	+	$2H_2O(g)$
ΔH_f^0, kJ/mol	–74.85		0.00		–393.51		–238.92
S^0, J/K · mol	186.2		205.03		213.64		188.72

The ΔH_f^0 of the products and the reactants are first used to calculate ΔH^0 for the reaction:

$$\Delta H^0 = \left[(1\ \text{mol}\ CO_2)\left(\frac{-393.51\ \text{kJ}}{\text{mol}\ CO_2}\right) + (2\ \text{mol}\ H_2O)\left(\frac{-238.92\ \text{kJ}}{\text{mol}\ H_2O}\right)\right]$$

$$-\left[(1\ \text{mol}\ CH_4)\left(\frac{-74.85\ \text{kJ}}{\text{mol}\ CH_4}\right) + (2\ \text{mol}\ O_2)\left(\frac{0.00\ \text{kJ}}{\text{mol}\ O_2}\right)\right] = -796.50\ \text{kJ}$$

The S^0 of the products and the reactants are next used to calculate ΔS^0 for the reaction:

$$\Delta S^0 = \left[(1\ \text{mol}\ CO_2)\left(\frac{213.64\ \text{J}}{\text{K}\cdot\text{mol}\ CO_2}\right) + (2\ \text{mol}\ H_2O)\left(\frac{188.72\ \text{J}}{\text{K}\cdot\text{mol}\ H_2O}\right)\right]$$

$$-\left[(1\ \text{mol}\ CH_4)\left(\frac{186.2\ \text{J}}{\text{K}\cdot\text{mol}\ CH_4}\right) + (2\ \text{mol}\ O_2)\left(\frac{205.03\ \text{J}}{\text{K}\cdot\text{mol}\ O_2}\right)\right] = -5.2\ \frac{\text{J}}{\text{K}}$$

The Celsius temperature must be converted to kelvin:

$$\text{K} = (t_C + 273.2)\ ^\circ\text{C}\left(\frac{1\ \text{K}}{1\ ^\circ\text{C}}\right) = (25.0 + 273.2)\ ^\circ\text{C}\left(\frac{1\ \text{K}}{1\ ^\circ\text{C}}\right) = 298.2\ \text{K}$$

Next convert ΔS^0 from J/K to kJ/K so that the units will be consistent with those of ΔH^0:

$$\left(\frac{-5.2\ \text{J}}{\text{K}}\right)\left(\frac{1\ \text{kJ}}{10^3\ \text{J}}\right) = -5.2 \times 10^{-3}\ \frac{\text{kJ}}{\text{K}}$$

Finally, calculate ΔG^0 using the following equation:

$$\Delta G^0 = \Delta H^0 - T\Delta S^0 = -796.50\ \text{kJ} - (298.2\ \text{K})\left(-5.2 \times 10^{-3}\ \frac{\text{kJ}}{\text{K}}\right)$$

$$= -796.50\ \text{kJ} + 1.6\ \text{kJ} = -794.9\ \text{kJ}$$

17.8 –800.77 kJ

Equation	$CH_4(g)$	+	$2O_2(g)$	→	$CO_2(g)$	+	$2H_2O(g)$
ΔG_f^0, kJ/mol	–50.79		0.00		–394.38		–228.59

$$\Delta G^0_{rxn} = \text{Sum of } \Delta G_f^0 \text{ for products} - \text{Sum of } \Delta G_f^0 \text{ of reactants}$$

$$= \left[(1 \text{ mol } CO_2)\left(\frac{-394.38 \text{ kJ}}{\text{mol } CO_2}\right) + (2 \text{ mol } H_2O)\left(\frac{-228.59 \text{ kJ}}{\text{mol } H_2O}\right)\right]$$

$$- \left[(1 \text{ mol } CH_4)\left(\frac{-50.79 \text{ kJ}}{\text{mol } CH_4}\right) + (2 \text{ mol } O_2)\left(\frac{0.00 \text{ kJ}}{\text{mol } O_2}\right)\right]$$

$$= [-394.38 \text{ kJ} + (-457.18 \text{ kJ})] - (-50.79 \text{ kJ} + 0.00 \text{ kJ}) = -800.77 \text{ kJ}$$

17.9 −793.8 kJ

From Practice Problem 17.7, $\Delta H^0 = -796.50$ kJ and $\Delta S^0 = -5.2 \times 10^{-3}$ kJ/K at 25.0 °C. If we assume that ΔH^0 and ΔS^0 do not change significantly with temperature, then we can estimate ΔG^0 at temperatures other than 25.0 °C.

First convert 250 °C to kelvin:

$$K = (t_C + 273.2)\ °C\left(\frac{1 \text{ K}}{1\ °C}\right) = (250\ °C + 273.2\ °C)\left(\frac{1 \text{ K}}{1\ °C}\right) = 523 \text{ K}$$

Now substitute temperature, ΔH^0 from Practice Problem 17.7, and ΔS^0 from Practice Problem 17.7 into the following equation:

$$\Delta G^0 = \Delta H^0 - T\Delta S^0 = -796.50 \text{ kJ} - (523 \text{ K})\left(\frac{-5.2 \times 10^{-3} \text{ kJ}}{\text{K}}\right)$$

$$= -796.50 \text{ kJ} + 2.7 \text{ kJ} = -793.8 \text{ kJ}$$

17.10 77 °C

Equation	$CH_3CH_2OH(l)$	$\rightleftarrows$	$CH_3CH_2OH(g)$
ΔH_f^0, kJ/mol	−277.63		−235.3
S^0, J/K · mol	161		282

$$\Delta H^0 = \left[(1 \text{ mol } CH_3CH_2OH(g))\left(\frac{-235.3 \text{ kJ}}{\text{mol } CH_3CH_2OH(g)}\right)\right] - \left[(1 \text{ mol } CH_3CH_2OH(l))\left(\frac{-277.63 \text{ kJ}}{\text{mol } CH_3CH_2OH(l)}\right)\right]$$

$$= (-235.3 \text{ kJ}) - (-277.63 \text{ kJ}) = 42.3 \text{ kJ}$$

$$\Delta S^0 = \left[(1 \text{ mol } CH_3CH_2OH(g))\left(\frac{282 \text{ J}}{\text{K} \cdot \text{mol } CH_3CH_2OH(g)}\right)\right] - \left[(1 \text{ mol } CH_3CH_2OH(l))\left(\frac{161 \text{ J}}{\text{K} \cdot \text{mol } CH_3CH_2OH(l)}\right)\right]$$

$$= \left(282 \frac{\text{J}}{\text{K}}\right) - \left(161 \frac{\text{J}}{\text{K}}\right) = 121 \frac{\text{J}}{\text{K}} \text{ or } 0.121 \frac{\text{kJ}}{\text{K}}$$

$\Delta G^0 = \Delta H^0 - T\Delta S^0$ and $T = \dfrac{\Delta H^0 - \Delta G^0}{\Delta S^0}$ At equilibrium, $\Delta G^0 = 0$.

$$\text{Therefore, } T_K = \frac{(42.3 \text{ kJ}) - (0 \text{ kJ})}{\left(0.121 \frac{\text{kJ}}{\text{K}}\right)} = 350 \text{ K}$$

$$t_C = (350\ \cancel{K} - 273\ \cancel{K})\left(\frac{1\ °C}{1\ \cancel{K}}\right) = 77\ °C$$

(From the CRC Handbook, the normal boiling point of ethyl alcohol is 78.5 °C.)

17.11 $88.9\ \frac{J}{K \cdot mol}$

$Br_2(l) \rightleftarrows Br_2(g)$

$$\Delta G^0_{vap} = \Delta H^0_{vap} - T\Delta S^0_{vap} \text{ and } \Delta S^0_{vap} = \frac{\Delta H^0_{vap} - \Delta G^0_{vap}}{T}$$

At equilibrium, $\Delta G^0 = 0$

$$\text{Therefore, } \Delta S^0_{vap} = \frac{\left(29.5\ \frac{kJ}{mol}\right) - \left(0\ \frac{kJ}{mol}\right)}{(58.8 + 273.2)\ K} = \frac{\left(29.5\ \frac{kJ}{mol}\right)}{332.0\ K} = 0.0889\ \frac{kJ}{K \cdot mol} \text{ or } 88.9\ \frac{J}{K \cdot mol}$$

17.12 –770.5 kJ/mol

For the reaction

$$CH_4(g) + 2O_2(g) \rightarrow CO_2(g) + 2H_2O(g)$$

use the information given to calculate the reaction quotient, Q:

$$Q = \frac{p_{CO_2} \cdot p_{H_2O}{}^2}{p_{CH_4} \cdot p_{O_2}{}^2} = \frac{(2.10)(3.40)^2}{(0.56)(0.45)^2} = 2.1 \times 10^2$$

In Practice Problem 17.9, ΔG^0 at 250 °C was estimated to be –793.8 kJ. Calculate ΔG for the reaction at 250 °C (523 K) using the following equation:

$$\Delta G = \Delta G^0 + RT \ln Q$$

$$= -793.8\ kJ/mol + \left(0.008\ 315\ \frac{kJ}{\cancel{K} \cdot mol}\right)(523\ \cancel{K}) \ln (2.1 \times 10^2)$$

$$= -793.8\ kJ/mol + \left(0.008\ 315\ \frac{kJ}{mol}\right)(523)\ (5.35)$$

$$= -793.8\ kJ/mol + 23.3\ kJ/mol = -770.5\ kJ/mol$$

17.13 1×10^{14}

$$\Delta G^0 = -RT \ln K \quad \text{Therefore, } \frac{\Delta G^0}{-RT} = \ln K$$

$$\ln K = \frac{-80.0\ \cancel{\frac{kJ}{mol}}}{\left(\frac{-0.008\ 315\ \cancel{kJ}}{\cancel{K} \cdot \cancel{mol}}\right)(298\ \cancel{K})} = 32.3$$

$K = 1 \times 10^{14}$

17.14 –25.3 kJ/mol

$\Delta G^0 = -RT \ln K$

$$\Delta G^0 = \left(\frac{-0.008\ 315\ \text{kJ}}{\cancel{\text{K}}\cdot\text{mol}}\right)(298\ \cancel{\text{K}})\ \ln(2.7\times 10^4)$$

$$= -25.3\ \frac{\text{kJ}}{\text{mol}}$$

Additional Practice Problems

17.41 No.

For Br_2:

Equation	$CH_4(g)$	+	$Br_2(g)$	$\rightarrow$	$CH_3Br(g)$	+	$HBr(g)$
ΔG_f^0, kJ/mol	–50.79		0		–26		–53.43

$$\Delta G^0_{rxn} = \left[(1\ \cancel{\text{mol CH}_3\text{Br}})\left(\frac{-26\ \text{kJ}}{\cancel{\text{mol CH}_3\text{Br}}}\right) + (1\ \cancel{\text{mol HBr}})\left(\frac{-53.43\ \text{kJ}}{\cancel{\text{mol HBr}}}\right)\right]$$

$$-\left[(1\ \cancel{\text{mol CH}_4})\left(\frac{-50.79\ \text{kJ}}{\cancel{\text{mol CH}_4}}\right) + (1\ \cancel{\text{mol Br}_2})\left(\frac{0\ \text{kJ}}{\cancel{\text{mol Br}_2}}\right)\right]$$

$$= [-26\ \text{kJ} + (-53.43\ \text{kJ})] - (-50.79\ \text{kJ} + 0\ \text{kJ})$$

$$= -29\ \text{kJ}$$

Because ΔG^0_{rxn} is negative, the reaction is spontaneous.

For I_2:

Equation	$CH_4(g)$	+	$I_2(g)$	$\rightarrow$	$CH_3I(g)$	+	$HI(g)$
ΔG_f^0, kJ/mol	–50.79		19.37		22		1.30

$$\Delta G^0_{rxn} = \left[(1\ \cancel{\text{mol CH}_3\text{I}})\left(\frac{22\ \text{kJ}}{\cancel{\text{mol CH}_3\text{I}}}\right) + (1\ \cancel{\text{mol HI}})\left(\frac{1.30\ \text{kJ}}{\cancel{\text{mol HI}}}\right)\right]$$

$$-\left[(1\ \cancel{\text{mol CH}_4})\left(\frac{-50.79\ \text{kJ}}{\cancel{\text{mol CH}_4}}\right) + (1\ \cancel{\text{mol I}_2})\left(\frac{19.37\ \text{kJ}}{\cancel{\text{mol I}_2}}\right)\right]$$

$$= (22\ \text{kJ} + 1.30\ \text{kJ}) - (-50.79\ \text{kJ} + 19.37\ \text{kJ})$$

$$= +55\ \text{kJ}$$

Because ΔG^0_{rxn} is positive, reaction is not spontaneous.

17.43 **(a)** 68.69 kJ

Equation	$2CH_4(g)$	$\rightarrow$	$CH_3CH_3(g)$	+	$H_2(g)$
ΔG^0, kJ/mol	–50.79		–32.89		0

$$\Delta G^0_{rxn} = \left[(1\ \text{mol}\ CH_3CH_3)\left(\frac{-32.89\ \text{kJ}}{\text{mol}\ CH_3CH_3}\right) + (1\ \text{mol}\ H_2)\left(\frac{0\ \text{kJ}}{\text{mol}\ H_2}\right)\right] - (2\ \text{mol}\ CH_4)\left(\frac{-50.79\ \text{kJ}}{\text{mol}\ CH_4}\right)$$

$\Delta G^0 = 68.69$ kJ

(b) No.

ΔG^0_{rxn} is positive.

(c) No.

Equation	$2CH_4(g)$	$\rightarrow$	$CH_3CH_3(g)$	+	$H_2(g)$
ΔH_f^0, kJ/mol	–74.85		–84.67		0
S^0, J/K · mol	186.2		229.5		130.68

$$\Delta H^0_{rxn} = \left[(1\ \text{mol}\ CH_3CH_3)\left(\frac{-84.67\ \text{kJ}}{\text{mol}\ CH_3CH_3}\right) + (1\ \text{mol}\ H_2)\left(\frac{0\ \text{kJ}}{\text{mol}\ H_2}\right)\right]$$

$$-\left[(2\ \text{mol}\ CH_4)\left(\frac{-74.85\ \text{kJ}}{\text{mol}\ CH_4}\right)\right]$$

$$= +65.03\ \text{kJ}$$

$$\Delta S^0_{rxn} = \left[(1\ \text{mol}\ CH_3CH_3)\left(\frac{229.5\ \text{J}}{\text{K}\cdot\text{mol}\ CH_3CH_3}\right) + (1\ \text{mol}\ H_2)\left(\frac{130.68\ \text{J}}{\text{K}\cdot\text{mol}\ H_2}\right)\right]$$

$$-\left[(2\ \text{mol}\ CH_4)\left(\frac{186.2\ \text{J}}{\text{K}\cdot\text{mol}\ CH_4}\right)\right]$$

$$= -12.2\ \frac{\text{J}}{\text{K}}$$

Because ΔH^0_{rxn} is positive and ΔS^0_{rxn} is negative, ΔG^0_{rxn} is positive at all temperatures. Reaction is nonspontaneous under all conditions.

(d) No.

Looking for a catalyst would be a waste of time. The reaction is not spontaneous under any conditions.

17.45 **(a)** left; larger volume

Each circle has 10 molecules, but the one on the left provides more space. Therefore there are more ways to arrange the molecules on the left and more randomness or entropy.

(b) left; larger volume

Same as (a) above.

(c) left; more molecules in same volume

The circle on the left has more molecules, so there are more possible ways to arrange them within the circle.

Stop and Test Yourself

1. (a) the same for paths 1 through 5

Entropy is a state function; it does not depend on path.

2. (b) formation of hydrogen and oxygen by boiling water

The reverse reaction

$$2H_2(g) + O_2(g) \rightarrow 2H_2O(g \text{ or } l)$$

is spontaneous.

3. (b) $2HF(g) \rightarrow H_2(g) + F_2(g)$

Two moles of gaseous reactants yield two moles of gaseous products.

4. (c) 258.8 J/K

Equation	$2NH_3(g)$	+	$3Cl_2(g)$	$\rightarrow$	$N_2(g)$	+	$6HCl(g)$
S^0, J/K · mol	192.3		222.96		191.50		186.80

$$\Delta S^0 = \left[(1\ \cancel{\text{mol N}_2})\left(\frac{191.50\ \text{J}}{\text{K}\cdot\cancel{\text{mol N}_2}}\right) + (6\ \cancel{\text{mol HCl}})\left(\frac{186.80\ \text{J}}{\text{K}\cdot\cancel{\text{mol HCl}}}\right)\right]$$

$$-\left[(2\ \cancel{\text{mol NH}_3})\left(\frac{192.3\ \text{J}}{\text{K}\cdot\cancel{\text{mol NH}_3}}\right) + (3\ \cancel{\text{mol Cl}_2})\left(\frac{222.96\ \text{J}}{\text{K}\cdot\cancel{\text{mol Cl}_2}}\right)\right]$$

$$= \left(191.50\frac{\text{J}}{\text{K}} + 1120.80\frac{\text{J}}{\text{K}}\right) - \left(384.6\frac{\text{J}}{\text{K}} + 668.88\frac{\text{J}}{\text{K}}\right)$$

$$= 258.8\frac{\text{J}}{\text{K}}$$

5. (e) gives off thermal energy

The negative ΔH value indicates an exothermic reaction.

6. (b) $CS_2(g) + 3Cl_2(g) \rightarrow CCl_4(g) + S_2Cl_2(g) \quad \Delta H^0_{rxn} = -238$ kJ

To be spontaneous at low temperatures but nonspontaneous at high temperatures, ΔH and ΔS must both be negative. Of the three reactions for which ΔH is negative [(b), (c), (d)], ΔS will be negative only for (b), in which 4 mol gas $\rightarrow$ 2 mol gas.

7. **(a)** 8.07 kJ

Equation	$3NO_2(g)$	+	$H_2O(l)$	→	$2HNO_3(l)$	+	$NO(g)$
ΔG_f^0, kJ/mol	51.30		−237.18		−80.89		86.57

$$\Delta G^0 = \left[(2\ \text{mol HNO}_3)\left(\frac{-80.89\ \text{kJ}}{\text{mol HNO}_3}\right) + (1\ \text{mol NO})\left(\frac{86.57\ \text{kJ}}{\text{mol NO}}\right)\right]$$

$$-\left[(3\ \text{mol NO}_2)\left(\frac{51.30\ \text{kJ}}{\text{mol NO}_2}\right) + (1\ \text{mol H}_2\text{O})\left(\frac{-237.18\ \text{kJ}}{\text{mol H}_2\text{O}}\right)\right]$$

$$= (-161.78\ \text{kJ} + 86.57\ \text{kJ}) - (153.90\ \text{kJ} - 237.18\ \text{kJ})$$

$$= 8.07\ \text{kJ}$$

8. **(d)** 160.1 kJ

Equation	$N_2(g)$	+	$O_2(g)$	→	$2NO(g)$
ΔH_f^0, kJ/mol	0		0		90.25
S^0, J/K · mol	191.50		205.03		210.65

$$\Delta H^0_{\text{rxn}} = \left[(2\ \text{mol NO})\left(\frac{90.25\ \text{kJ}}{\text{mol NO}}\right)\right] - \left[(1\ \text{mol N}_2)\left(\frac{0\ \text{kJ}}{\text{mol N}_2}\right) + (1\ \text{mol O}_2)\left(\frac{0\ \text{kJ}}{\text{mol O}_2}\right)\right] = 180.50\ \text{kJ}$$

$$\Delta S^0 = \left[(2\ \text{mol NO})\left(\frac{210.65\ \text{J}}{\text{K}\cdot\text{mol NO}}\right)\right] - \left[(1\ \text{mol N}_2)\left(\frac{191.50\ \text{J}}{\text{K}\cdot\text{mol N}_2}\right) + (1\ \text{mol O}_2)\left(\frac{205.03\ \text{J}}{\text{K}\cdot\text{mol O}_2}\right)\right]$$

$$= +24.77\ \frac{\text{J}}{\text{K}}\ \text{or}\ 0.024\ 77\ \frac{\text{kJ}}{\text{K}}$$

$$\Delta G^0 = \Delta H^0 - T\Delta S^0 = 180.50\ \text{kJ} - (552 + 273)\ \text{K}\left(\frac{0.024\ 77\ \text{kJ}}{\text{K}}\right) = 160.1\ \text{kJ}$$

This answer is an estimate because its calculation assumes that ΔH^0 and ΔS^0 are the same at 552 °C as at 25 °C.

9. **(c)** the reaction mixture is at equilibrium but the concentration of products is small

The fact that ΔG is zero shows that the reaction is at equilibrium. The positive value of ΔG^0 shows that the forward reaction is nonspontaneous. Thus, at equilibrium the concentrations of the reactants are high and the concentrations of the products are low.

10. **(c)** ii and iii

Entropy increases with increasing temperature. Under the same conditions, the more complex molecule (HCl) has greater entropy than the simpler molecule (Ar). More dilute solutions allow for more disorder.

11. **(a)** The entropy of the universe increases.

12. **(c)** 549 °C

$$\Delta G^0 = \Delta H^0 - T\Delta S^0$$

If $\Delta G^0 = 0$, then $T = \frac{\Delta H^0}{\Delta S^0} = \frac{-108.28 \text{ kJ}}{\frac{-0.131\,63 \text{ kJ}}{\text{K}}} = 823 \text{ K}$

(Answer is an estimate because calculation assumes that ΔH^0 and ΔS^0 are the same at 823 K as at 25 °C. Therefore, three significant figures are enough.)

823 K is 550 °C (823 – 273 = 550)

(Best answer, without intermediate round off, is 549 °C.)

13. (b) 1×10^{-5}

$\Delta G^0 = -RT \ln K$

Solving for K

$$\text{antiln}\left(\frac{\Delta G^0}{-RT}\right) = K = \text{antiln}\frac{28.5 \text{ kJ} \cdot \text{mol}^{-1}}{\left(-0.008\,315 \frac{\text{kJ}}{\text{K} \cdot \text{mol}}\right)(298 \text{ K})} = 1 \times 10^{-5}$$

14. (a) Thermal energy cannot be completely converted to work. There is always some thermal energy that is transferred to the surroundings.

Putting Things Together

17.47 –46.3 kJ

$\Delta G = \Delta G^0 + RT \ln Q$

To calculate the value of ΔG, we first need to find ΔG^0 and Q.

We can estimate ΔG^0 at 525 K from ΔH^0 and ΔS^0 at 25°C.

Equation	$CoO(s)$	+	$H_2(g)$	→	$Co(s)$	+	$H_2O(g)$
H_f^0, kJ/mol	–211.3		0		0		–238.92
S^0, J/K · mol	43.9		130.68		28.5		188.72

$$\Delta H^0_{rxn} = \left[(1 \text{ mol Co})\left(\frac{0 \text{ kJ}}{\text{mol Co}}\right) + (1 \text{ mol } H_2O)\left(\frac{-238.92 \text{ kJ}}{\text{mol } H_2O}\right)\right]$$

$$- \left[(1 \text{ mol CoO})\left(\frac{-211.3 \text{ kJ}}{\text{mol CoO}}\right) + (1 \text{ mol } H_2)\left(\frac{0 \text{ kJ}}{\text{mol } H_2}\right)\right]$$

$= [0 \text{ kJ} + (-238.92 \text{ kJ})] - (-211.3 \text{ kJ} + 0 \text{ kJ})$

$= -27.6 \text{ kJ}$

$$\Delta S^0_{rxn} = \left[(1\ \text{mol Co})\left(\frac{28.5\ \text{J}}{\text{K}\cdot\text{mol Co}}\right) + (1\ \text{mol H}_2\text{O})\left(\frac{188.72\ \text{J}}{\text{K}\cdot\text{mol H}_2\text{O}}\right)\right]$$

$$-\left[(1\ \text{mol CoO})\left(\frac{43.9\ \text{J}}{\text{K}\cdot\text{mol CoO}}\right) + (1\ \text{mol H}_2)\left(\frac{130.68\ \text{J}}{\text{K}\cdot\text{mol H}_2}\right)\right]$$

$$= 42.6\frac{\text{J}}{\text{K}}\ \text{or}\ 0.0426\frac{\text{kJ}}{\text{K}}$$

$$\Delta G^0 = -27.6\ \text{kJ} - (525\ \text{K})\left(\frac{0.0426\ \text{kJ}}{\text{K}}\right) = -50.0\ \text{kJ}$$

$$Q = \left(\frac{p_{H_2O}}{p_{H_2}}\right) = \frac{4.9\ \text{atm}}{2.1\ \text{atm}} = 2.3$$

$$\Delta G = -50.0\ \text{kJ} + \left(\frac{0.008\,315\ \text{kJ}}{\text{K}}\right)(525\ \text{K})\ (\ln 2.3) = -46.4\ \text{kJ}$$

(Pressures are in atmospheres and quantities in moles.)

(Best answer, without intermediate round off, is –46.3 kJ.)

17.49 methyl alcohol

The negative ΔG^0 value indicates a spontaneous reaction.

For $CO(g) + H_2(g) \rightarrow H_2CO(g)$

$$\Delta G^0_{rxn} = \left[(1\ \text{mol H}_2\text{CO})\left(\frac{-110\ \text{kJ}}{\text{mol H}_2\text{CO}}\right)\right] - \left[(1\ \text{mol CO})\left(\frac{-137.27\ \text{kJ}}{\text{mol CO}}\right) + (1\ \text{mol H}_2)\left(\frac{0\ \text{kJ}}{\text{mol H}_2}\right)\right] = +27\ \text{kJ}$$

For $CO(g) + 2H_2(g) \rightarrow CH_3OH(l)$

$$\Delta G^0_{rxn} = \left[(1\ \text{mol CH}_3\text{OH})\left(\frac{-166.3\ \text{kJ}}{\text{mol CH}_3\text{OH}}\right)\right] - \left[(1\ \text{mol CO})\left(\frac{-137.27\ \text{kJ}}{\text{mol CO}}\right) + (2\ \text{mol H}_2)\left(\frac{0\ \text{kJ}}{\text{mol H}_2}\right)\right] = -29.0\ \text{kJ}$$

The negative value of ΔG^0 for the production of $CH_3OH(l)$ indicates that $CH_3OH(l)$ will be formed (provided that the reaction takes place at a reasonable rate). The positive value of ΔG^0 for the production of $H_2CO(g)$ indicates that this product will not be formed.

17.51 4.0×10^{-4} atm or 0.31 mmHg

Equation	$I_2(s)$	$\rightarrow$	$I_2(g)$
ΔG_f^0, kJ/mol	0		19.37

$$\Delta G^0 = (1\ \text{mol I}_2)\left(\frac{19.37\ \text{kJ}}{\text{mol I}_2}\right) - (1\ \text{mol I}_2)\left(\frac{0\ \text{kJ}}{\text{mol I}_2}\right) = 19.37\ \text{kJ}$$

$$\Delta G^0 = -RT \ln K_p$$

$$\text{antiln}\left(\frac{\Delta G^0}{-RT}\right) = K_p = \text{antiln}\frac{19.37\ \text{kJ}\cdot\text{mol}^{-1}}{\left(-0.008\,315\ \frac{\text{kJ}}{\text{K}\cdot\text{mol}}\right)(25+273)\text{K}} = \text{antiln}\ (-7.82)$$

$K_p = 4.0 \times 10^{-4}$ atm

$p_{I_2} = K_p = 4.0 \times 10^{-4}$ atm or 0.30 mmHg

(Best answer, without intermediate round off, is 4.0×10^{-4} atm or 0.31 mmHg.)

17.53 **(a)** positive

Melting is an endothermic process.

(b) positive

Entropy increases because liquids are more random than solids.

(c) zero

At equilibrium, the free energy is zero.

(d) positive

Melting is not spontaneous at temperatures below 5.5 °C.

(e) negative

Melting is spontaneous at temperatures above 5.5 °C.

17.55 –233.54 kJ

Equation	$2H_2O_2(l)$	$\rightarrow$	$2H_2O(l)$	+	$O_2(g)$
ΔG_f^0, kJ/mol	–120.41		–237.18		0

$$\Delta G^0 = \left[(2\ \cancel{\text{mol } H_2O})\left(\frac{-237.18\text{ kJ}}{\cancel{\text{mol } H_2O}}\right) + (1\ \cancel{\text{mol } O_2})\left(\frac{0\text{ kJ}}{\cancel{\text{mol } O_2}}\right)\right] - (2\ \cancel{\text{mol } H_2O_2})\left(\frac{-120.41\text{ kJ}}{\cancel{\text{mol } H_2O_2}}\right) = -233.54\text{ kJ}$$

17.57 reaction is nonspontaneous at all temperatures ($\Delta H > 0, \Delta S < 0$)

Equation	$CH_4(g)$	+	$H_2O(g)$	$\rightarrow$	$CH_3OH(g)$	+	$H_2(g)$
ΔH_f^0, kJ/mol	–74.85		–238.92		–201.2		0
S^0, J/K · mol	186.2		188.72		238		130.68

$$\Delta H^0_{rxn} = \left[(1\ \cancel{\text{mol } CH_3OH})\left(\frac{-201.2\text{ kJ}}{\cancel{\text{mol } CH_3OH}}\right) + (1\ \cancel{\text{mol } H_2})\left(\frac{0\text{ kJ}}{\cancel{\text{mol } H_2}}\right)\right]$$

$$-\left[(1\ \cancel{\text{mol } CH_4})\left(\frac{-74.85\text{ kJ}}{\cancel{\text{mol } CH_4}}\right) + (1\ \cancel{\text{mol } H_2O})\left(\frac{-238.92\text{ kJ}}{\cancel{\text{mol } H_2O}}\right)\right]$$

$$= +112.6\text{ kJ}$$

$$\Delta S^0 = \left[(1\ \text{mol CH}_3\text{OH})\left(\frac{238\ \text{J}}{\text{K}\cdot\text{mol CH}_3\text{OH}}\right) + (1\ \text{mol H}_2)\left(\frac{130.68\ \text{J}}{\text{K}\cdot\text{mol H}_2}\right)\right]$$

$$-\left[(1\ \text{mol CH}_4)\left(\frac{186.2\ \text{J}}{\text{K}\cdot\text{mol CH}_4}\right) + (1\ \text{mol H}_2\text{O})\left(\frac{188.72\ \text{J}}{\text{K}\cdot\text{mol H}_2\text{O}}\right)\right]$$

$$= -6\ \frac{\text{J}}{\text{K}}$$

Because ΔH is positive and ΔS is negative, the reaction is not spontaneous at any temperature (Table 17.2).

17.59 1.7×10^6

Calculate ΔG^0:

Equation	$SO_2(g)$	+	$NO_2(g)$	$\rightleftharpoons$	$NO(g)$	+	$SO_3(g)$
ΔG_f^0, kJ/mol	−300.19		51.30		86.57		−371.08

$$\Delta G^0 = \left[(1\ \text{mol NO})\left(\frac{86.57\ \text{kJ}}{\text{mol NO}}\right) + (1\ \text{mol SO}_3)\left(\frac{-371.08\ \text{kJ}}{\text{mol SO}_3}\right)\right]$$

$$-\left[(1\ \text{mol SO}_2)\left(\frac{-300.19\ \text{kJ}}{\text{mol SO}_2}\right) + (1\ \text{mol NO}_2)\left(\frac{51.30\ \text{kJ}}{\text{mol NO}_2}\right)\right]$$

$$= -35.62\ \text{kJ}$$

Use $\Delta G^0 = RT \ln K$ to calculate K_p.

$$\text{antiln}\left(\frac{\Delta G^0}{-RT}\right) = K_p = \text{antiln}\ \frac{-35.62\ \text{kJ}\cdot\text{mol}^{-1}}{\left(\frac{-0.008\,315\ \text{kJ}}{\text{K}\cdot\text{mol}}\right)(298.2\ \text{K})} = \text{antiln}\ 14.37 = 1.7 \times 10^6$$

Because there are two moles of gas on each side of the equation for the reaction, $K_c = K_p = 1.7 \times 10^6$

17.61 **(a)** −1.7 kJ

To estimate the value of ΔG^0 at 250 °C, use the relationship

$$\Delta G^0 = \Delta H^0 - T\Delta S^0$$

Assume that the values of ΔH^0 and ΔS^0 are the same at 250 °C as they are at 25 °C.

Equation	$PCl_5(g)$	$\rightleftharpoons$	$PCl_3(g)$	+	$Cl_2(g)$
ΔH_f^0, kJ/mol	−374.5		−287.0		0
S^0, J/K · mol	364.1		311.78		222.96

$$\Delta H^0_{rxn} = \left[(1\ \text{mol PCl}_3)\left(\frac{-287.0\ \text{kJ}}{\text{mol PCl}_3}\right) + (1\ \text{mol Cl}_2)\left(\frac{0\ \text{kJ}}{\text{mol Cl}_2}\right)\right]$$

$$-\left[(1\ \text{mol PCl}_5)\left(\frac{-374.5\ \text{kJ}}{\text{mol PCl}_5}\right)\right]$$

$$= 87.5\ \text{kJ}$$

$$\Delta S^0_{rxn} = \left[(1\ \cancel{mol\ PCl_3})\left(\frac{311.78\ J}{K \cdot \cancel{mol\ PCl_3}}\right) + (1\ \cancel{mol\ Cl_2})\left(\frac{222.96\ J}{K \cdot \cancel{mol\ Cl_2}}\right)\right]$$

$$-\left[(1\ \cancel{mol\ PCl_5})\left(\frac{364.1\ J}{K \cdot \cancel{mol\ PCl_5}}\right)\right]$$

$$= 170.6\,\frac{J}{K} = 0.1706\,\frac{kJ}{K}$$

$$\Delta G^0 = 87.5\ kJ - (250 + 273)\ \cancel{K}\left(\frac{0.1706\ kJ}{\cancel{K}}\right) = -1.7\ kJ$$

(b) 1.5

$$\Delta G^0 = -RT \ln K_p$$

$$\text{antiln}\left(\frac{\Delta G^0}{-RT}\right) = K_p = \text{antiln}\,\frac{-1.7\ \cancel{kJ} \cdot \cancel{mol^{-1}}}{\left(-0.008\ 315\,\frac{\cancel{kJ}}{\cancel{K} \cdot \cancel{mol}}\right)(250 + 273)\ \cancel{K}} = \text{antiln}\ 0.39$$

$$K_p = 1.5$$

(c) $p_{PCl_3} = p_{Cl_2} = 0.68$ atm, $p_{PCl_5} = 0.31$ atm

	$PCl_5(g)$	$\rightleftarrows$	$PCl_3(g)$	+	$Cl_2(g)$
Init. press., atm	0.99		0.00		0.00
Change, Δatm	$-x$		$+x$		$+x$
Eq. press., atm	$0.99 - x$		x		x

$$K = 1.49 = \frac{p_{PCl_3}\,p_{Cl_2}}{p_{PCl_5}} = \frac{x^2}{0.99 - x}$$

$$1.49(0.99 - x) = x^2$$

$$x^2 + 1.49x - 1.5 = 0$$

$$x = p_{PCl_3} = p_{Cl_2} = 0.69\ \text{atm}$$

$$p_{PCl_5} = 0.99 - 0.69 = 0.30\ \text{atm}$$

(Best answer, without intermediate round off, is 0.68 atm and 0.31 atm.)

Check: $\dfrac{(0.68)^2}{0.31} = 1.49$

(d) to the right

According to Le Chatelier's principle, the reaction will proceed to the right to reestablish equilibrium.

(e) At equilibrium, $\Delta G = 0$. For the forward (to the right) reaction to occur spontaneously, ΔG must be negative.

$\Delta G = \Delta G^0 + RT \ln Q$, where for the reaction in this problem, $Q = \dfrac{p_{PCl_3} p_{Cl_2}}{p_{PCl_5}}$

If p_{PCl_3} is lowered by removing p_{PCl_3}, Q will be lowered, $RT \ln Q$ will be lowered, and ΔG will be lower (i.e. negative).

17.63 ΔH = zero; ΔS = positive; ΔG = negative

There is no enthalpy change when an ideal solution forms. Entropy increases because the volume of the solution is greater than the volume of either component. Free energy decreases because the solution forms spontaneously. This last conclusion can also be reached by applying mathematics to the equation $\Delta G = \Delta H - T\Delta S$.

Applications

17.65 When small molecules are joined to form large ones, the entropy of the system decreases. However, the energy released as a result of bond formation increases the entropy of the surroundings more, so that the entropy of the universe increases as required by the second law of thermodynamics.

17.67 No. The reaction is nonspontaneous at all temperatures ($\Delta H^0 > 0$, $\Delta S^0 < 0$).

Equation	$N_2(g)$	+	$2H_2(g)$	$\rightarrow$	$H_2NNH_2(l)$
ΔH_f^0, kJ/mol	0		0		50.63
S^0, J/K · mol	191.50		130.68		121.2

$$\Delta S^0 = \left[(1\ \cancel{\text{mol } H_2NNH_2})\left(\frac{121.2\ \text{J}}{\text{K}\cdot\cancel{\text{mol } H_2NNH_2}}\right)\right] - \left[(1\ \cancel{\text{mol } N_2})\left(\frac{191.50\ \text{J}}{\text{K}\cdot\cancel{\text{mol } N_2}}\right) + (2\ \cancel{\text{mol } H_2})\left(\frac{130.68\ \text{J}}{\text{K}\cdot\cancel{\text{mol } H_2}}\right)\right] = -331.7\ \frac{\text{J}}{\text{K}}$$

Thus, ΔS^0 is negative as predicted by the equation for the formation of hydrazine in which 3 mol gas $\rightarrow$ 1 mol gas.

Because ΔH is positive and ΔS is negative, the reaction is not spontaneous at any temperature (under standard conditions) and hydrazine probably cannot be prepared by this reaction. [Industrially hydrazine is prepared by the reaction $2NH_3(aq) + NaOCl(aq) \rightarrow H_2NNH_2(aq) + NaCl(aq) + H_2O(l)$].

17.69 **(a)** 3.06 g

$(CH_3)_2NNH_2(l) + 2N_2O_4(l) \rightarrow 2CO_2(g) + 3N_2(g) + 4H_2O(g)$

$$(1.00\ \cancel{\text{g DMH}})\left(\frac{1\ \cancel{\text{mol DMH}}}{60.10\ \cancel{\text{g DMH}}}\right)\left(\frac{2\ \cancel{\text{mol } N_2O_4}}{1\ \cancel{\text{mol DMH}}}\right)\left(\frac{92.01\ \text{g } N_2O_4}{1\ \cancel{\text{mol } N_2O_4}}\right) = 3.06\ \text{g } N_2O_4$$

(b) Reaction will be spontaneous because it is exothermic (ΔH is negative) and entropy increases. Three moles of liquid reactants yield nine moles of gaseous products.

17.71 positive

For $\Delta G = \Delta H - T\Delta S$, if $\Delta H \approx 0$, ΔG will only be negative if ΔS is positive (since $T \geq 0$). A negative ΔG value indicates a spontaneous change.

17.73 **(a)** $\Delta H^0 = 180.50$ kJ for $N_2(g) + O_2(g) \rightarrow 2NO(g)$

$\Delta H^0 = -114.1$ kJ for $2NO(g) + O_2(g) \rightarrow 2NO_2(g)$

$\Delta H^0 = 306.2$ kJ for $NO_2(g) \xrightarrow{\text{light}} NO(g) + O(g)$

$\Delta H^0 = -106.5$ kJ for $O(g) + O_2(g) \rightarrow O_3(g)$

Equation	$N_2(g)$	+	$O_2(g)$	$\rightarrow$	$2NO(g)$
ΔH_f^0, kJ/mol	0		0		90.25

$$\Delta H^0_{rxn} = \left[(2\ \cancel{\text{mol NO}})\left(\frac{90.25\ \text{kJ}}{\cancel{\text{mol NO}}}\right)\right] - \left[(1\ \cancel{\text{mol } N_2})\left(\frac{0\ \text{kJ}}{\cancel{\text{mol } N_2}}\right) + (1\ \cancel{\text{mol } O_2})\left(\frac{0\ \text{kJ}}{\cancel{\text{mol } O_2}}\right)\right] = 180.50\ \text{kJ}$$

Equation	$2NO(g)$	+	$O_2(g)$	$\rightarrow$	$2NO_2(g)$
ΔH_f^0, kJ/mol	90.25		0		33.2

$$\Delta H^0_{rxn} = \left[(2\ \cancel{\text{mol } NO_2})\left(\frac{33.2\ \text{kJ}}{\cancel{\text{mol } NO_2}}\right)\right] - \left[(2\ \cancel{\text{mol NO}})\left(\frac{90.25\ \text{kJ}}{\cancel{\text{mol NO}}}\right) + (1\ \cancel{\text{mol } O_2})\left(\frac{0\ \text{kJ}}{\cancel{\text{mol } O_2}}\right)\right] = -114.1\ \text{kJ}$$

Equation	$NO_2(g)$	$\xrightarrow{\text{light}}$	$NO(g)$	+	$O(g)$
ΔH_f^0, kJ,mol	33.2		90.25		249.17

$$\Delta H^0_{rxn} = \left[(1\ \cancel{\text{mol NO}})\left(\frac{90.25\ \text{kJ}}{\cancel{\text{mol NO}}}\right) + (1\ \cancel{\text{mol O}})\left(\frac{249.17\ \text{kJ}}{\cancel{\text{mol O}}}\right)\right] - \left[(1\ \cancel{\text{mol } NO_2})\left(\frac{33.2\ \text{kJ}}{\cancel{\text{mol } NO_2}}\right)\right] = 306.2\ \text{kJ}$$

Equation	$O(g)$	+	$O_2(g)$	$\rightarrow$	$O_3(g)$
ΔH_f^0, kJ/mol	249.17		0		142.7

$$\Delta H^0_{rxn} = \left[(1\ \cancel{\text{mol } O_3})\left(\frac{142.7\ \text{kJ}}{\cancel{\text{mol } O_3}}\right)\right] - \left[(1\ \cancel{\text{mol O}})\left(\frac{249.17\ \text{kJ}}{\cancel{\text{mol O}}}\right) + (1\ \cancel{\text{mol } O_2})\left(\frac{0\ \text{kJ}}{\cancel{\text{mol } O_2}}\right)\right] = -106.5\ \text{kJ}$$

(b) The sign can be predicted for each reaction except $N_2(g) + O_2(g) \rightarrow 2NO(g)$.

ΔS should be negative for

$2NO(g) + O_2(g) \rightarrow 2NO_2(g)$

$O(g) + O_2(g) \rightarrow O_3(g)$

because the number of moles of gas decreases. ΔS should be positive for $NO_2(g) \rightarrow NO(g) + O(g)$ because the number of moles of gas increases. For $N_2(g) + O_2(g) \rightarrow 2NO(g)$, there is no change in the number of moles of gas, and thus ΔS may be either positive or negative.

(c) 24.77 J/K

Equation	$N_2(g)$	+	$O_2(g)$	$\rightarrow$	$2NO(g)$
S^0, J/K · mol	191.50		205.03		210.65

$$\Delta S^0 = \left[(2\ \cancel{\text{mol NO}})\left(\frac{210.65\ \text{J}}{\text{K}\cdot\cancel{\text{mol NO}}}\right)\right] - \left[(1\ \cancel{\text{mol } N_2})\left(\frac{191.50\ \text{J}}{\text{K}\cdot\cancel{\text{mol } N_2}}\right) + (1\ \cancel{\text{mol } O_2})\left(\frac{205.03\ \text{J}}{\text{K}\cdot\cancel{\text{mol } O_2}}\right)\right] = 24.77\ \frac{\text{J}}{\text{K}}$$

(d)

Reaction	Signs ΔH	ΔS	Conclusion
$N_2(g) + O_2(g) \rightarrow 2NO(g)$	+	+	Nonspontaneous at low temperatures, spontaneous at high temperatures
$2NO(g) + O_2(g) \rightarrow 2NO_2(g)$	–	–	Spontaneous at low temperatures, nonspontaneous at high temperatures
$NO_2(g) \xrightarrow{\text{light}} NO(g) + O(g)$	+	+	Nonspontaneous at low temperatures, spontaneous at high temperatures
$O(g) + O_2(g) \rightarrow O_3(g)$	–	–	Spontaneous at low temperatures, nonspontaneous at high temperatures

(e)

O_3:

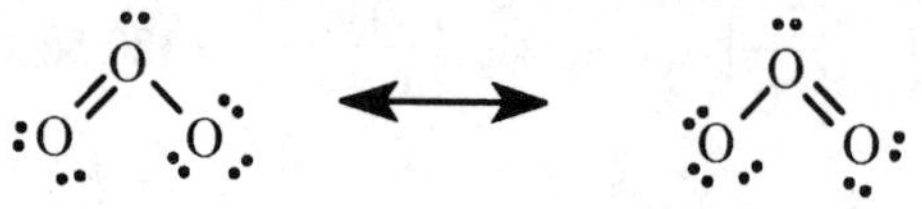

NO:

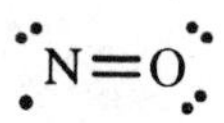

NO_2:

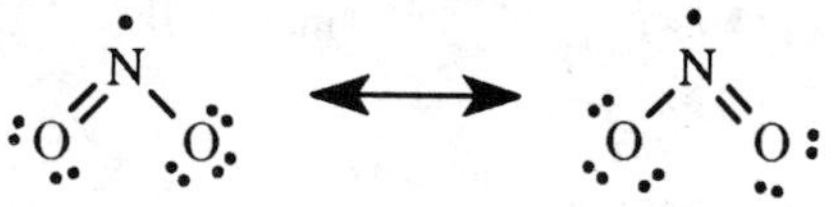

(f) Both NO and NO_2 have an unpaired electron. Species with unpaired electrons (free radicals) are very reactive.

17.75 **(a)** –918 kJ/mol

Equation	$C_6H_{12}O_6(aq) \rightarrow$	$3CH_4(g)$	+	$3CO_2(g)$	$\Delta G^0 = -418$ kJ
ΔG_f^0, kJ/mol	x	–50.79		–394.38	

$$\Delta G^0 = -418 \text{ kJ} = \left[(3 \text{ mol } CH_4)\left(\frac{-50.79 \text{ kJ}}{\text{mol } CH_4}\right) + (3 \text{ mol } CO_2)\left(\frac{-394.38 \text{ kJ}}{\text{mol } CO_2}\right)\right] - (1 \text{ mol } C_6H_{12}O_6)\left(\frac{x \text{ kJ}}{\text{mol } C_6H_{12}O_6}\right)$$

$x = \Delta G_f^0 = -918$ kJ/mol

(b) –2453.85 kJ

Since three moles of CH_4 are produced from one mole of glucose, write the equation for the combustion of three moles of CH_4:

Equation	$3CH_4(g)$	+	$9O_2(g)$	$\rightarrow$	$3CO_2(g)$	+	$6H_2O(l)$
ΔG_f^0, kJ/mol	–50.79		0		–394.38		–237.18

$$\Delta G^0 = \left[(3 \text{ mol } CO_2)\left(\frac{-394.38 \text{ kJ}}{\text{mol } CO_2}\right) + (6 \text{ mol } H_2O)\left(\frac{-237.18 \text{ kJ}}{\text{mol } H_2O}\right)\right]$$

$$-\left[(3 \text{ mol } CH_4)\left(\frac{-50.79 \text{ kJ}}{\text{mol } CH_4}\right) + (9 \text{ mol } O_2)\left(\frac{0 \text{ kJ}}{\text{mol } O_2}\right)\right]$$

$$= -2453.85 \text{ kJ}$$

17.77 spontaneous, because $\Delta G°$ is negative

Equation	$24H_2S(aq)$	+	$6CO_2(aq)$	+	$6O_2(g)$	→	$C_6H_{12}O_6(aq)$	+ $18H_2O(l)$	+	$24S(s)$
ΔG_f^0, kJ/mol	–27.87		–386.2		16.3		–919	–237.18		0

$$\Delta G^0 = \left[(1\ \cancel{\text{mol } C_6H_{12}O_6})\left(\frac{-919\text{ kJ}}{\cancel{\text{mol } C_6H_{12}O_6}}\right) + (18\ \cancel{\text{mol } H_2O})\left(\frac{-237.18\text{ kJ}}{\cancel{\text{mol } H_2O}}\right)\right]$$

$$-\left[(24\ \cancel{\text{mol } H_2S})\left(\frac{-27.87\text{ kJ}}{\cancel{\text{mol } H_2S}}\right) + (6\ \cancel{\text{mol } CO_2})\left(\frac{-386.2\text{ kJ}}{\cancel{\text{mol } CO_2}}\right) + (6\ \cancel{\text{mol } O_2})\left(\frac{16.3\text{ kJ}}{\cancel{\text{mol } O_2}}\right)\right]$$

$$= -2.300 \times 10^3\text{ kJ}$$

To determine the free energy of formation for aqueous glucose:

Equation	$6CO_2(g)$ +	$6H_2O(l)$	→	$C_6H_{12}O_6(aq)$	+	$6O_2(g)$	$\Delta G^0 = -2870$ kJ
ΔG_f^0, kJ/mol	–394.38	–237.18		x		0	

$$\Delta G^0 = -2870\text{ kJ} = (1\ \cancel{\text{mol } C_6H_{12}O_6})\left(\frac{x\text{ kJ}}{\cancel{\text{mol } C_6H_{12}O_6}}\right) - \left[(6\ \cancel{\text{mol } CO_2})\left(\frac{-394.38\text{ kJ}}{\cancel{\text{mol } CO_2}}\right) + (6\ \cancel{\text{mol } H_2O})\left(\frac{-237.18\text{ kJ}}{\cancel{\text{mol } H_2O}}\right)\right]$$

$$x = \Delta G_f^0 = -919\text{ kJ/mol}$$

CHAPTER 18: Kinetics: A Closer Look at Reaction Rates

Practice Problems

18.1 **(a)** 0.000 30 mL/s

Average rate of formation of oxygen $= \dfrac{\Delta Vol_{O_2}}{\Delta t}$

$$= \frac{10.53\text{ mL} - 10.17\text{ mL}}{7800\text{ s} - 6600\text{ s}}$$

$$= \frac{0.36\text{ mL}}{1200\text{ s}} = 0.000\,30\text{ mL/s or } 3.0 \times 10^{-4}\text{ mL/s.}$$

(b) Yes.

Sample Problem 18.1 showed that the average rate of formation over the first 300 s of the reaction was 0.0038 mL/s, whereas the average rate of formation was only 0.000 30 mL/s, less than one tenth as high over the time range from 6600 s to 7800 s.

18.2 $0.0013\ \frac{\text{mL}}{\text{s}}$

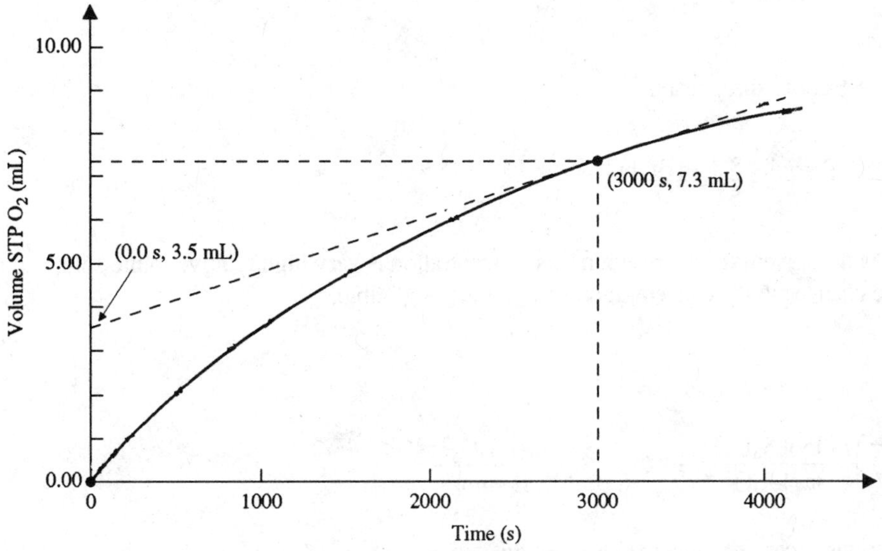

$$\text{Slope of the tangent} = \lambda = \frac{(y_2 - y_1)}{(x_2 - x_1)} = \frac{7.3\text{ mL} - 3.5\text{ mL}}{3000\text{ s} - 0\text{ s}} = 0.0013 \text{ or } 1.3 \times 10^{-3}\ \frac{\text{mL}}{\text{s}}$$

18.3 (a) $-7.1 \times 10^{-5}\ \frac{\text{M}}{\text{s}}$

$$\text{Initial rate of disappearance of } N_2O_5 = \lambda_{N_2O_5} = \frac{0.000\text{ M} - 0.200\text{ M}}{2.8 \times 10^3\text{ s} - 0\text{ s}} = -7.1 \times 10^{-5}\ \frac{\text{M}}{\text{s}}$$

(b) $3.6 \times 10^{-5}\ \frac{\text{M}}{\text{s}}$

$$\text{Initial rate of appearance of } O_2 = \lambda_{O_2} = \frac{0.125\text{ M} - 0.000\text{ M}}{3.5 \times 10^3\text{ s} - 0\text{ s}} = 3.6 \times 10^{-5}\ \frac{\text{M}}{\text{s}}$$

(c) It is about two times the initial rate of formation of O_2.

$$\frac{7.1 \times 10^{-5}\ \frac{\cancel{\text{M}}}{\cancel{\text{s}}}}{3.6 \times 10^{-5}\ \frac{\cancel{\text{M}}}{\cancel{\text{s}}}} = 2.0$$

18.4 (a) $\text{rate} = -\frac{1}{5}\frac{\Delta[Br^-]}{\Delta t} = -\frac{\Delta[BrO_3^-]}{\Delta t} = -\frac{1}{6}\frac{\Delta[H^+]}{\Delta t} = \frac{1}{3}\frac{\Delta[Br_2]}{\Delta t} = \frac{1}{3}\frac{\Delta[H_2O]}{\Delta t}$

(b) $\text{rate} = -\frac{\Delta[H_2O_2]}{\Delta t} = -\frac{1}{3}\frac{\Delta[I^-]}{\Delta t} = -\frac{1}{2}\frac{\Delta[H^+]}{\Delta t} = \frac{\Delta[I_3^-]}{\Delta t} = \frac{1}{2}\frac{\Delta[H_2O]}{\Delta t}$

18.5 (a) Br_2 is formed three times as fast as BrO_3^- disappears.

$$\frac{1}{3}\frac{\Delta[Br_2]}{\Delta t} = -\frac{\Delta[BrO_3^-]}{\Delta t}\text{; therefore, } \frac{\Delta[Br_2]}{\Delta t} = 3\left(-\frac{\Delta[BrO_3^-]}{\Delta t}\right)$$

(b) Br_2 is formed $\frac{3}{5}$ as fast as Br^- disappears.

$$\frac{1}{3}\frac{\Delta[Br_2]}{\Delta t} = -\frac{1}{5}\frac{\Delta[Br^-]}{\Delta t}; \frac{\Delta[Br_2]}{\Delta t} = \frac{3}{5}\left(-\frac{\Delta[Br^-]}{\Delta t}\right)$$

(c) No, because water is the solvent for the reaction. Its concentration is very high relative to the solute concentrations. This means that the changes in the concentration of water are negligible.

18.6 $1.28 \times 10^{-4} \frac{M}{min}$

$$\left(\frac{2.56 \times 10^{-4}\ \text{mol } S_2O_3^{2-}}{L \cdot min}\right)\left(\frac{1\ \text{mol } S_4O_6^{2-}}{2\ \text{mol } S_2O_3^{2-}}\right) = 1.28 \times 10^{-4} \frac{\text{mol } S_4O_6^{2-}}{L \cdot min}$$

18.7 (a) No. All of the reactions except reaction (3) illustrate this point.

(b) No. In some cases, such as reactions (4), (6), and (9), one or more of the reactants are not even included in the rate law.

18.8 (1) The reaction is first order with respect to N_2O_5.

(2) The reaction is first order with respect to NO_2 and F_2.

(3) The reaction is second order with respect to NO and first order with respect to O_2.

(4) The reaction is zero order with respect to NO and first order with respect to N_2O_5.

(5) The reaction is second order with respect to NO.

18.9 (a) extremely rare

None of the twelve reactions is third order with respect to a species.

(b) About 25% of reactions are zero order with respect to a species. Reaction (4) is zero order with respect to NO, reaction (6) is zero order with respect to Tl^+, and reaction (9) is zero order with respect to CO.

18.10 rate = $k\,[I^-][H_3AsO_4][H^+]$

18.11 (a) first order

Comparing experiment 2 with experiment 3, $[HgCl_2]$ in experiment 2 is twice $[HgCl_2]$ in experiment 3. The initial rate in experiment 2 is 1.9 times the initial rate in experiment 3. From equation 18.5

$$\frac{1.2 \times 10^{-4}}{6.2 \times 10^{-5}} = \left(\frac{0.100}{0.050}\right)^x \text{ or } 1.9 = (2)^x$$

By inspection, $x = 1$ and the reaction is first order with respect to $HgCl_2$.

(b) second order

Comparing Experiment 2 with Experiment 1, the $[C_2O_4^{2-}]$ was doubled while $[HgCl_2]$ remained the same. The initial rate increased fourfold. From equation 18.5

$$\frac{1.2 \times 10^{-4}}{3.1 \times 10^{-5}} = \frac{0.40}{0.20} \text{ or } (3.9) = (2.0)^y$$

By inspection, $y = 2$ and the reaction is second order with respect to $C_2O_4^{2-}$.

(c) $\text{rate} = k\,[HgCl_2][C_2O_4^{2-}]^2$

(d) $7.8 \times 10^{-3}\ s^{-1} \cdot M^{-2}$

In parts (a) and (b), the rate law was found to be $\text{rate} = k[HgCl_2][C_2O_4^{2-}]^2$. Solving for k and substituting the concentrations and initial rate from experiment 1 gives

$$k = \frac{\text{rate}}{[HgCl_2][C_2O_4^{2-}]^2} = \frac{3.1 \times 10^{-5}\ M/s}{(0.100\ M)(0.20\ M)^2} = 7.8 \times 10^{-3}\ s^{-1} \cdot M^{-2}$$

(e) Initial rate will increase.

Since the rate is proportional to $[HgCl_2][C_2O_4^{2-}]^2$, as these concentrations are increased, the rate is increased.

(f) $1.4 \times 10^{-4} \frac{M}{s}$

$$\text{rate} = k\,[HgCl_2][C_2O_4^{2-}]^2 = (7.8 \times 10^{-3}\ s^{-1} \cdot M^{-2})(0.20\ M)(0.30\ M)^2 = 1.4 \times 10^{-4} \frac{M}{s}$$

18.12 **(a)** increase fourfold

If a reaction is second order with respect to a reactant A,

$$\text{rate} = k\,[A]^2$$

If the concentration of A is doubled,

$$\text{rate} = k\,\{2[A]\}^2 = k\,2^2[A]^2 = 4k[A]^2$$

(b) no change

If a reaction is zero order with respect to reactant A,

$$\text{rate} = k[A]^0$$

Any number to the zeroth power equals 1. Therefore,

$$\text{rate} = k$$

and doubling the concentration of A has no effect on the rate.

(c) increase threefold

If a reaction is first order with respect to a reactant A,

rate = k [A]

If the concentration of A is tripled,

rate = k {3[A]} = 3k[A]

18.13 increase eightfold

rate = $k\{2[A]\}\{2[B]\}^2 = 8k[A][B]^2$

18.14 **(a)** zero order

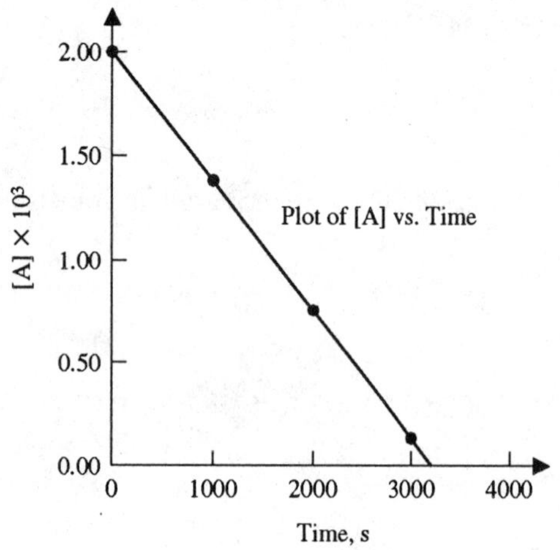

The graph of concentration against time is a straight line. The reaction is zero order with respect to A.

(b) 6.15×10^{-7} M/s

$$\text{slope} = \frac{0.000\text{ M} - (2.00 \times 10^{-3})\text{ M}}{(3.25 \times 10^{3})\text{ s} - 0.00\text{ s}} = \frac{-2.00 \times 10^{-3}\text{ M}}{3.25 \times 10^{3}\text{ s}} = -6.15 \times 10^{-7}\text{ M/s} = -k$$

$k = 6.15 \times 10^{-7}$ M/s

18.15 0.102 M

The time 702 days is three half-lives.

Time, days	M
0	0.816
234	0.408
468	0.204
702	0.102

After 702 days the concentration is 0.102 M.

18.16 304.8 min

It will take 4 half–lives to decrease to $\frac{1}{16}\left(\frac{1}{2}\times\frac{1}{2}\times\frac{1}{2}\times\frac{1}{2}\right)$ of its original value. Therefore, 4(76.2 min.) = 304.8 min.

18.17 (a) and (b) reaction with $t_{1/2} = 273$ s

(a) Reactant concentration decreases by one half in 273 s if $t_{1/2} = 273$ s. If $t_{1/2} = 541$ s, 541 s are required. Thus, the former reaction takes place in less time and is faster.

(b) For first order reactions, rate = k[A]. For the same concentration, for rate to be larger, k must be larger. Therefore, the reaction with $t_{1/2} = 273$ s, which is faster, has the larger value for k.

18.18 2×10^{-2} M

The time 9.9 years is not a whole number multiple of the half–life. The integrated rate law must be used.

First find the value of k using equation 18.7:

$$t_{1/2} = \frac{0.693}{k}$$

$$k = \frac{0.693}{t_{1/2}} = \frac{0.693}{2.6\ \text{yr}} = 0.27\ \text{yr}^{-1}$$

The integrated rate law for a first–order reaction is

$$\ln\,[\text{A}]_t = -kt + \ln\,[\text{A}]_0$$

Substitution gives

$$\ln[\text{A}]_{9.9\ \text{yr}} = -(0.27\ \cancel{\text{yr}^{-1}})(9.9\ \cancel{\text{yr}}) + \ln 0.25$$

$$= -2.7 + (-1.39) = -4.1$$

$$[\text{A}]_{9.9\ \text{yr}} = 2 \times 10^{-2}\ \text{M}$$

18.19 (a) 1.0×10^2 kJ/mol

Construct a data table using the information given in the problem to calculate values of ln k and $\frac{1}{T}$.

Temp. °C	T, K	$1/T$, $\times 10^3$	$k \times 10^4$, s^{-1}	ln k
20.00	293.15	3.4112	0.235	–10.659
25.00	298.15	3.3540	0.469	–9.967
30.00	303.15	3.2987	0.933	–9.280
35.00	308.15	3.2452	1.820	–8.6115
40.00	313.15	3.1934	3.62	–7.924
45.00	318.15	3.1432	6.29	–7.371

Next, construct a graph of ln k vs. $\frac{1}{T}$.

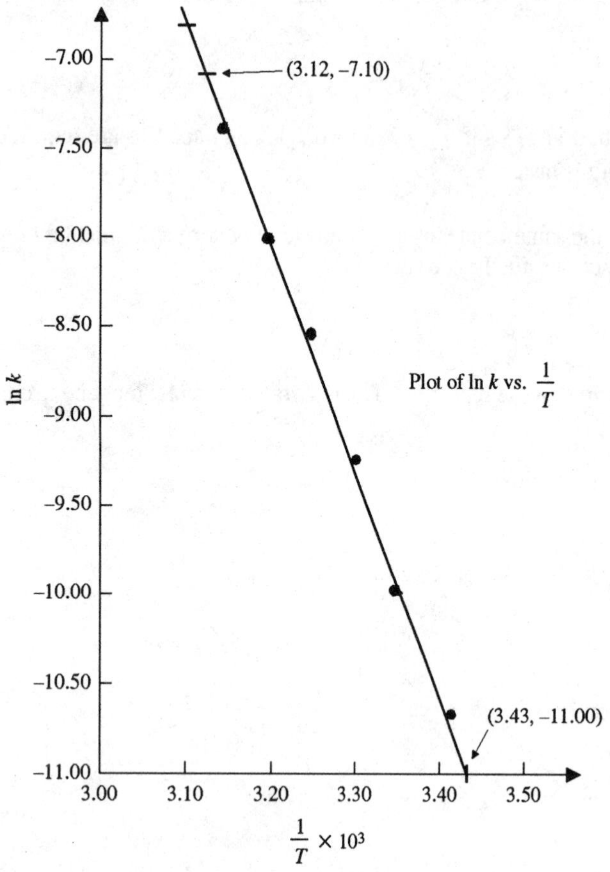

Determine the slope of the line.

$$\lambda = \frac{-7.10 - (-11.00)}{(3.12 \times 10^{-3}) - (3.43 \times 10^{-3})} = \frac{3.90}{-0.31 \times 10^{-3}} = -1.3 \times 10^{4}$$

From the Arrhenius equation (equation 18.8 b), the slope of the line is $\frac{-E_a}{R}$, where E_a is the activation energy. Therefore,

$$\frac{-E_a}{R} = \lambda \text{ and } E_a = -R\lambda = -\left(\frac{0.008\ 315 \text{ kJ}}{\text{mol} \cdot \cancel{\text{K}}}\right)(-1.3 \times 10^{4}\ \cancel{\text{K}}) = 1.1 \times 10^{2} \text{ kJ/mol}$$

(Best answer, without intermediate round off, is 1.0×10^{2} kJ/mol.)

(b) 1.1×10^{-3}

$$T_K = (50.00\ \cancel{°C} + 273.15\ \cancel{°C})\left(\frac{1\text{ K}}{1\ \cancel{°C}}\right) = 323.15 \text{ K}$$

$$\frac{1}{T} = 3.0945 \times 10^{-3} \text{ K}^{-1}$$

$\ln k_{50.00\,°C} = -6.82$ (from the graph)

$k_{50.00\,°C} = 1.1 \times 10^{-3}$

18.20 $102.0\,\dfrac{\text{kJ}}{\text{mol}}$

Equation 18.9 is

$$\ln\left(\frac{k_2}{k_1}\right) = \frac{E_a}{R}\left(\frac{1}{T_1} - \frac{1}{T_2}\right)$$

Solving this equation for E_a gives

$$\frac{R \ln\left(\dfrac{k_2}{k_1}\right)}{\dfrac{1}{T_1} - \dfrac{1}{T_2}} = E_a,$$

Use the data for the lowest and highest temperatures in the table in Practice Problem 18.19

$T_1 = 20.00\ °\text{C} = 293.15\ \text{K};\quad \dfrac{1}{T_1} = 3.4112 \times 10^{-3}$ and $k = 0.235 \times 10^{-4}$

$T_2 = 45.00\ °\text{C} = 318.15\ \text{K};\quad \dfrac{1}{T_2} = 3.1432 \times 10^{-3}$ and $k = 6.29 \times 10^{-4}$

$$E_a = \frac{\left(\dfrac{0.008\,315\ \text{kJ}}{\cancel{\text{K}} \cdot \text{mol}}\right) \ln\left(\dfrac{6.29 \times 10^{-4}}{0.235 \times 10^{-4}}\right)}{(3.4112 \times 10^{-3} - 3.1432 \times 10^{-3})\ \cancel{\text{K}}^{-1}} = 102.0\,\frac{\text{kJ}}{\text{mol}}$$

18.21

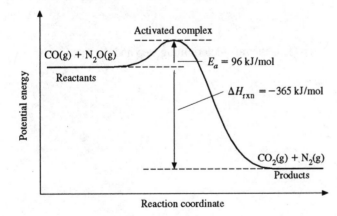

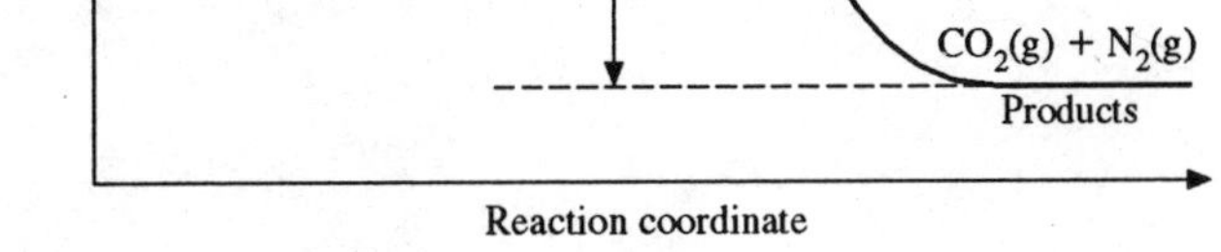

18.22 (3) $2NO(g) + O_2(g) \rightarrow 2NO_2(g)$

The only reaction that has exponents in the rate law equal to coefficients in the equation is (3). Therefore, the only reaction that *can* take place in a single step is (3):

$$2NO(g) + O_2(g) \rightarrow 2NO_2(g)$$

for which the rate law is rate = $k[NO]^2[O_2]$. This rate law indicates that all three reactant molecules are involved in the activated complex. However, for this reaction to take place in a single step, three molecules must collide at the same time and this is very improbable.

18.23 (1) N_2O_5

The rate law is rate = $k[N_2O_5]$.

(3) N_2O_4

The rate law is $k[NO]^2[O_2]$.

18.24 **(a)** $2NO_2 + O_3 \rightarrow N_2O_5 + O_2$

The equation for the overall reaction is the sum of the equations for the steps.

$$\begin{array}{l} NO_2 + O_3 \xrightarrow{k_1} \cancel{NO_3} + O_2 \\ + \cancel{NO_3} + NO_2 \xrightarrow{k_2} N_2O_5 \\ \hline 2NO_2 + O_3 \rightarrow N_2O_5 + O_2 \end{array}$$

(b) step (1): rate = $k_1\,[NO_2][O_3]$

step (2): rate = $k_2\,[NO_3][NO_2]$

(c) overall reaction: rate = $k_1\,[NO_2][O_3]$

Since step (1) is a slower reaction than step (2), step (1) is the rate–determining step and the rate law is rate = $k_1[NO_2][O_3]$.

(d) step (1): NO_5

step (2): N_2O_5

overall reaction: NO_5

(e) NO_3

NO_3 is formed in Step (1) and used up in Step (2). Therefore, it is an intermediate.

18.25 No. The rate law would have to be rate = $k[NO][O_2]$, whereas the experimentally determined rate law, as shown in the reactions for Practice Problems 18.7 – 18.9, is rate = $k[NO]^2[O_2]$.

18.26

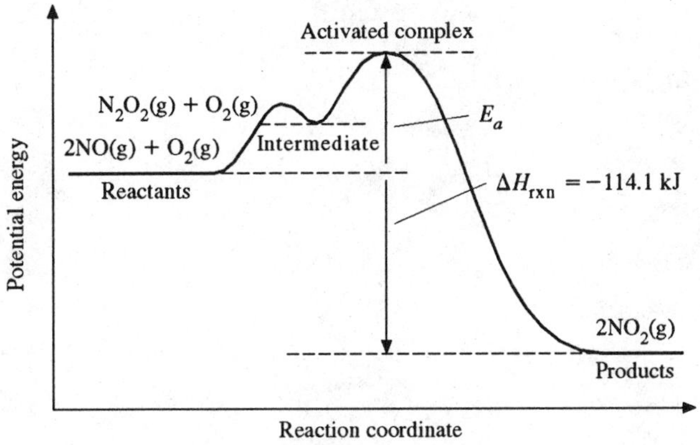

18.27 **(a)** 1 and 3

Mechanism 2 would lead to the rate law rate = k[NO][Br_2]. The experimental rate law is rate = $k[NO]^2[Br_2]$.

(b) 3

Mechanism 1 is very improbable because threee-body collisions are unlikely.

18.28 **(a)** $2Ce^{4+} + Tl^{+} \rightarrow 2Ce^{3+} + Tl^{3+}$

The equation for the overall reaction is the sum of the equations for the steps.

$$Ce^{4+} + \cancel{Mn^{2+}} \xrightarrow{k_1} Ce^{3+} + \cancel{Mn^{3+}}$$
$$Ce^{4+} + \cancel{Mn^{3+}} \xrightarrow{k_2} Ce^{3+} + \cancel{Mn^{4+}}$$
$$\cancel{Mn^{4+}} + Tl^{+} \xrightarrow{k_3} \cancel{Mn^{2+}} + Tl^{3+}$$
$$2Ce^{4+} + Tl^{+} \rightarrow 2Ce^{3+} + Tl^{3+}$$

(b) rate = k_1 [Ce^{4+}][Mn^{2+}]

This is the rate law for the first step, which is the slowest and, therefore, the rate–determining step.

(c) Mn^{2+}

Mn^{2+} is used up in Step (1) and re–formed in Step (3).

(d) Mn^{3+} and Mn^{4+}

Mn^{3+} is formed in Step (1) and used up in Step (2). Mn^{4+} is formed in Step (2) and used up in Step (3).

18.29 77 kJ/mol

Sketching the reaction profile helps determine E_a for the reverse reaction.

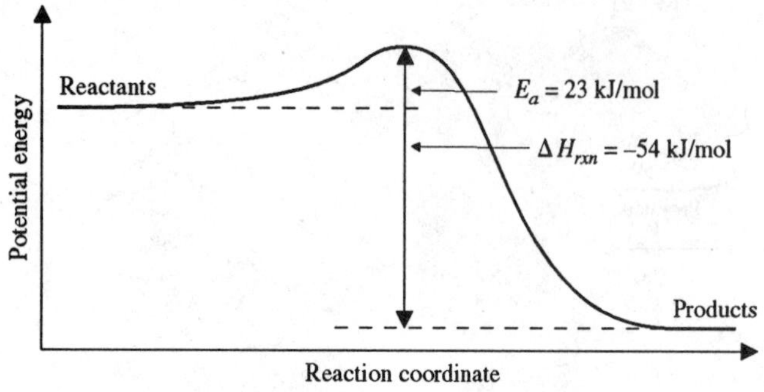

From the reaction profile, one can see that E_a for the reverse reaction must be equal to the sum of E_a for the forward reaction and the negative of ΔH for the forward reaction.

$E_{a\ \text{(reverse)}} = 23\ \text{kJ/mol} + 54\ \text{kJ/mol} = 77\ \text{kJ/mol}$

Additional Practice Problems

18.77 0.020 M

The rate law is rate = $k[\text{NOCl}]^2$. The reaction is second order with respect to NOCl. From Table 18.3 the integrated rate law for a second order reaction is

$$\frac{1}{[\text{A}]_t} = kt + \frac{1}{[\text{A}]_0}$$

Substitution of the information given in the problem in the integrated rate law gives

$$\frac{1}{[\text{NOCl}]_{5.00\ \text{hr}}} = (5.9 \times 10^{-4}\ \text{M}^{-1}\cancel{\text{s}}^{-1})\ (5.00\ \cancel{\text{hr}})\left(\frac{60\ \cancel{\text{min}}}{\cancel{\text{hr}}}\right)\left(\frac{60\ \cancel{\text{s}}}{\cancel{\text{min}}}\right) + \frac{1}{0.025}$$

Solving for $[\text{NOCl}]_{5.00\ \text{hr}}$ gives 0.020 M.

18.79 **(a)** For a zero–order reaction, the second half–life is shorter than the first half–life.

The rate of a zero–order reaction does not depend on the concentration of the reactant – it is constant. After half the reactant has reacted, only half the original concentration remains. Reaction of half this smaller quantity at the same rate takes less time.

(b) For a second–order reaction, the second half–life is longer than the first half–life.

The rate of a second–order reaction is more sensitive to concentration than the rate of a first–order reaction. Because successive half–lives are constant for first order reactions, successive half–lives for higher order reactions must get shorter and shorter.

18.81 For zero–order reactions, rate = k; for first–order reactions, rate = k [A]. The rates of zero–order reactions are independent of concentration. Zero–order reactions do not slow down as the reactant is used up. However, because the rates of first order reactions depend on concentration, reaction slows down as the reactant is used up. If the values of k are similar, the time required for A to react will be less if the reaction is zero order than if it is first order.

18.83 **(a)** $M^{-2}\ s^{-1}$

If rate (M/s) = $k\ [A]^3$, then $k = \frac{M/s}{M^3} = M^{-2}\ s^{-1}$

(b) $\frac{1}{[A]_t^2}$

The equation is in the form of $y = mx + b$, where $y = \frac{1}{[A]_t^2}$, $m = 2k$, $x = t$, $b = \frac{1}{[A]_0^2}$.

A plot of $y\left(\frac{1}{[A]_t^2}\right)$ vs. x (t) should give a straight line.

18.85 **(a)**

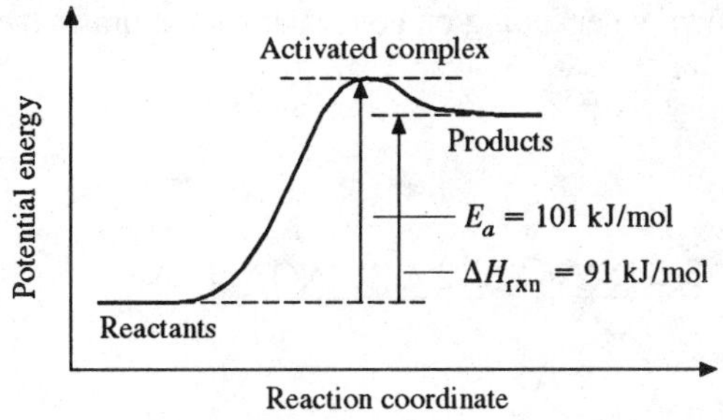

(b) The minimum activation energy for an endothermic reaction is ΔH_{rxn}.

(c) No. A very endothermic reaction must have a very high activation energy.

The higher the activation energy, the slower a reaction will be. The activation energy must be at least ΔH_{rxn} for an endothermic reaction. The more endothermic, the greater the activation energy. Therefore, such reactions are *not* likely to be very fast.

18.87 **(a)**

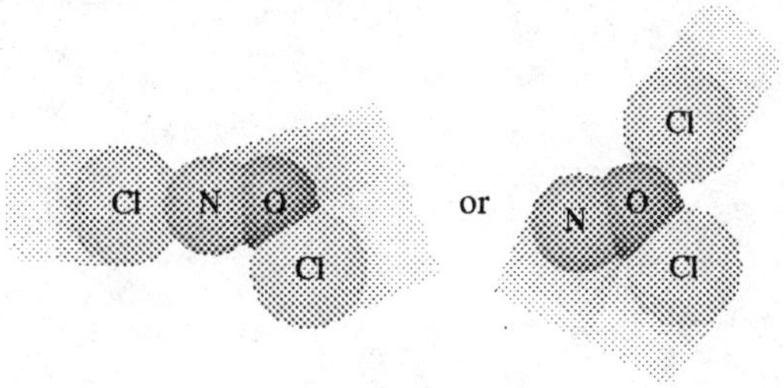

(b)

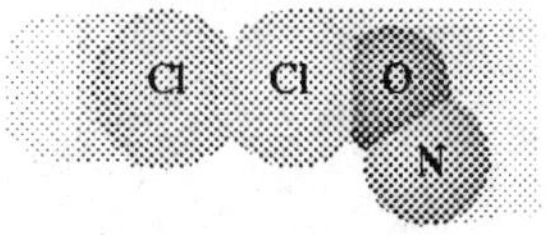

Stop and Test Yourself

1. (a) $A + B \rightarrow 2C$

Notice, for example, that at 8 min, 0.6 mol/L of A and B had reacted to make 1.2 mol/L of C.

2. (c) 6.0 min

After 6.0 min, [A] = [B] = 0.50 M, half the original concentration of 1.00 M.

3. (b) +0.25 M/min

Drawing a tangent to the curve at time 0 and using the formula for the slope of a line,

$\lambda = \frac{y_2 - y_1}{x_2 - x_1}$, where x_1 and y_1

represent the origin ($y_1 = 0$, $x_1 = 0$), gives positive values of rates of formation of C close to 0.25 M/min. For example, $\lambda = \frac{(2.20 - 0)\text{ M}}{(8.80 - 0)\text{ min}} = 0.25$ M/min. Values will vary depending on how each person draws the tangent line.

4. (c) $\frac{-3\,\Delta[KMnO_4]}{\Delta t} = \frac{2\,\Delta[KCNO]}{\Delta t}$

$-\frac{1}{2}\frac{[KMnO_4]}{\Delta t} = \frac{1}{3}\frac{[KCNO]}{\Delta t}$; Clearing of fractions, $\frac{-3\,\Delta[KMnO_4]}{\Delta t} = \frac{2\,\Delta[KCNO]}{\Delta t}$

5. (d) $k\,[NO_2]^2$

Since the reaction is zero order in CO, a [CO] term does not appear in the rate expression. The exponent of the concentration term shows the order; thus, the exponent for NO_2 is 2 because the reaction is second order in NO_2.

6. (d) 2.69×10^{-4}

rate = $k\,[NH_4^+][NO_2^-]$

$2.15 \times 10^{-6}\,\frac{\text{M}}{\text{s}} = k\,(0.200\text{ M})(0.0400\text{ M})$

$k = 2.69 \times 10^{-4}\text{ M}^{-1}\text{ s}^{-1}$

7. (b) 3.4×10^{-5}

rate = $k[NO_2]^2$

rate = $0.543\text{ M}^{-1}\text{ s}^{-1}\,(0.0079\text{ M})^2$

rate = $3.4 \times 10^{-5}\text{ M} \cdot \text{s}^{-1}$

8. (c) 2

When the concentration is doubled (multiplied by a factor of 2), the rate increases by a factor of 4, which is the square of the concentration increase factor (2^2).

9. **(a)** ln [B] vs. time

The reaction is first order.

10. **(b)** 0.050 M

The time 2.6×10^4 s is two half–lives. In 1.3×10^4 s, the concentration is reduced by half to 0.100 M and in another 1.3×10^4 s (total time = 2.6×10^4 s), the concentration is again reduced by half to 0.050 M.

11. **(a)** the fraction of the molecules with kinetic energy greater than the activation energy increases very rapidly as temperature increases.

12. **(a)** $k\,[H_2][NO]^2$

According to the mechanism given, the first step is the slow rate–determining step. Its rate law is the rate law for the overall reaction. For elementary processes, the exponents in the rate law are the same as the coefficients in the equation. Thus, the rate law for the elementary process $H_2 + 2NO \rightarrow N_2 + H_2O_2$ and for the overall reaction is rate = $k[H_2][NO]^2$.

13. **(c)** This is the energy necessary to form the activated complex from the reactants.

14. **(c)** Activation energy is lower, rate is faster, ΔH is the same.

A catalyst lowers the activation energy but ΔH remains the same.

Putting Things Together

18.89 **(a)** $Cl\cdot + CH_4 \rightarrow HCl + \cdot CH_3$

This reaction has greater probability of a collision occurring with the correct orientation because all faces of the CH_4 molecule are the same. If the Cl· collides with the molecule, it must collide near a hydrogen atom. If Cl· collides with $CHCl_3$, it has only one chance in four of bumping into a hydrogen atom.

(b) (*i*) and (*iii*)

In (*i*), no bond needs to be broken. Bond breaking always requires energy. In (*iii*), opposite charges attract, and also no bond needs to be broken.

18.91 **(a)** rate = $k\,[Cl_2][H_2S]$, rate = $k\,[Cl_2]^2$, rate = $k\,[H_2S]^2$

For rate to be in ($M \cdot s^{-1}$), when $k = 3.5 \times 10^{-5}\ M^{-1}\ s^{-1}$, concentration must be squared

$$k = \frac{M \cdot s^{-1}}{M^2} = M^{-1}\ s^{-1}$$

(b) See what happens to the initial rate if $[Cl]_0$ is kept constant and $[H_2S]_0$ is doubled and see what happens to the initial rate if $[H_2S]_0$ is kept constant and $[Cl]_0$ is doubled.

18.93 **(a)** The fraction of molecules having energy greater than the activation energy increases rapidly as temperature increases.

(b) If the reaction involves an intermediate that is the product of an exothermic equilibrium, a temperature increase will decrease the concentration of the intermediate (Le Châtelier's principle) and slow the reaction.

18.95 (a) The rate of disappearance of the red–brown color of $NO_2(g)$ could be measured with a spectrophotometer.

(b) The rate of formation of $O_2(g)$ could be measured by measuring the volume of the gas at different times.

(c, d) The volume of $CO_2(g)$ or the mass of $Hg_2Cl_2(s)$ could be measured. Also, small samples could be removed from the reaction mixture as the reaction progressed in order to determine (1) the decrease in $C_2O_4^{2-}$ concentration vs. time or (2) the increase in concentration of Cl^- vs. time.

(e) Titration of samples from the mixture would show the rate of formation of lactic acid or pH measurements vs. time might be used to determine the increase in acid concentration.

18.97 Reaction (a) fails to take place for kinetic reasons ($\Delta G^0 < 0$), and reaction (b) fails to occur for thermodynamic reasons ($\Delta G^0 > 0$.)

Reaction (a) is favored thermodynamically so it must fail to take place for kinetic reasons.

	(1)	$H_2(g)$	+	$Cl_2(g)$	→	$2HCl(g)$
ΔG_f^0, kJ/mol		0		0		-95.30

$$\Delta G^0_{rxn} = \left[(2\ \cancel{\text{mol HCl}})\left(\frac{-95.30\ \text{kJ}}{\cancel{\text{mol HCl}}}\right)\right] - \left[(1\ \cancel{\text{mol H}_2})\left(\frac{0\ \text{kJ}}{\cancel{\text{mol H}_2}}\right) + (1\ \cancel{\text{mol Cl}_2})\left(\frac{0\ \text{kJ}}{\cancel{\text{mol Cl}_2}}\right)\right] = -190.60\ \text{kJ}$$

Because ΔG^0 is negative, reaction (a) is a spontaneous reaction thermodynamically under standard conditions.

	(2)	$N_2(g)$	+	$O_2(g)$	→	$2NO(g)$
ΔG_f^0, kJ/mol		0		0		86.57

$$\Delta G^0_{rxn} = \left[(2\ \cancel{\text{mol NO}})\left(\frac{86.57\ \text{kJ}}{\cancel{\text{mol NO}}}\right)\right] - \left[(1\ \cancel{\text{mol N}_2})\left(\frac{0\ \text{kJ}}{\cancel{\text{mol N}_2}}\right) + (1\ \cancel{\text{mol O}_2})\left(\frac{0\ \text{kJ}}{\cancel{\text{mol O}_2}}\right)\right] = 173.14\ \text{kJ}$$

With a positive value for ΔG, reaction (b) fails for thermodynamic reasons.

18.99 (a) bromine is limiting

Since bromine and acetone react in a 1:1 mole ratio, it is clear from the data that a small amount of bromine is used compared with a large amount of acetone.

(b) The concentration of acetone does not change significantly because it is so much higher than the concentration of bromine. The concentration of HCl does not change because HCl is the catalyst.

(c) zero order

The experimental design is to time the reaction until the color of the bromine disappears. If the reaction proceeds at the same rate (i.e., is independent of the concentration of bromine), then it should take twice as long for twice as much bromine to disappear as was observed in experiment #2. If the reaction rate had depended on the initial concentration of bromine, the time for experiment #2 would have been different.

(d) first order for both acetone and HCl

Comparing experiments #1 and 3, doubling the inital concentration of acetone halves the time required for bromine to react. The rate is doubled. A similar conclusion can be drawn for HCl by comparing experiments #1 and 4.

(e) rate = k [acetone][HCl]

(f) 1.5×10^2 s

As concentration of bromine is halved, the time is halved.

(g) 7.0×10^1 s

Compare experiment #6 with #1 and #4.

[HCl]	0.20 M	0.40 M	0.80 M
time (s)	2.9×10^2	1.4×10^2	7.0×10^1

As the concentration of HCl is doubled, the time is halved.

18.101 **(a)** rate = k [$(CH_3)_3CBr$]

The rate of formation of $(CH_3)_3COH$ depends on the rate of the first step. Once the cations, $(CH_3)_3C^+$, are formed, they react rapidly with OH^-.

rate = k [$(CH_3)_3CBr$]

(b) The rate would double.

(c) no effect

Doubling [OH^-] has no effect on the rate of the reaction because OH^- is not a factor in the rate–controlling step.

(d) Water is a much more polar solvent than alcohol and solvates the ions better.

18.103 **(a)** $2NO(g) + Cl_2(g) \rightarrow 2ONCl(g)$

(b) $2NO(g) + F_2(g) \rightarrow 2ONF(g)$

(c) The rate law differences indicate clearly that even though a similar overall reaction occurs, the mechanistic route to product in (a) differs from that in (b).

18.105 **(a)** The solution process was exothermic and, at the microscopic level, the acetone and water molecules occupied less volume when mixed than the sum of the volumes of the pure solvents. The attractive forces among the different molecular species were great enough to cause a shrinkage in total volume and a release of thermal energy. Acetone can accept hydrogen bonds from water.

(b) 29.5 M

$$M = \frac{\text{mol solute}}{\text{vol. soln in L}}$$

Assume 1.0 mL H_2O = 1.0 g.

$$(75.0 \text{ g } \cancel{H_2O})\left(\frac{1 \text{ mol } H_2O}{18.02 \text{ g } \cancel{H_2O}}\right) = 4.16 \text{ mol } H_2O$$

$$M = \frac{4.16 \text{ mol}}{0.141 \text{ L}} = 29.5$$

(c) No.

Although water is a reactant, the concentration of water (29.5 M) is so high compared to the concentration of *tertiary*–butyl chloride that the change in concentration of water is negligible.

18.107 (a) *Allotropes* are different forms of the same element that exist in the same physical state under the same conditions of temperature and pressure but differ in bonding.

(b) red

The negative ΔH_f^0 value for the red form ($\Delta H_f^0 = -17.56$ kJ/mol; see Appendix D) shows that the change from white to red is exothermic. Therefore, red phosphorus is lower in energy and is more stable.

(c)

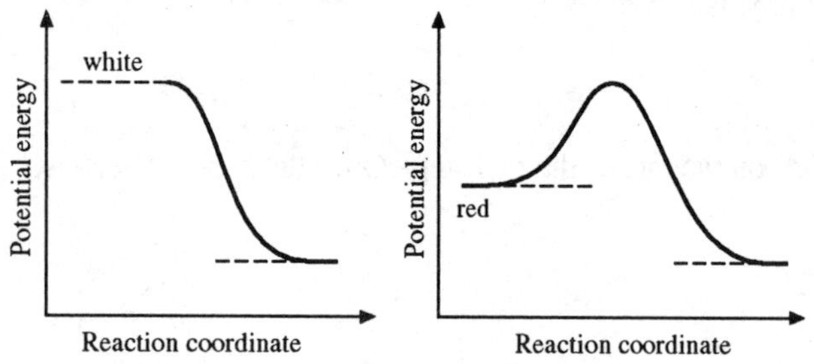

The energy of activation for the reaction of oxygen with white phosphorus is very small. Assuming the same activated complex the energy of activation for the reaction of red phosphorus with oxygen must be at least 17.56 kJ/mol higher; energy must be supplied for red phosphorus to begin to burn.

(d) tetraphosphorus decoxide, $P_4(g) + 5O_2(g) \rightarrow P_4O_{10}(s)$

18.109 *Advantages:* (1) It is easier to mix reactants. (2) It is easier to control concentrations of reactants without losses due to leakage of gases. (3) Rates can be controlled by adjusting concentrations. (4) Volatile organic solvents or water boil at convenient temperatures. The temperature of the solution cannot exceed the boiling point. (5) Flammable reactants can be made less flammable by dilution with nonflammable solvent.

Disadvantages: (1) Solvents cost money. (2) Volatile organic solvents contribute to air pollution. (3) The solvent might participate in the reaction and/or change the mechanism of the reaction.

Applications

18.111 At constant volume and temperature, seven times the normal amount of gas resulted in seven times the normal pressure. Even more important, the increased concentration would increase the reaction rate. The reaction must be spontaneous and therefore is probably exothermic. Thermal energy released by greater quantities of reactant will further increase the reaction rate.

18.113 Only $\frac{1}{64}$ of what you learned will be retained.

There are six 6–month periods of forgetting, each with a loss of one half of the previously retained learning $\left(\left[\frac{1}{2}\right]^6\right) = \frac{1}{64}$.

18.115 Assuming that physical activity is based on chemical reactions, the rates of these reactions will decrease with decreasing temperature.

18.117 **(a)** rate = $k\,[NO]^2[O_2]$

When [NO] is doubled, the rate increases by a factor of four (2^2). When the $[O_2]$ doubles, the rate doubles.

(b) $7.0 \times 10^3\ M^{-2}\ s^{-1}$

$2.8 \times 10^{-5}\ M \cdot s^{-1} = k\,(0.0020\ M)^2(0.0010\ M)$

$k = 7.0 \times 10^3\ M^{-2}\ s^{-1}$

(c) $9.5 \times 10^{-5}\ M \cdot s^{-1}$

rate = $7.0 \times 10^3\ M^{-2}\ s^{-1}\ (0.0030\ M)^2(0.0015\ M) = 9.5 \times 10^{-5}\ M \cdot s^{-1}$

18.119 about 90%

$$t_{1/2} = \frac{0.693}{k}$$

$$k = \frac{0.693}{6\ h} = 0.1\ h^{-1}$$

From the equation for a first–order reaction:

$\ln\,[A]_t = -kt + \ln\,[A]_0$

$\ln\,conc_t = (-0.1\ \cancel{h}^{-1})(1\ \cancel{h}) + \ln\,(100)$

$\ln\,conc_t = 4.5$

antiln 4.5 = 9×10^1 = 90%

18.121 **(a)** ≈ 70 °C, ≈ 158 °F

Converting 75 °F to °C,

$$t_C = \frac{5\ °C}{9\ \cancel{°F}}(\cancel{°F} - 32\ \cancel{°F}) = 24\ °C$$

Thus, if the rate of aging increased threefold for every 10 °C increase in temperature,

24 °C → 100 y	54 °C → 3.7 y
34 °C → 33 y	64 °C → 1.2 y
44 °C → 11 y	74 °C → 0.40 y

The necessary temperature is between 64 °C and 74 °C. If we assume 70 °C, then

$$70\ °C = \frac{5\ °C}{9\ °F}(t_F - 32\ °F)$$

$$158\ °F = t_F$$

(b) The ln time vs. $\frac{1}{T}$ plot would be linear, and thus extrapolation would be more accurate.

18.123 **(a)** No. Sunlight is not a catalyst because it is not a substance, but rather is a form of energy.

(b) Cl acts as a catalyst. It is present at the beginning and then is re–formed. ClO acts as an intermediate. It is formed and used up in the reaction. The overall reaction is $O_3 + O \rightarrow 2O_2$.

$$\begin{array}{l} \not{Cl} + O_3 \rightarrow \not{ClO} + O_2 \\ \underline{\not{ClO} + O \rightarrow \not{Cl} + O_2} \\ O_3 + O \rightarrow 2O_2 \end{array}$$

18.125 **(a)** about 6 h

(b) about 21 h

27 h – 6 h = 21 h

(c) about 27 h

(d) No, because some of the first dose remains.

(e) increase the size of the first dose

(f) By taking two capsules to start, time required to reach minimum effective relative concentration is reduced. After this level is reached, a smaller dose is all that is needed to maintain it.

(g) first order

The half life is appoximately constant.

18.127 $6 \times 10^1\ \text{kJ} \cdot \text{mol}^{-1}$

$$\ln\frac{k_2}{k_1} = \frac{E_a}{R}\left(\frac{1}{T_1} - \frac{1}{T_2}\right)$$

Let the ratio of k values represent the relative speeds of reaction (faster at higher temperature).

$$\ln\frac{4.5}{3.0} = \frac{E_a}{8.315\ \text{J} \cdot \text{mol}^{-1} \cdot \text{K}^{-1}}\left(\frac{1}{365\ \text{K}} - \frac{1}{373\ \text{K}}\right)$$

$$E_a = (6 \times 10^4\ \not{\text{J}} \cdot \text{mol}^{-1})\left(\frac{1\ \text{kJ}}{10^3\ \not{\text{J}}}\right) = 6 \times 10^1\ \text{kJ} \cdot \text{mol}^{-1}$$

18.129 **(a)** $63\ \text{kJ} \cdot \text{mol}^{-1}$

$$\ln\frac{k_2}{k_1} = \frac{E_a}{R}\left(\frac{1}{T_1} - \frac{1}{T_2}\right)$$

$$\ln\frac{3.0 \times 10^3}{7.9 \times 10^1} = \frac{E_a}{8.315\ \text{J} \cdot \text{mol}^{-1} \cdot \text{K}^{-1}}\left(\frac{1}{(273 + 25)\ \text{K}} - \frac{1}{(273 + 75)\ \text{K}}\right)$$

$$3.64 = \frac{E_a\,(4.8 \times 10^{-4}\,\text{K}^{-1})}{8.315\ \text{J}\cdot\text{mol}^{-1}\cdot\text{K}^{-1}}$$

$$E_a = 6.3 \times 10^4\ \text{J}\cdot\text{mol}^{-1} = 6.3 \times 10^1\ \text{kJ}\cdot\text{mol}^{-1}$$

(b) 10^{13}

$$\ln k_1 = \frac{-E_a}{R}\left(\frac{1}{T_1}\right) + \ln A$$

$$\ln A = \ln k_1 + \frac{E_a}{R}\left(\frac{1}{T_1}\right)$$

$$\ln A = \ln 7.9 \times 10^1 + \frac{6.3 \times 10^4\ \text{J}\cdot\text{mol}^{-1}}{8.315\ \text{J}\cdot\text{mol}^{-1}\cdot\text{K}^{-1}}\left(\frac{1}{298\ \text{K}}\right)$$

$$\ln A = 4.37 + 25 = 29$$

$$A = \text{antiln } 29 = 10^{12}$$

(Best answer, without intermediate round off, is 10^{13}.)

(c) Yes. The rate law has exponents equal to the coefficients in the equation. The reaction may be an elementary process involving only the collision of NO and O_3.

(d) $7.9 \times 10^{-16}\ \text{M}\cdot\text{s}^{-1}$ at 25 °C, $3.0 \times 10^{-14}\ \text{M}\cdot\text{s}^{-1}$ at 75 °C

$$\text{rate} = k\,[\text{NO}][\text{O}_3]$$

At 25 °C rate = $(7.9 \times 10^1\ \text{M}^{-1}\ \text{s}^{-1})(1.0 \times 10^{-8}\ \text{M})(1.0 \times 10^{-9}\ \text{M}) = 7.9 \times 10^{-16}\ \text{M}\cdot\text{s}^{-1}$

At 75 °C rate = $(3.0 \times 10^3\ \text{M}^{-1}\ \text{s}^{-1})(1.0 \times 10^{-8}\ \text{M})(1.0 \times 10^{-9}\ \text{M}) = 3.0 \times 10^{-14}\ \text{M}\cdot\text{s}^{-1}$

(e) about 38 times as fast

$$\text{Ratio of rates} = \frac{3.0 \times 10^{-14}}{7.9 \times 10^{-16}} = 3.8 \times 10^1$$

(f) –199.8 kJ

	NO(g)	+	O_3(g)	→	NO_2(g)	+	O_2(g)
ΔH_f^0, kJ/mol	90.25		142.7		33.2		0

$$\Delta H^0_{\text{rxn}} = \left[(1\ \text{mol NO}_2)\left(\frac{33.2\ \text{kJ}}{\text{mol NO}_2}\right) + (1\ \text{mol O}_2)\left(\frac{0\ \text{kJ}}{\text{mol O}_2}\right)\right] - \left[(1\ \text{mol NO})\left(\frac{90.25\ \text{kJ}}{\text{mol NO}}\right) + (1\ \text{mol O}_3)\left(\frac{142.7\ \text{kJ}}{\text{mol O}_3}\right)\right]$$

$$= -199.8\ \text{kJ}$$

(g)

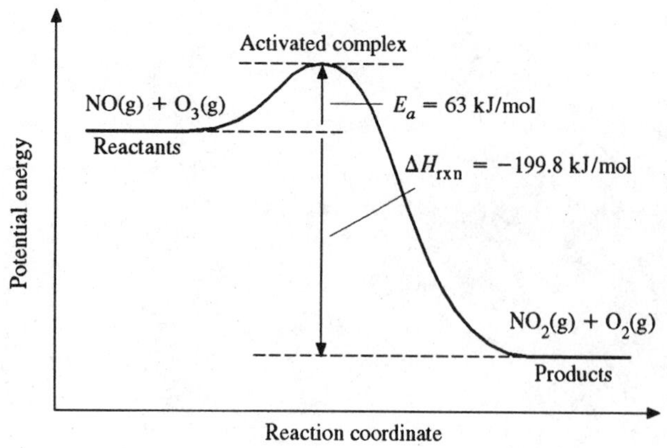

(h) $263\ \text{kJ} \cdot \text{mol}^{-1}$

$$E_{a\ (\text{reverse})} = 199.8\ \text{kJ} \cdot \text{mol}^{-1} + E_{a\ (\text{forward})} = 199.8\ \text{kJ} \cdot \text{mol}^{-1} + 63\ \text{kJ} \cdot \text{mol}^{-1} = 263\ \text{kJ} \cdot \text{mol}^{-1}$$

(i) No.

The equation for the reaction is

$$2NO_2(g) + O_3(g) \rightarrow N_2O_5(g) + O_2(g)$$

This cannot be an elementary process because the exponents in the rate law do not equal the coefficients in the equation. It requires intermediate reaction steps.

18.131 **(a)** $CH_3CH_2OH(g) + 3O_2(g) \rightarrow 2CO_2(g) + 3H_2O(l \text{ or } g)$

(b) The mechanism probably takes place by a series of steps. A single–step mechanism based on the molecularity of the equation would require instantaneous collision of four particles. This is highly improbable.

18.133 a steady state

18.135 **(a)** transfer of an electron from NO to O^+

If there were formation of a new N—O bond, some of the O–18 would have been incorporated into the NO and the NO^+ would have become enriched in O–18.

(b) 8 protons, 10 neutrons, and 7 electrons

CHAPTER 19:
Electrochemistry

Practice Problems

19.1 **(a)**

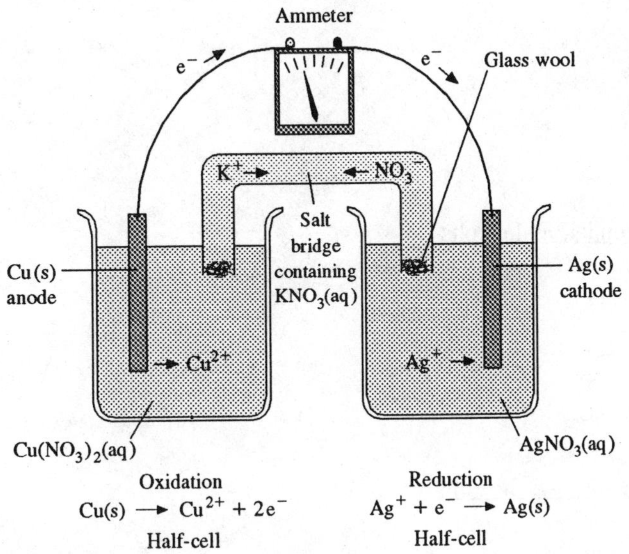

KNO_3 is used in the salt bridge because Cl^- from KCl would form a precipitate with Ag^+.

(b) Net ionic equation: $2Ag^+ + Cu(s) \rightarrow Cu^{2+} + 2Ag(s)$

Molecular equation: $2AgNO_3(aq) + Cu(s) \rightarrow Cu(NO_3)_2(aq) + 2Ag(s)$

$$Cu(s) \rightarrow Cu^{2+} + 2e^-$$
$$\underline{2(Ag^+ + e^- \rightarrow Ag(s))}$$
$$2Ag^+ + Cu(s) \rightarrow Cu^{2+} + 2Ag(s)$$

19.2 $Cu(s)\ |\ Cu^{2+}(aq)\ ||\ Ag^+(aq)\ |\ Ag(s)$

In this cell diagram, Cu(s) is the anode and Ag(s) is the cathode.

19.3 (a) oxidizing agent

(b) oxidizing agent

(c) reducing agent

(d) reducing agent

(e) oxidizing agent

(f) both

19.4 (a) $Fe^{2+} < Ag^+ < NO_3^- < Br_2(l)$

Oxidizing Agent	Std. Reduction Potential, E^0 in Volts
Fe^{2+}	–0.44
Ag^+	+0.80
NO_3^-	+0.96
$Br_2(l)$	+1.07

(b) $Fe^{2+} < I^- < Al(s)$

Reducing Agent	Std. Oxidation Potential, $-E^0$ in Volts
Fe^{2+}	–0.77
I^-	–0.54
Al(s)	+1.67

19.5 Yes. H^+ has a higher standard reduction potential than Al^{3+}.

H^+ has a higher standard reduction potential, 0.00 V, than Al^{3+}, –1.67 V. This means that H^+ is reduced more readily than Al^{3+} and is therefore a stronger oxidizing agent. It will, thus, readily oxidize Al(s) to Al^{3+}, dissolving the Al(s) in the process.

19.6 (a) +0.04 V, spontaneous

$$E^0_{cell} = E^0_{\text{half-cell of reduction}} - E^0_{\text{half-cell of oxidation}}$$

$$= E^0_{H^+/H_2(g,\ 1\ atm)} - E^0_{Fe^{3+}/Fe(s)}$$

$$= 0.00\ V - (-0.04\ V) = +0.04\ V$$

(b) +0.46 V, spontaneous

$$E^0_{cell} = E^0_{O_2(g,\ 1\ atm)/H_2O(l)} - E^0_{Fe^{3+}/Fe^{2+}}$$

$$= 1.23\ V - 0.77\ V = +0.46\ V$$

19.7 –0.14 V

$$E^0_{cell} = E^0_{half\text{–}cell\ of\ reduction} - E^0_{half\text{–}cell\ of\ oxidation}$$

$$E^0_{half\text{–}cell\ of\ reduction} = E^0_{cell} + E^0_{half\text{–}cell\ of\ oxidiation}$$

$$E^0_{Sn^{2+}/Sn(s)} = E^0_{cell} + E^0_{Zn^{2+}/Zn(s)}$$

$$= +0.62\ V + (-0.76\ V) = -0.14\ V$$

19.8 (1) $F_2(g) + 2Cl^- \rightarrow 2F^- + Cl_2(g)$

By inspection, it is seen that $F_2(g)$ will oxidize Cl^- as well as Br^- and I^-.

(2) $F_2(g) + 2Br^- \rightarrow 2F^- + Br_2(l)$

(3) $F_2(g) + 2I^- \rightarrow 2F^- + I_2(s)$

(4) $Cl_2(g) + 2Br^- \rightarrow 2Cl^- + Br_2(l)$

By inspection, it is seen that $Cl_2(g)$ will oxidize Br^- and I^-.

(5) $Cl_2(g) + 2I^- \rightarrow 2Cl^- + I_2(s)$

(6) $Br_2(l) + 2I^- \rightarrow 2Br^- + I_2(s)$

By inspection, it is seen that $Br_2(l)$ will oxidize I^-.

19.9 Ni^{2+} will oxidize Zn to Zn^{2+}. E^0_{cell} for $Zn(s) + Ni^{2+} \rightarrow Zn^{2+} + Ni(s)$ is +.

The standard reduction potential for $Ni^{2+} + 2e^- \rightarrow Ni(s)$ is higher than the standard reduction potential for $Zn^{2+} + 2e^- \rightarrow Zn(s)$. This means that Ni^{2+} will *spontaneously* oxidize Zn(s) to Zn^{2+}. This is demonstrated quantitatively by calculating E^0_{cell} for the reaction $Zn(s) + Ni^{2+} \rightarrow Zn^{2+} + Ni(s)$:

$$E^0_{cell} = E^0_{reduction} - E^0_{oxidation} = -0.26\ V - (-0.76\ V) = +0.50\ V$$

19.10 +0.37 V

For the cell $Cu(s)\ |\ Cu^{2+}\ (aq,\ 0.100\ M)\ ||\ Ag^+\ (aq,\ 0.0100\ M)\ |\ Ag(s)$

the equation for the cell reaction is

$$Cu(s) + 2Ag^+ \rightarrow Cu^{2+} + 2Ag(s)$$

For this reaction $n = 2$ and $Q = \dfrac{[Cu^{2+}]}{[Ag^{+}]^2}$

$E^0 = +0.46$ V from the reverse of the reaction in Sample Problem 19.1.

Substitution in the equation $E = E^0 - \dfrac{0.0257\ \text{V}}{n} \ln Q$ gives

$$E = 0.46\ \text{V} - \frac{0.0257\ \text{V}}{2} \ln \frac{(0.100)}{(0.000\ 1000)}$$

$$= 0.46\ \text{V} - (0.0129\ \text{V})(6.91)$$

$$= (0.46\ \text{V}) - (8.91 \times 10^{-2}\ \text{V}) = 0.37\ \text{V}$$

Although both concentrations are lower than 1 M, the concentration of reactant, Ag^{+}, is lowered much more than the concentration of product Cu^{2+}. Also recall that in the equation for the reaction, Ag^{+} has a coefficient of 2 and Q depends on $[Ag^{+}]^2$. Thus, according to Le Chatelier's principle, the equilibrium should shift to the left, lowering the driving force for the reaction. Since 0.37 V is less than 0.46 V, the answer is reasonable.

19.11 -2.12×10^5 J or -212 kJ

$$\Delta G^0 = -nFE^0 = (-2\ \cancel{\text{mol}})\left(9.649 \times 10^4 \frac{\text{C}}{\cancel{\text{mol}}}\right)(1.10\ \text{V})$$

$$= -2.12 \times 10^5\ \text{J} = -212\ \text{kJ}$$

(Remember that $\text{C} \cdot \text{V} = \text{J}$.)

19.12 +0.162 V

$$\Delta G^0 = -nFE^0$$

$$\text{and } E^0 = \frac{\Delta G^0}{-nF} = \frac{-31.2\ \cancel{\text{kJ}}}{(-2\ \cancel{\text{mol}})\left(9.649 \times 10^4 \frac{\text{C}}{\cancel{\text{mol}}}\right)}\left(\frac{10^3\ \text{J}}{1\ \cancel{\text{kJ}}}\right)$$

$$= 0.162 \frac{\text{J}}{\text{C}} = 0.162\ \text{V}$$

19.13 $K = \dfrac{[Au^{3+}]}{[Au^{+}]^3} = 10^{16}$

For the reaction

$$3Au^{+} \rightarrow Au^{3+} + 2Au(s)$$

the equilibrium constant expression is

$$K = \frac{[Au^{3+}]}{[Au^{+}]^3}$$

The value of the equilibrium constant can be calculated by means of the Nernst equation.

$$E = E^0 - \frac{0.0257\text{ V}}{n}\ln Q$$

When the system is at equilibrium, $E = 0$ and $Q = K$. Thus,

$$0 = E^0 - \frac{0.0257\text{ V}}{n}\ln K,$$

$$E^0 = \frac{0.0257\text{ V}}{n}\ln K, \quad \text{and} \quad \frac{nE^0}{0.0257\text{ V}} = \ln K$$

The half–reactions are

oxidation	$Au^+ \rightarrow Au^{3+} + \cancel{2e^-}$
reduction	$\cancel{2e^-} + 2Au^+ \rightarrow 2Au(s)$
	$3Au^+ \rightarrow Au^{3+} + 2Au(s)$

and $n = 2$. Substitution gives

$$\frac{n\,E^0}{0.0257\text{ V}} = \ln K = \frac{2\,(0.47\,\cancel{\text{V}})}{(0.0257\,\cancel{\text{V}})} = 37, \quad K = 10^{16}$$

19.14 Eventually the gutter will fall from the house as holes form in the aluminum. In contact with the iron nails, the aluminum in the gutter is like zinc on galvanized iron or a sacrificial anode.

19.15 (a) Anode: $2Br^- \rightarrow Br_2(l) + 2e^-$

Cathode: $2H_2O(l) + 2e^- \rightarrow H_2(g) + 2OH^-\ (10^{-7}\text{ M})$

Overall: $2Br^- + 2H_2O(l) \rightarrow Br_2(l) + H_2(g) + 2OH^-\ (10^{-7}\text{ M})$

The C electrodes are inert. Therefore, only reactions of solute ions and solvent need be considered.

The possible anode reactions are:

	E
$2H_2O(l) \rightarrow O_2(g) + 4H^+(10^{-7}\text{ M}) + 4e^-$	–0.82 V
$2Br^- \rightarrow Br_2(l) + 2e^-$	–1.07 V

From the half–cell potentials, the oxidation of water would be predicted. However, since Cl^- is preferentially oxidized in aqueous solution, $2Cl^- \rightarrow Cl_2(g) + 2e^-$, $E^0 = -1.36$ V, the oxidation of Br^- in this case is predicted to be the predominant reaction.

The possible cathode reactions are:

	E
$Li^+ + e^- \rightarrow Li(s)$	–3.05 V
$2H_2O(l) + 2e^- \rightarrow H_2(g) + 2OH^-(10^{-7}\ M)$	–0.42 V

Of all the species in Table 19.1, Li^+ is the most difficult to reduce. Thus, the reduction of H_2O is predicted.

The overall reaction is the sum of the anode and cathode reactions.

$2Br^- + 2H_2O \rightarrow Br_2(l) + H_2(g) + 2OH^-(10^{-7}\ M)$

(b) Anode: $2H_2O(l) \rightarrow O_2(g) + 4H^+(10^{-7}\ M) + 4e^-$

Cathode: $Ag^+ + e^- \rightarrow Ag(s)$

Overall: $4Ag^+ + 2H_2O(l) \rightarrow 4Ag(s) + O_2(g) + 4H^+(10^{-7}\ M)$

The platinum electrodes are inert. Therefore, only reactions of solute ions and solvent need be considered.

The possible anode reaction is:

	E
$2H_2O(l) \rightarrow O_2(g) + 4H^+(10^{-7}\ M) + 4e^-$	–0.82 V

The NO_3^- will not be oxidized, since N in NO_3^- is already in the +5 oxidation state, which is the highest state possible for N.

The possible cathode reactions are:

	E
$Ag^+ + 1e^- \rightarrow Ag(s)$	+0.80 V
$2H_2O(l) + 2e^- \rightarrow H_2(g) + 2OH^-(10^{-7}\ M)$	–0.42 V

From the half–cell potentials, Ag^+ would be predicted to be reduced preferentially.

The overall reaction is the sum of the anode and cathode reactions:

$4Ag^+ + 2H_2O(l) \rightarrow 4Ag(s) + O_2(g) + 4H^+(10^{-7}\ M)$

(c) Anode: $2H_2O(l) \rightarrow O_2(g) + 4H^+(10^{-7}\ M) + 4e^-$

Cathode: $2H_2O(l) + 2e^- \rightarrow H_2(g) + 2OH^-(10^{-7}\ M)$

Overall: $2H_2O(l) \rightarrow O_2(g) + 2H_2(g)$

The platinum electrodes are inert. Therefore, only reactions of solute ions and solvent need be considered.

The possible anode reaction is:

	E
$2H_2O(l) \rightarrow O_2(g) + 4H^+(10^{-7}\ M) + 4e^-$	–0.82 V

The nitrate ion will not be oxidized because N in nitrate ion is already in its highest oxidation state.

The possible cathode reactions are:

	E
$K^+ + e^- \rightarrow K(s)$	–2.93 V
$2H_2O + 2e^- \rightarrow H_2(g) + 2OH^-(10^{-7}\ M)$	–0.42 V

The half–cell potentials indicate that water will be reduced rather than K^+. This is not surprising, since K(s) reacts spontaneously and rapidly with water to form K^+ and $H_2(g)$.

The overall reaction is the sum of the anode and cathode reactions (the electrolysis of water):

$$2H_2O(l) \rightarrow O_2(g) + 2H_2(g)$$

(d) Anode: $Ni(s) \rightarrow Ni^{2+} + 2e^-$

Cathode: $Ni^{2+} + 2e^- \rightarrow Ni(s)$

No net reaction occurs; nickel is transferred from anode to cathode.

In this reaction, the electrode is *not* inert.

The possible anode reactions are:

	E
$Ni(s) \rightarrow Ni^{2+} + 2e^-$	+0.26 V
$2H_2O(l) \rightarrow O_2(g) + 4H^+(10^{-7}\ M) + 4e^-$	–0.82 V

From the half–cell potentials, the Ni(s) is predicted to be oxidized. The HSO_4^- ion would not be oxidized, because S is already in its highest oxidation state (+6) in the hydrogen sulfate ion.

The possible cathode reactions are:

	E
$Ni^{2+} + 2e^- \rightarrow Ni(s)$	–0.26 V
$2H_2O + 2e^- \rightarrow H_2(g) + 2OH^-(10^{-7}\ M)$	–0.42 V

The half–cell potentials indicate that Ni^{2+} will be preferentially reduced.

No net reaction occurs; nickel is transferred from anode to cathode.

19.16 0.771 g

The equation for the reduction of Ni^{2+} to Ni(s) is:

$Ni^{2+} + 2e^- \rightarrow Ni(s)$

Two moles of electrons are needed to reduce one mole of Ni^{2+} ions to one mole of Ni(s).

Use equation 19.11

coulombs = (amperes)(seconds) or C = (A)(s)

and the relationship 1 mol $e^- \approx 9.649 \times 10^4$ C (the Faraday constant, F)

$$\left(\frac{0.563\ \text{C}}{\text{s}}\right)(1.25\ \text{h})\left(\frac{60\ \text{min}}{1\ \text{h}}\right)\left(\frac{60\ \text{s}}{\text{min}}\right)\left(\frac{1\ \text{mol e}^-}{9.649 \times 10^4\ \text{C}}\right) = 2.63 \times 10^{-2}\ \text{mol e}^-$$

From the number of moles of electrons used and the equation for the half–reaction, calculate the grams of nickel deposited:

$$(2.63 \times 10^{-2}\ \text{mol e}^-)\left(\frac{1\ \text{mol Ni}}{2\ \text{mol e}^-}\right)\left(\frac{58.69\ \text{g Ni}}{\text{mol Ni}}\right) = 0.772\ \text{g Ni}$$

(Best answer, without intermediate round off, is 0.771 g.)

19.17 0.966 A

The equation is

$Ag^+ + e^- \rightarrow Ag(s)$

First convert mass of Ag deposited to moles of e^- used:

$$(0.324\ \text{g Ag})\left(\frac{1\ \text{mol Ag}}{107.9\ \text{g Ag}}\right)\left(\frac{1\ \text{mol e}^-}{\text{mol Ag}}\right) = 3.00 \times 10^{-3}\ \text{mol e}^-$$

Next calculate the number of coulombs represented by the moles of electrons using the Faraday constant:

$$(3.00 \times 10^{-3}\ \text{mol e}^-)\left(\frac{9.649 \times 10^4\ \text{C}}{\text{mol e}^-}\right) = 2.89 \times 10^2\ \text{C}$$

Using the relationship $A = \frac{C}{s}$, calculate the amperes necessary:

$$A = \frac{C}{s} = \left(\frac{289\ \text{C}}{5.00\ \text{min}}\right)\left(\frac{1\ \text{min}}{60\ \text{s}}\right) = 0.963\ \frac{C}{s}$$

(Best answer, without intermediate round off, is 0.966 A.)

Additional Practice Problems

19.55 1.08 g Cu, 2.24 g Au

$$(3.68\ \text{g Ag})\left(\frac{1\ \text{mol Ag}}{107.9\ \text{g Ag}}\right)\left(\frac{1\ \text{mol Cu}}{2\ \text{mol Ag}}\right)\left(\frac{63.55\ \text{g Cu}}{1\ \text{mol Cu}}\right) = 1.08\ \text{g Cu}$$

$$(3.68\ \text{g Ag})\left(\frac{1\ \text{mol Ag}}{107.9\ \text{g Ag}}\right)\left(\frac{1\ \text{mol Au}}{3\ \text{mol Ag}}\right)\left(\frac{197.0\ \text{g Au}}{1\ \text{mol Au}}\right) = 2.24\ \text{g Au}$$

19.57 **(a)** stable

From Table 19.1,

$F_2(g) + 2e^- \rightarrow 2F^-$ $E^0 = +2.87$ V
$Ag^+ + e^- \rightarrow Ag(s)$ $E^0 = +0.80$ V

From their positions in the table, Ag^+ and F^- will not spontaneously undergo a redox reaction with each other. (For the reaction, $2Ag^+ + 2F^- \rightarrow F_2(g) + Ag(s)$, $E^0 = +0.80\ \text{V} - (+2.87\ \text{V}) = -2.07$ V.)

(b) stable

From Table 19.1,

$Br_2(l) + 2e^- \rightarrow 2Br^-$ $E^0 = +1.07$ V
$Fe^{3+} + 3e^- \rightarrow Fe(s)$ $E^0 = -0.04$ V

From their positions in the table, Fe^{3+} and Br^- will not spontaneously undergo a redox reaction with each other. (For the reaction, $2Fe^{3+} + 6Br^- \rightarrow 3Br_2(l) + 2Fe(s)$, $E^0 = -0.04\ \text{V} - (+1.07\ \text{V}) = -1.11$ V.)

(c) redox reaction

From Table 19.1,

$Fe^{3+} + e^- \rightarrow Fe^{2+}$ $E^0 = +0.77$ V
$I_2(s) + 2e^- \rightarrow 2I^-$ $E^0 = +0.54$ V

From their positions in the table, Fe^{3+} and I^- will spontaneously undergo a redox reaction with each other. (For the reaction, $2Fe^{3+} + 2I^- \rightarrow I_2(s) + 2Fe^{2+}$, $E^0 = +0.77\ \text{V} - (+0.54\ \text{V}) = +0.23$ V.)

19.59 neodymium (Nd)

$$(0.3001\ \text{A})(0.7500\ \text{h})\left(\frac{60\ \text{min}}{1\ \text{h}}\right)\left(\frac{60\ \text{s}}{1\ \text{min}}\right) = 810.3\ \text{coulombs (C)}$$

$$(810.3\ \text{C})\left(\frac{1\ \text{mol e}^-}{9.6485 \times 10^4\ \text{C}}\right)\left(\frac{1\ \text{mol metal}}{3\ \text{mol e}^-}\right) = 2.799 \times 10^{-3}\ \text{mol metal}$$

$$\frac{0.4036\ \text{g metal}}{2.799 \times 10^{-3}\ \text{mol metal}} = 144.2\ \frac{\text{g}}{\text{mol}}$$

The atomic mass of the metal is 144.2 u. From the periodic table, the element with atomic mass 144.2 u is Nd, neodymium.

19.61 (1) $B^+ + e^- \rightarrow B$
(2) $A^+ + e^- \rightarrow A$
(3) $C^+ + e^- \rightarrow C$

From the first reaction, A is more reactive, i.e., more easily oxidized than B. Therefore, B^+ is more easily reduced than A^+. From the second reaction, A^+ is more easily reduced than C^+. The third reaction confirms that B^+ is more easily reduced than C^+. Therefore, $B^+ > A^+ > C^+$.

Stop and Test Yourself

1. **(d)** Zn(s)

From Table 19.1 the half–reactions that involve the species in this question are:

$Cl_2(g) + 2e^- \rightarrow 2Cl^-$	$E^0 = 1.36$ V
$I_2(s) + 2e^- \rightarrow 2I^-$	$E^0 = 0.54$ V
$2H^+ + 2e^- \rightarrow H_2(g)$	$E^0 = 0.00$ V
$Zn^{2+} + 2e^- \rightarrow Zn(s)$	$E^0 = -0.76$ V

In Table 19.1, reducing agents are in the right–hand column and the strongest reducing agents are at the bottom. Therefore, Zn(s) is the strongest reducing agent.

2. **(d)** Mn^{2+}

Of the species listed, only Mn^{2+} is to the right and above $Cr_2O_7^{2-}$ in Table 19.1.

3. **(c)** 0.44

half–cell of reduction: $MnO_4^- + 8H^+ + 5e^- \rightarrow Mn^{2+} + 4H_2O(l)$ $E^0 = 1.51$ V

half–cell of oxidation: $Br_2(l) + 2e^- \rightarrow 2Br^-$ $E^0 = 1.07$ V

$$E^0_{cell} = E^0_{\text{half-cell of reduction}} - E^0_{\text{half-cell of oxidation}} = E^0_{MnO_4^-/Mn^{2+}} - E^0_{Br_2(l)/Br^-}$$

$$= +1.51\text{ V} - (+1.07\text{ V}) = +0.44\text{ V}$$

4. **(e)** increasing $[Br^-]$

(a) no shift in equilibrium – concentration of *liquid* is constant

(b) shift to left – decreases emf

(c) shift to left – decreases emf

(d) shift to left – decreases emf

(e) shift to right – *increases* emf

5. **(b)** 0.0257

In 0.025 M $Au(s) \rightarrow Au^{3+}(aq) + 3e^-$ (anode, oxidation)
In 0.50 M $Au^{3+}(aq) + 3e^- \rightarrow Au(s)$ (cathode, reduction)

$E^0_{cell} = E^0_{half\text{–}cell\ red} - E^0_{half\text{–}cell\ ox} = 0$ (since both half–cells are the same)

Reaction is spontaneous, so E^0_{cell} must be +. For E to be +, ln Q must be – and Q must be < 1.

Therefore, $E = \frac{-0.0257}{3} \ln\left(\frac{0.025}{0.50}\right) = 0.0257$ V

6. **(b)** 1.64 V

$\Delta G = -nFE^0$ $\qquad 1\text{V} = 1\frac{\text{J}}{\text{C}}$

$$\left(\frac{10^3\ \text{J}}{\cancel{\text{kJ}}}\right)(-1268.2\ \cancel{\text{kJ}}) = (-8\ \cancel{\text{mol}})\left(\frac{9.65 \times 10^4\ \text{C}}{\cancel{\text{mol}}}\right)E^0$$

$$E^0 = 1.64\frac{\text{J}}{\text{C}} = 1.64\ \text{V}$$

7. **(c)** 6×10^{-10}

$$E^0 = \left(\frac{0.0257\ \text{V}}{n}\right)\ln K$$

$$-0.272\ \text{V} = \left(\frac{0.0257\ \text{V}}{2}\right)\ln K$$

$$-21.2 = \ln K$$

$$6 \times 10^{-10} = K$$

8. **(a)** $H_2(g)$ is produced at the cathode and $Cl_2(g)$ at the anode.

The possible anode reactions are:

$2Cl^- \rightarrow Cl_2(g) + 2e^-$ $\quad E^0 = -1.36$ V

$2H_2O \rightarrow O_2(g) + 4H^+(10^{-7}\ M) + 4e^-$ $\quad E^0 = -0.82$ V

Predicting the formation of $O_2(g)$ from these voltages is correct *but* $O_2(g)$ formation has an unusually high overpotential, so $Cl_2(g)$ is produced.

At the cathode the two possible half–cell reactions are:

$2H_2O(l) + 2e^- \rightarrow H_2(g) + 2OH^-(10^{-7}\ M)$ $\quad E^0 = -0.42$ V

$Ca^{2+} + 2e^- \rightarrow Ca(s)$ $\quad E^0 = -2.84$ V

Because the formation of $H_2(g)$ at the cathode has the less negative potential, water, not Ca^{2+}, will be reduced.

9. (c) 1.02 V

$Cu^{2+} + 2e^- \rightarrow Cu(s) \quad E^0 = +0.34\ V$

Copper(II) ion is reduced at the cathode.

$2Cl^- \rightarrow Cl_2(g) + 2e^- \quad E^0 = -1.36\ V$

Chlorine gas is produced at the anode because of the high overpotential for $O_2(g)$. The voltage to be applied must be greater than the sum of the two half–cell voltages: $(0.34 + (-1.36)) = -1.02\ V$

10. (c) 0.17 g

$$(0.25\ A)\ (37\ min)\left(\frac{60\ s}{1\ min}\right) = 5.6 \times 10^2\ C$$

$\left(\text{Remember that } A = \frac{C}{s}.\right)$

$$(5.6 \times 10^2\ \cancel{C})\left(\frac{1\ \cancel{mol\ e^-}}{9.65 \times 10^4\ \cancel{C}}\right)\left(\frac{1\ \cancel{mol\ Ni}}{2\ \cancel{mol\ e^-}}\right)\left(\frac{58.7\ g\ Ni}{1\ \cancel{mol\ Ni}}\right) = 0.17\ g\ Ni$$

Putting Things Together

19.63 (a)

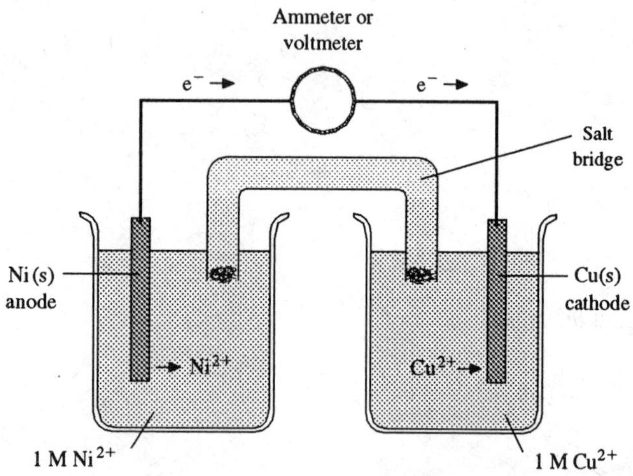

anode: $Ni(s) \rightarrow Ni^{2+} + 2e^-$ cathode: $Cu^{2+} + 2e^- \rightarrow Cu(s)$

(b) $Ni(s) + Cu^{2+} \rightarrow Ni^{2+} + Cu(s)$

19.65 (a) -1.5×10^4 J ($Cu^{2+} \rightarrow Cu^+$), -5.0×10^4 J ($Cu^+ \rightarrow Cu$)

$\Delta G^0 = -nFE^0$

$= -(1)(9.65 \times 10^4\ \cancel{C})(0.16\ J/\cancel{C})$

$= -1.5 \times 10^4\ J\ (Cu^{2+} \rightarrow Cu^+)$

$\Delta G^0 = -(1)(9.65 \times 10^4 \text{ C})(0.52 \text{ J/C})$

$= -5.0 \times 10^4 \text{ J } (Cu^+ \rightarrow Cu)$

(b) 0.34 V

For the reaction

$Cu^{2+} + 2e^- \rightarrow Cu(s)$,

$\Delta G^0 = (-1.5 \times 10^4 \text{ J}) + (-5.0 \times 10^4 \text{ J}) = -6.5 \times 10^4 \text{ J/mol}$

E^0 can be calculated from ΔG^0:

$\Delta G^0 = -nFE^0$

$-6.5 \times 10^4 \text{ J} = -(2)(9.65 \times 10^4 \text{ C})\, E^0$

$E^0 = 0.34 \frac{\text{J}}{\text{C}} = 0.34 \text{ V}$

(c) Take the weighted average of the two half–cell potentials $E^0{}_1$ and $E^0{}_2$: $E^0 = \frac{n_1E^0{}_1 + n_2E^0{}_2}{n_1 + n_2}$

Given half–reactions 1 and 2,

$\Delta G^0{}_1 = -n_1 FE^0{}_1$

$\Delta G^0{}_2 = -n_2 FE^0{}_2$.

Since, overall, $\Delta G^0 = -nFE^0$,

$\Delta G^0 = \Delta G^0{}_1 + \Delta G^0{}_2 = F(-n_1E^0{}_1 - n_2E^0{}_2)$

Thus, $-nFE^0 = (-n_1E^0{}_1 - n_2E^0{}_2)F$. Since $n = n_1 + n_2$

$\frac{n_1E^0{}_1 + n_2E^0{}_2}{n_1 + n_2} = E^0$

19.67 –48 kJ

First, calculate the standard cell potential, $E^0{}_{cell}$. Then, use the Nernst equation to find E. Finally, calculate ΔG from E.

For the reaction

$Cl_2(g) + 2Br^- \rightarrow Br_2(l) + 2Cl^-$

$E^0{}_{cell} = 1.36 \text{ V} - (1.07 \text{ V}) = 0.29 \text{ V}$

$E = 0.29 \text{ V} - \frac{0.0257 \text{ V}}{2} \ln Q$

where $Q = \frac{[Cl^-]^2}{p_{Cl_2} \cdot [Br^-]^2} = \frac{(0.50)^2}{(1.00)(0.100)^2} = 25$

$$E = 0.29\text{ V} - \frac{0.0257\text{ V}}{2}\ln 25$$

$$= 0.29\text{ V} - 0.0414\text{ V} = 0.25\text{ V}$$

$$\Delta G = -2(9.65 \times 10^4 \cancel{C})(0.25\text{ J}/\cancel{C})$$

$$= -4.8 \times 10^4\text{ J} = -48\text{ kJ}$$

19.69 (a) 0.00 V

For the reaction

$Cr_2O_7^{2-} + Cl^- \rightarrow Cr^{3+} + Cl_2$

$E^0_{cell} = 1.36\text{ V} - (1.36\text{ V}) = 0.00\text{ V}$

(b) No. No reaction will take place.

Because $E^0 = 0.00$ V, no reaction will take place under standard conditions. Therefore, this reaction is not a good way to oxidize Cl^- to Cl_2 under standard conditions.

(c) remove $Cl_2(g)$ as formed or increase $[H^+]$

The best ways to increase the percent conversion of Cl^- to Cl_2 would be to remove $Cl_2(g)$ as formed and to increase the concentration of H^+. In the equation for the reaction

$$Cr_2O_7^{2-} + 14H^+ + 6Cl^- \rightarrow 2Cr^{3+} + 7H_2O(l) + 3Cl_2(g)$$

the coefficient of H^+ is very high. Increasing the concentration of Cl^- would also help shift the equilibrium to the right. An increase in the concentration of $Cr_2O_7^{2-}$ would be relatively ineffective because the coefficient of this species in the equation is only one.

19.71 (a) +0.828 V

Consider the voltaic cell as a concentration cell,

anode: $H_2(g) \rightarrow 2H^+ + 2e^-$ $[H^+] = 1.00 \times 10^{-14}$

cathode: $2H^+ + 2e^- \rightarrow H_2(g)$ $[H^+] = 1.00$

$$E = 0 - \left(\frac{0.0257}{2}\right)\ln\left[\frac{(1.00 \times 10^{-14})^2}{(1.00)^2}\right] = +0.828\text{ V}$$

(b) E_{cell} will be lower.

NH_3 is a weak base. Therefore, $[OH^-]$ will be lower and $[H^+]$ greater. Thus, there will be less driving force for reaction and E_{cell} will be lower.

(c) +0.688 V

Find [OH^-] in 1.00 M NH_3

$NH_3 + H_2O \rightleftharpoons NH_4^+ + OH^-$ $K_b = 1.8 \times 10^{-5}$

Let x = [OH^-] = [NH_4^+], [NH_3] = 1.00 – x

$$\frac{x^2}{1.00 - x} = 1.8 \times 10^{-5}$$

Assume x negligible compared with 1.00

$x^2 = 1.8 \times 10^{-5}$

$x = 4.2 \times 10^{-3}$

$[H^+](4.2 \times 10^{-3}) = 1.00 \times 10^{-14}$

$[H^+] = 2.4 \times 10^{-12}$

$$E = 0.00 - \frac{0.0257}{2} \ln\left[\frac{(2.4 \times 10^{-12})^2}{(1.00)^2}\right] = +0.688\ \text{V}$$

19.73 $1.602\ 177 \times 10^{-19}$ C/e^-

$$\left(\frac{9.648\ 531 \times 10^4\ \text{C}}{1\ \text{mol}\ e^-}\right)\left(\frac{1\ \text{mol}\ e^-}{6.022\ 137 \times 10^{23}\ e^-}\right) = 1.602\ 177 \times 10^{-19}\ \frac{\text{C}}{e^-}$$

19.75 **(a)** 1.23 V

$O_2(g) + 4H^+ + 4e^- \rightarrow 2H_2O(l)$ $E^0 = 1.23$ V

(b) 1.0 atm

(c) 1.22 V

$$E = 1.23 - \frac{0.0257}{4} \ln\left(\frac{1}{0.21}\right)$$

$E = 1.23 - 0.0100 = 1.22$ V

(d) $CH_3COOH(aq) \rightleftharpoons CH_3COO^- + H^+$

$H_2O(l) + H^+ \rightleftharpoons H_3O^+$

(e)

	Table 15.3	*Table 19.1*
Top left:	most easily ionized acids	most easily reduced species
Bottom right:	strongest bases (best H^+ acceptors)	strongest reducing agents (best electron donors)
Reactions:	acid–base involve transfer of H^+	redox involve transfer of e^-

A similar diagonal relationship can be used to make predictions from both tables.

19.77 **(a)** $E^0 = (T\Delta S^0 - \Delta H^0)/nF$

$\Delta G^0 = \Delta H^0 - T\Delta S^0$
$\Delta G^0 = -nFE^0$ ⟶ $\Delta H^0 - T\Delta S^0 = -nFE^0$

$$\frac{(\Delta H^0 - T\Delta S^0)}{-nF} = E^0$$

$$\text{or} \quad \frac{T\Delta S^0 - \Delta H^0}{nF} = E^0$$

(b) small value of ΔS^0

ΔH^0 is approximately constant with respect to temperature. Thus, as T changes, a small value of ΔS^0 will minimize the change in E^0.

19.79 **(a)** 7×10^{-7}

Cell is a concentration cell, so $E^0 = 0$

Let $x = [Ag^+] = [Br^-]$

$$0.305 \text{ V} = 0 - \frac{0.0257}{1} \ln\left(\frac{x}{0.100}\right)$$

$$-11.9 = \ln x - \ln (0.100)$$

$$= \ln x + 2.303$$

$$-14.2 = \ln x$$

$x = 7 \times 10^{-7} = [Ag^+] = [Br^-]$ in saturated AgBr

(b) 5×10^{-13}

The solubility equilibrium for AgBr(s) is $AgBr(s) \rightleftharpoons Ag^+ + Br^-$ and

$$K_{sp} = [Ag^+][Br^-] = (7 \times 10^{-7})^2$$

$$= 5 \times 10^{-13}$$

(From Table 16.6, K_{sp} for AgBr is 5×10^{-13} at 25 °C.)

(c)

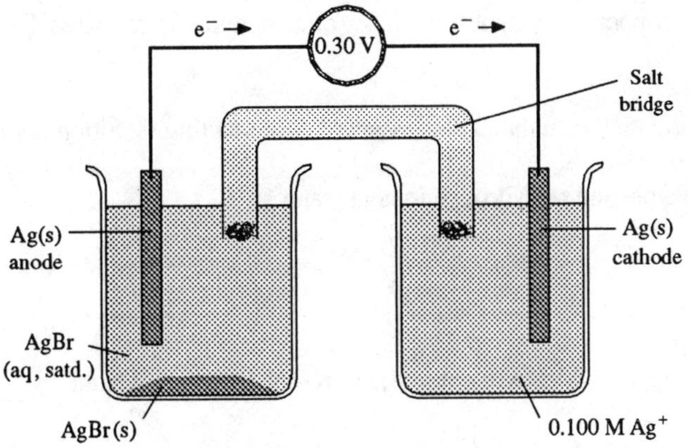

19.81 (a) –3164.6 kJ for $4Al(s) + 3O_2(g) \rightarrow 2Al_2O_3(s)$, –636.64 kJ for $2Zn(s) + O_2(g) \rightarrow 2ZnO(s)$

Equation	$4Al(s)$	+	$3O_2(g)$	$\rightarrow$	$2Al_2O_3(s)$
ΔG_f^0, kJ/mol	0		0		–1582.3

$$\Delta G^0_{rxn} = (2 \text{ mol } Al_2O_3)\left(\frac{-1582.3 \text{ kJ}}{\text{mol } Al_2O_3}\right) = -3164.6 \text{ kJ}$$

Equation	$2Zn(s)$	+	$O_2(g)$	$\rightarrow$	$2ZnO(s)$
ΔG_f^0, kJ/mol	0		0		–318.32

$$\Delta G^0_{rxn} = (2 \text{ mol } ZnO)\left(\frac{-318.32 \text{ kJ}}{\text{mol } ZnO}\right) = -636.64 \text{ kJ}$$

(b) 1.33×10^4 kJ/lb Al, 2.21×10^2 kJ/lb Zn

$$(454 \text{ g Al})\left(\frac{1 \text{ mol Al}}{26.98 \text{ g Al}}\right)\left(\frac{2 \text{ mol } Al_2O_3}{4 \text{ mol Al}}\right)\left(\frac{-3164.6 \text{ kJ}}{2 \text{ mol } Al_2O_3}\right) = -1.33 \times 10^4 \text{ kJ}$$

$$(454 \text{ g Zn})\left(\frac{1 \text{ mol Zn}}{65.39 \text{ g Zn}}\right)\left(\frac{2 \text{ mol ZnO}}{2 \text{ mol Zn}}\right)\left(\frac{-636.64 \text{ kJ}}{2 \text{ mol ZnO}}\right) = -2.21 \times 10^2 \text{ kJ}$$

(c) $E^0 = 2.73$ V for the Al battery, $E^0 = 1.65$ V for the Zn battery

$$\Delta G^0 = -nFE^0$$

and for the Al battery, $$E^0 = \frac{\Delta G^0}{-nF} = \frac{-3164.6 \text{ kJ}}{(-12 \text{ mol})\left(9.65 \times 10^4 \frac{\text{C}}{\text{mol}}\right)}\left(\frac{10^3 \text{ J}}{1 \text{ kJ}}\right)$$

$$= 2.73 \frac{\text{J}}{\text{C}} = 2.73 \text{ V}$$

and for the Zn battery, $$E^0 = \frac{\Delta G^0}{-nF} = \frac{-636.64 \text{ kJ}}{(-4 \text{ mol})\left(9.65 \times 10^4 \frac{\text{C}}{\text{mol}}\right)}\left(\frac{10^3 \text{ J}}{1 \text{ kJ}}\right)$$

$$= 1.65 \frac{\text{J}}{\text{C}} = 1.65 \text{ V}$$

19.83 (a) Bubbles are formed on the surface of both copper and magnesium. The solution remains colorless. Part of the magnesium strip is eaten away.

(b) Conclusions: a reaction is taking place (formation of bubbles) but copper is not reacting (solution not blue).

(c) Copper simply conducts electrons from magnesium to hydrogen ions in water.

Applications

19.85 The iron will act as a sacrificial anode for the copper, which will cause the iron pipe to spring a leak.

19.87 2.8×10^6 C

$$(2.36\ \text{g Zn})\left(\frac{1\ \text{mol Zn}}{65.39\ \text{g Zn}}\right)\left(\frac{2\ \text{mol } e^-}{1\ \text{mol Zn}}\right)\left(\frac{9.649 \times 10^4\ \text{C}}{1\ \text{mol } e^-}\right) = 6.96 \times 10^3\ \text{C}$$

$$0.0025 = \frac{6.96 \times 10^3\ \text{C}}{?\ \text{C}}$$

$$?\ \text{C} = 2.8 \times 10^6\ \text{C}$$

19.89 (a) $Cd(s) \rightarrow Cd(OH)_2(s)$ (anode), $NiO_2(s) \rightarrow Ni(OH)_2(s)$ (cathode)

Oxidation takes place at the anode, so $Cd(s) \rightarrow Cd(OH)_2(s)$ is the anode reaction. Cd(s) is oxidized to Cd^{2+}.

(b)
$$\begin{array}{ll} NiO_2(s) + 2H_2O(l) + 2e^- \rightarrow Ni(OH)_2(s) + 2OH^- & E^0 = -0.49\ \text{V} \\ Cd(s) + 2OH^- \rightarrow Cd(OH)_2(s) + 2e^- & E^0 = 0.82\ \text{V} \\ \hline Cd(s) + NiO_2(s) + 2H_2O(l) \rightarrow Cd(OH)_2(s) + Ni(OH)_2(s) & \end{array}$$

(c) +0.33 V

$$E_{\text{cell}} = E_{\text{half-cell of reduction}} - E_{\text{half-cell of oxidation}}$$

where $E_{\text{half-cell of oxidation}}$ is the *reduction* potential.

Thus, $E_{\text{cell}} = -0.49\ \text{V} - (-0.82\ \text{V}) = 0.33\ \text{V}$

19.91 (a) 11.4 s

The half-reaction of reduction is

$$Al^{3+} + 3e^- \rightarrow Al(s)$$

$$(106\ \text{g Al})\left(\frac{1\ \text{mol Al}}{26.98\ \text{g Al}}\right)\left(\frac{3\ \text{mol } e^-}{1\ \text{mol Al}}\right)\left(\frac{9.649 \times 10^4\ \text{C}}{1\ \text{mol } e^-}\right) = 1.14 \times 10^6\ \text{C}$$

$$1.14 \times 10^6\ \text{C} = (1.00 \times 10^5\ \text{A}) \cdot \text{s}$$

$$\text{s} = 11.4$$

(b) 35.4 g C (all to CO_2), 70.8 g C (all to CO)

$2Al_2O_3(\text{cryolite}) + 3C(s) \rightarrow 4Al(l) + 3CO_2(g)$

$$(106\ \text{g Al})\left(\frac{1\ \text{mol Al}}{26.98\ \text{g Al}}\right)\left(\frac{3\ \text{mol C}}{4\ \text{mol Al}}\right)\left(\frac{12.01\ \text{g C}}{1\ \text{mol C}}\right) = 35.4\ \text{g C}$$

$A_2O_3(\text{cryolite}) + 3C(s) \rightarrow 2Al(l) + 3CO(g)$

$$(106\ \text{g Al})\left(\frac{1\ \text{mol Al}}{26.98\ \text{g Al}}\right)\left(\frac{3\ \text{mol C}}{2\ \text{mol Al}}\right)\left(\frac{12.01\ \text{g C}}{1\ \text{mol C}}\right) = 70.8\ \text{g C}$$

(c) 67.9%

Let x = fraction C to CO_2; then $1 - x$ = fraction C to CO and

$35.4x + 70.8(1 - x) = 45.0$

$x = 0.729$ and $1 - x = 0.271$

Check:

$$\begin{aligned} (35.4\ \text{g})(0.729) &= 25.8\ \text{g C to } CO_2 \\ (70.8\ \text{g})(0.271) &= \underline{19.2\ \text{g C to CO}} \\ &\ 45.0\ \text{g C} \end{aligned}$$

$$(25.8\ \text{g C})\left(\frac{44.01\ \text{g } CO_2}{12.01\ \text{g C}}\right) = 94.5\ \text{g } CO_2$$

$$(19.2\ \text{g C})\left(\frac{28.01\ \text{g CO}}{12.01\ \text{g C}}\right) = \underline{44.8\ \text{g CO}}$$

$$139.3\ \text{g total}$$

$$\frac{94.5\ \text{g } CO_2}{139.3\ \text{g total}} \times 100 = 67.8\%\ CO_2 \text{ in gas}$$

(Best answer, without intermediate round off, is 67.9%.)

(d) 98.7%

The theoretical yield of aluminum is

$$(203\ \text{g } Al_2O_3)\left(\frac{1\ \text{mol } Al_2O_3}{102.0\ \text{g } Al_2O_3}\right)\left(\frac{4\ \text{mol Al}}{2\ \text{mol } Al_2O_3}\right)\left(\frac{26.98\ \text{g Al}}{1\ \text{mol Al}}\right) = 107\ \text{g Al}$$

The percent yield of aluminum is

$$\frac{106\ \text{g Al}}{107\ \text{g Al}} \times 100 = 99.1\%\ \text{yield}$$

(Best answer, without intermediate round off, is 98.7%.)

19.93 **(a)** 1.01×10^4 A

$$(15.0\ \text{kg NaOH})\left(\frac{10^3\ \text{g}}{1\ \text{kg}}\right)\left(\frac{1\ \text{mol NaOH}}{40.00\ \text{g NaOH}}\right)\left(\frac{2\ \text{mol } e^-}{2\ \text{mol NaOH}}\right)\left(\frac{9.649 \times 10^4\ \text{C}}{1\ \text{mol } e^-}\right) = 3.62 \times 10^7\ \text{C}$$

$$(1\ \cancel{h})\left(\frac{60\ \cancel{min}}{1\ \cancel{h}}\right)\left(\frac{60\ s}{1\ \cancel{min}}\right) = 3.600 \times 10^3\ s$$

coulombs (C) = amps (A) × seconds (s)

$$A = \frac{C}{s} = \frac{3.62 \times 10^7\ C}{3.600 \times 10^3\ s} = 1.01 \times 10^4\ A$$

(b) 9.8 kg/h

$$65\% = \frac{x}{15.0\ kg} \times 100$$

$x = 0.65 \times 15.0\ kg = 9.8\ kg/h$

(c) The rest of the electrical energy is probably converted to thermal energy.

19.95 **(a)** 0.76 V

$$(0.050\ 00\ \cancel{L})\left(\frac{0.010\ 00\ mol\ Fe^{2+}}{1\ \cancel{L}}\right) = 5.000 \times 10^{-4}\ mol\ Fe^{2+}$$

$$(0.010\ 00\ \cancel{L})\left(\frac{0.020\ 00\ mol\ Ce^{4+}}{1\ \cancel{L}}\right) = 2.000 \times 10^{-4} mol\ Ce^{4+}$$

$$\overline{3.000 \times 10^{-4}\ mol\ Fe^{2+}\ left}$$

2.000×10^{-4} mol Fe^{3+} are formed, since the reaction is rapid and complete. (Otherwise, it would not be used for titration.)

$$E = 0.77\ V - \frac{0.257}{1}\ \ln\left[\frac{\left(\frac{3.000 \times 10^{-4}\ \cancel{mol\ Fe^{2+}}}{6.000 \times 10^{-2}\ \cancel{L}}\right)}{\left(\frac{2.000 \times 10^{-4}\ \cancel{mol\ Fe^{3+}}}{6.000 \times 10^{-2}\ \cancel{L}}\right)}\right] = 0.76\ V$$

Note: Before the equivalence point, there is no significant amount of Ce^{4+} left. This is why $Fe^{3+} + e^- \rightarrow Fe^{2+}$, $E^0 = 0.77$ V, is used.

(b) 12.50 mL

$$(0.050\ 00\ \cancel{L})\left(\frac{0.010\ 00\ mol\ Fe^{2+}}{1\ \cancel{L}}\right) = 5.000 \times 10^{-4}\ mol\ Fe^{2+}\ at\ start$$

At the half–equivalence point, 2.500×10^{-4} mol Fe^{2+} will be left, and 2.500×10^{-4} mol Fe^{2+} has reacted to form 2.500×10^{-4} mol Fe^{3+}. To react with 2.500×10^{-4} mol Fe^{2+}, we need 2.500×10^{-4} mol Ce^{4+}, since the ratio is 1:1 according to the equation.

$$(2.500 \times 10^{-4}\ \cancel{mol\ Ce^{4+}})\left(\frac{1\ L}{2.000 \times 10^{-2}\ \cancel{mol\ Ce^{4+}}}\right) = 0.012\ 50\ L\ or\ 12.50\ mL\ of\ 2.000 \times 10^{-2}\ M\ Ce^{4+}\ added$$

(c) 0.77 V

Since mol Fe^{2+} left = mol Fe^{3+} formed, and ln 1.000 = 0, $E = 0.77$ V.

(d) Solutions of reducing agents are oxidized by O_2 in air, and thus their concentrations change.

19.97 **(a)** ultraviolet region

$$(352\ \cancel{nm})\left(\frac{10^{-9}\ m}{\cancel{nm}}\right) = 3.52 \times 10^{-7}\ m$$

According to Figure 7.15 (a), a wavelength of 3.52×10^{-7} m is in the ultraviolet region.

(b) $O_3(g) + 3I^- + 2H^+ \rightarrow O_2(g) + I_3^- + H_2O(l)$

$$\begin{array}{c} O_3(g) + 2H^+ + \cancel{2e^-} \rightarrow O_2(g) + H_2O(l) \\ 3I^- \rightarrow I_3^- + \cancel{2e^-} \\ \hline O_3(g) + 3I^- + 2H^+ \rightarrow O_2(g) + I_3^- + H_2O(l) \end{array}$$

(c) 1.54 V

$$E^0_{cell} = E^0_{half\text{-}cell\ of\ reduction} - E^0_{half\text{-}cell\ of\ oxidation} = 2.08\ V - (0.54\ V) = 1.54\ V$$

(d) 1.14 V

$$E = E^0 - \frac{0.0257}{n} \ln Q, \qquad Q = \frac{p_{O_2}\,[I_3^-]}{p_{O_3}\,[H^+]^2[I^-]^3}$$

$$E = 1.54\ V - \frac{0.0257}{2} \ln\left[\frac{1 \cdot 1}{1 \cdot (1.58 \times 10^{-7})^2 \cdot 1^3}\right]$$

$$E = 1.54\ V - 0.402\ V = 1.14\ V$$

(e) $H_2O_2(aq) + 3I^- + 2H^+ \rightarrow 2H_2O(l) + I_3^-$

$Cl_2(g) + 3I^- \rightarrow 2Cl^- + I_3^-$

$$\begin{array}{c} 2H^+ + H_2O_2(aq) + \cancel{2e^-} \rightarrow 2H_2O(l) \\ 3I^- \rightarrow I_3^- + \cancel{2e^-} \\ \hline H_2O_2(aq) + 3I^- + 2H^+ \rightarrow 2H_2O(l) + I_3^- \end{array}$$

$$\begin{array}{c} Cl_2(g) + \cancel{2e^-} \rightarrow 2Cl^- \\ 3I^- \rightarrow I_3^- + \cancel{2e^-} \\ \hline Cl_2(g) + 3I^- \rightarrow 2Cl^- + I_3^- \end{array}$$

CHAPTER 20:
Nuclear Chemistry

Practice Problems

20.1 $^{222}_{86}\text{Rn} \rightarrow {}^{218}_{84}\text{Po} + {}^{4}_{2}\text{He}$

$^{222}_{86}\text{Rn} \rightarrow {}^{A}_{Z}X + {}^{4}_{2}\text{He}$

$222 = A + 4 \qquad A = 222 - 4 = 218$

$86 = Z + 2 \qquad Z = 86 - 2 = 84$

Z = 84 corresponds to polonium.

Therefore, $^{222}_{86}\text{Rn} \rightarrow {}^{218}_{84}\text{Po} + {}^{4}_{2}\text{He}$

20.2 **(a)** beta decay – an electron is given off.

(b) positron emission – a positron is given off.

(c) spontaneous fission – two pieces are formed from the radionuclide and neutrons are given off.

(d) alpha decay – a helium nucleus is given off.

(e) electron capture – an electron is captured by the nucleus.

(f) beta decay – an electron is given off.

20.3 **(a)** $^{139}_{57}\text{La}$

$^{139}_{58}\text{Ce} + ^{0}_{-1}\text{e} \rightarrow ^{139}_{57}X, \quad X = ^{139}_{57}\text{La}$ – electron capture

(b) $^{23}_{11}\text{Na}$

$^{23}_{12}\text{Mg} \rightarrow ^{23}_{11}X + ^{0}_{+1}\text{e}, \quad ^{23}_{11}X = ^{23}_{11}\text{Na}$ – positron emission

(c) $^{176}_{77}\text{Ir}$

$^{176}_{77}X \rightarrow ^{172}_{75}\text{Re} + ^{4}_{2}\text{He}, \quad ^{176}_{77}X = ^{176}_{77}\text{Ir}$ – alpha decay

20.4 **(a)** $^{245}_{100}\text{Fm} \rightarrow ^{241}_{98}\text{Cf} + ^{4}_{2}\text{He}$

$^{245}_{100}\text{Fm} \rightarrow ^{241}_{98}X + ^{4}_{2}\text{He}, \quad ^{241}_{98}X = ^{241}_{98}\text{Cf}$

(b) $^{251}_{100}\text{Fm} + ^{0}_{-1}\text{e} \rightarrow ^{251}_{99}\text{Es}$

$^{251}_{100}\text{Fm} + ^{0}_{-1}\text{e} \rightarrow ^{251}_{99}X, \quad ^{251}_{99}X = ^{251}_{99}\text{Es}$

(c) $^{18}_{7}\text{N} \rightarrow ^{18}_{8}\text{O} + ^{0}_{-1}\text{e}$

$^{18}_{7}\text{N} \rightarrow ^{18}_{8}X + ^{0}_{-1}\text{e}, \quad ^{18}_{8}X = ^{18}_{8}\text{O}$

(d) $^{14}_{8}\text{O} \rightarrow ^{14}_{7}\text{N} + ^{0}_{+1}\text{e}$

$^{14}_{8}\text{O} \rightarrow ^{14}_{7}X + ^{0}_{+1}\text{e}, \quad ^{14}_{7}X = ^{14}_{7}\text{N}$

(e) $^{258}_{102}\text{No} \rightarrow ^{120}_{47}\text{Ag} + ^{135}_{55}\text{Cs} + 3^{1}_{0}\text{n}$

20.5 target: $^{27}_{13}\text{Al}$, particle ejected: $^{1}_{0}\text{n}$ (neutron),

projectile: $^{4}_{2}\text{He}$ (alpha ray), product nuclide: $^{30}_{15}\text{P}$

20.6 **(a)** (i) $^{239}_{94}\text{Pu} + ^{2}_{1}\text{H} \rightarrow ^{240}_{95}\text{Am} + ^{1}_{0}\text{n}$

(iv) $^{139}_{58}\text{Ce} + ^{1}_{0}\text{n} \rightarrow ^{140}_{58}\text{Ce} + \gamma$

(v) $^{239}_{94}\text{Pu} + ^{4}_{2}\text{He} \rightarrow ^{242}_{96}\text{Cm} + ^{1}_{0}\text{n}$

(b) (iv) $^{139}_{58}\text{Ce} + ^{1}_{0}\text{n} \rightarrow ^{140}_{58}\text{Ce} + \gamma$

20.7 **(a)** 0.17 g

The time, 221.4 d, is three half–lives

Time, d	g
0	1.36
73.8	0.68
147.6	0.34
221.4	0.17

(b) 0.22 g

$$t_{1/2} = \frac{0.693}{k}, \quad k = \frac{0.693}{t_{1/2}} = \frac{0.693}{73.8\text{ d}} = 0.009\ 39\text{ d}^{-1}$$

$$\ln\frac{N_t}{N_0} = -kt, \quad \ln N_t - \ln N_0 = -kt$$

$$\ln N_t = \ln N_0 - kt = \ln 1.36 - (0.009\ 39\ \cancel{\text{d}^{-1}})(192.4\ \cancel{\text{d}})$$

$$= \ln 1.36 - 1.81$$

$$= -1.50$$

$$N_t = 0.22\text{ g}$$

20.8 **(a)** $93\ \frac{\text{events}}{\text{s}}$

$$(6.0\ \cancel{\text{y}})\left(\frac{365\ \cancel{\text{d}}}{\cancel{\text{y}}}\right)\left(\frac{24\ \cancel{\text{h}}}{\cancel{\text{d}}}\right)\left(\frac{60\ \cancel{\text{min}}}{\cancel{\text{h}}}\right)\left(\frac{60\text{ s}}{\cancel{\text{min}}}\right) = 1.9 \times 10^8\text{ s}$$

$$k = \frac{0.693}{1.9 \times 10^8\text{ s}} = 3.6 \times 10^{-9}\text{ s}^{-1} \cdot \text{atom}^{-1}$$

$$(8.2\ \cancel{\text{pg Os–194}})\left(\frac{10^{-12}\ \cancel{\text{g}}}{1\ \cancel{\text{pg}}}\right)\left(\frac{1\ \cancel{\text{mol Os–194}}}{194\ \cancel{\text{g Os–194}}}\right)\left(\frac{6.02 \times 10^{23}\text{ atom Os–194}}{1\ \cancel{\text{mol Os–194}}}\right) = 2.5 \times 10^{10}\text{ atom}$$

$$\text{rate} = \left(\frac{3.6 \times 10^{-9}\text{ events}}{\text{s} \cdot \cancel{\text{atom}}}\right)(2.5 \times 10^{10}\ \cancel{\text{atom}}) = 9.0 \times 10^1\ \frac{\text{events}}{\text{s}}$$

(Best answer, without intermediate round off, is 93 events/s.)

(b) 1.4×10^{16} atoms

$$\text{rate} = kN, \quad N = \frac{\text{rate}}{k} = \frac{5.0 \times 10^7\ \frac{\cancel{\text{disintegrations}}}{\cancel{\text{s}}}}{3.6 \times 10^{-9}\ \cancel{\text{disintegrations}} \cdot \cancel{\text{s}^{-1}} \cdot \text{atom}^{-1}}$$

$$N = 1.4 \times 10^{16}\text{ atoms}$$

(c) 4.4 μg

$$(1.4 \times 10^{16} \text{ \sout{atoms}})\left(\frac{1 \text{ \sout{mol}}}{6.02 \times 10^{23} \text{ \sout{atoms}}}\right)\left(\frac{194 \text{ g}}{1 \text{ \sout{mol}}}\right) = 0.000\,045 \text{ g}$$

$$(0.000\,0045 \text{ g})\left(\frac{10^{6}\ \mu\text{g}}{\text{g}}\right) = 4.5\ \mu\text{g}$$

(Best answer, without intermediate round off in parts (a) and (b), is 4.4 μg.)

20.9 (a) probably stable

The atomic mass of Ba–138 is nearly identical to the atomic mass of naturally occurring barium (137.327 u). Since Ba–138 has an even number of both protons and neutrons, it is probably stable.

(b) probably radioactive, beta emission

The atomic mass of Ne–24, which is close to 24, is higher than the atomic mass of naturally occurring neon (20.1797). Ne–24 has too many neutrons and will probably decay by beta emission.

(c) probably radioactive, electron capture

The atomic mass of Pb–196, which is close to 196, is lower than the atomic mass of naturally occurring lead (207.2). Pb–196 has too few neutrons and, since it has a high atomic number, it will probably undergo electron capture.

(d) probably radioactive, beta decay or alpha decay

The atomic mass of Pb–214, which is close to 214, is higher than the atomic mass of naturally occurring lead (207.2). Thus, Pb–214 has too many neutrons and is probably radioactive. Beta decay would reduce the neutron:proton ratio. Since Pb–214 is very near the top of the stability band, it may also decay by alpha emission.

(e) probably radioactive, alpha decay

The atomic number of radium (88) is beyond the band of stability, and thus Ra–226 will probably decay by alpha emission.

(According to the Table of the Isotopes in the CRC Handbook, all these predictions are correct except that lead–214 does not alpha decay.)

20.10 fluorine–17

It is farther from the band of stability than fluorine–18. The only stable isotope of fluorine is fluorine–19.

20.11 (a) –0.000 167 u

Assume nuclear masses = atomic masses.

$$\Delta m = m_{\text{products}} - m_{\text{reactants}}$$

$$= 14.003\,074 \text{ u} - 14.003\,241 \text{ u} = -0.000\,167 \text{ u}$$

(b) Yes. ΔE is negative

Since Δm is – and $\Delta E = \Delta mc^2$, ΔE is –.

(c) $-1.50 \times 10^7 \dfrac{\text{kJ}}{\text{mol}}$

$\Delta E = \Delta mc^2$

$$= \left(\frac{-1.67 \times 10^{-4}\,\cancel{\text{u}}}{\cancel{\text{atom}}}\right)\left(\frac{1.661 \times 10^{-27}\text{ kg}}{\cancel{\text{u}}}\right)\left(\frac{6.022 \times 10^{23}\,\cancel{\text{atoms}}}{\text{mol}}\right)\left(\frac{2.998 \times 10^8\text{ m}}{\text{s}}\right)^2$$

$$= \left(\frac{-1.50 \times 10^{10}\,\cancel{\text{kg}} \cdot \cancel{\text{m}^2}}{\cancel{\text{s}^2} \cdot \text{mol}}\right)\left(\frac{1\,\cancel{\text{J}}}{\cancel{\text{kg}} \cdot \cancel{\text{m}^2} \cdot \cancel{\text{s}^{-2}}}\right)\left(\frac{1\text{ kJ}}{1000\,\cancel{\text{J}}}\right) = -1.50 \times 10^7 \frac{\text{kJ}}{\text{mol}}$$

20.12 **(a)** $-0.528\ 46$ u

Mass number – number of protons = number of neutrons
56 – 26 = 30

mass of H-1 = 1.007 825 u, mass of neutron = 1.008 665 u

mass defect = atomic mass – (mass of protons + mass of neutrons)

$$\text{mass defect} = 55.934\ 939\text{ u} - \left[26\,\cancel{\text{H}}\left(\frac{1.007\ 825\text{ u}}{\cancel{\text{H}}}\right) + 30\,\cancel{\text{n}}\left(\frac{1.008\ 665\text{ u}}{\cancel{\text{n}}}\right)\right]$$

$$= 55.934\ 939\text{ u} - 26.203\ 45\text{ u} - 30.259\ 95\text{ u}$$

$$= -0.528\ 46\text{ u}$$

(b) -7.8868×10^{-11} J

$\Delta E = \Delta mc^2$

$$= (-0.528\ 46\,\cancel{\text{u}})\left(\frac{1.660\ 54 \times 10^{-27}\text{ kg}}{\cancel{\text{u}}}\right)\left(\frac{2.997\ 92 \times 10^8\text{ m}}{\text{s}}\right)^2$$

$$= -7.8868 \times 10^{-11}\text{ J}$$

(c) $-1.4084 \times 10^{-12} \dfrac{\text{J}}{\text{nucleon}}$

$$\frac{-7.8868 \times 10^{-11}\text{ J}}{56\text{ nucleons}} = -1.4084 \times 10^{-12} \frac{\text{J}}{\text{nucleon}}$$

20.13 2.50×10^3 y

$$\ln\frac{N_t}{N_0} = -kt, \quad t_{1/2} = \frac{0.693}{k}, \quad k = \frac{0.693}{t_{1/2}}$$

$$k = \frac{0.693}{5715\text{ y}} = 1.21 \times 10^{-4}\text{ y}^{-1}$$

$$\frac{N_t}{N_0} = \frac{11.0\ \cancel{\text{disintegrations}}}{14.9\ \cancel{\text{disintegrations}}} = 0.738$$

$$t = \frac{\ln\frac{N_t}{N_0}}{-k} = \frac{\ln 0.738}{-(1.21 \times 10^{-4}\text{ y}^{-1})} = \frac{-0.304}{-(1.21 \times 10^{-4}\text{ y}^{-1})} = 2.51 \times 10^3\text{ y}$$

(Best answer, without intermediate round off, is 2.50×10^3 y.)

Additional Practice Problems

20.59 $^{9}_{4}\text{Be} + ^{4}_{2}\text{He} \rightarrow ^{12}_{6}\text{C} + ^{1}_{0}\text{n}$

Since we know the atomic and mass numbers of both reactants and that one of the products is a neutron, we can determine the atomic and mass number of the second product.

$^{9}_{4}\text{Be} + ^{4}_{2}\text{He} \rightarrow ^{A}_{Z}X + ^{1}_{0}\text{n}$

atomic number: $4 + 2 = Z + 0, \quad Z = 6$

mass number: $9 + 4 = A + 1, \quad A = 12$

The second product is $^{12}_{6}\text{C}$.

20.61

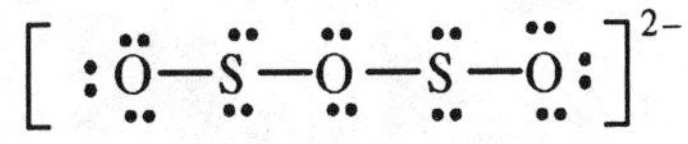

The chain structure below is eliminated by the studies with S–35. If it had been the correct structure, the radionuclide would have been evenly distributed between the two sulfur positions:

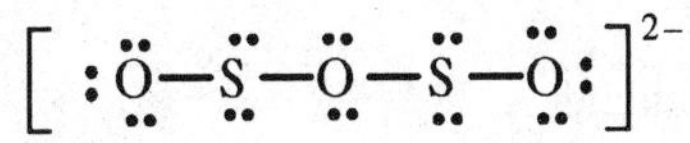

O—S*—O—S—O and O—S—O—S*—O

(where * indicates radioactive)

Upon treatment with acid, the ion would have an equal probability (or nearly equal) of fragmenting in either direction. This would result in 50% of the ^{35}S in the sulfur and 50% in the SO_2.

Stop and Test Yourself

1. **(d)** The energy changes that accompany nuclear reactions are of the order of $10–10^3$ kJ/mol.

Even very small changes in mass produce (or require) quantities of energy on the order of 10^8 kJ/mol.

2. **(e)** $^{194}_{82}Pb$

$$^{198}_{84}Po \xrightarrow[\alpha\ decay]{} {}^{A}_{Z}X + {}^{4}_{2}He$$

mass number of X: $198 = A + 4$, $A = 194$

$X = {}^{194}_{82}Pb$

atomic number of X: $84 = Z + 2$, $Z = 82$

3. **(d)** $^{185}_{79}Au$

	$^{20}_{12}Mg$	$^{45}_{19}K$	$^{52}_{26}Fe$	$^{185}_{79}Au$	$^{197}_{79}Au$
No. of neutrons	too few	too many	too few	too few	stable
Mode of decay	positron or electron capture	beta	positron or electron capture	positron, electron capture, or *alpha*	

4. **(c)** $^{204}_{83}Bi$

	$^{8}_{5}B$	$^{13}_{5}B$	$^{204}_{83}Bi$	$^{209}_{83}Bi$	$^{215}_{83}Bi$
No. of neutrons	too few	too many	too few	stable	too many
Mode of decay	positron or electron capture	beta	positron or electron capture		beta

Electron capture is more likely for $^{204}_{83}Bi$ than for $^{8}_{5}B$ because of Bi's high atomic number.

5. **(c)** d

Note from Table 20.1 that d (for deuteron) is the particle symbol for the $^{2}_{1}H$ nuclide.

6. **(e)** 2.34

The time, 7.86 h, is three half–lives.

Time, h	µg Si–31	µg P–31
0	2.68	0
2.62	1.34	1.34
5.24	0.67	2.01
7.86	0.34	2.34

7. **(d)** 301

When 85.0% of sample has decayed, 15%, or 15 atoms of 100, remain.

$$k = \frac{0.693}{109.8 \text{ min}} = 6.31 \times 10^{-3} \text{ min}^{-1}$$

$$\ln\left(\frac{15}{100}\right) = -(6.31 \times 10^{-3})t$$

$$\frac{-1.90}{-6.31 \times 10^{-3}} = t = 301 \text{ min}$$

8. **(b)** 1.01×10^3

To solve this problem, we must convert the mass of K–40 to atoms and the $t_{1/2}$ to seconds.

$$(3.91 \text{ mg K–40})\left(\frac{1 \text{ mmol K–40}}{40.0 \text{ mg K–40}}\right)\left(\frac{6.022 \times 10^{20} \text{ atoms}}{1 \text{ mmol K–40}}\right) = 5.89 \times 10^{19} \text{ atoms}$$

$$(1.28 \times 10^9 \text{ y})\left(\frac{365 \text{ d}}{1 \text{ y}}\right)\left(\frac{24 \text{ h}}{1 \text{ d}}\right)\left(\frac{60 \text{ min}}{1 \text{ h}}\right)\left(\frac{60 \text{ s}}{1 \text{ min}}\right) = 4.04 \times 10^{16} \text{ s}$$

Next, we must determine the rate constant k:

$$k = \frac{0.693}{4.04 \times 10^{16} \text{ s}} = 1.72 \times 10^{-17} \text{ s}^{-1}$$

Now, the rate can be calculated:

$$\text{rate} = Nt$$

$$= (5.89 \times 10^{19} \text{ atoms})(1.72 \times 10^{-17} \text{ events/s} \cdot \text{atom})$$

$$= 1.01 \times 10^3 \text{ events/s}$$

9. **(b)** $^{63}_{29}Cu$

	$^{58}_{29}Cu$	$^{63}_{29}Cu$	$^{64}_{29}Cu$	$^{68}_{29}Cu$	$^{72}_{36}Kr$
No. of neutrons	too few	stable n/p	stable n/p	too many	too few
No. of protons/ No. of neutrons		odd/even	odd/odd		

Since isotopes with odd numbers of both protons *and* neutrons are extremely rare, $^{63}_{29}Cu$ is more likely to be stable than $^{64}_{29}Cu$.

10. **(c)** 0.186 953

	mass number	–	number of protons	=	number of neutrons
For Na–22	22	–	11	=	11

The mass deficit is

$$21.994\,437\text{ u} - [11\text{ H}(1.007\,825\text{ u}\cdot\text{H}^{-1}) + 11\text{ n}(1.008\,665\text{ u}\cdot\text{n}^{-1})]$$

$$= 21.994\,437\text{ u} - (11.086\,075\text{ u} + 11.095\,315\text{ u})$$

$$= -0.186\,953\text{ u}$$

11. **(b)** -1.82×10^{-12} J, spontaneous

$$\Delta m = 51.9405\text{ u} - (50.9440\text{ u} + 1.0087\text{ u})$$

$$= -0.0122\text{ u}$$

$$\Delta E = \Delta mc^2 = (-0.0122\text{ u})(1.661 \times 10^{-27}\text{ kg}\cdot\text{u}^{-1})(2.998 \times 10^8\text{ m}\cdot\text{s}^{-1})^2$$

$$= -1.82 \times 10^{-12}\text{ kg}\cdot\text{m}^2\cdot\text{s}^{-2} = -1.82 \times 10^{-12}\text{ J}$$

Since ΔE is negative, reaction is spontaneous.

12. **(d)** ${}^{3}_{2}\text{He} + {}^{4}_{2}\text{He} \rightarrow {}^{7}_{4}\text{Be} + \gamma$

Both reacting nuclides have a relatively low mass number. Fusion occurs to form ${}^{7}_{4}\text{Be}$.

13. **(e)** all of the above

14. **(c)** 4.5×10^3 y

$$\ln\left(\frac{N_t}{N_0}\right) = -kt, \quad \frac{N_t}{N_0} = \frac{8.6}{14.9} = 0.58$$

$$\ln(0.58) = -1.21 \times 10^{-4}\text{ y}^{-1}(t) \quad k = \frac{0.693}{5715\text{ y}} = 1.21 \times 10^{-4}\text{ y}^{-1}$$

$$t = \frac{\ln(0.58)}{-1.21 \times 10^{-4}\text{ y}^{-1}} = 4500\text{ y or } 4.5 \times 10^3\text{ y}$$

15. **(c)** H

93.4% of all atoms in the universe are estimated to be H.

Putting Things Together

20.63 Nuclear reactions involve protons and neutrons within the nucleus, whereas chemical reactions involve electrons outside the nucleus. In nuclear reactions, elements are transmuted into other elements, unlike chemical reactions, which have the same number of each kind of atom as both reactants and products. Different isotopes of an element will undergo different nuclear reactions, but behave similarly in chemical reactions. Unlike chemical reactions, nuclear reactions are independent of state of chemical combination. The energy changes of nuclear reactions are of the order of 10^8–10^9 kJ/mol; for chemical reactions, however, the energies are much smaller (10–10^3 kJ/mol). Changes in mass are detectable in nuclear reactions but undetectable in chemical reactions.

20.65 **(a)** In electron capture, one of the electrons, usually one from the first shell, becomes part of the nucleus. This leaves a hole in the first energy shell into which an electron from an outer shell falls. The transition from an outer energy level to $n = 1$ corresponds to radiation in the X–ray region of the spectrum.

(b) Nuclear reactions involve the release of energy from an unstable nucleus. Energy levels for nucleons exist in the nucleus similar to the energy levels of electrons outside the nucleus. Transitions between these nuclear energy levels correspond in energy to radiation in the gamma-ray region of the electromagnetic spectrum. The emission of γ rays is one way a nucleus with excess energy can relax to its ground state.

20.67 **(a)** ${}^{14}_{7}\text{N} + {}^{1}_{0}\text{n} \rightarrow {}^{3}_{1}\text{H} + {}^{12}_{6}\text{C}$

(b) 17.7 y

average lifetime $= \dfrac{1}{k}$, $\quad k = \dfrac{0.693}{t_{1/2}} = \dfrac{0.693}{12.26 \text{ y}} = 0.0565 \text{ y}^{-1}$

Therefore, average lifetime $= \dfrac{1}{0.0565 \text{ y}^{-1}}$

$= 17.7$ y

(c) ${}^{3}_{1}\text{H} \rightarrow {}^{3}_{2}\text{He} + {}^{0}_{-1}\text{e}$

(d) 9.80×10^{13} disintegrations

Calculate the # of ${}^{3}_{1}\text{H}$ atoms in the sample:

$$(1.00 \text{ g } {}^{3}\text{H}_2\text{O})\left(\frac{1 \text{ mol } {}^{3}\text{H}_2\text{O}}{22.03 \text{ g } {}^{3}\text{H}_2\text{O}}\right)\left(\frac{2 \text{ mol } {}^{3}_{1}\text{H}}{1 \text{ mol } {}^{3}\text{H}_2\text{O}}\right)\left(\frac{6.022 \times 10^{23} \text{ atoms}}{1 \text{ mol } {}^{3}_{1}\text{H}}\right) = 5.47 \times 10^{22} \text{ atoms}$$

Convert $t_{1/2}$ to seconds, calculate k:

$$(12.26 \text{ y})\left(\frac{365 \text{ d}}{1 \text{ y}}\right)\left(\frac{24 \text{ h}}{1 \text{ d}}\right)\left(\frac{60 \text{ min}}{1 \text{ h}}\right)\left(\frac{60 \text{ s}}{1 \text{ min}}\right) = 3.87 \times 10^8 \text{ s}$$

$$k = \frac{0.693}{3.87 \times 10^8 \text{ s}} = 1.79 \times 10^{-9} \text{ s}^{-1}$$

Calculate the rate:

rate $= k$ (# atoms)

$= (1.79 \times 10^{-9} \text{ s}^{-1})(5.47 \times 10^{22} \text{ atoms})$

$= 9.79 \times 10^{13}$ atoms disintegrated

(Best answer, without intermediate round off, is 9.80×10^{13} disintegrations.)

(e) 2.65×10^3 Ci

$$(9.80 \times 10^{13}\ \text{events s}^{-1})\left(\frac{1\ \text{Bq}}{1\ \text{event s}^{-1}}\right)\left(\frac{1\ \text{Ci}}{3.70 \times 10^{10}\ \text{Bq}}\right) = 2.65 \times 10^3\ \text{Ci}$$

(f) 1.20×10^{-4} disintegrations

Convert sample to number of ${}^3_1\text{H}$ atoms:

$$(1.00\ \text{g H}_2\text{O})\left(\frac{1\ \text{mol H}_2\text{O}}{18.02\ \text{g H}_2\text{O}}\right)\left(\frac{2\ \text{mol H}}{1\ \text{mol H}_2\text{O}}\right)\left(\frac{6.022 \times 10^{23}\ \text{H atoms}}{1\ \text{mol H}}\right)\left(\frac{1\ {}^3_1\text{H atom}}{1 \times 10^{18}\ \text{H atoms}}\right) = 6.68 \times 10^4\ {}^3_1\text{H atoms}$$

Calculate the rate:

$$\text{rate} = (1.79 \times 10^{-9}\ \text{s}^{-1})(6.68 \times 10^4\ \text{atoms})$$

$$= 1.20 \times 10^{-4}\ \text{atoms disintegrated}$$

20.69 2.88×10^9 y old

Since 1 U decays to 1 Pb, there must have been 10 627 atoms of U at $t = 0$.

$$k = \frac{0.693}{t_{1/2}} = \frac{0.693}{7.04 \times 10^8\ \text{y}} = 9.84 \times 10^{-10}\ \text{y}^{-1}$$

$$\ln\left(\frac{N_t}{N_0}\right) = -kt$$

$$\ln\left(\frac{627}{10\ 627}\right) = -(9.84 \times 10^{-10}\ \text{y}^{-1})\ t$$

$$2.88 \times 10^9\ \text{y} = t$$

20.71 NO-30: 99.39%; NO-31: 0.41%; NO-32: 0.20%; NO-33: 0.00%

Possible molecular formulas:

NO-30:	${}^{14}\text{N}^{16}\text{O}$
NO-31:	${}^{14}\text{N}^{17}\text{O}$ and ${}^{15}\text{N}^{16}\text{O}$
NO-32:	${}^{14}\text{N}^{18}\text{O}$ and ${}^{15}\text{N}^{17}\text{O}$
NO-33:	${}^{15}\text{N}^{18}\text{O}$

Calculate probabilities for each molecular combination:

NO-30: $(0.9963)(0.9976) = 0.993$ or 99.39%

NO-31: $(0.9963)(0.0004) = 0.0004$
$(0.0037)(0.9976) = \underline{0.0037}$
0.0041 or 0.41%

NO-32: $(0.9963)(0.0020) = 0.0020$
$(0.0037)(0.0004) = \underline{0.000\,0015}$
0.0020 or 0.20%

NO-33: $(0.0037)(0.0020) = 0.000\,0074$ or 0.00%

20.73 **(a)** metal

(b) IIB

(c) mercury

Placing elements in the periodic table in order of increasing atomic number, one continues across the seventh period from 109, which puts element–112 in Group IIB under mercury.

(d) highest state: +2, lowest state: +1

20.75 Prepare saturated solutions of labeled and unlabeled barium sulfate by stirring one milligram of barium sulfate with 100 mL of water overnight. From the labeled solution, remove the undissolved solid by filtration. Add the filtrate to the saturated solution of unlabeled barium sulfate and stir the mixture overnight. Then separate the undissolved solid by filtration, wash thoroughly to remove the mother liquor, and measure the radioactivity of the solid. If the solubility equilibrium is indeed dynamic, the solid will be radioactive because it will have exchanged some of its unlabeled sulfate ions for radioactive sulfate ions from the labeled solution.

20.77 **(a)** A *nuclide* is any atom defined by a specific atomic and mass number. A *nucleon* is a general term for the building blocks of the nucleus: neutrons and protons.

(b) *Gamma rays* are electromagnetic radiation of very high energy, typically released in nuclear reactions as a means for the nucleus to stabilize. *X–rays* are electromagnetic radiation of slightly less energy than gamma rays, typically released when electrons make transitions from higher to lower energy levels.

(c) *Atomic number* represents the number of protons in a nucleus. *Mass number* represents the total number of nucleons.

(d) *Fission* is the splitting of one heavy nucleus into two nuclei with smaller mass numbers. *Fusion* is the combining of two light nuclei to form a heavier, more stable nucleus.

(e) A *beta particle* is an electron and is negatively charged. A *positron* is a particle with the same mass as a beta particle but opposite in charge.

(f) A *rad* is a measure of the amount of energy deposited in tissue as radiation passes. A *rem* is a rad times a correction factor that takes into account the effectiveness of the radiation in causing biological damage.

(g) *Slow neutrons* have average kinetic energies similar to those that they would have if they existed as a gas at room temperature. *Fast neutrons* have higher kinetic energies.

20.79 reduction of IO_4^-

If IO_3^- were formed by the oxidation of $^{128}I^-$, the I in the IO_3^- would be ^{128}I, i.e. $^{128}IO_3^-$.

Applications

20.81 **(a)** 9.57×10^6 y

Assume a 10 000 atom sample to calculate *t*. One atom will be left:

$$\ln\left(\frac{1}{10\,000}\right) = -kt \quad k = \frac{0.693}{7.20 \times 10^5\ \text{y}} = 9.63 \times 10^{-7}\ \text{y}^{-1}$$

$$\frac{\ln(0.000\,100)}{-9.63 \times 10^{-7}\ \text{y}^{-1}} = t = 9.56 \times 10^6\ \text{y}$$

(Best answer, without intermediate round off, is 9.57×10^6 y.)

(b) ≈0%; none of the orginally formed Al is present

We have k and are given t, solve for $\frac{N_t}{N_0}$:

$$\ln\left(\frac{N_t}{N_0}\right) = -(9.63 \times 10^{-7}\ \cancel{\text{y}^{-1}})(4.6 \times 10^9\ \cancel{\text{y}}) = -4.4 \times 10^3$$

$$\frac{N_t}{N_0} = e^{-4400} \approx 0$$

20.83 **(a)** 7.7×10^{-13} %

Calculate the number of atoms ^2_1H available in seawater:

$$(1.36 \times 10^{21}\ \cancel{\text{kg } H_2O})\left(\frac{1000\ \cancel{\text{g}}}{1\ \cancel{\text{kg}}}\right)\left(\frac{1\ \cancel{\text{mol } H_2O}}{18.02\ \cancel{\text{g } H_2O}}\right)\left(\frac{2\ \cancel{\text{mol H}}}{1\ \cancel{\text{mol } H_2O}}\right)\left(\frac{6.022 \times 10^{23}\ \text{atoms}}{1\ \cancel{\text{mol H}}}\right) = 9.09 \times 10^{46}\ \text{atoms H}$$

$$9.09 \times 10^{46}\ \text{atoms H}\ (0.000\,15) = 1.4 \times 10^{43}\ \text{atoms}\ ^2_1H$$

Each fusion requires two ^2_1H atoms:

$$^2_1H + ^2_1H \rightarrow ^4_2He$$

$$E = \Delta mc^2$$

$$= [4.002\,60 - 2(2.0140)]\ \cancel{\text{u}}\left(\frac{1.661 \times 10^{-27}\ \text{kg}}{1\ \cancel{\text{u}}}\right)\left(\frac{2.998 \times 10^8\ \text{m}}{\text{s}}\right)^2$$

$$= -3.79 \times 10^{-12}\ \frac{\text{kg} \cdot \text{m}^2}{\text{s}^2}/\text{event} = -3.79 \times 10^{-12}\ \text{J/fusion event}$$

$$\text{Need } (1.98 \times 10^{14}\ \cancel{\text{kJ}})\left(\frac{10^3\ \cancel{\text{J}}}{1\ \cancel{\text{kJ}}}\right)\left(\frac{1\ \text{event}}{3.79 \times 10^{-12}\ \cancel{\text{J}}}\right) = 5.22 \times 10^{28}\ \text{fusion events}$$

To calculate the %:

$$\frac{2(5.22 \times 10^{28} \text{ atoms } {}^{2}_{1}\text{H required})}{1.4 \times 10^{43} \text{ atoms } {}^{2}_{1}\text{H available}} \times 100 = 7.5 \times 10^{-13}\ \%$$

(Best answer, without intermediate round off, is 7.7×10^{-13} %.)

(b) 1.3×10^{14} y

$$\frac{1.4 \times 10^{43} \text{ atoms available}}{2(5.22 \times 10^{28} \text{ atoms needed /y})} = 1.3 \times 10^{14} \text{ y}$$

20.85 **(a)**

Positron emission	β emission
P–28	P–32
P–29	P–33
P–30	P–34

The stable ratio of neutrons to protons in P–31 is $\frac{16}{15}$. In P–28 through P–30, the n/p ratio is less than $\frac{16}{15}$. Positron emission increases the n/p ratio to achieve stability. With P–32 through P–34, the n/p ratio is greater than $\frac{16}{15}$. β emission decreases the n/p ratio to achieve stability.

(b) P–32 and P–33

The half–lives of all the other radionuclides are too short for them to be useful as diagnostic tools in a biological system.

20.87 Gamma radiation is detected because it penetrates through more material than alpha or beta radiation can. Thus, the prospector can detect uranium buried deeply in the ground by detecting gamma radiation.

20.89 12 μg B–12

Let x = mg B–12 in 60 tablets.

$$\left(\frac{1.282\ \mu\text{Ci}}{\text{mg}}\right)(x + 0.395) \text{ mg} = \left(\frac{3.620\ \mu\text{Ci}}{\text{mg}}\right)(0.395 \text{ mg})$$

$$x = 0.720 \text{ mg}$$

$$\frac{0.720 \text{ mg}}{60 \text{ tablets}} = 0.012 \text{ mg/tablet or } 12\ \mu\text{g/tablet}$$

20.91 Fuel for fusion is plentiful and large quantities of radioactive wastes are not produced. However, release of useful amounts of energy by controlled fusion has not yet been achieved because of the extremely high temperatures required to initiate fusion. The major problem with fission as a source of energy is the radioactive wastes produced. Neither fission nor fusion produce harmful oxides of nitrogen and sulfur or large quantities of carbon dioxide as combustion does. Another problem with combustion is limited fuel supplies. However, fossil fuels are convenient and equipment for using them is readily available. Substitution of plant materials for fossil fuels would use up carbon dioxide produced by combustion in photosynthesis and provide a renewable energy source.

20.93 **(a)** lower

If Ar–40 escapes, less will be detected. Thus, when interpreting the age of a sample, the assumption will be that less K–40 has decayed because less Ar–40 is present. The actual age is much older then the age found.

(b) electron capture: ${}^{40}_{19}K + {}^{0}_{-1}e \rightarrow {}^{40}_{18}Ar$

positron emission: ${}^{40}_{19}K \rightarrow {}^{40}_{18}Ar + {}^{0}_{+1}e$

(c) beta decay: ${}^{40}_{19}K \rightarrow {}^{40}_{20}Ca + {}^{0}_{-1}e$

(d) K–40 has 21 (an odd number) neutrons and 19 (an odd number) protons. K–39 and K–41 both have even numbers of neutrons.

(e) All three types yield stable nuclides.

20.95 (a) ${}^{57}_{27}Co + {}^{0}_{-1}e \rightarrow {}^{57}_{26}Fe$, ${}^{60}_{27}Co \rightarrow {}^{60}_{28}Ni + {}^{0}_{-1}e$

	${}^{57}_{27}Co$	${}^{59}_{27}Co$	${}^{60}_{27}Co$
No. of neutrons	too few	stable	too many
Mode of decay	probably electron capture (possibly positron)		beta

(b) Co–57

Cobalt–57 probably decays by electron capture, and thus no particles are emitted. It also has a shorter half–life and emits lower energy gamma rays than Co–60. All of these properties are more desirable for purposes of diagnosis.

(c) Co–60

Cobalt–60 will be used because the gamma radiation produced is of a higher energy.

20.97 (a) 4.00 mCi

$$(1.481 \times 10^{8}\ \cancel{Bq})\left(\frac{1\ \cancel{Ci}}{3.70 \times 10^{10}\ \cancel{Bq}}\right)\left(\frac{10^{3}\ mCi}{1\ \cancel{Ci}}\right) = 4.00\ mCi$$

(b) 98.9% remains

Convert $t_{1/2}$ to minutes, and calculate k:

$$(6.02\ \cancel{h})\left(\frac{60\ min}{1\ \cancel{h}}\right) = 361\ min$$

$$k = \frac{0.693}{361\ min} = 1.92 \times 10^{-3}\ min^{-1}$$

Solve for ratio $\frac{N_t}{N_0}$:

$$\ln\left(\frac{N_t}{N_0}\right) = -(1.92 \times 10^{-3}\ \cancel{min^{-1}})(6.00\ \cancel{min}) = -0.0115$$

$$\frac{N_t}{N_0} = 0.989 \text{ or } 98.9\%$$

(c) 5.25×10^3 mL

Activity after 6 min if no dilution is

$$\left(\frac{1.481 \times 10^8 \text{ Bq}}{\text{mL}}\right)(0.989) = 1.46 \times 10^8 \frac{\text{Bq}}{\text{mL}}$$

Let x = volume person's blood in mL

$$\left(\frac{2.79 \times 10^4 \text{ Bq}}{\cancel{\text{mL}}}\right)(x\ \cancel{\text{mL}}) = 1.46 \times 10^8 \text{ Bq}$$

$x = 5.23 \times 10^3$ mL

(Best answer, without intermediate round off, is 5.25×10^3 mL.)

(d) 1.43×10^{-9} g $NaTcO_4$

rate = kN

$$1.481 \times 10^8 \text{ s}^{-1} = \left(\frac{1.92 \times 10^{-3}}{\cancel{\text{m}}}\right)\left(\frac{1\ \cancel{\text{m}}}{60 \text{ s}}\right)N$$

4.63×10^{12} atoms = N

$$(4.63 \times 10^{12}\ \cancel{\text{atoms Tc}})\left(\frac{1\ \cancel{\text{mol Tc}}}{6.022 \times 10^{23}\ \cancel{\text{atoms Tc}}}\right)\left(\frac{1\ \cancel{\text{mol NaTcO}_4}}{1\ \cancel{\text{mol Tc}}}\right)\left(\frac{185.9 \text{ g NaTcO}_4}{1\ \cancel{\text{mol NaTcO}_4}}\right) = 1.43 \times 10^{-9} \text{ g NaTcO}_4$$

(e) Prepare a solution of 1.00×10^{-5} g $NaTcO_4$ in 1.000 L.

$$(1.43 \times 10^{-9}\ \cancel{\text{g NaTcO}_4})\left(\frac{1\ \cancel{\text{L}}}{1.00 \times 10^{-5}\ \cancel{\text{g NaTcO}_4}}\right)\left(\frac{10^3 \text{ mL}}{\cancel{\text{L}}}\right) = 0.143 \text{ mL or } 143\ \mu\text{L}$$

Measure 143 μL of the solution with a 250–μL syringe and dilute to 1.00 mL. The new solution will have the desired concentration.

20.99 **(a)** as a catalyst

C–12 is acting as a catalyst. It is used up in the first step and re–formed in the last step. It is not consumed by the process

(b) ${}^{13}_{7}\text{N}$, ${}^{13}_{6}\text{C}$, ${}^{14}_{7}\text{N}$, ${}^{15}_{8}\text{O}$, ${}^{15}_{7}\text{N}$

(c) The energy released should be the same as that released when four H–1 nuclei fuse by the proton–proton chain, since the overall equation is the same. The difference in energy between reactants and products is independent of the path from reactants to products.

20.101 **(a)** ${}^{99}_{43}\text{Tc}^* \rightarrow {}^{99}_{43}\text{Tc} + {}^{0}_{0}\gamma$

${}^{99}_{43}\text{Tc} \rightarrow {}^{99}_{44}\text{Ru} + {}^{0}_{-1}\text{e}$

(b) 0.480 events/s

From Problem 20.97(d), 4.63×10^{12} atoms of Tc are administered.

$$2.12 \times 10^5 \text{ y} = \frac{0.693}{k}, \quad k = 3.27 \times 10^{-6} \text{ y}^{-1}$$

Rate = kN

$$= (3.27 \times 10^{-6} \text{ y}^{-1})(4.63 \times 10^{12} \text{ atoms})$$

$$= 1.51 \times 10^7 \text{ events/y}$$

$$1 \text{ y} = 3.154 \times 10^7 \text{ s}$$

$$\text{Rate} = \left(\frac{1.51 \times 10^7 \text{ events}}{\cancel{\text{y}}}\right)\left(\frac{1 \cancel{\text{y}}}{3.154 \times 10^7 \text{ s}}\right) = 0.479 \frac{\text{events}}{\text{s}} \text{ or } \frac{1}{0.479 \text{ events/s}} = 2.09 \text{ s/event}$$

(Best answer, without intermediate round off, is 0.480 events/s.)

(c) (i) 1.36×10^7 kJ/mol (ii) 8.79×10^{-12} m (iii) 1.51×10^{-4} g

(i) $$(0.141 \cancel{\text{MeV}})\left(\frac{6.022 \times 10^{23} \text{ photon}}{1 \text{ mol}}\right)\left(\frac{1.602 \times 10^{-13} \cancel{\text{J}}}{1 \cancel{\text{MeV}}}\right)\left(\frac{1 \text{ kJ}}{10^3 \cancel{\text{J}}}\right) = 1.36 \times 10^7 \text{ kJ/mol}$$

(ii) $E_{\text{photon}} = \dfrac{hc}{\lambda}$ (Section 7.6)

$$\lambda = \frac{hc}{E_{\text{photon}}}$$

$$= \frac{(6.626 \times 10^{-34} \text{ kg} \cdot \text{m}^2/\text{s})(2.998 \times 10^8 \text{ m/s})}{(0.141 \text{ MeV})(1.602 \times 10^{-13} \text{ J})}$$

$$= 8.79 \times 10^{-12} \text{ m}$$

$$= 8.79 \times 10^{-3} \text{ nm or } 8.79 \text{ pm}$$

(iii) $\Delta E = \Delta mc^2$

$$\frac{\Delta E}{c^2} = \Delta m = (m_{\text{Tc}} - m_{\text{Tc*}}) = \frac{1.36 \times 10^{10} \text{ kg} \cdot \text{m}^2 \cdot \text{s}^{-2}}{(2.998 \times 10^8 \text{ m} \cdot \text{s}^{-1})^2} = 1.51 \times 10^{-7} \text{ kg or } 1.51 \times 10^{-4} \text{ g}$$

The mass of a mole of $^{99}_{43}$Tc atoms is 1.51×10^{-4} g less than the mass of a mole of $^{99}_{43}$Tc* atoms.

CHAPTER 21:
A Closer Look at Inorganic Chemistry: Nonmetals and Semimetals and Their Compounds

Practice Problems

21.1 $CH_3CH_2CH_3(g) + 3H_2O(g) \xrightarrow[\text{heat}]{\text{catalyst}} 7H_2(g) + 3CO(g)$

21.2 **(a)** $[Ar]4s^2 + \begin{matrix} 1s^1 \\ 1s^1 \end{matrix} \rightarrow [Ar] + \begin{matrix} 1s^2 \text{ or } [He] \\ 1s^2 \text{ or } [He] \end{matrix}$

$Ca + 2H \rightarrow Ca^{2+} + 2H^-$

(b) $Ca(s) + H_2(g) \rightarrow CaH_2(s)$

(c) Net ionic: $H^- + H_2O(l) \rightarrow H_2(g) + OH^-$

Complete ionic: $Ca^{2+} + 2H^- + 2H_2O(l) \rightarrow 2H_2(g) + Ca^{2+} + 2OH^-$

Molecular: $CaH_2(s) + 2H_2O(l) \rightarrow 2H_2(g) + Ca(OH)_2(s)$

21.3 **(a)** H_2O, proton donor

(b) H^-, proton acceptor

21.4 (a) hydrogen in H^-

(b) one hydrogen in $H_2O(l)$

(c) H_2O

(d) H^-

21.5

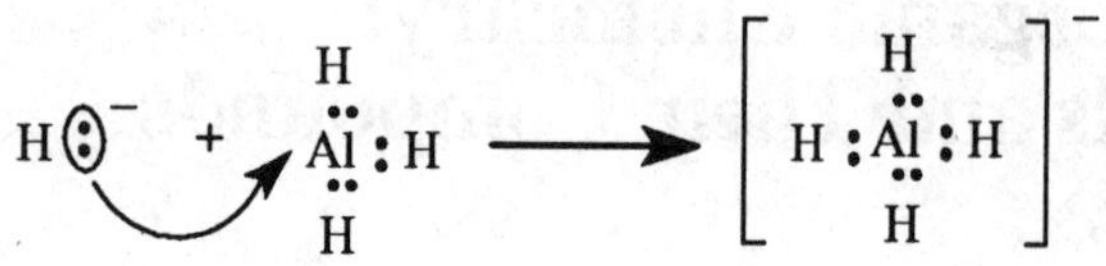

Lewis base   Lewis acid

21.6 (a) $CH_3CH{=}CH_2(g) + H_2(g) \xrightarrow{\text{catalyst}} CH_3CH_2CH_3(g)$

The product is propane, $CH_3CH_2CH_3$.

(b) $MoO_3(s) + 3H_2(g) \rightarrow Mo(s) + 3H_2O(g)$

Molybdenum is in the same group in the periodic table as tungsten and would, therefore, be expected to behave similarly. A high temperature would probably be required.

21.7

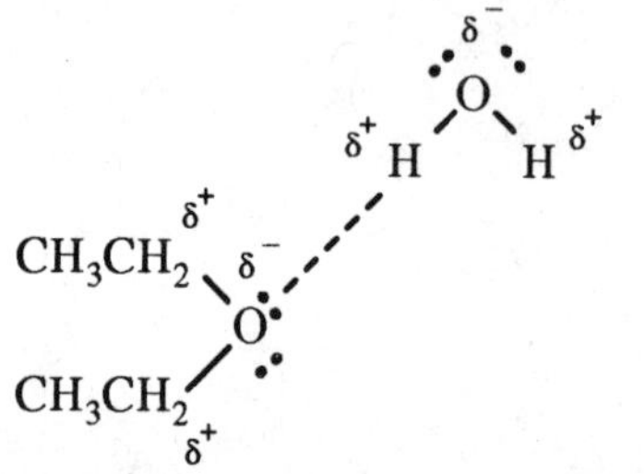

The critical parts of the sketch are that the hydrogen bond (represented by the dashed line) is in a straight line with the covalent bond connecting hydrogen to oxygen in the water molecule and that the hydrogen bond is longer than the covalent bond. Therefore, the following sketch is equally correct:

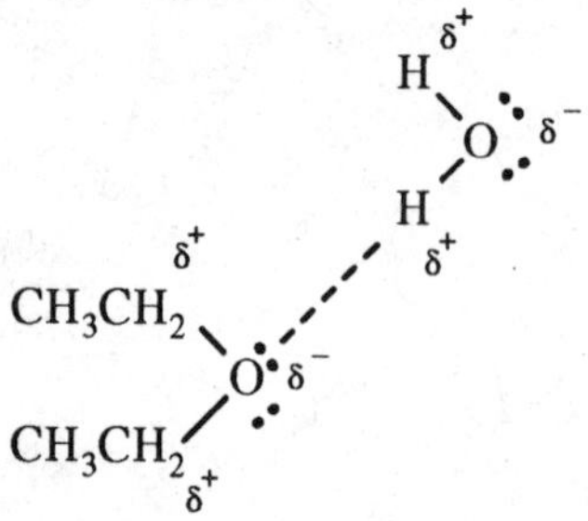

21.8 37%

The equilibrium vapor pressure for H_2O vapor at 20 °C = 17.5 mmHg according to Appendix C and Table 5.4.

The relative humidity at 20 °C = $\frac{6.5 \text{ mmHg}}{17.5 \text{ mmHg}} \times 100 = 37\%$

21.9 (a) $CaO(s) + H_2O(l) \rightarrow Ca(OH)_2(s)$

With base, CaO (a metal oxide), water reacts as an acid.

(b) $HNO_3(l) + H_2O(l) \rightarrow H_3O^+ + NO_3^-$

With strong acid, HNO_3, water reacts as a base.

(c) $2Na(s) + 2H_2O(l) \rightarrow 2NaOH(aq) + H_2(g)$

With strong reducing agent, Na, water reacts as an oxidizing agent.

(d) $CH_3COOH(l) + H_2O(l) \rightleftharpoons H_3O^+ + CH_3COO^-$

With weak acid, CH_3COOH, water reacts as a base; mostly molecules are present at equilibrium.

21.10 (a)

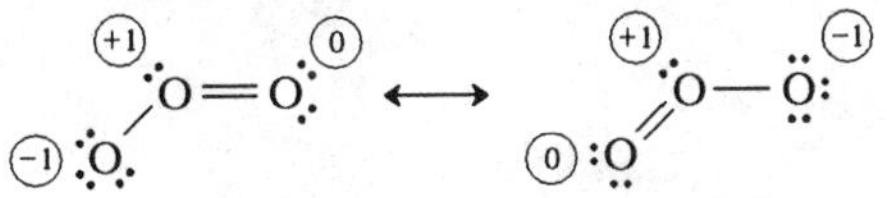

(b)

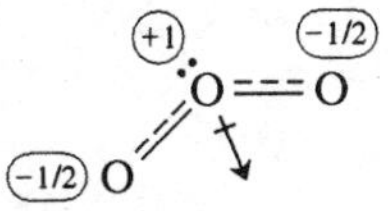

21.11 (a) $Pb^{2+} + H_2S(g) \rightarrow PbS(s) + 2H^+$

(b) $ZnS(s) + 2HCl(aq) \rightarrow ZnCl_2(aq) + H_2S(g)$

21.12 (a) potassium hydrogen sulfate

(b) iron(II) sulfide

(c) sulfite ion

21.13 (a) $MgSO_4 \cdot 7H_2O$

(b) $Ca(HSO_3)_2$

(c) K_2SO_4

21.14 (a) $2Al(s) + N_2(g) \xrightarrow{\text{heat}} 2AlN(s)$

(b) $NH_4Cl(s) + NaOH(aq) \rightarrow NaCl(aq) + NH_3(g) + H_2O(l)$

(c) $NH_3(g) + HCl(g) \rightarrow NH_4Cl(s)$

(d) $NH_4Cl(s) \xrightarrow{\text{heat}} NH_3(g) + HCl(g)$

(e) $Fe(s) + 4HNO_3(6\text{ M}) \rightarrow Fe(NO_3)_3(aq) + NO(g) + 2H_2O(l)$

(f) $CaCO_3(s) + 2HNO_3(6\ M) \rightarrow Ca(NO_3)_2(aq) + CO_2(g) + H_2O(l)$

21.15 (a) +5

$$\overset{+5\ -2}{N_2O_5}$$
$+10 - 10 = 0$

(b) +5

$$\overset{+5\ -2}{NO_3^-}$$
$+5 - 6 = -1$

(c) +4

$$\overset{+4\ -2}{N_2O_4}$$
$+8 - 8 = 0$

(d) +3

$$\overset{+3\ -2}{NO_2^-}$$
$+3 - 4 = -1$

(e) −3

$$\overset{-3\ +1\ -1}{NH_4Cl}$$
$-3 + 4 - 1 = 0$

21.16 (a) $\underset{\text{excess}}{2H_3PO_4(aq)} + K_2CO_3(aq) \rightarrow 2KH_2PO_4(aq) + CO_2(g) + H_2O(l)$

(b) $H_3PO_4(aq) + \underset{\text{excess}}{K_2CO_3(aq)} \rightarrow K_2HPO_4(aq) + CO_2(g) + H_2O(l)$

(c) $H_3PO_4(aq) + \underset{\text{excess}}{3KOH(aq)} \rightarrow K_3PO_4(aq) + H_2O(l)$

(d) $3Zn(s) + 2H_3PO_4(aq) \rightarrow Zn_3(PO_4)_2(s) + 3H_2(g)$

21.17 (a) $C(\text{graphite}) + 2S(g) \xrightarrow{\text{heat}} CS_2(g)$

(b) $MnO_2(s) + C(\text{graphite}) \xrightarrow{\text{heat}} Mn(l) + CO_2(g)$

(c) $CH_3COCH_3(l) + \underset{\text{excess}}{4O_2(g)} \xrightarrow{\text{spark}} 3CO_2(g) + 3H_2O(l)$
or $C_3H_6O(l)$

(d) $Na_2CO_3(s) + 2HNO_3(aq) \rightarrow 2NaNO_3(aq) + CO_2(g) + H_2O(l)$

Additional Practice Problems

21.19 The density of $H_2(g)$ is very low. Therefore, hydrogen gas was not held by the Earth's gravitational field and escaped into space.

21.23 **(a, b)** $2Na(s) + 2D_2O(l) \rightarrow 2NaOD(soln) + D_2(g)$

(c) $Cl_2(g) + D_2(g) \rightarrow 2DCl(g)$

21.27 **(a)** The *greenhouse effect* is caused by substances that are transparent to visible radiation but absorb infrared radiation. They let the sun warm the Earth during the day but prevent loss of thermal energy during the night.

(b) water, H_2O; carbon dioxide, CO_2; methane, CH_4

21.31 **(a)** Nitric acid is colorless but turns yellow as the result of the formation of $NO_2(g)$.

(b) Copper and silver are soluble in nitric acid because the nitrate ion in acidic solution is a powerful oxidizing agent.

21.35 The HPO_4^{2-} ion can behave as an acid or as a base:

addition of H^+: $HPO_4^{2-} + H^+ \rightleftharpoons H_2PO_4^-$

addition of OH^-: $HPO_4^{2-} + OH^- \rightleftharpoons PO_4^{3-} + H_2O$

21.39 two

Putting Things Together

21.43 **(a)** *Allotropes* are different forms of the same element that differ in bonding.

(b) O, S, P, C

(c) oxygen: $O_2(g)$, $O_3(g)$
sulfur: rhombic, monoclinic
phosphorus: white, red
carbon: diamond, graphite, buckyballs, and $\sim C\equiv C-C\equiv C-C\equiv C\sim$

(d) Silicon does not readily form double bonds.

In graphite, carbon is doubly bonded. Because silicon is in the third period and has a larger radius, it does not form double bonds well.

21.47 **(a)**

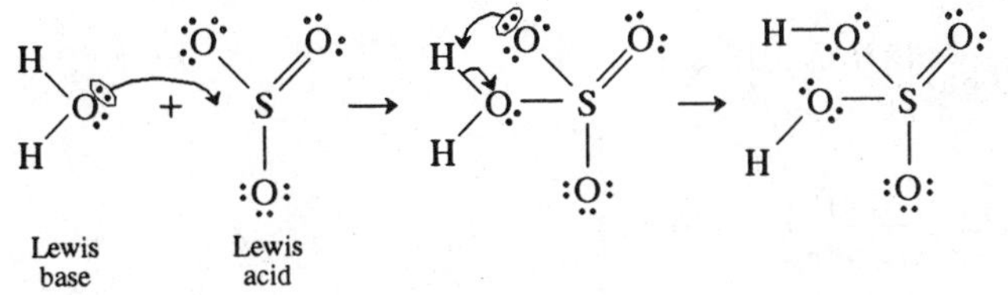

(b) NH_2^- + H_2O $\longrightarrow$ NH_3 + OH^-
B–L base   B–L acid

21.51 **(a)** covalent bonds

Silicon forms a covalent network crystal like diamond that is held together by covalent bonds.

(b) electrostatic attraction

Magnesium oxide forms ionic crystals held together by strong electrostatic attractive forces.

(c) London forces

Rhombic sulfur is composed of molecules consisting of rings of eight sulfur atoms. These molecules are nonpolar and are held together by London forces in a molecular crystal.

(d) hydrogen bonds

Ice is composed of water molecules held in fixed positions by hydrogen bonding.

(e) London forces

Molecules of carbon disulfide are nonpolar and are held together by London forces.

21.55 **(a)** one

(b) sodium hypophosphite monohydrate

(c) +1

$$\overset{+1}{H_3}\overset{+1}{P}\overset{-2}{O_2}$$

$$+3 + 1 - 4 = 0$$

$$\overset{+1}{Na}\overset{+1}{H_2}\overset{+1}{P}\overset{-2}{O_2} \cdot \overset{+1}{H_2}\overset{-2}{O}$$

$$+1 + 2 + 1 - 4 + 2 - 2 = 0$$

21.59 **(a)** $T = 937.5$ K for $SnO_2(s) + 2C(s) \rightarrow Sn(s) + 2CO(g)$,
$T = 2706$ K for $CaO(s) + C(s) \rightarrow Ca(s) + CO(g)$

Determine the values for ΔH^0 and ΔS^0 in order to find the temperature at which the reaction is spontaneous.

	$SnO_2(s)$	+	$2C(s)$	$\rightarrow$	$Sn(s)$	+	$2CO(g)$
ΔH_f^0, kJ/mol	–580.7		0		0		–110.52
S^0, J/K · mol	52.3		5.69		51.5		197.91

$$\Delta H^0 = \left[(2\ \cancel{\text{mol CO}})\left(\frac{-110.52\ \text{kJ}}{\cancel{\text{mol CO}}}\right) + (1\ \cancel{\text{mol Sn}})\left(\frac{0\ \text{kJ}}{\cancel{\text{mol Sn}}}\right)\right]$$

$$-\left[(1\ \cancel{\text{mol SnO}_2})\left(\frac{-580.7\ \text{kJ}}{\cancel{\text{mol SnO}_2}}\right) + (2\ \cancel{\text{mol C}})\left(\frac{0\ \text{kJ}}{\cancel{\text{mol C}}}\right)\right] = 359.7\ \text{kJ}$$

$$\Delta S^0 = \left[(1\ \text{mol Sn})\left(\frac{51.5\ \text{J}}{\text{K}\cdot\text{mol Sn}}\right) + (2\ \text{mol CO})\left(\frac{197.91\ \text{J}}{\text{K}\cdot\text{mol CO}}\right)\right]$$

$$-\left[(1\ \text{mol SnO}_2)\left(\frac{52.3\ \text{J}}{\text{K}\cdot\text{mol SnO}_2}\right) + (2\ \text{mol C})\left(\frac{5.69\ \text{J}}{\text{K}\cdot\text{mol C}}\right)\right] = 383.6\ \frac{\text{J}}{\text{K}} = 0.3836\ \frac{\text{kJ}}{\text{K}}$$

Set $\Delta G^0 = 0$ to determine the temperature at which the reaction becomes spontaneous.

$$\Delta G^0 = \Delta H^0 - T\Delta S^0 = 0$$

$$T = \frac{(\Delta H^0 - \Delta G^0)}{\Delta S^0} = \frac{(359.7 - 0)\ \text{kJ}}{0.3836\ \text{kJ/K}} = 937.7\ \text{K}$$

(Best answer, without intermediate round off, is 937.5.)

Similarly, for the second reaction:

	CaO(s)	+	C(s)	→	Ca(s)	+	CO(g)
ΔH_f^0, kJ/mol	–635.09		0		0		–110.52
S^0, J/K · mol	39.75		5.69		41.4		197.91

$$\Delta H^0 = \left[(1\ \text{mol Ca})\left(\frac{0\ \text{kJ}}{\text{mol Ca}}\right) + (1\ \text{mol CO})\left(\frac{-110.52\ \text{kJ}}{\text{mol CO}}\right)\right]$$

$$-\left[(1\ \text{mol CaO})\left(\frac{-635.09\ \text{kJ}}{\text{mol CaO}}\right) + (1\ \text{mol C})\left(\frac{0\ \text{kJ}}{\text{mol C}}\right)\right] = 524.57\ \text{kJ}$$

$$\Delta S^0 = \left[(1\ \text{mol Ca})\left(\frac{41.4\ \text{J}}{\text{K}\cdot\text{mol Ca}}\right) + (1\ \text{mol CO})\left(\frac{197.91\ \text{J}}{\text{K}\cdot\text{mol CO}}\right)\right]$$

$$-\left[(1\ \text{mol CaO})\left(\frac{39.75\ \text{J}}{\text{K}\cdot\text{mol CaO}}\right) + (1\ \text{mol C})\left(\frac{5.69\ \text{J}}{\text{K}\cdot\text{mol C}}\right)\right] = 193.9\ \frac{\text{J}}{\text{K}} = 0.1939\ \frac{\text{kJ}}{\text{K}}$$

When $\Delta G^0 = 0$,

$$T = \frac{(\Delta H^0 - \Delta G^0)}{\Delta S^0} = \frac{(524.57 - 0)\ \text{kJ}}{0.1939\ \text{kJ/K}} = 2705\ \text{K}$$

(Best answer, without intermediate round off, is 2706 K.)

(b) The first reaction requires a temperature of 937.5 K, which can be reached relatively easily. The second reaction requires a temperature of 2706 K, which is very high and probably not reachable in ancient times. Thus, the first reaction was probably the one used in ancient times.

21.63 **(a)**

$$\text{Cl} + H_2 \rightarrow \text{HCl} + \text{H}$$
$$+ (\text{H} + Cl_2 \rightarrow \text{HCl} + \text{Cl})$$
$$H_2 + Cl_2 \rightarrow 2\text{HCl}$$

(b) Cl and H are radicals.

Step (1) $Cl_2 \xrightarrow{\text{light (or thermal energy)}} 2Cl\cdot$

Step (2) $Cl\cdot + H_2 \rightarrow HCl + H\cdot$

Step (3) $H\cdot + Cl_2 \rightarrow HCl + Cl\cdot$

(c) one, one

(d) $Cl\cdot + Cl\cdot \rightarrow Cl_2$ or $H\cdot + Cl\cdot \rightarrow HCl$

(e) In Steps (2) and (3), $H_2 + Cl_2$ are reactants; their concentration is high, at least until reaction approaches completion. $H\cdot$ and $Cl\cdot$, on the other hand, are reactive intermediates, and their concentration is low.

(f) –92.31 kJ/mol

The equation is

$\frac{1}{2}H_2(g) + \frac{1}{2}Cl_2(g) \rightarrow HCl(g)$, the formation equation for HCl(g).

From Table 6.2, ΔH_f^0 for HCl(g) is 92.31 kJ/mol.

(g) Once the reaction begins, much thermal energy is released. Thermal energy, as well as light, can cause chains to start. The reaction gets faster and faster, yielding an explosion.

21.67 30.3 g $H_2(g)$

Equation for production of $H_2(g)$ by electrolysis of water is

$2H_2O(l) + 2e^- \rightarrow H_2(g) + 2OH^-$

$1\ V = \frac{1\ J}{C}, \quad C = \frac{J}{V}$

$$C = \frac{3.60 \times 10^6\ J}{1.24\ V}$$

$C = 2.90 \times 10^6$ coulombs

$$2.90 \times 10^6\ \cancel{C} \times \frac{1\ \text{mol } e^-}{9.649 \times 10^4\ \cancel{C}} = 30.1\ \text{mol } e^- \rightarrow 15.1\ \text{mol } H_2(g)$$

$$(15.1\ \cancel{\text{mol } H_2(g)})\left(\frac{2.016\ g\ H_2}{\cancel{\text{mol } H_2}}\right) = 30.4\ g\ H_2(g)$$

(Best answer, without intermediate round off, is 30.3 g $H_2(g)$.)

21.71 **(a)** NO

(b) SO_2

(c) O_3 and NO

(d) hydrocarbons containing double bonds such as isoprene $[H_2C{=}C(CH_3)CH{=}CH_2]$

21.75 3.11×10^3 kJ

$S(s) + O_2(g) \rightarrow SO_2(g)$

$\Delta H^0_{\text{combustion}}[S(s)] = \Delta H_f^0[SO_2(g)] = -296.83$ kJ/mol

$$(672 \text{ g } SO_2)\left(\frac{1 \text{ mol } SO_2}{64.07 \text{ g } SO_2}\right)\left(\frac{-296.8 \text{ kJ}}{1 \text{ mol } SO_2}\right) = -3.11 \times 10^3 \text{ kJ}$$

21.79 **(a)** Yes. H_2O_2 appears in both left and right columns of a table of standard reduction potentials. From the diagonal relationship in the table, the disproportionation reaction is spontaneous.

(b) 1.078 V

$E^0_{\text{cell}} = 1.763 \text{ V} - 0.685 \text{ V} = 1.078 \text{ V}$

(c) $3H_2O_2(aq) + 2Cr^{3+} + H_2O(l) \rightarrow Cr_2O_7^{2-} + 8H^+$; hydrogen peroxide is acting as an oxidizing agent.

(d) $Cr_2O_7^{2-} + 8H^+ + 3H_2O_2(aq) \rightarrow 2Cr^{3+} + 7H_2O(l) + 3O_2(g)$; hydrogen peroxide is acting as a reducing agent.

(e) HO_2^-

Applications

21.83 **(a)** 33% P_2O_5

$$\left(\frac{71 \text{ g BPL}}{100 \text{ g rock}}\right)\left(\frac{1 \text{ mol BPL}}{310 \text{ g BPL}}\right)\left(\frac{1 \text{ mol } P_2O_5}{1 \text{ mol BPL}}\right)\left(\frac{142 \text{ g } P_2O_5}{1 \text{ mol } P_2O_5}\right) \times 100 = 33\% \ P_2O_5$$

(b) 14% P

$$\left(\frac{33 \text{ g } P_2O_5}{100 \text{ g rock}}\right)\left(\frac{61.9 \text{ g P}}{142 \text{ g } P_2O_5}\right) \times 100 = 14\% \text{ P}$$

21.87 $KAlSi_3O_8$

Atoms can't be divided. For orthoclase to have one-fourth of the silicon atoms replaced by Al, there must be four Si:

$4(SiO_2) = Si_4O_8$

Replacing 1 Si by Al gives $AlSi_3O_8^-$. One K^+ is needed for charge balance. Therefore, the empirical formula is $KAlSi_3O_8$.

21.91 **(a)** 5.85 atm

$P_{\text{total}} = p_{O_2} + p_{He}$

$P_{\text{total}} - p_{O_2} = p_{He} = 6.06 \text{ atm} - 0.21 \text{ atm} = 5.85 \text{ atm}$

(b) 96.5% He, 3.5% O_2

Consider 100 mL of the mixture.

$$n_{He} = \frac{(5.85\ \cancel{atm})(0.100\ \cancel{L})}{\left(0.082\ 06\ \frac{\cancel{L}\cdot\cancel{atm}}{\cancel{K}\cdot mol}\right)(277.6\ \cancel{K})} = 0.0257\ mol$$

$$n_{O_2} = \frac{(0.21\ \cancel{atm})(0.100\ \cancel{L})}{\left(0.082\ 06\ \frac{\cancel{L}\cdot\cancel{atm}}{\cancel{K}\cdot mol}\right)(277.6\ \cancel{K})} = 0.000\ 92\ mol$$

According to Avogadro's law, the volume of a gas is proportional to the number of moles. Therefore,

$$\left(\frac{0.0257}{0.0257 + 0.0009}\right) \times 100 = 96.6\% \text{ by volume He}$$

$$\left(\frac{0.000\ 92}{0.0257 + 0.0009}\right) \times 100 = 3.5\% \text{ by volume } O_2$$

(Best answers, without intermediate round off, are 96.5% He and 3.5% O_2.)

(c) 0.222 g/L

$$(0.0257\ \cancel{mol\ He})\left(\frac{4.003\ g\ He}{\cancel{mol\ He}}\right) = 0.103\ g\ He$$

$$(0.000\ 92\ \cancel{mol\ O_2})\left(\frac{32.0\ g\ O_2}{\cancel{mol\ O_2}}\right) = \underline{0.029\ g\ O_2}$$

0.132 g total

Need to find volume at STP.

	Volume, mL	Temperature, K	Pressure, atm
Old	100	277.6	6.06
New (STP)	?	273.2	1.00

$$(100\ mL)\left(\frac{273.2\ \cancel{K}}{277.6\ \cancel{K}}\right)\left(\frac{6.06\ \cancel{atm}}{1.00\ \cancel{atm}}\right) = 596\ mL = 0.596\ L$$

$$d_{STP} = \frac{0.132\ g}{0.596\ L} = 0.221\ g/L$$

(Best answer, without intermediate round off, is 0.222 g/L.)

(d) 1.32 g/L

$$d_{4.4\ °C,\ 6.06\ atm} = \frac{0.132\ g}{0.100\ L} = 1.32\ g/L$$

21.95 Ice is a thermal insulator. Once the ice has formed, it is a poor conductor of thermal energy and does not transfer thermal energy from the plant to the outside air. Also, freezing of the water releases thermal energy that warms the trees and fruit.

21.99 **(a)** ice: melting or fusion; dry ice: sublimation; liquid nitrogen: boiling or evaporation or vaporization

(b) ice: freezing or crystallization; dry ice: deposition; liquid nitrogen: condensation

21.103 **(a)** 5×10^9 kg

From data in Section 21.2,

Amount of seawater = 1.4×10^{21} kg $\times$ 0.973 = 1.4×10^{21} kg seawater

Assume $d = 1.00$. Then 1 kg water = 1 L water:

$$(1.4 \times 10^{21}\,\cancel{\text{L}})\left(\frac{4 \times 10^{-6}\,\cancel{\text{mg Au}}}{\cancel{\text{L}}}\right)\left(\frac{1\,\cancel{\text{g Au}}}{10^3\,\cancel{\text{mg Au}}}\right)\left(\frac{1\text{ kg Au}}{10^3\,\cancel{\text{g Au}}}\right) = 6 \times 10^9\text{ kg Au}$$

(Best answer, without intermediate round off, is 5×10^9 kg.)

(b) 3×10^{11} L

$$(1\,\cancel{\text{kg Au}})\left(\frac{1\text{ L H}_2\text{O}}{4 \times 10^{-6}\,\cancel{\text{mg Au}}}\right)\left(\frac{10^3\,\cancel{\text{mg Au}}}{1\,\cancel{\text{g Au}}}\right)\left(\frac{10^3\,\cancel{\text{g Au}}}{1\,\cancel{\text{kg Au}}}\right) = 3 \times 10^{11}\text{ L}$$

21.107 **(a)** Mass of $CaCl_2$ is greater, probably making transportation and labor costs greater.

$$\frac{110.99\text{ g CaCl}_2}{\text{mol}}, \quad \frac{58.443\text{ g NaCl}}{\text{mol}}$$

To get 0.0300 mol ions, need (choosing any convenient number of ions)

0.0100 mol $CaCl_2$ and 0.0150 mol NaCl

$$(0.0100\,\cancel{\text{mol CaCl}_2})\left(\frac{111.0\text{ g}}{\cancel{\text{mol CaCl}_2}}\right) = 1.11\text{ g CaCl}_2$$

$$(0.0150\,\cancel{\text{mol NaCl}})\left(\frac{58.44\text{ g}}{\cancel{\text{mol NaCl}}}\right) = 0.877\text{ g NaCl}$$

(b) $CaCl_2$ costs almost twice as much as NaCl.

$$(1.11\,\cancel{\text{g CaCl}_2})\left(\frac{1.50\text{ cost}}{\cancel{\text{g CaCl}_2}}\right) = 1.67$$

$$(0.877\,\cancel{\text{g NaCl}})\left(\frac{1.00\text{ cost}}{\cancel{\text{g NaCl}}}\right) = 0.877$$

$$\frac{1.67}{0.877} = 1.90$$

21.111 **(a)** Water extinguishes burning paper by lowering the temperature of the paper and by excluding oxygen, a necessary reactant for combustion.

(b) Water can't be used to extinguish a sodium fire because sodium reacts violently with water. Furthermore, one of the products of this reaction is the highly flammable gas H_2:

$$2\text{Na(s)} + 2\text{H}_2\text{O(l)} \rightarrow 2\text{NaOH(aq)} + \text{H}_2\text{(g)}.$$

(c) Water can't be used to extinguish burning oil because oil is less dense than water. As a result, the oil floats on the surface of the water and the fire is spread out over a larger area.

(d) If you lie down and roll, the supply of oxygen will be cut off and the fire will be extinguished. Running to a faucet will replenish the oxygen consumed by the fire, and the fire will burn more vigorously.

(e) Ammonium bromide is water soluble and washes out when the fabric is laundered.

21.115 **(a)** 1×10^3 ppm

$$\frac{0.1 \text{ L CO}}{10^2 \text{ L air}} = \frac{x}{10^6 \text{ L air}}$$

$x = 1 \times 10^3$ ppm

(b) 49 g CO

$$\text{volume of room} = (13.0 \text{ ft})(14.5 \text{ ft})(8.0 \text{ ft})\left(\frac{12 \text{ in}}{\text{ft}}\right)^3\left(\frac{2.54 \text{ cm}}{\text{in}}\right)^3\left(\frac{10^{-2} \text{ m}}{\text{cm}}\right)^3\left(\frac{1 \text{ L}}{10^{-3} \text{ m}^3}\right) = 4.3 \times 10^4 \text{ L air}$$

$$m = \frac{PV(\text{MM})}{RT} = \frac{(755.0 \text{ mmHg})(4.3 \times 10^4 \text{ L air})\left(\frac{0.1 \text{ L CO}}{100 \text{ L air}}\right)\left(\frac{28.0 \text{ g CO}}{\text{mol CO}}\right)}{\left(\frac{62.36 \text{ L} \cdot \text{mmHg}}{\text{mol} \cdot \text{K}}\right)(294.3 \text{ K})} = 50 \text{ g CO}$$

(Best answer, without intermediate round off, is 49 g CO.)

(c) 8.6×10^{-3} M

The number of moles of oxygen in 1 L of air is

$$n_{O_2} = \frac{PV}{RT} = \frac{(760 \text{ mmHg})(0.21 \text{ L } O_2)}{\left(\frac{62.36 \text{ L} \cdot \text{mmHg}}{\text{mol} \cdot \text{K}}\right)(298 \text{ K})} = 8.6 \times 10^{-3} \text{ mol } O_2$$

$$\text{molarity } O_2 = \frac{8.6 \times 10^{-3} \text{ mol } O_2}{1 \text{ L air}} = 8.6 \times 10^{-3} \text{ M}$$

(d) 4×10^{-5} M

$$n_{CO} = \frac{(760 \text{ mmHg})(0.001 \text{ L})}{\left(\frac{62.36 \text{ mmHg} \cdot \text{L}}{\text{mol} \cdot \text{K}}\right)(298 \text{ K})} = 4 \times 10^{-5} \text{ mol CO}$$

$$\text{molarity CO} = \frac{4 \times 10^{-5} \text{ mol CO}}{1 \text{ L air}} = 4 \times 10^{-5} \text{ M}$$

(e) 2×10^2

$$\frac{[O_2]}{[CO]} = \frac{8.6 \times 10^{-3}}{4 \times 10^{-5}} = 2 \times 10^2$$

(f) 1

From Related Topic 14 (p. 538), we have

$$K_c = \frac{[\text{Hb(CO)}][\text{O}_2]}{[\text{Hb(O}_2)][\text{CO}]} = 2.1 \times 10^2 = \frac{[\text{Hb(CO)}]}{[\text{Hb(O}_2)]} \cdot 2 \times 10^2$$

$$\frac{[\text{Hb(CO)}]}{[\text{Hb(O}_2)]} = \frac{2.1 \times 10^2}{2 \times 10^2} = 1$$

(g) 50% of the hemoglobin is tied up as Hb(CO).

CHAPTER 22:
A Closer Look at Organic Chemistry

Practice Problems

22.1 **(a)**

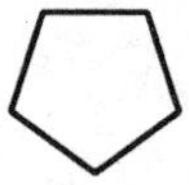

(b)

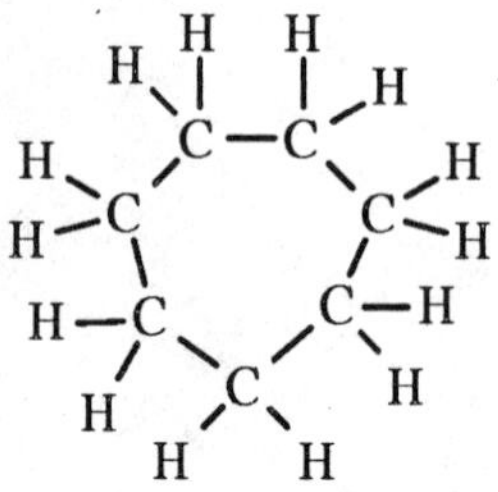

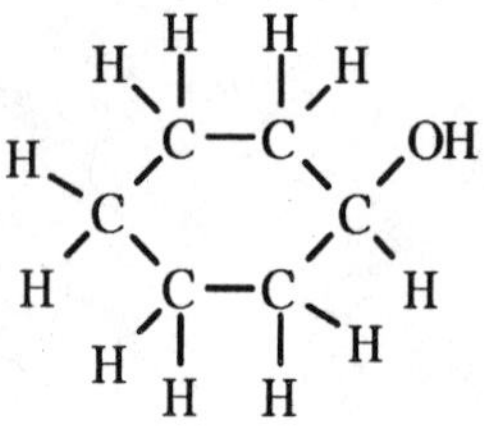

22.2 **(a)**

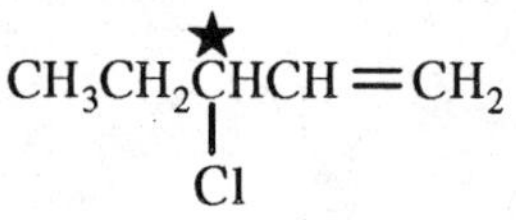

$$CH_3CH_2\overset{\bigstar}{C}HCH = CH_2$$
$$\underset{|}{}Cl$$

This carbon has 4 different groups attached to it:

CH_3CH_2—, H—, Cl—, and —CH=CH_2.

(b)

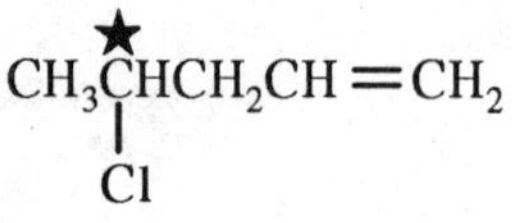

$$CH_3\overset{\bigstar}{C}HCH_2CH = CH_2$$
$$\underset{|}{}Cl$$

This carbon has 4 different groups attached to it: CH_3—, H—, Cl—, and —CH_2CH=CH_2.

(c) No stereocenter is present.

$ClCH_2CH_2CH_2CH{=}CH_2$

22.3 

This is the *trans*–isomer because the methyl groups (CH_3—) are on opposite sides of the plane of the ring.

22.4 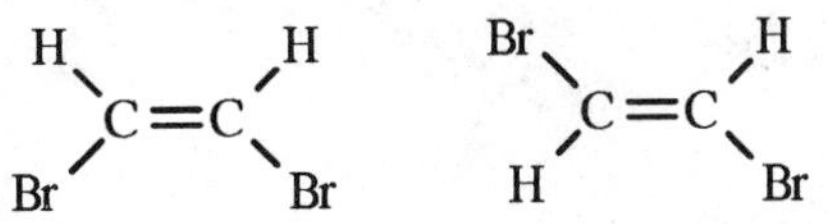

cis *trans*

The "*cis*" isomer has both bromines on the same side of the C═C double bond.

22.5 No. The two CH_3— groups are identical and interchangeable.

22.6 pentane, hexane, heptane, octane, nonane, and decane

These all have a liquid range (the temperature range from the melting point to the boiling point) that includes "ordinary" conditions or room temperature of about 20 °C.

22.7 (a)

$(CH_3)_2C{=}CH_2$, 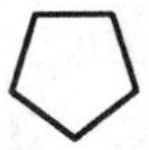, $(CH_3)_2CHCH_2C(CH_3)_3$, $CH_3C{\equiv}CCH_3$, 

All of the others contain elements other than C and H.

(b) $(CH_3)_2CHCH_2C(CH_3)_3$

The other hydrocarbons contain either rings and/or multiple bonds.

(c) 

The other ring hydrocarbon contains a double bond.

22.8 C_nH_{2n}

22.9 If O_2 is present, complete or incomplete combustion will be the predominant reaction.

22.10 $CH_3CH_2CH_2CH_3(g) + Cl_2(g) \xrightarrow{\text{thermal energy and/or light}} CH_3CH_2CH_2CH_2Cl(l) + HCl(g)$

$CH_3CH_2CH_2CH_3(g) + Cl_2(g) \xrightarrow{\text{thermal energy and/or light}} CH_3CH_2CH(Cl)CH_3(l) + HCl(g)$

22.11 propyl

This name is derived from propane, the corresponding alkane, by changing the ending –ane to –yl.

22.12 No. The two methyl groups will be located on the third carbon regardless of which end carbon is number 1.

22.13 (b) 2–methylpentane

The longest continuous chain is five carbons long. Therefore, (a) 1, 1–dimethylbutane and (d) 2–propylpropane are incorrect (Rule 1). (c) 4–methylpentane is wrong because the position of the methyl group has a higher number than necessary (Rule 3).

22.14 (i)

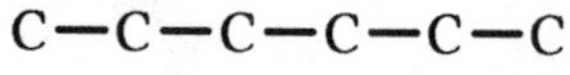

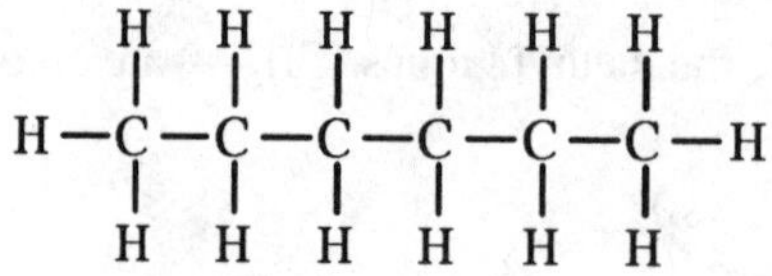

$CH_3CH_2CH_2CH_2CH_2CH_3$, hexane

(ii)

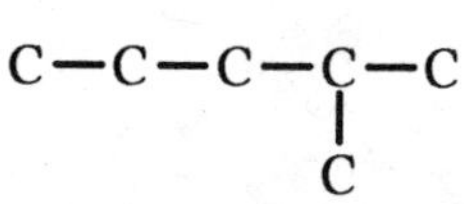

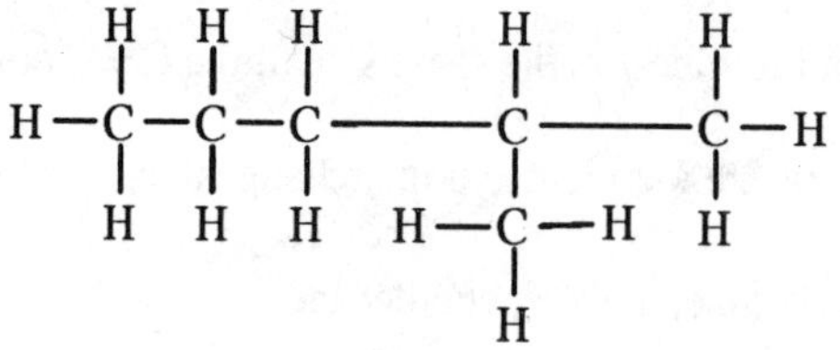

$CH_3CH_2CH_2CH(CH_3)_2$, 2-methylpentane

(iii

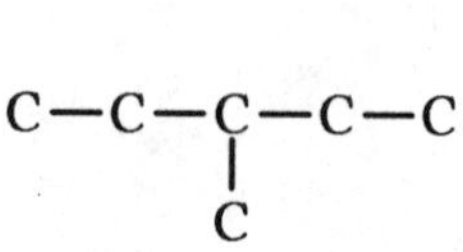

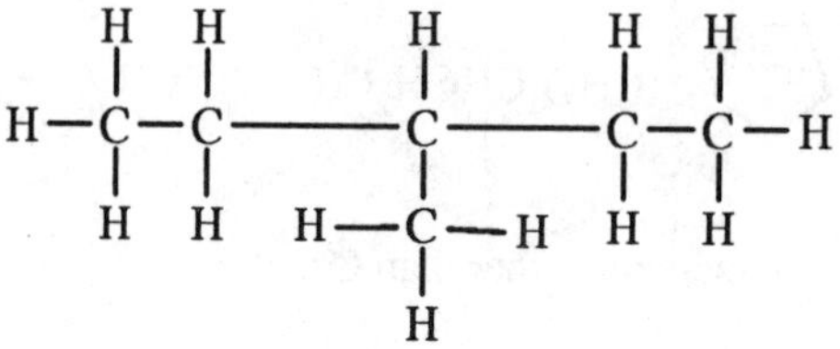

$CH_3CH_2CH(CH_3)CH_2CH_3$, 3-methylpentane

(iv)

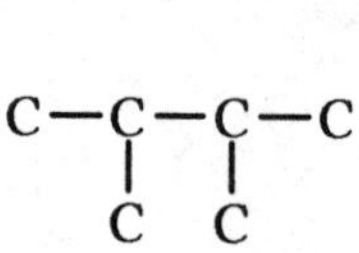

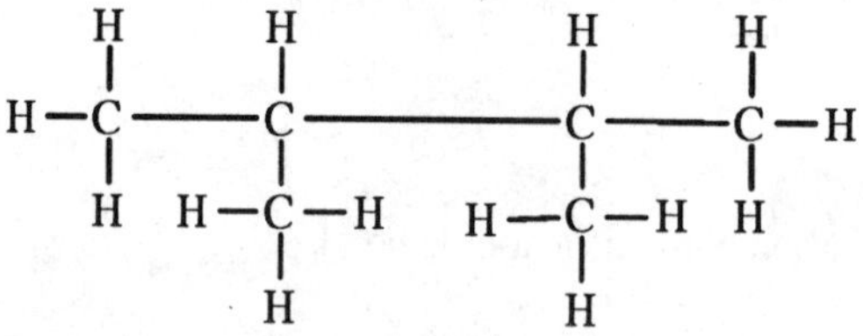

$(CH_3)_2CHCH(CH_3)_2$, 2, 3-dimethylbutane

(v)

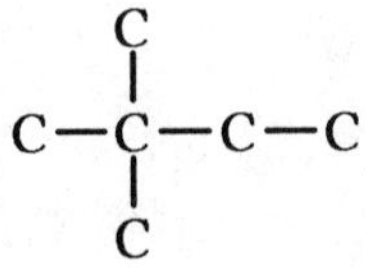

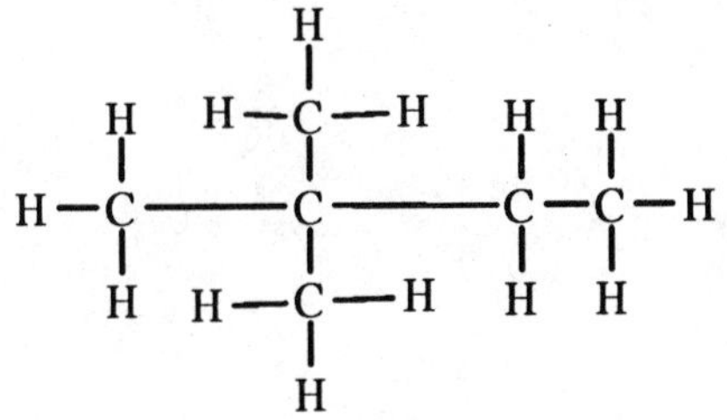

$(CH_3)_3CCH_2CH_3$, 2, 2-dimethylbutane

22.15 (a)

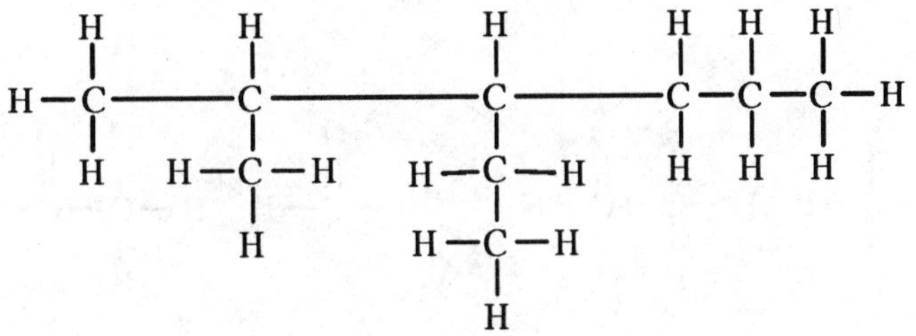

(b)

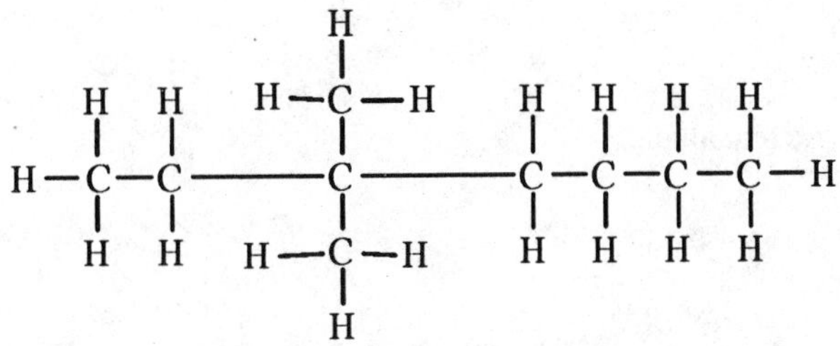

22.16 (a) $CH_3CH_2COOH(aq) + NaOH(aq) \rightarrow CH_3CH_2COONa(aq) + H_2O(l)$

(b) $CH_3CH_2NH_2(aq) + HCl(aq) \rightarrow CH_3CH_2NH_3Cl(aq)$

(c) $CH_3CH{=}CH_2(g) + H_2(g) \xrightarrow{\text{catalyst}} CH_3CH_2CH_3(g)$

(d) $2CH_3OH(l) + 2Na(s) \rightarrow 2CH_3ONa(soln) + H_2(g)$

22.17

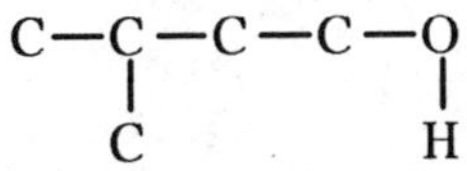

3–methyl–1–butanol

22.18 (a)

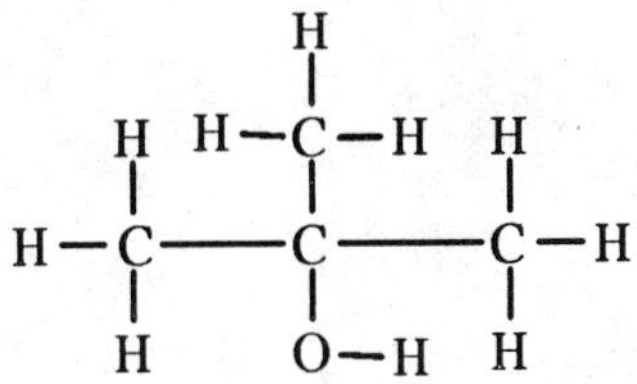

$(CH_3)_3COH$ or $(CH_3)_2C(OH)CH_3$

(b)

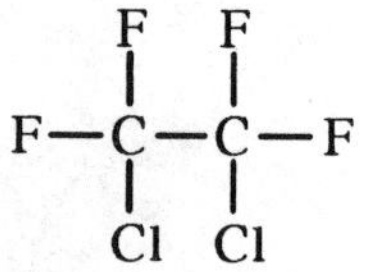

F—C—C—F with F, F on top and Cl, Cl on bottom $CClF_2CClF_2$

22.19 **(a)** $(CH_3)_2C\!=\!CH_2$

(b)

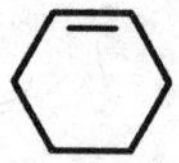

(c) $CH_3C\equiv CCH_3$

22.20

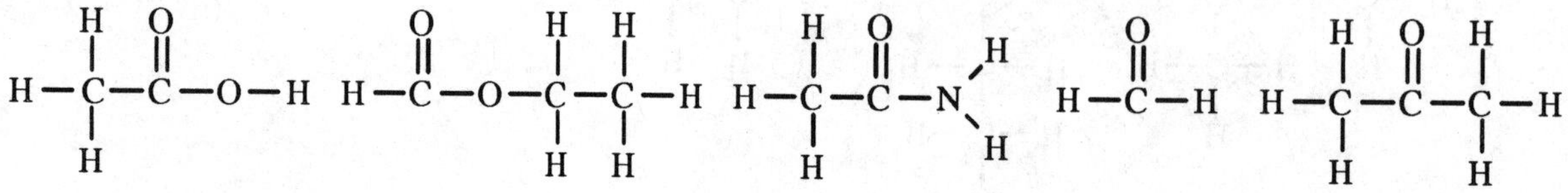

22.21 **(a)**

$\boxed{Cl}CH_2CH_2\boxed{OH}$ alkyl chloride and alcohol

(b)   aromatic and alkene

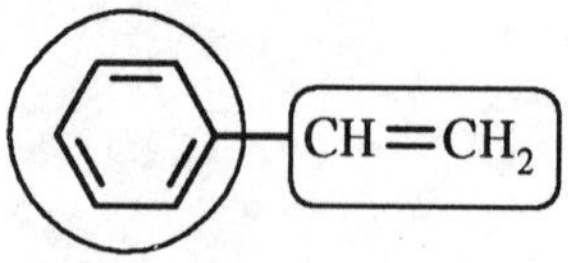

(c) $CH_3CH_2\boxed{COOC}H_2CH_3$ ester

(d) $CH_3CH_2\boxed{Br}$ alkyl bromide

(e) $H\boxed{CON}(CH_3)_2$ amide

22.22 **(a) – (c)**

$\boxed{CH_3CH_2CH_2\,Br}$ + CH_3O^- ⟶ $CH_3CH_2CH_2OCH_3$ + Br^-

substrate   leaving group   nucleophile

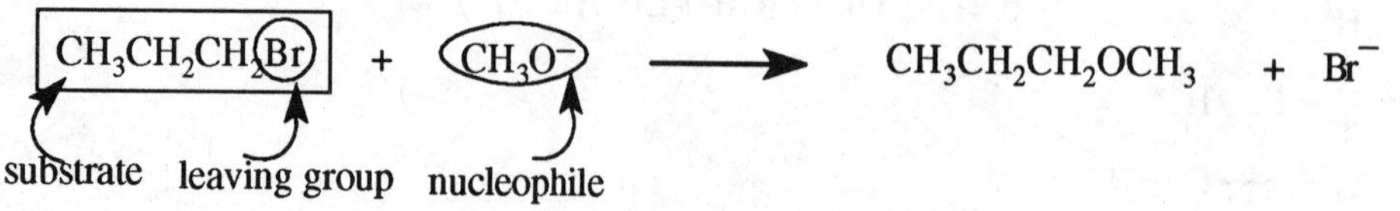

(d) $CH_3CH_2CH_2-$ (propyl group)

22.23

nucleophile   substrate   activated complex

product   leaving group

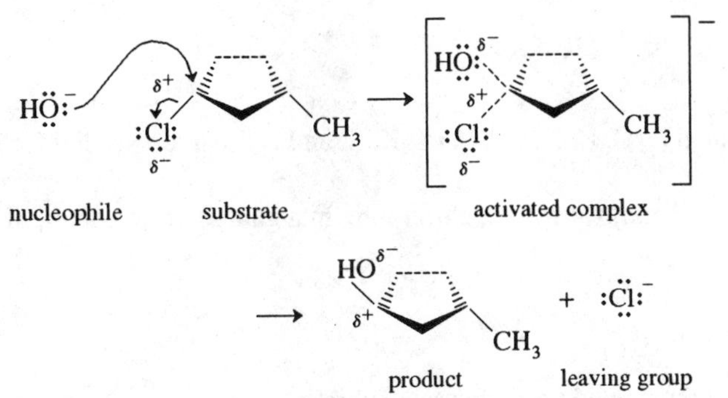

22.24   (a) $CH_3CH_2CH_2CH_2C \equiv CH + Cl^-$

$HC \equiv C^-$ + $CH_3CH_2CH_2 \!-\! Cl$ $\longrightarrow$ $CH_3CH_2CH_2 \!-\! C \equiv CH$ + Cl^-

Nuc:$^-$ R   X   R   Nuc: X$^-$

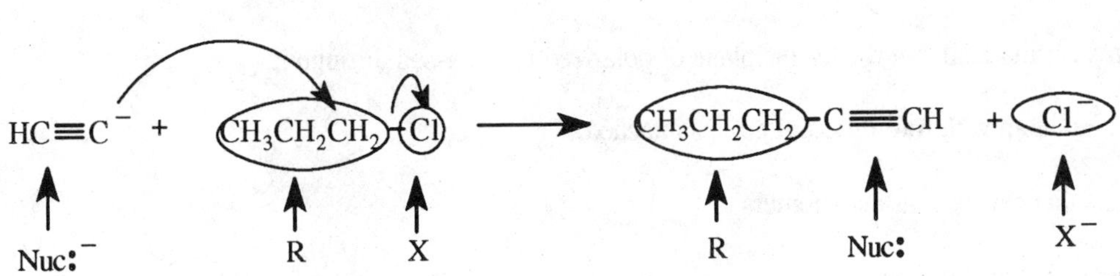

(b) $H_2C \!=\! CHCH_2OOCCH_3$ (or $H_2C \!=\! CHCH_2O\underset{\underset{O}{\|}}{C}CH_3$) + Br$^-$

$CH_3\underset{\underset{O}{\|}}{C}O^-$ + $H_2C \!=\! CHCH_2 \!-\! Br$ $\longrightarrow$ $H_2C \!=\! CHCH_2 \!-\! O \!-\! \underset{\underset{O}{\|}}{C}CH_3$ + Br$^-$

Nuc:$^-$ R   X   R   Nuc: X$^-$

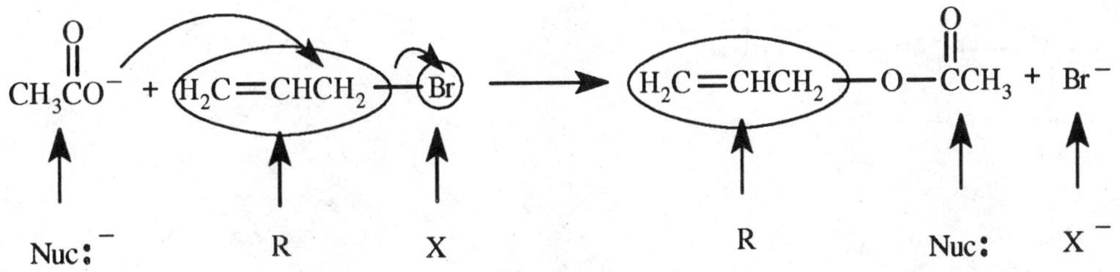

22.25   (a) Reaction will occur; CN$^-$ is a stronger base than Br$^-$.

CN$^-$ is the conjugate base of a weak acid, while Br$^-$ is the conjugate base of a strong acid. Thus, CN$^-$ is a stronger base than Br$^-$.

(b) Reaction will not occur; OH$^-$ is a stronger base than Br$^-$.

OH$^-$ is the conjugate base of a very weak acid, while Br$^-$ is the conjugate base of a strong acid. In S_N^2 reactions, the stronger base displaces the weaker base. In this case Br$^-$ is the weaker base, so no reaction will occur.

22.26

$H_2C \!=\! CHCH_2Br$ + $^-\!:\!\overset{\cdot\cdot}{\underset{\cdot\cdot}{O}}\!-\!H$

substrate   nucleophile

OH$^-$ is a strong base compared to Br$^-$ and should displace it in an S_N^2 type reaction. $H_2C\!=\!CHCH_2\!-\!$ has the correct structure for R.

Additional Practice Problems

22.27 (a) Elements termed "essential" are elements other than O, C, H, N, Ca, P, Cl, K, S, and Na that are needed for life.

These elements are: boron, fluorine, magnesium, silicon, vanadium, chromium, manganese, iron, cobalt, nickel, copper, zinc, selenium, molybdenum, and iodine.

(b) aluminum and titanium

(c) At the pH of biological systems, the oxides of aluminum and titanium are insoluble.

22.31 (a) *plane polarized light*: a light beam with waves vibrating in only one direction or plane.

(b) *optically active*: a material that rotates the plane of polarized light passed through it.

(c) *stereocenter*: an atom with four different groups attached.

(d) *chiral*: objects that can exist as enantiomers.

(e) *achiral*: objects that are not chiral.

(f) *racemic mixture:* a mixture of exactly 50% of one form of an optically active material with 50% of its enantiomer.

22.35 (a) two

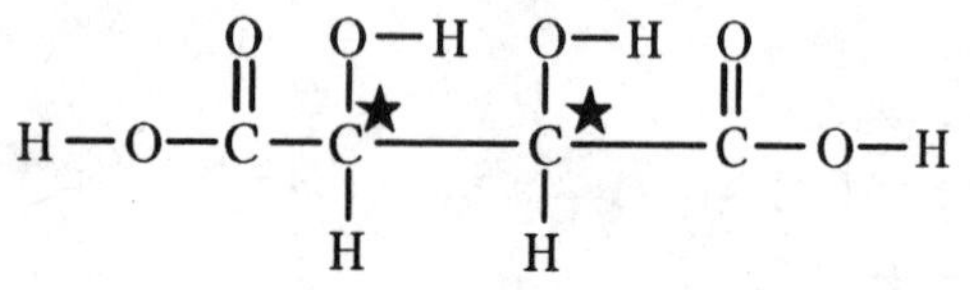

2 stereocenters

(b) three

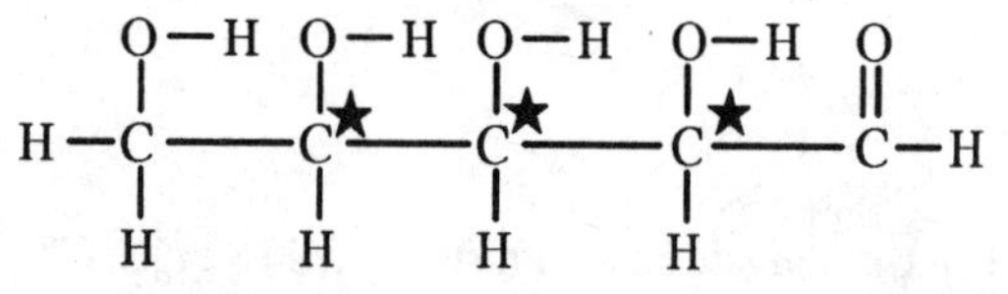

3 stereocenters

(c) one

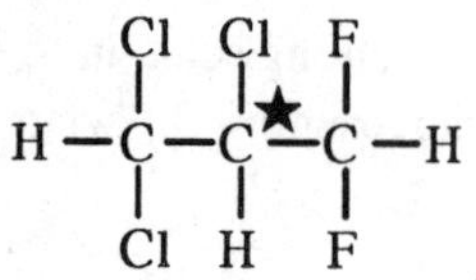

1 stereocenter

(d) one

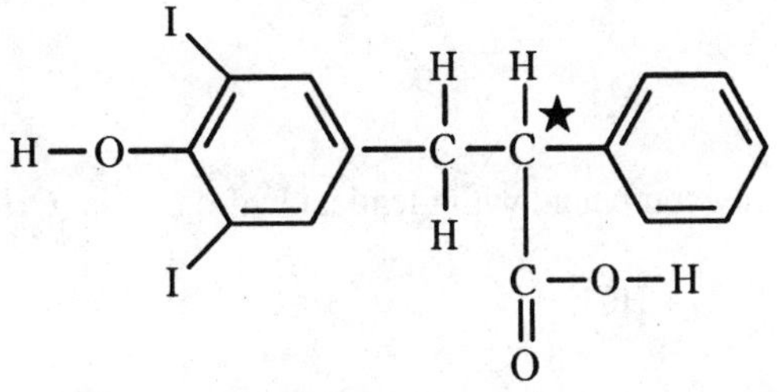

1 stereocenter

(e) zero

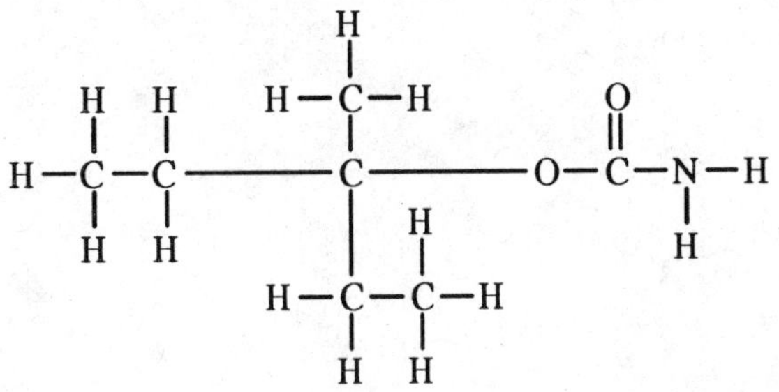

0 stereocenters

(f) two

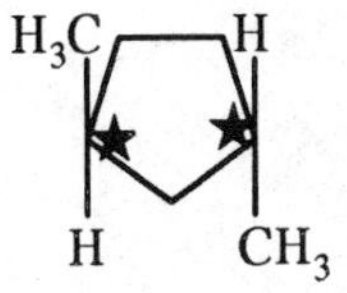

2 stereocenters

22.39 **(a)** Butane is a gas under ordinary conditions; pentane is a liquid.

(b) 0.740 g/mL

The incremental increase in density from pentane through decane is decreasing:

Formula	Density at 20 °C, g/mL	Δ density
$CH_3(CH_2)_3CH_3$	0.626	
		+0.034
$CH_3(CH_2)_4CH_3$	0.660	
		+0.024
$CH_3(CH_2)_5CH_3$	0.684	
		+0.018
$CH_3(CH_3)_6CH_3$	0.702	
		+0.016
$CH_3(CH_2)_7CH_3$	0.718	
		+0.012
$CH_3(CH_2)_8CH_3$	0.730	
$CH_3(CH_2)_9CH_3$	?	

The pattern suggests that the next change in density will be about 0.010. Therefore, we would predict the density to be 0.730 + 0.010 g/mL = 0.740 g/mL. (According to the CRC Handbook, the density of $CH_3(CH_2)_9CH_3$ is 0.7402 g/mL.)

22.43 **(a)** $(CH_3)_4C$

An alkane of molecular mass 72 must be C_5H_{12}. The only compound with identical hydrogens is $(CH_3)_4C$.

(b) $\underset{\text{excess}}{(CH_3)_4C(g)} + Cl_2(g) \xrightarrow{\text{thermal energy or light}} (CH_3)_3CCH_2Cl(l) + HCl(g)$

(c) An excess of the alkane is required. At a lower ratio of alkane to chlorine, disubstitution and higher substitutions occur.

(d) Thermal energy or light is required to initiate the reaction.

22.47 **(a)** alcohol

$(CH_3)_2CHCH_2\textcircled{OH}$ - alcohol

(b) ether

$CH_3\textcircled{O}CH_2CH_3$ - ether

(c) alkyl chloride

$(CH_3)_2CH\textcircled{Cl}$ - alkyl chloride

(d) alkyl iodide

$CH_3CH_2\textcircled{I}$ - alkyl iodide

(e) carboxylic acid

$CH_3CH_2CH_2\textcircled{COOH}$ - carboxylic acid

(f) aldehyde

$CH_3CH_2\textcircled{CHO}$ - aldehyde

(g) phenol

HO—⟨⟩—CH₃ —phenol

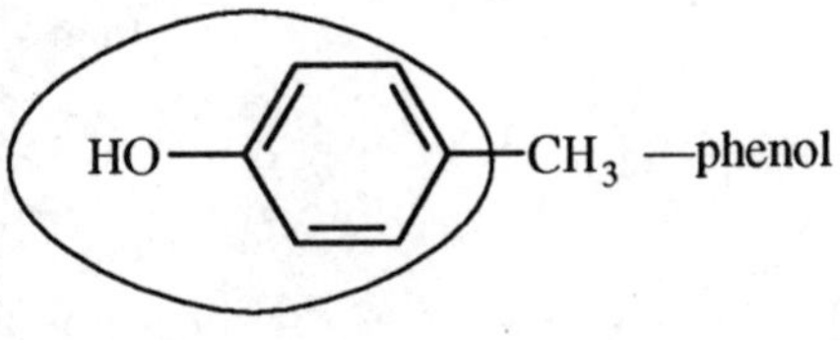

(h) aromatic alcohol

—CH₂—(OH) —aromatic alcohol

22.51 (a) – (c)

$CH_3CH_2CH_2CH_2$(Br) + NH_3 ⟶ $CH_3CH_2CH_2CH_2\overset{+}{N}H_3$ + Br^-

substrate

leaving group

nucleophile

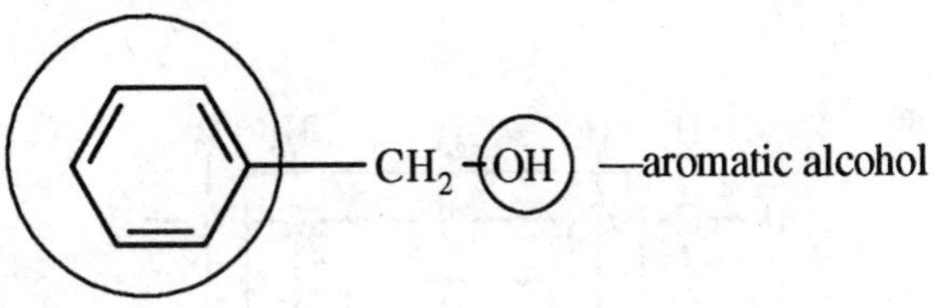

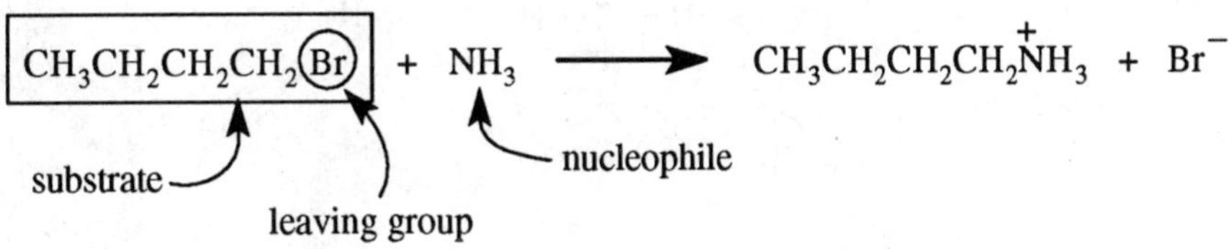

(d) $CH_3CH_2CH_2CH_2-$

22.55

nucleophile   substrate   activated complex

product   leaving group

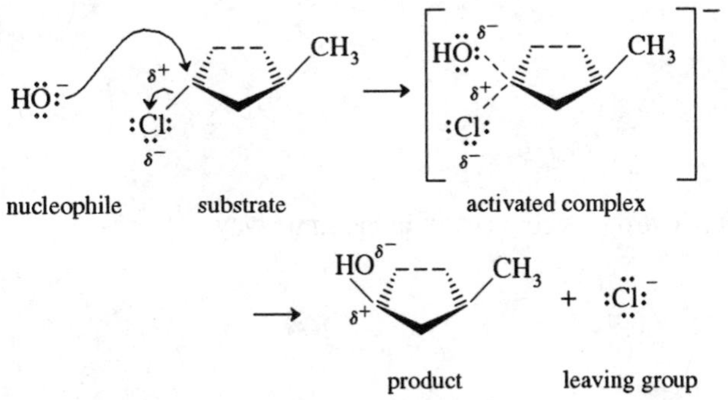

22.59 (a) $2(CH_3)_3COH(l) + 2K(s) \rightarrow H_2(g) + 2(CH_3)_3COK(soln)$

(b) $2(CH_3)_2NH(aq) + H_2SO_4(aq) \rightarrow [(CH_3)_2NH_2]_2SO_4(aq)$ or $(CH_3)_2NH(aq) + H_2SO_4(aq) \rightarrow [(CH_3)_2NH_2]HSO_4(aq)$

(c) $2CH_3COOH(aq) + Ba(OH)_2(aq) \rightarrow (CH_3COO)_2Ba(aq) + 2H_2O(l)$

(d)

⟨⟩ (l) + H_2(g) $\xrightarrow{\text{catalyst}}$ ⟨⟩ (l)

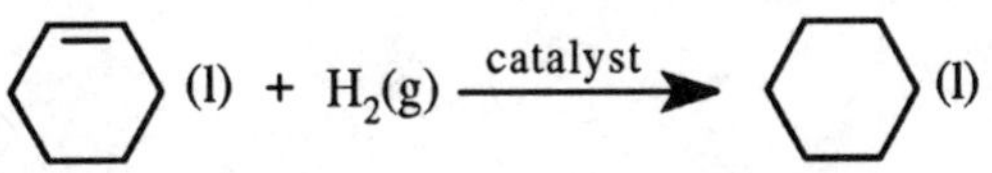

(e)

$$\langle \text{benzene ring} \rangle - OH(aq) + NaOH(aq) \longrightarrow \langle \text{benzene ring} \rangle - ONa(aq) + H_2O(l)$$

Stop and Test Yourself

1.   (a) CoBr$_2$

Organic compounds must contain carbon.

2.   (e) 2, 4–dimethylhexane

Longest carbon chain consists of six carbons:

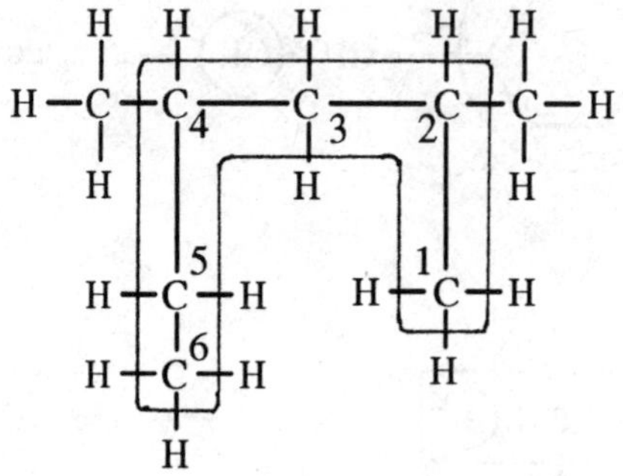

3.   (b)

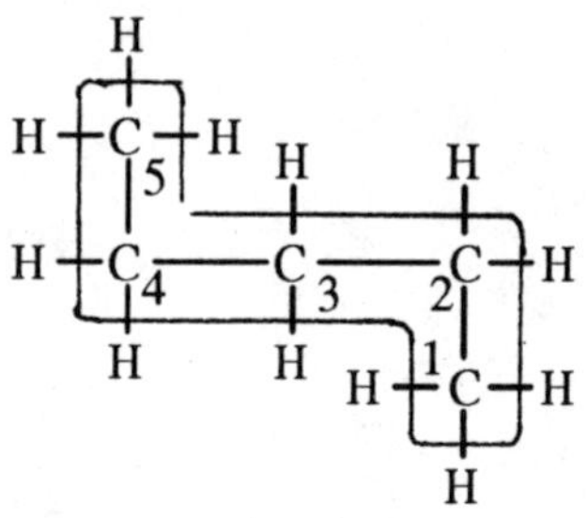

This structure has the same number of carbons and hydrogens connected in the same way.

4.   (c)

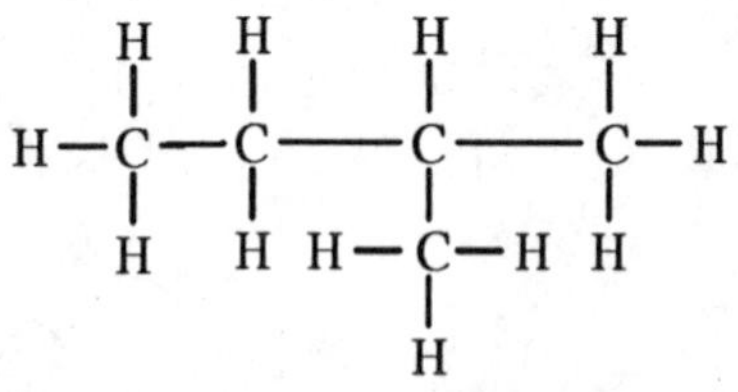

This structure has the same number of carbons and hydrogens connected differently.

5.   **(e)** 7

CCCC—OH, CCCC, $\overset{\overset{\text{C}}{|}}{\text{CCC}}$—OH, $\overset{\overset{\text{C}}{|}}{\underset{\underset{\text{OH}}{|}}{\text{CCC}}}$ = four alcohols
                   $\underset{\underset{\text{OH}}{|}}{}$

COCCC, $\text{COC}\overset{\diagup\text{C}}{\diagdown\text{C}}$, CCOCC = three ethers

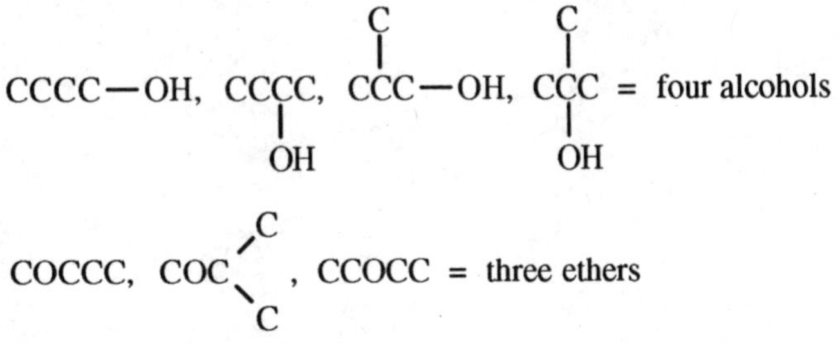

6.   **(d)** Carbon can expand its octet.

This statement is false. Carbon is in the second period and cannot expand its octet.

7.   **(e)** a ketone

CH_3CH_2⟨CO⟩CH_2CH_3. The general formula for a ketone is: $R-\overset{\overset{\text{O}}{\|}}{C}-R$

8.   **(d)** ii, iii, and v

9.   **(c)** 3

The carbons marked with an asterisk each
have four different groups attached.

$$
\begin{array}{c}
CH_2OH \\
| \\
C{=}O \\
| \\
H{-}\overset{\star}{C}{-}OH \\
| \\
H{-}\overset{\star}{C}{-}OH \\
| \\
H{-}\overset{\star}{C}{-}OH \\
| \\
CH_2OH
\end{array}
$$

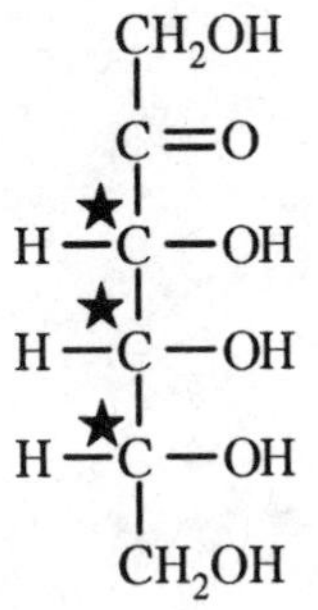

10.   **(a)** D < C < B < A

Note that D, C, and B are constitutional isomers. The most highly branched, D, has the lowest bp. The least branched, B, has the highest bp. Compound A has one more carbon and is straight chain; thus, it has the highest bp.

11.   **(d)** $\overset{\text{H}}{\diagdown}\underset{\text{H}}{\diagup}C{=}C\overset{\diagup\text{H}}{\diagdown\text{H}}$ This molecule has equal differences in electronegativity for every C—H bond. The four C—H bonds are arranged symmetrically. All other molecules listed have some asymmetry and are thus polar.

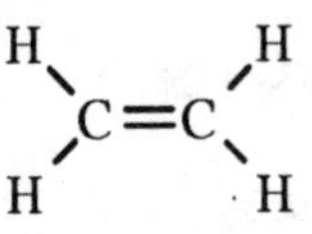

12. **(d)** i, ii, and iv

i) Pyrolysis will take place.

ii) Bromination will take place.

iv) Combustion will take place.

13. **(e)**

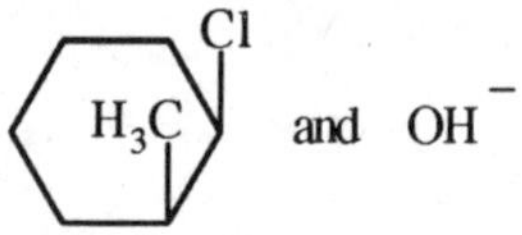

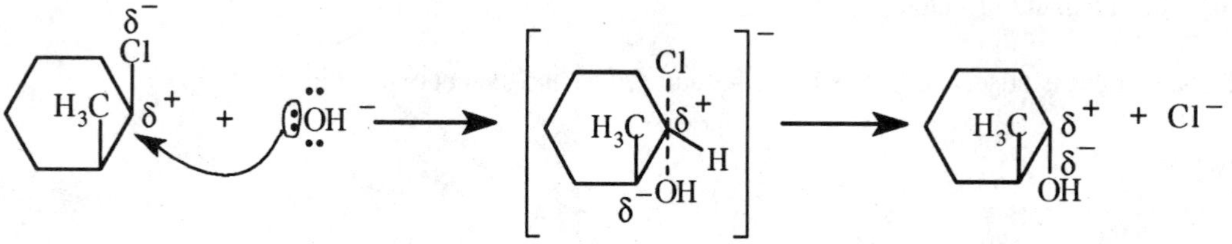

Note that combination (d) would put the Cl on the wrong side of the ring. (*cis* – to the methyl group instead of *trans*.)

Putting Things Together

22.63 **(a)** ① $CH_3CH_2CH_2CH_2CHO$ ② $CH_3CH_2CH(CH_3)CHO$

③ $(CH_3)_2CHCH_2CHO$ ④ $(CH_3)_3CCHO$

(b) ① $CH_3CH_2CH_2CH_2OH$ ② $CH_3CH_2CH(OH)CH_3$

③ $(CH_3)_2CHCH_2OH$ ④ $(CH_3)_3COH$

(c) $CH_3(CH_2)_{10}CH_3$

An aldehyde has 2 less hydrogens than an alkane.

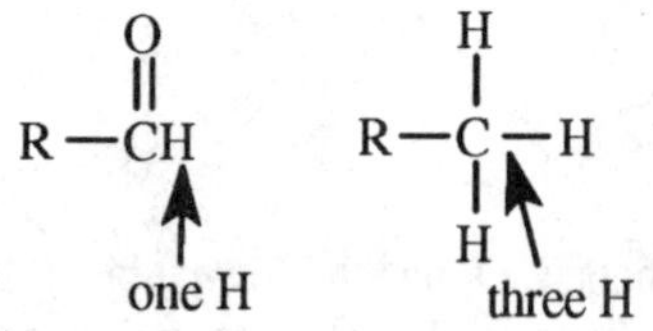

The formula given is: $C_5H_{10}O$ or $C_nH_{2n}O$

Thus, we know that the rest of the aldehyde cannot have a ring or double bond.

An alcohol has the same number of hydrogens as an alkane and has a general formula: $C_nH_{2n+2}O$

The formula given is $C_4H_{10}O$, which satisfies the general alcohol formula. Thus, we know that the rest of the alcohol is saturated without a ring.

A continuous chain alkane has the general formula

C_nH_{2n+2}. Solve for n: $2n + 2 = 26$

$2n = 24$

$n = 12$

Thus, this molecule has 12 carbons.

22.67 **(a)**

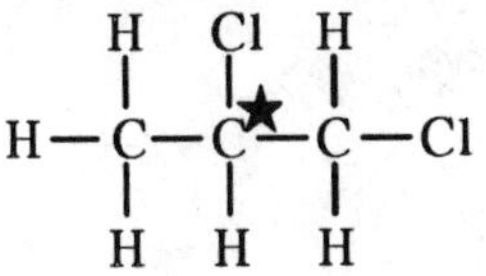

$CH_3CH(Cl)CH_2Cl$

(b) 1, 2–dichloropropane

(c) Three others are possible:

① 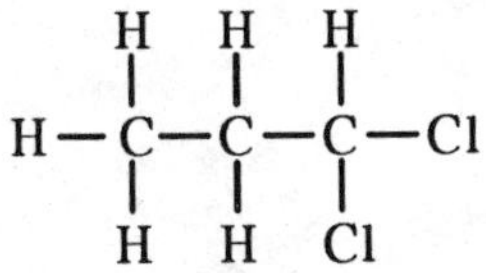 $CH_3CH_2CHCl_2$ 1,1-dichloropropane

② 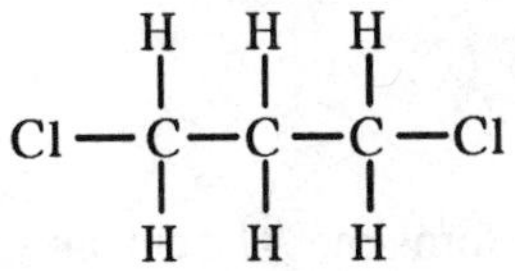 $CH_2ClCH_2CH_2Cl$ 1,3-dichloropropane

③ 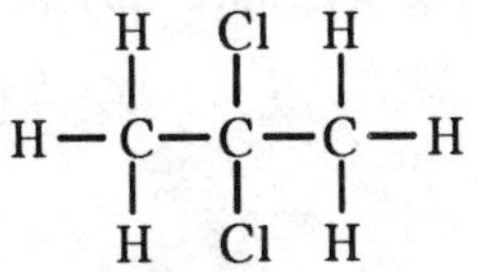 $CH_3CCl_2CH_3$ 2,2-dichloropropane

22.71 (a) CH_3CN, (b) CH_3COCH_3, (d) $(CH_3)_2S{=}O$

None of these solvents has H bonded to N or O.

22.75 Double bonds are shorter and stronger than single bonds.

22.79 **(a)**

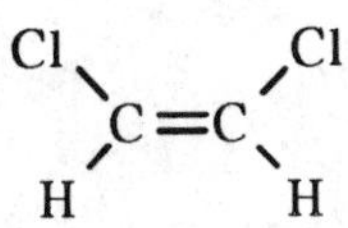

cis

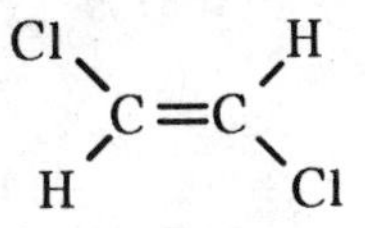

trans

(b) *trans*–1, 2–dichloroethene is nonpolar as a result of its symmetry.

The differences in electronegativity between Cl and C create polar bonds. However, in *trans*–1,2–dichloroethene the chlorines are attached symmetrically so that no molecular dipole is created. The polarities of the C—Cl bonds cancel each other, resulting in a nonpolar molecule.

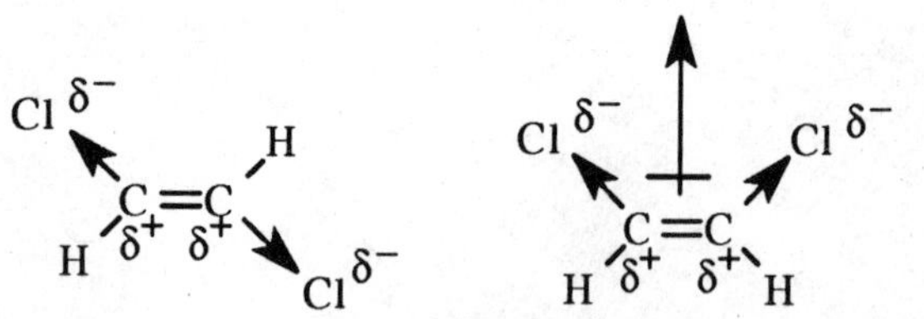

Dipoles cancel   Dipoles add. One side of the molecule is slightly negative.

cis–1, 2–dichloroethene is polar.

22.83 **(a)** Step 1 is a Brønsted-Lowry reaction. Step 2 is a Lewis acid-base reaction.

(b)

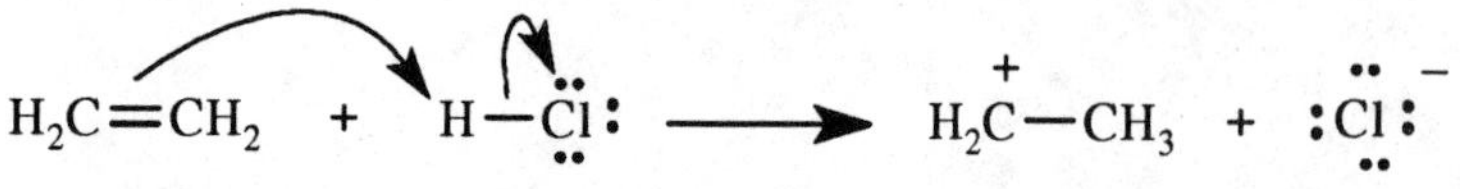

Brønsted-Lowry base   Brønsted-Lowry acid

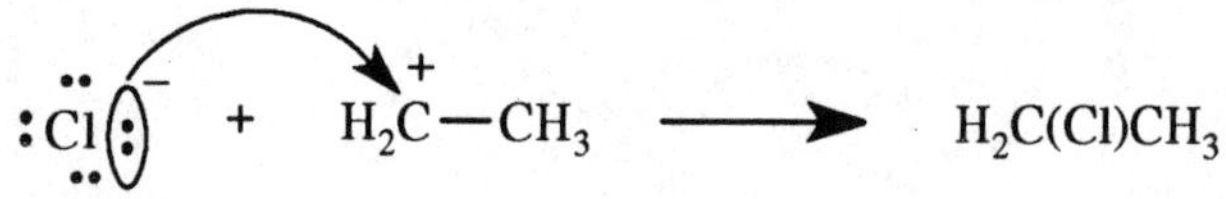

Lewis base   Lewis acid

(c) The entropy of the system decreases, as two molecules of gas combine to form one. The enthalpy of the system decreases; thermal energy is released, and the reaction is exothermic (see below). Since the reaction takes place spontaneously, free energy must decrease.

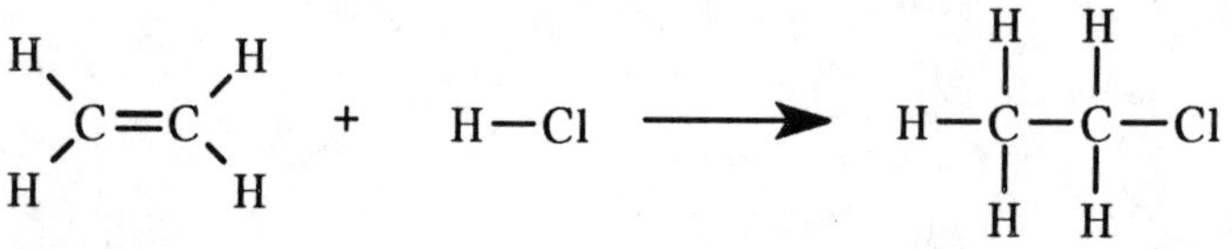

Break	C=C	615 kJ/mol	Form	C–C	347 kJ/mol
	H–Cl	432		C–H	414
				C–Cl	326
		1047			1087

ΔH = +1047 kJ – 1087 kJ = –40 kJ, exothermic

(d) $H_2C{=}CH_2(g) + H_2(g) \xrightarrow{\text{catalyst}} H_3C{-}CH_3(g)$ (catalytic hydrogenation)

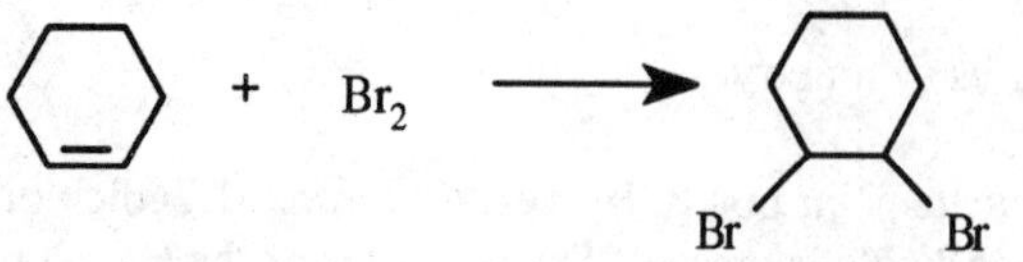

(Section 10.7)

22.87 **(a)**

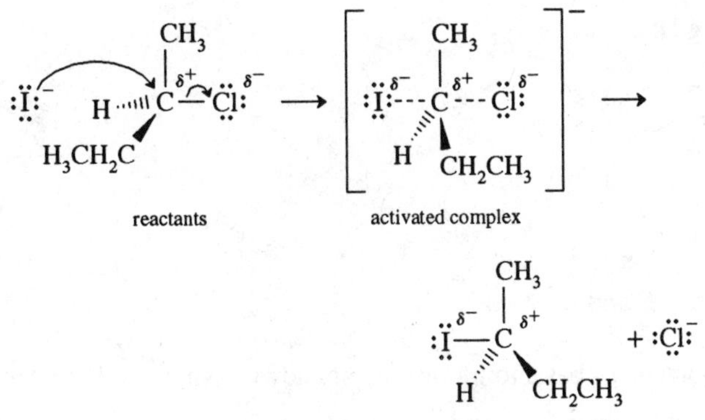

(b)

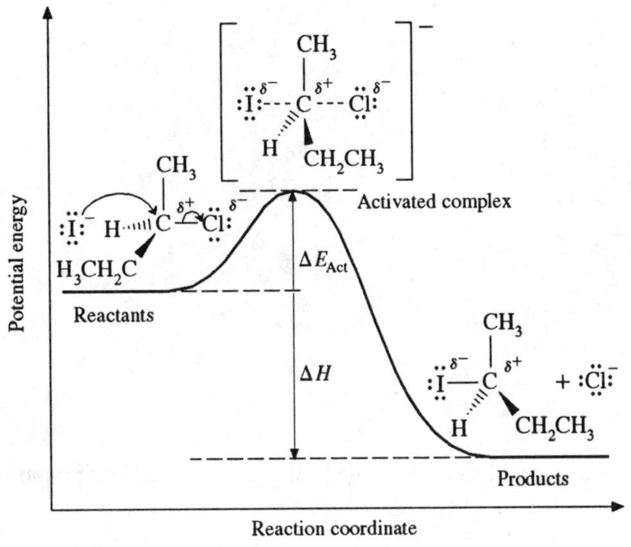

(c) rate = $k[I^-][CH_3CH_2CHClCH_3]$

22.91 CH_3NH_2, methylamine

Use data from combustion to calculate the empirical formula and other data to calculate the molecular mass.

$$n = \frac{PV}{RT} = \frac{(765\ \text{mmHg})(0.615\ \text{L})}{(62.36\ \text{mmHg} \cdot \text{L} \cdot \text{mol}^{-1} \cdot \text{K}^{-1})(293.4\ \text{K})} = 0.0257\ \text{mol}$$

$$\frac{0.825\ \text{g}}{0.0257\ \text{mol}} = 32.1\ \frac{\text{g}}{\text{mol}}$$

The molecular mass is 32.1 u.

$$(0.032\ 68\ \text{g}\ CO_2)\left(\frac{12.011\ \text{g C}}{44.010\ \text{g}\ CO_2}\right) = 0.008\ 919\ \text{g C}$$

$$\frac{0.008\ 919\ \text{g C}}{0.023\ 06\ \text{g sample}} \times 100 = 38.68\%\ \text{C}$$

$$(0.033\ 45\ \text{g H}_2\text{O})\left(\frac{2.0159\ \text{g H}}{18.015\ \text{g H}_2\text{O}}\right) = 0.003\ 743\ \text{g H}$$

$$\frac{0.003\ 743\ \text{g H}}{0.023\ 06\ \text{g sample}} \times 100 = 16.23\%\ \text{H}$$

$$38.68\% + 16.23\% = 54.91\%$$

The compound must contain an element other than C and H.

Because a water solution of the unknown compound was basic to litmus, the unknown compound is probably an amine and contains N.

$$100.00\% - 54.91\% = 45.09\%\ \text{N}$$

Assume 100–g sample

$$(38.68\ \text{g C})\left(\frac{1\ \text{mol C}}{12.011\ \text{g C}}\right) = 3.220 \qquad \frac{3.220}{3.220} = 1$$

$$(16.23\ \text{g H})\left(\frac{1\ \text{mol H}}{1.0079\ \text{g H}}\right) = 16.10 \qquad \frac{16.10}{3.220} = 5$$

$$(45.09\ \text{g N})\left(\frac{1\ \text{mol N}}{14.0067\ \text{g N}}\right) = 3.220 \qquad \frac{3.220}{3.220} = 1$$

The empirical formula is CH_5N, mass 31 u.

Because the experimental molecular mass is 32.1 u, the molecular formula is also CH_5N. The compound is CH_3NH_2, methylamine.

22.95 The two atoms on either side of a triple bond lie on a straight line with the triply bonded atoms

C—C≡C—C

This leaves only two more carbon atoms to complete the 6–membered ring. They can't reach.

Applications

22.99 $CH_3I + {}^-OCH_2CH_2CH_3 \rightarrow CH_3OCH_2CH_2CH_3 + I^-$

$CH_3CH_2CH_2Br + {}^-OCH_3 \rightarrow CH_3CH_2CH_2OCH_3 + Br^-$

22.103 (a) A *homogeneous catalyst* is in the same phase as the reactants.

(b) $CH_3CH_2CH_2CH_2CHO$, $CH_3CH_2CH(CHO)CH_3$

(c) 2.3×10^3 lb $CH_3CH_2CH_2CHO$, 5.1×10^2 lb $(CH_3)_2CHCHO$

Problem	1 ton = 2000 lb		? lb (67% yield)		? lb (15% yield)
Equation	$CH_3CH{=}CH_2$	$\rightarrow$	$CH_3CH_2CH_2CHO$	+	$(CH_3)_2CHCHO$
Formula mass, u	42.08		72.10		72.10
Recipe, mol	1		1		1

$$(2000\ \text{lb CH}_3\text{CH=CH}_2)\left(\frac{1\ \text{lb-mol CH}_3\text{CH=CH}_2}{42.08\ \text{lb CH}_3\text{CH=CH}_2}\right)\times \left(\frac{1\ \text{lb-mol CH}_3\text{CH}_2\text{CH}_2\text{CHO}}{1\ \text{lb-mol CH}_3\text{CH=CH}_2}\right)\left(\frac{72.10\ \text{lb CH}_3\text{CH}_2\text{CH}_2\text{CHO}}{1\ \text{lb-mol CH}_3\text{CH}_2\text{CH}_2\text{CHO}}\right)$$

$$= 3.4\times 10^3\ \text{lb CH}_3\text{CH}_2\text{CH}_2\text{CHO theory}$$

$$\text{Actual yield} = \left(\frac{\text{(Theoretical yield)(percent yield)}}{100}\right) = \left(\frac{(3.4\times 10^3\ \text{lb CH}_3\text{CH}_2\text{CH}_2\text{CHO})(67)}{100}\right)$$

$$= 2.3\times 10^3\ \text{lb CH}_3\text{CH}_2\text{CH}_2\text{CHO}$$

Since the second product is an isomer of the first product, the theoretical yield will be the same, 3.4×10^3 lb $(CH_3)_2CHCHO$.

$$\text{Actual yield} = \left(\frac{\text{(Theoretical yield)(percent yield)}}{100}\right) = \left(\frac{(3.4\times 10^3\ \text{lb (CH}_3)_2\text{CHCHO})(15)}{100}\right)$$

$$= 5.1\times 10^2\ \text{lb (CH}_3)_2\text{CHCHO}$$

22.107 **(a)** 2878 kJ

The equation for photosynthesis is

	$6CO_2(g)$	+	$6H_2O(l)$	$\xrightarrow{\text{light}}$	$C_6H_{12}O_6(aq)$	+	$6O_2(g)$
ΔG_f^0, kJ/mol	–394.38		–237.18		–911		0

$$\Delta G^0 = \left[(1\ \text{mol C}_6\text{H}_{12}\text{O}_6)\left(\frac{-911\ \text{kJ}}{\text{mol C}_6\text{H}_{12}\text{O}_6}\right) + (6\ \text{mol O}_2)\left(\frac{0\ \text{kJ}}{\text{mol O}_2}\right)\right]$$

$$-\left[(6\ \text{mol CO}_2)\left(\frac{-394.38\ \text{kJ}}{\text{mol CO}_2}\right) + (6\ \text{mol H}_2\text{O})\left(\frac{-237.18\ \text{kJ}}{\text{mol H}_2\text{O}}\right)\right]$$

$$= -911\ \text{kJ} + 0\ \text{kJ} + 2366.28\ \text{kJ} + 1423.08\ \text{kJ} = +2878\ \text{kJ}$$

(b) 2956 kJ

$$\Delta G = \Delta G^0 + RT\ln Q \qquad Q = \frac{[C_6H_{12}O_6]\,p_{O_2}{}^6}{p_{CO_2}{}^6}$$

$$\Delta G = 2878\ \text{kJ} + \left(\frac{0.008\,315\ \text{kJ}}{\text{K}}\right)(298\ \text{K})\ln\frac{(1.0\times 10^{-3})(0.21)^6}{(3.4\times 10^{-4})^6} = 2956\ \text{kJ}$$

(c) No – ΔG is positive.

(d) Sunlight provides the energy required to make the nonspontaneous process take place.

(e) –2956 kJ

The metabolism of glucose is the reverse of the photosynthesis of glucose.

CHAPTER 23: Polymers: Synthetic and Natural

Practice Problems

23.1

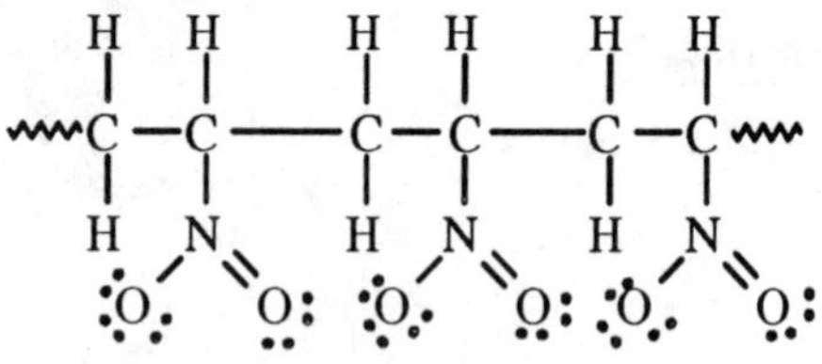

23.2 **(a)** $CH_2{=}CCl_2$

(b)

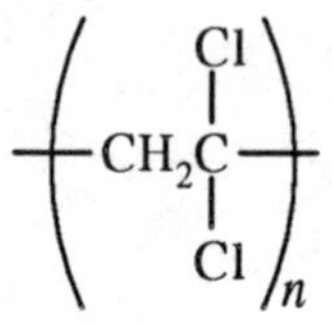

23.3 Chain initiation:

Rad· + CH_2::$\dot{C}HOCH_3$ ⟶ Rad··CH_2··$\dot{C}HOCH_3$

Chain propagation:

Rad··CH_2··$\dot{C}HOCH_3$ + CH_2::$\dot{C}HOCH_3$ →
Rad··CH_2··$CH(OCH_3)$··CH_2··$\dot{C}HOCH_3$

and then

Rad··CH_2··$CH(OCH_3)$··CH_2··$CHOCH_3$ +
CH_2::$\dot{C}HOCH_3$ →
Rad : CH_2 : $CH(OCH_3)$: CH_2 : $CH(OCH_3)$: CH_2 : $\dot{C}HOCH_3$, etc.

Chain termination:

Rad· + $CH_3O\dot{C}H$··CH_2··$CH(OCH_3)$··CH_2··Rad or
Rad··CH_2··$\dot{C}HOCH_3$ + $CH_3O\dot{C}H$··CH_2··Rad or
Rad· + ·Rad or any other step in which two radicals combine.

23.4 **(a)**

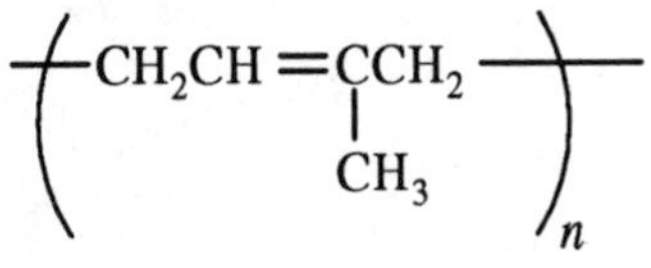

(b)

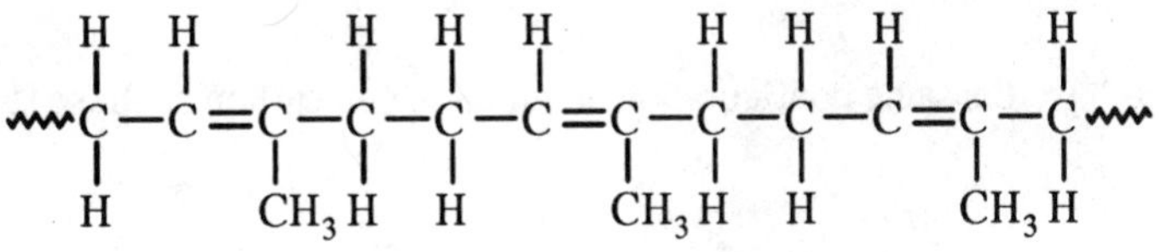

These structures can be determined by analogy to those for polychloroprene (Neoprene) in Table 23.1.

23.5 **(a)**

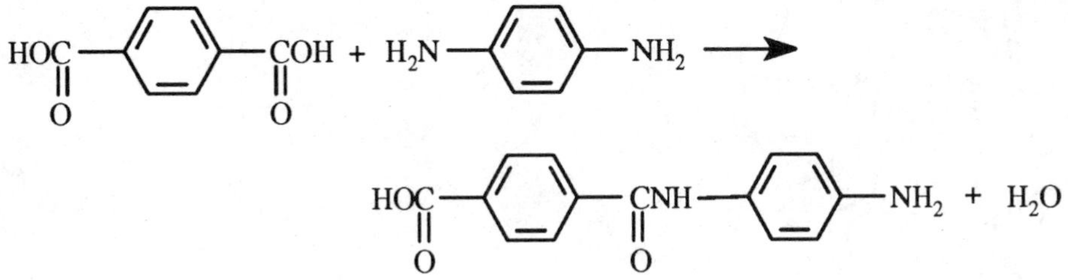

(b)

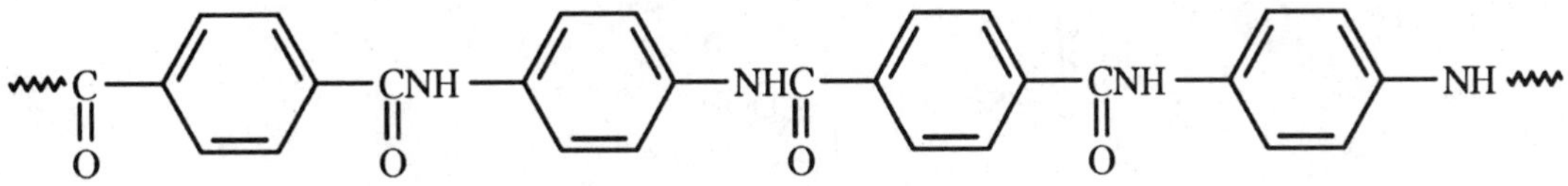

(c)

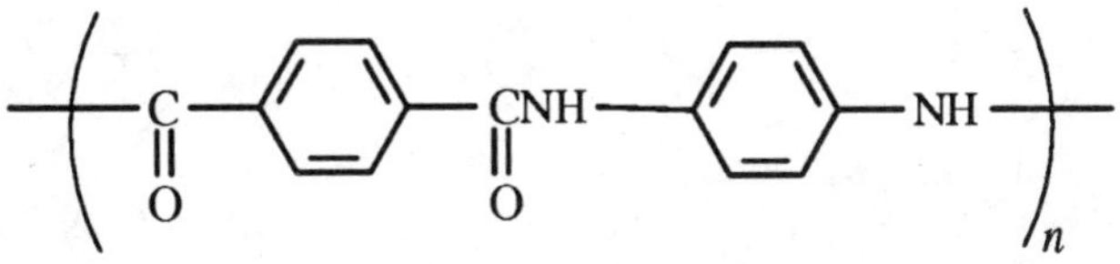

23.6 $H_3\overset{+}{N}CH(CH_3)COO^-$

23.7 **(a)** leucine

(b) hydroxylysine

The group in the asparagine side chain is an amide, which is neutral, not an amine.

(c) aspartic acid

(d) hydroxylysine and threonine

$H_2NCH[CH_2CH_2\overset{*}{C}H(OH)CH_2NH_2]COOH$ and $H_2NCH[\overset{*}{C}H(OH)CH_3]COOH$

23.8

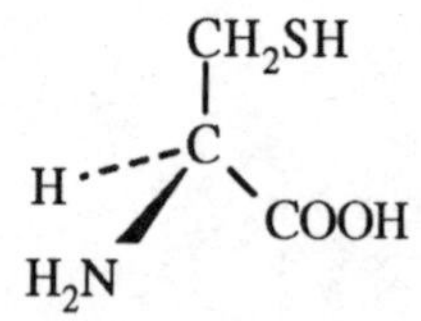

Most of the alpha-amino acids in proteins have the same configuration about the alpha carbon as those shown in Figure 23.4.

23.9 **(a)** VF

See Table 23.2 for abbreviations.

(b)

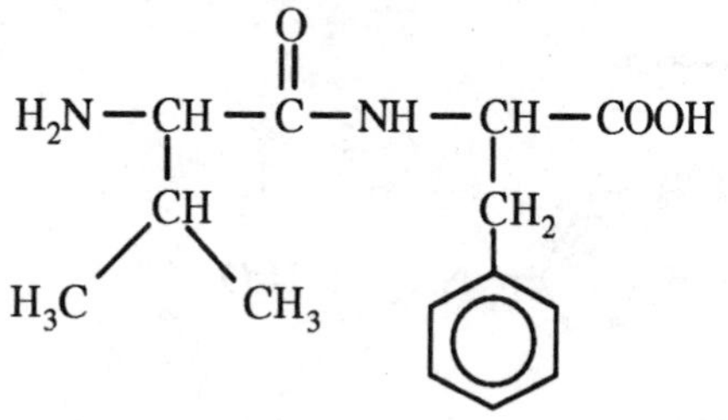

(c) Phe-Val

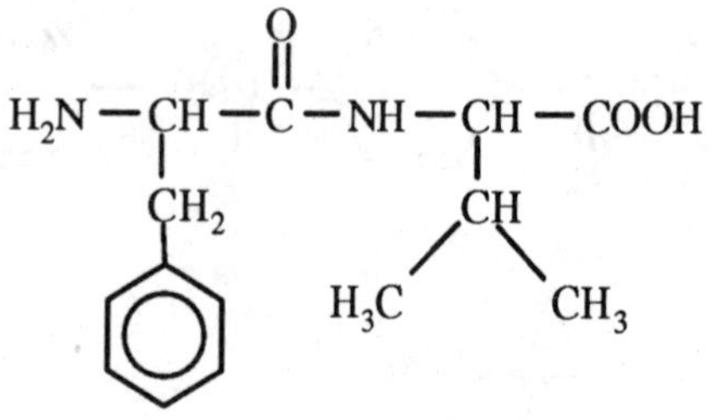

23.10 Cys-Gly and Gly-Glu

Cleavage occurs at either end of the two peptide links:

Cleavage here gives Cys-Gly (and Glu).

↓

Cys—Gly—Glu

↑

Cleavage here gives Gly-Glu (and Cys).

23.11 Gly-Glu-Cys

By overlapping the two dipeptides, one gets Gly-Glu-Cys as the tripeptide:

Glu - Cys
Gly - Glu
Gly - Glu - Cys

23.12 **(a)** $Ala_2Gly_1Ser_1$

One mole peptide gives two moles Ala, one mole Gly, and one mole Ser, a total of four moles. The molecular formula is $Ala_2Gly_1Ser_1$.

(b) tetra–

(c) Ala-Gly-Ser-Ala

Using the overlap technique,

Gly -Ser
Ala - Gly
Ser - Ala

Ala - Gly - Ser- Ala is the tetrapeptide.

23.13 (a) During cyclization, the second-to-the-last —OH group reacts with the carbonyl (aldehyde) group.

(b) The ring form has one more stereocenter than the straight-chain form.

Since the original carbonyl group is converted into a hydroxyl group during cyclization, a new stereocenter is formed. Thus, the cyclic sugar has one more stereocenter than the straight-chain form.

(c) In aqueous solution, the cyclic forms are in equilibrium with the straight-chain form.

23.14 (a) Polysaccharides are macromolecules made up of hundreds or thousands of simple sugar units.

(b) The two most important polysaccharides are cellulose and starch. Cellulose is the chief structural material of plants. It is used as wood for building, cotton for clothing, and paper. Acetate rayon, viscose rayon, and cellophane are modified forms of cellulose. Starch functions as a reserve food supply in plants. Starch in potatoes, corn, and rice is used as food.

23.15 TTC

23.16 DNA is missing the oxygen at the 2′ position that is present in RNA. DNA contains thymine; RNA contains uracil.

23.17 (a) two

(b) three

Adenine and thymine are connected by two hydrogen bonds. Guanine and cytosine are linked by three hydrogen bonds.

(c) guanine and cytosine

The more hydrogen bonds, the stronger the linkage between the two bases.

23.18

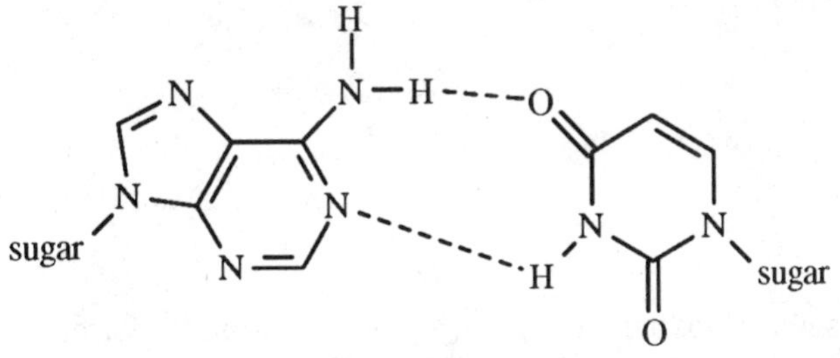

23.19 The complementary triplets of RNA to the glycine-coding CCA, CCG, CCT, and CCC sequences are GGU, GGC, GGA, and GGG, respectively.

23.21 **(a)** $\sim\sim CH(CH_2CH_3)-CH_2-CH(CH_2CH_3)-CH_2 \sim\sim$ $\quad -\!\!\left(CH(CH_2CH_3)CH_2\right)_n\!\!-$

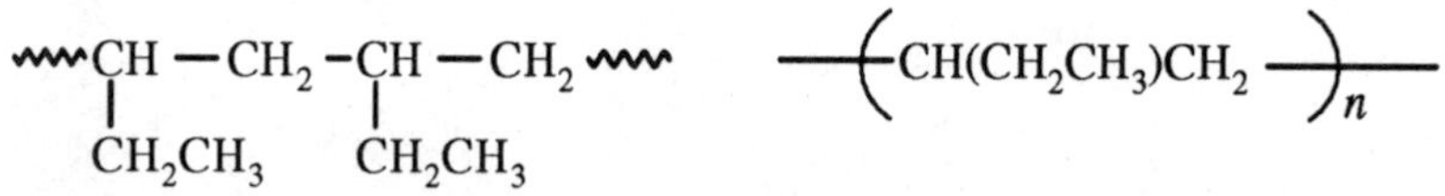

(b) $\sim\sim CH_2CH{=}CHCH_2-CH_2CH{=}CHCH_2 \sim\sim$ $\quad -\!\!\left(CH_2CH{=}CHCH_2\right)_n\!\!-$

(c) $\sim\sim OC(CH_2)_{10}CONH(CH_2)_6NHOC(CH_2)_{10}CONH(CH_2)_6NH \sim\sim$

$-\!\!\left(OC(CH_2)_{10}CONH(CH_2)_6NH\right)_n\!\!-$

(d)

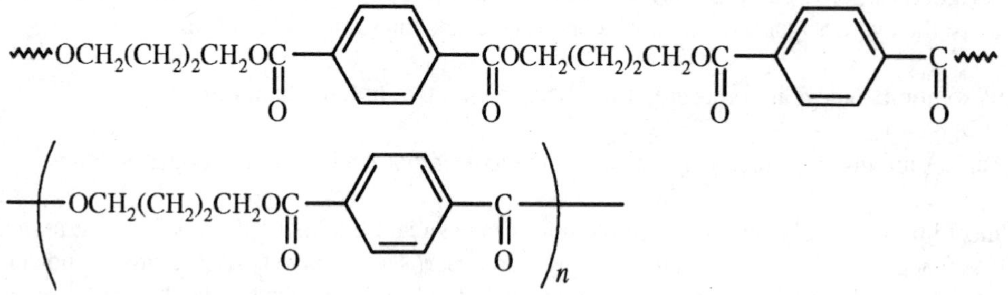

23.25 **(a)** a C═C double bond

(b) The monomers must have two functional groups (either two different molecules each with two identical functional groups or one molecule with two different groups).

23.29 **(a)** Molecules of *fibers* are regularly shaped and have strong intermolecular forces. Molecules of *elastomers* are irregularly shaped and have weak intermolecular forces. Molecules of *plastics* are intermediate between those of fibers and those of elastomers.

(b) *Thermoplastic plastics* soften when heated, whereas *thermosetting plastics* do not soften when heated.

Cross-linking holds the molecules of a thermosetting plastic in position so that it will not soften when heated.

23.33 **(a)** $H_2NCH(CH_2CH_2CONH_2)COOH$

(b)

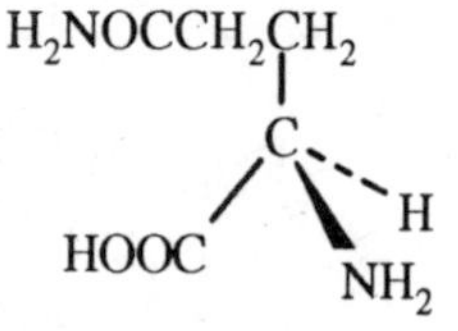

(c) amide

23.37 **(a)** The *primary structure* is the order in which the amino acids are arranged in a protein or peptide molecule. The *secondary structure* is the way that the chains of amino acids are coiled or folded. The *tertiary structure* is the way that the secondary structures fold and coil up (see Figure 23.7). *Quaternary structure* only exists in proteins containing more than one peptide chain and is the way that peptide chains pack together (see Figure 23.8).

(b) hydrolysis

Hydrolysis yields the amino acids contained in the protein. Partial hydrolysis cleaves the protein into smaller segments. The overlap technique is used to determine the sequence of the original protein.

(c) X-ray crystallography and a knowledge of the primary structure

(d) alpha helix and pleated sheet

(e) a disorderly arrangement (see Figure 23.7)

23.41 **(a)** *Nucleic acids* are acidic substances present in the nuclei of cells that store genetic information and direct the biological synthesis of proteins.

(b) *Nucleotides* are the repeating units of nucleic acids and are composed of one unit each of phosphate, sugar, and one of five heterocyclic bases.

(c) *Heterocyclic compounds* are cyclic organic compounds in which one or more of the ring atoms is not carbon.

23.45 CCTAGG

23.49 **(a)**
$$CH_3COOH + CH_3CH_2CH_2CH_2OH \xrightarrow{H^+} CH_3COOCH_2CH_2CH_2CH_3 + H_2O$$

(b)
$$HOOCCH_2CH_2COOH + HOCH_2CH_2OH \xrightarrow{H^+} HOOCCH_2CH_2COOCH_2CH_2OH + H_2O$$

(c)
$$2CH_3CH(OH)COOH \xrightarrow{H^+} CH_3CH(OH)COOCH(CH_3)COOH + H_2O$$

(d)

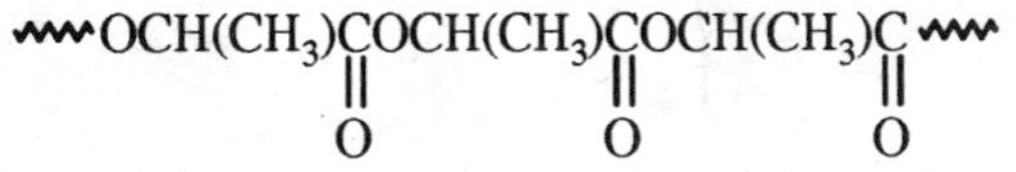

$$\text{www}OCH(CH_3)\underset{O}{\overset{||}{C}}OCH(CH_3)\underset{O}{\overset{||}{C}}OCH(CH_3)\underset{O}{\overset{||}{C}}\text{www}$$

(e)

$$\left(OCH(CH_3)\underset{O}{\overset{||}{C}}\right)_n$$

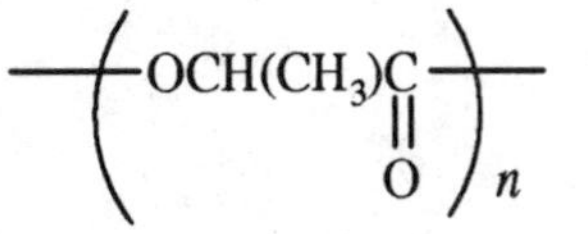

23.53 One pure enantiomer of an essential amino acid occurs naturally. A racemic mixture obtained by synthesis does *not* have to be separated to obtain this enantiomer.

23.57 **(a)** three **(b)** four

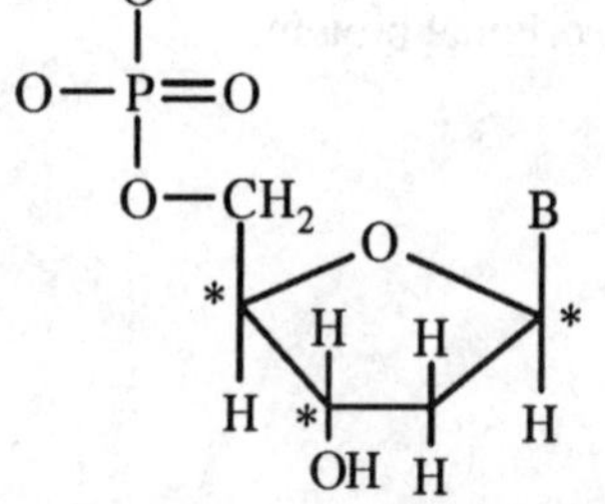

3 chiral carbons in DNA

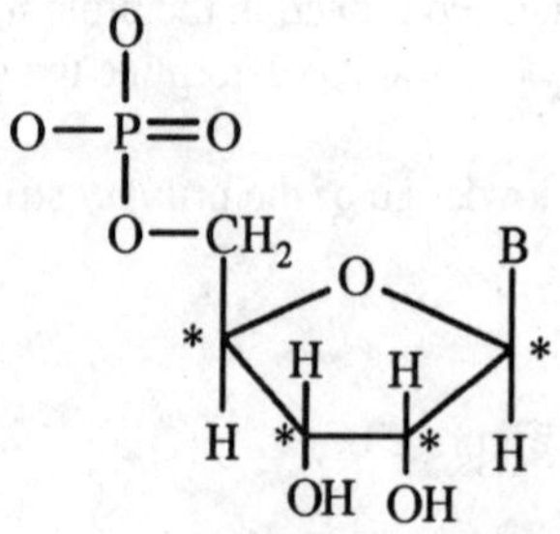

4 chiral carbons in RNA

23.61 **(a)**

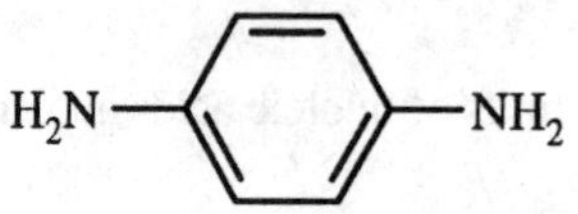

(b)

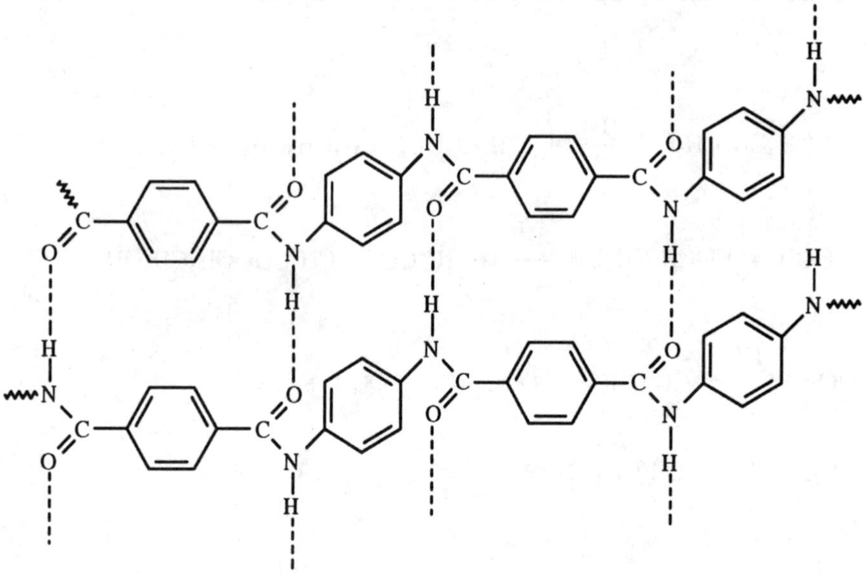

(c) all of them

(d) 400 (4×10^2)

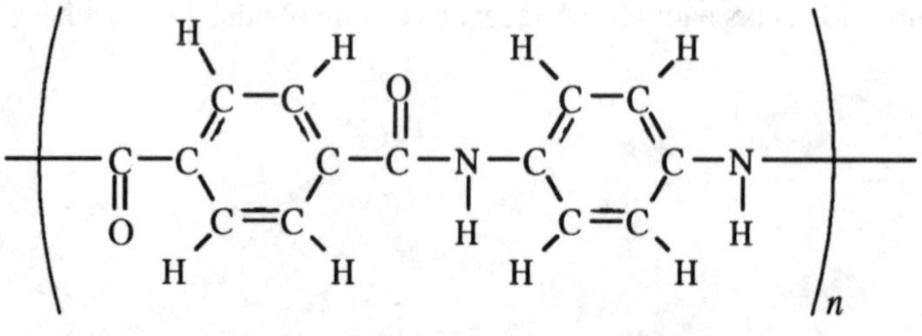

Repeating unit is $C_{14}H_{10}N_2O_2$, molar mass = 238 g/mol

$$\left(\frac{10^5 \text{ g}}{1 \text{ mol polymer}}\right)\left(\frac{1 \text{ mol repeating unit}}{238 \text{ g}}\right) = 4 \times 10^2 \text{ units per polymer}$$

(e) Kevlar does not corrode.

23.65 **(a)**

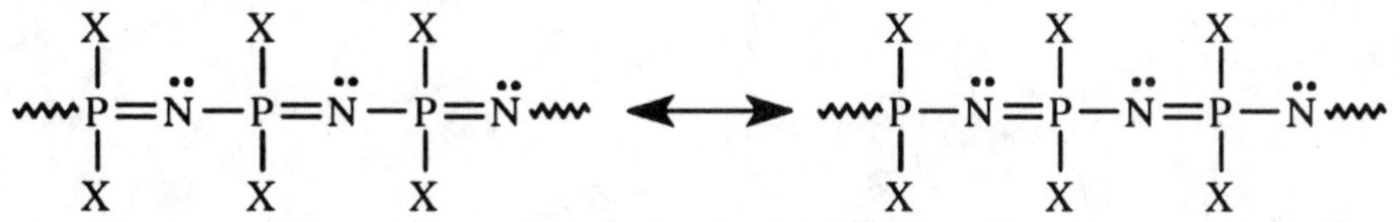

(b) P–N bond lengths in the polymer are shorter than P–N single bonds.

Resonance in the polymer leads to P–N bonds that are shorter than P–N single bonds. This prediction is correct.

(c) $-N(CH_3)_2$; its structure is similar to NH_3.

23.69 Gly – Ser – Gly – Ala – Gly – Ala

The overlap method gives:

```
Gly – Ser – Gly
      Ser – Gly – Ala
            Gly – Ala – Gly
                  Ala – Gly – Ala
-----------------------------------
Gly – Ser – Gly – Ala – Gly – Ala
```

23.73 Wood is strong because London forces between the long chains of cellulose are very large. There is also considerable hydrogen bonding between chains. The twisting together of the bundles of cellulose chains adds to wood's strength.

23.77 For the 21 amino acid chain, 69; for the 30 amino acid chain, 96

Polypeptide chain containing 21 amino acids requires 21 codons for amino acids, 1 codon for starting the synthesis, and 1 codon for stopping the synthesis. 23 codons × 3 nucleotides/codon = 69 nucleotides.

Polypeptide chain containing 30 amino acids requires 30 codons for amino acids, 1 codon for starting the synthesis, and 1 codon for stopping the synthesis. 32 codons × 3 nucleotides/codon = 96 nucleotides.

CHAPTER 24:
A Closer Look at Inorganic Chemistry: Transition Metals and Complexes

Practice Problems

24.1 **(a)** four

The electron configuration for iron is $[Ar]4s^2 3d^6$. Electrons in the $3d$ orbitals can be depicted as
[↑↓] [↑] [↑] [↑] [↑] for iron.

(b) five

The electron configuration for Fe^{3+} is $[Ar]3d^5$. Electrons in the $3d$ orbitals can be depicted as
[↑] [↑] [↑] [↑] [↑].

24.2 Rhenium would be most likely to resemble Tc because of their similarities in size and electron configuration.

Manganese, technetium, and rhenium all have similar electron configurations. However, technetium is very close in size to rhenium but is larger than manganese. Since technetium and rhenium are similar in size and electron configuration, they have similar properties.

24.3 two

24.4 **(a)** $[AuF_4]^-$

(b) $K[AuF_4]$

24.5 **(a)** +3

(b) +2

(c) +5

(d) 0

24.6 **(a)** four

In each case, four ligands are bonded directly to platinum.

(b) $[Pt(NH_3)_4]Cl_2$

$[PtCl(NH_3)_3]Cl$

$[PtCl_2(NH_3)_2]$

24.7 **(a)** two

(b)

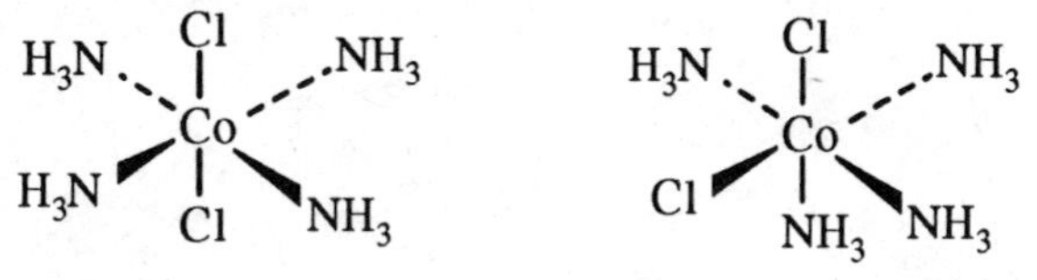

(c) No.

24.8 **(a)** six

(b) +3

24.9

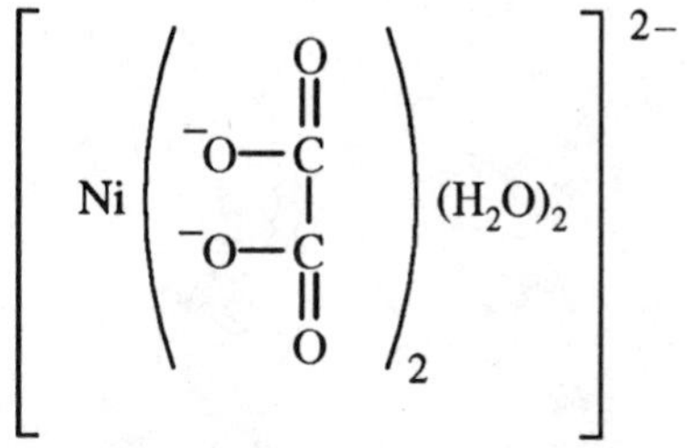

24.10 **(a)** $[CoBr(NH_3)_5]SO_4(aq) + Ba^{2+} \rightarrow BaSO_4(s) + [CoBr(NH_3)_5]^{2+}$

(b) no reaction

(c) no reaction

(d) $[CoSO_4(NH_3)_5]Br(aq) + Ag^+ \rightarrow AgBr(s) + [CoSO_4(NH_3)_5]^+$

24.11 **(a)** pentaaquahydroxoaluminum ion

(b) potassium hexacyanoferrate(II)

(c) hexacarbonylvanadium

24.12 **(a)** $[Cr(H_2O)_6]Cl_3$

(b) $[Cu(NH_3)_4][PtCl_4]$

24.13 — —

↑↓ ↑↓ ↑↓

The electron configuration for cobalt is [Ar]$3d^7 4s^2$; the electron configuration for Co^{3+} is [Ar]$3d^6$. The fact that the $[Co(NH_3)_6]^{3+}$ ion is diamagnetic tells us that this ion has no unpaired electrons. Thus, the ion is probably low spin, strong field with a relatively large separation between the two sets of *d* orbitals.

— —

↑ Δ_O relatively large ↓

↑↓ ↑↓ ↑↓

24.14 No. Only three electrons are available.

Cr^{3+} ion has the following electron configuration:

	3d					4s	4p		
[Ar]$3d^3$ or [Ar]	↑	↑	↑						

Since it has only three electrons to add to the split *d* orbitals, they will all remain unpaired in the lowest energy orbitals no matter what the Δ_O:

— —

↑ ↑ ↑

24.15 — —

↑ Δ_O relatively large (low spin, strong field splitting) ↓

↑↓ ↑↓ ↑↓

Ir^{3+} ion has six *d* electrons to place into orbitals split by the crystal field. Since the elements in the third transition series have relatively high nuclear charges, this produces strong field splitting and low spin.

24.16 **(a)** smaller

Water produces lower crystal field splitting than ammonia.

(b) increase

The wavelength of light absorbed will increase, which tells us that Δ_O is smaller.

(c) blue end

The wavelength of light absorbed is shifted toward the red end of the visible spectrum. Thus, the light reflected and transmitted will shift toward the blue end.

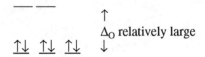

24.17 ↑↓ ↑ ↑

↑
Δ_t = weak field, high spin
↓

↑↓ ↑↓

Since tetrahedral crystal fields are always weak fields, Ni^{2+} with its eight *d* electrons will be in a high–spin configuration.

24.18 ___

↑↓

↑↓

↑↓ ↑↓

The ground–state electron configuration of the Pt atom is

$[Xe]4f^{14}5d^{9}6s^{1}$

The ground–state electron configuration of the Pt^{2+} ion is

$[Xe]4f^{14}5d^{8}$

Because the $[Pt(NH_3)_4]^{2+}$ ion is diamagnetic, the seventh and eighth electrons must be in the same orbital.

Additional Practice Problems

24.19 **(a)** three

V [Ar] [↑] [↑] [↑] [ ] [ ] (3*d*) [↑↓] (4*s*) 3 unpaired electrons

(b) three

V^{2+} [Ar] [↑] [↑] [↑] [ ] [ ] (3*d*) [ ] (4*s*) 3 unpaired electrons

24.23 **(a)** Characteristic properties of transition elements: metallic, high melting and boiling points, high densities, hard and strong. Most transition elements have more than one oxidation number in their common compounds; the oxidation numbers differ by one. Compounds in at least one oxidation state are usually colored. Compounds in at least one oxidation state are ususally paramagnetic. Most transition metal ions form complexes readily.

(b) Melting points, boiling points, and densities increase to a maximum toward the middle of a row in the periodic table and then decrease.

(c) tungsten

Moving down a group, transition elements in the fifth period (e.g., molybdenum) resemble those of the sixth period (e.g., tungsten) more than those of the fourth period (e.g., chromium).

(d) Outer electron configuration is ns^2 for most transition elements.

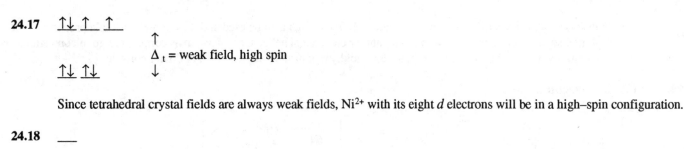

(e) Transition metals are sufficiently unreactive for the free elements to be used and have many desirable properties. For example, some, such as iron, are very strong and/or can be magnetized. Some, such as copper, are good conductors of electricity and thermal energy. Some, such as gold, are unusually unreactive and do not corrode.

24.27 **(a)** no geometric isomers

(b)

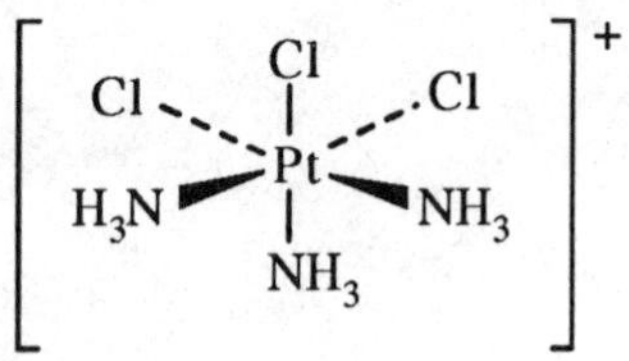

and

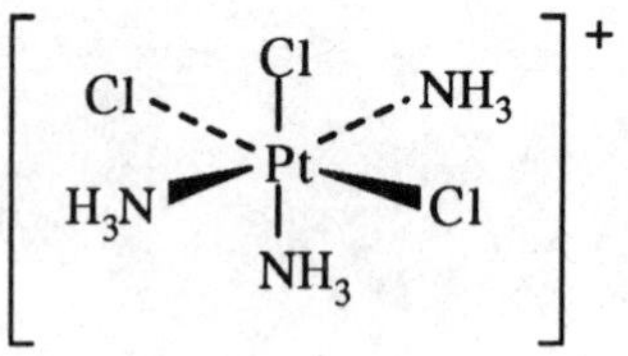

(c)

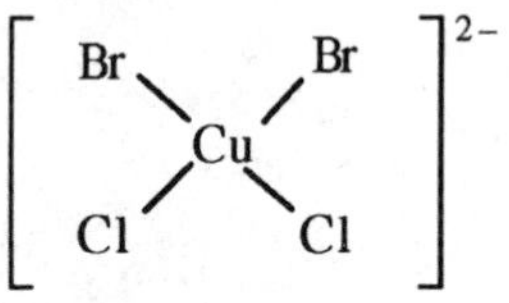

and

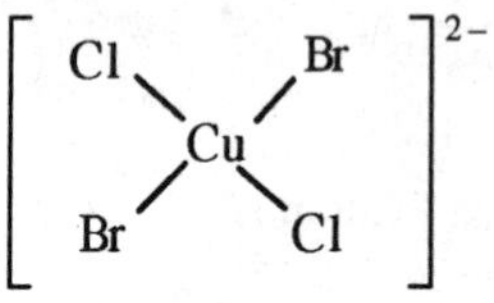

(d) no geometric isomers

24.31 **(a)** tridentate

(b) didentate

(c) tetradentate

(d) monodentate

(e) didentate

24.35

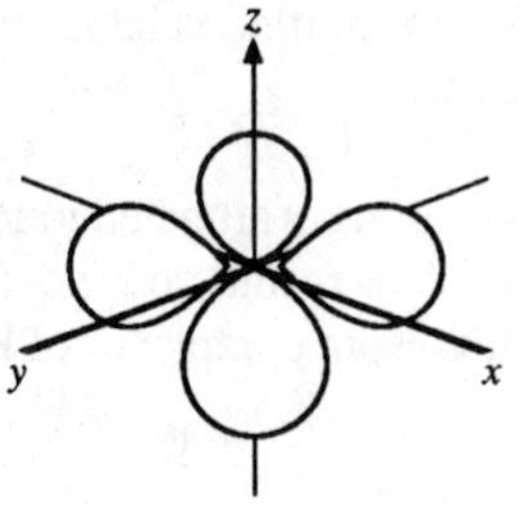

xy orbital

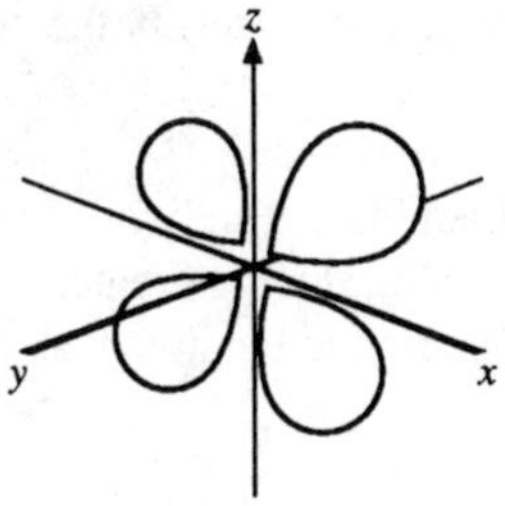

xz orbital

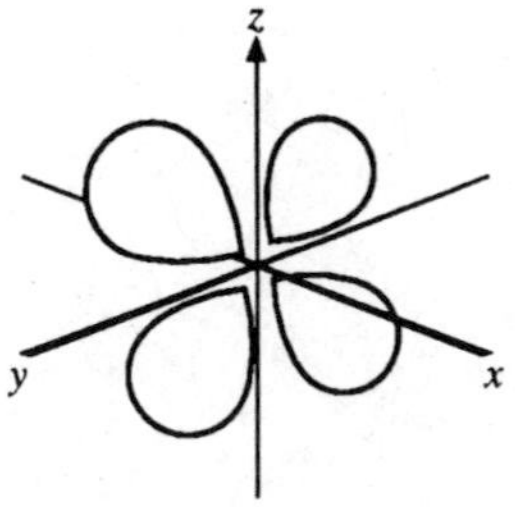

yz orbital

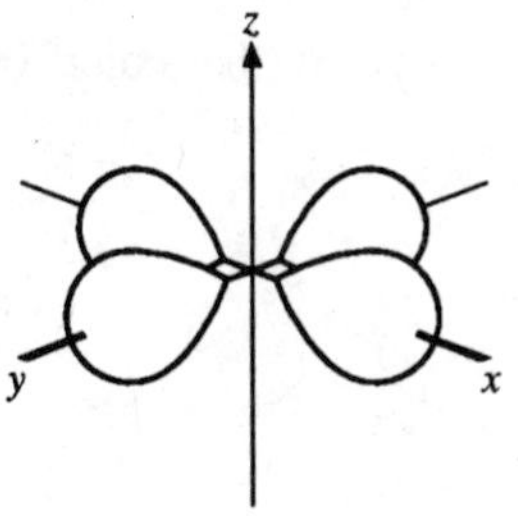

x^2-y^2 orbital

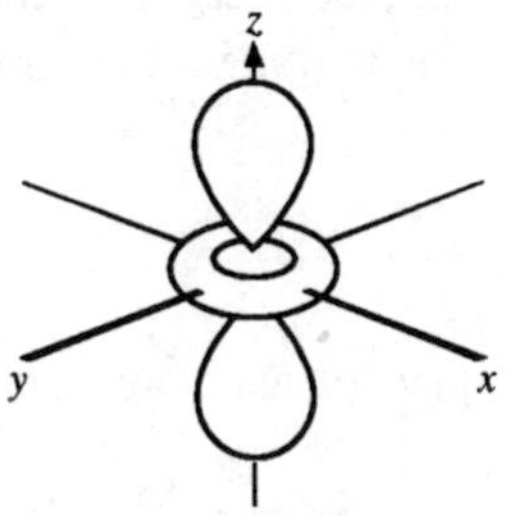

z^2 orbital

24.39 **(a)** The number of unpaired electrons is determined by measuring with a sensitive balance the effect of a strong magnetic field on a sample of the material suspended in the magnetic field. (Remember Figure 8.3.) The more unpaired electrons present, the stronger the attraction of the sample to the magnet.

(b) weak field, high spin = $[Fe(H_2O)_6]^{2+}$

strong field, low spin = $[Fe(CN)_6]^{4-}$

Fe^{2+} is d^6.

	__ __
↑ ↑	
↑↓ ↑ ↑	↑↓ ↑↓ ↑↓
paramagnetic	diamagnetic
$[Fe(H_2O)_6]^{2+}$	$[Fe(CN)_6]^{4-}$

(c) yes

$H_2O < CN^-$

24.43 **(a)** Cyanide ion is at the far right of the spectrochemical series. As a result, the crystal field splitting is large and $[Fe(CN)_6]^{3-}$ is low spin.

(b) Ruthenium and osmium have higher nuclear charges than iron. The ligands are more strongly attracted to Ru^{3+} and Os^{3+} than to Fe^{3+} and approach the nucleus more closely, producing greater splitting. As a result, no weak field, high spin complexes are known for Ru(III) and Os(III), although many weak field, high spin complexes of Fe(III) are known.

(c) one

Ru^{3+} and Os^{3+} are d^5. The electron configuration is

__ __

↑↓ ↑↓ ↑

with one unpaired electron.

24.47 The ligands vary in strength in the order $Br^- < H_2O <$ en. Thus, the energy absorbed increases from left to right. The wavelength of the light absorbed decreases. Thus, the wavelength of light transmitted or reflected (the "color" of the compound) increases from violet to blue to green.

24.51 square planar

Gold(III) ion is d^8. Splitting diagrams for tetrahedral and square planar complexes are

	__
	↑↓
↑↓ ↑ ↑	↑↓
↑↓ ↑↓	↑↓ ↑↓
tetrahedral	square planar
Paramagnetic with two unpaired electrons	Diamagnetic

Stop and Test Yourself

1. **(e)** pentacyanohydroxocobaltate(III) ion

2. **(c)** 3, 1

The complex dissociates to form one $[Co(NH_3)_5Cl]^{2+}$ ion and two Cl^- ions, a total of three ions. The chlorine contained in the brackets is tightly held to the cobalt and does not dissociate.

3. **(e)**

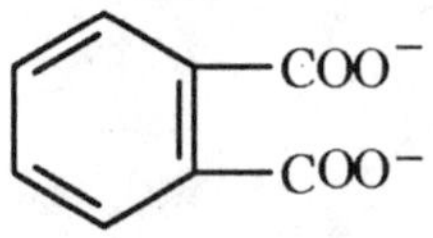

4. **(d)** Fe^{2+}

The ground state for Fe is [Ar]$3d^6 4s^2$. Removal of the two *s* electrons to form Fe^{2+} ion makes it a d^6 ion.

5. **(a)** d^3

The only way to fill the *d* orbitals in an octahedral complex with three electrons is independent of crystal field strength:

__ __

↑ ↑ ↑ ← three unpaired electrons

In an uncomplexed d^3 ion, the number of unpaired electrons is the same:

↑ ↑ ↑ __ __ ← three unpaired electrons

The d^4 – d^7 ions may have variable numbers of unpaired electrons, depending on the strength of the ionic field.

6. **(b)** octahedral with weak field and high spin

Octahedral complexes generate two sets of orbitals as shown in the diagram. The field must be weak because an electron was placed in a higher energy orbital instead of in a lower energy orbital.

7. **(e)** $[Zn(CN)_4]^{2-}$

Zinc has a completely filled *d* orbital. Thus, no visible wavelengths are absorbed because the lower energy electrons cannot be excited to the higher energy orbitals. As a result, the complex is colorless.

8. **(a)** blue–green

The absorbed wavelength range is in the red–orange region. Thus, the yellow, green, blue, and violet wavelengths are allowed to pass.

9. **(b)** $[CoF_6]^{3-}$

Since there are six ligands, we can assume that the ion is octahedral. The F^- ion is the farthest to the left in the spectrochemical series of the ligands shown and produces the least splitting. Thus, the fluoride complex ion is weak field and the electron configuration is

↑ ↑

↑↓ ↑ ↑ ← four unpaired electrons: paramagnetic

The other ligands result in electron configuration

__ __

↑↓ ↑↓ ↑↓

10. **(a)** $[CdBr_4]^{2-}$

The complex with the greatest K_f is the most thermodynamically stable.

Putting Things Together

24.55 Vanadium forms the oxide V_2O_5.

24.59 **(a)** $[Pt(NH_3)_4][PtCl_4]$: tetraammineplatinum(II) tetrachloroplatinate(II)

$PtCl_2(NH_3)_2$:	1 Pt @	195	=	195 u
	2 Cl @	35.5	=	71 u
	2 N @	14	=	28 u
	6 H @	1	=	6 u
				300 u

Since the compound has a formula mass of 600, the molecular formula must be $[PtCl_2(NH_3)_2]_2 = Pt_2Cl_4(NH_3)_4$

$Pt_2Cl_4(NH_3)_4 = [Pt(NH_3)_4][PtCl_4]$

(b) $[Pt(NH_3)_4]^{2+} + [PtCl_4]^{2-} \rightarrow [Pt(NH_3)_4][PtCl_4](s)$

(c) $[Pt(NH_3)_4]Cl_2$ (tetraammineplatinum(II) chloride) and $Na_2[PtCl_4]$ (sodium tetrachloroplatinate(II))

Complete ionic equation: $[Pt(NH_3)_4]^{2+} + 2Cl^- + 2Na^+ + [PtCl_4]^{2-} \rightarrow [Pt(NH_3)_4][PtCl_4](s) + 2Na^+ + 2Cl^-$

Molecular equation: $[Pt(NH_3)_4]Cl_2(aq) + Na_2[PtCl_4](aq) \rightarrow [Pt(NH_3)_4][PtCl_4](s) + 2NaCl(aq)$

24.63 **(a)** one with $K_f = 5 \times 10^5$

(b) one with $K_f = 3 \times 10^8$

(c) one with $K_f = 3 \times 10^{-5}$

(d) one with $pK_f = 5.2$

K_f is the equilibrium constant for the formation of a complex (see Section 16.4). The larger the value of K_f, the more stable the complex.

24.67 **(a)** +7 **(b)** +4

(a) $\overset{+1\ +7\ -2}{KMnO_4}$, $+1 +7 -8 = 0$

(b) $\overset{+4\ -2}{MnO_2}$, $+4 -4 = 0$

(c) +2

$MnSO_4$ — The sulfate ion has a 2– charge. Therefore, manganese ion in this compound must have a 2+ charge. Its oxidation number is therefore +2.

(d) +6

$$\overset{+1}{K_2}\overset{+6}{Mn}\overset{-2}{O_4}$$
+2 +6 −8 = 0

(e) +3

$$\overset{+3}{Mn_2}\overset{-2}{O_3}$$
+6 −6 = 0

24.71 **(a)** In the presence of cyanide ion, the spontaneous reaction is $Cu^{2+} + Cu(s) + 4CN^- \rightarrow 2[Cu(CN)_2]^-$. For this cell, $E^0_{cell} = +1.12\ V - (-0.44\ V) = +1.56\ V$. In the absence of cyanide ion, the spontaneous reaction is $2Cu^+ \rightarrow Cu^{2+} + Cu(s)$. For this cell, $E^0_{cell} = +0.52\ V - (+0.159\ V) = +0.36\ V$.

(b) disproportionation

24.75 The aluminum ion is small. Six small fluoride ions can approach the aluminum ion to form an octahedral complex. Chloride ions are larger than fluoride ions, and only four chloride ions can get close enough to form a complex, which is therefore tetrahedral.

24.79 The chemical properties of the transition elements vary from element to element because the configuration of *d* electrons varies and the sizes of the atoms vary across a period. The lanthanides are similar in chemical properties because the valence electron configuration of the elements is the same, $6s^2$, and they all form a 3+ ion with configuration $4f^n5d^06s^0$. In addition, the lanthanides all have similar radii.

24.83

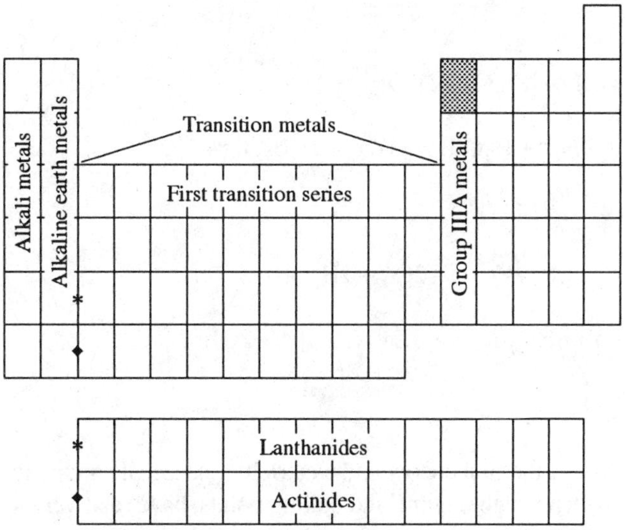

24.87 **(a)** 515 nm and 625 nm, respectively

Wavelengths may be calculated by converting cm^{-1} to nm:

$$19\,400\ cm^{-1} = (5.15 \times 10^{-5}\ \cancel{cm})\left(\frac{10^{-2}\ \cancel{m}}{\cancel{cm}}\right)\left(\frac{1\ nm}{10^{-9}\ \cancel{m}}\right) = 515\ nm \quad \text{strong}$$

$$16\,000\ cm^{-1} = (6.25 \times 10^{-5}\ \cancel{cm})\left(\frac{10^{-2}\ \cancel{m}}{\cancel{cm}}\right)\left(\frac{1\ nm}{10^{-9}\ \cancel{m}}\right) = 625\ nm \quad \text{weak}$$

(b) green

The strong absorption band at 515 nm is in the green region of the spectrum. Thus, green light is strongly absorbed.

(c) red to red–blue

The red, orange, blue, and violet wavelengths are allowed to pass through an aqueous solution of $[Co(H_2O)_6](NO_3)_2$. Thus, the solution appears red to red–blue. (See Figure 24.11.)

24.91 (a) The actual solubility of zinc hydroxide will only be about 1% greater than the solubility calculated from K_{sp}.

From K_{sp} the solubility of $Zn(OH)_2$ is about 10^{-4} M:

$$Zn(OH)_2(s) \rightarrow \underset{s}{Zn^{2+}} + \underset{2s}{2OH^-}$$

$$K_{sp} = [Zn^{2+}][OH^-]^2 \approx 10^{-11}$$

Let solubility of $Zn(OH)_2$ equal s:

$$(s)(2s)^2 \approx 10^{-11}$$

$$4s^3 \approx 10^{-11}$$

$$s \approx 10^{-4}$$

The concentration of undissociated $Zn(OH)_2$ is 10^{-6}.

(b) (i) Net ionic equation: $Zn(OH)_2(s) + OH^- + H_2O(l) \rightarrow [Zn(OH)_3H_2O]^-$

Complete ionic equation: $Zn(OH)_2(s) + OH^- + Na^+ + H_2O(l) \rightarrow [Zn(OH)_3H_2O]^- + Na^+$

(ii) Net ionic equation: $Zn(OH)_2(s) + 2H^+ \rightarrow Zn^{2+} + 2H_2O(l)$

Complete ionic equation: $Zn(OH)_2(s) + 2H^+ + 2Cl^- \rightarrow Zn^{2+} + 2Cl^- + 2H_2O(l)$

(iii) Net and complete ionic equation: $Zn(OH)_2(s) + 4NH_3(aq) \rightarrow [Zn(NH_3)_4]^{2+} + 2OH^-$

(c) amphoteric

24.95 Properties of metals: ductile, malleable, good conductors of thermal energy and electricity (electrical conductivity decreases with increasing temperature), lustrous, usually have high melting and boiling points, have relatively low ionization energies.

24.99 708 nm

$$\left(\frac{169\ \cancel{kJ}}{1\ \cancel{mol}}\right)\left(\frac{1000\ J}{1\ \cancel{kJ}}\right)\left(\frac{1\ \cancel{mol}}{6.022 \times 10^{23}}\right) = 2.81 \times 10^{-19}\ J$$

$$\lambda = \frac{hc}{\Delta E} = \frac{(6.626 \times 10^{-34}\ \cancel{J} \cdot \cancel{s})(2.998 \times 10^{8}\ m \cdot \cancel{s^{-1}})}{2.81 \times 10^{-19}\ \cancel{J}} = 7.07 \times 10^{-7}\ m \text{ or } 707\ nm$$

(Best answer, without intermediate round off, is 708 nm.)

24.103 (a) A *complex* is a species in which a central atom is surrounded by a set of Lewis bases that are covalently bonded to the central atom.

(b) *Coordination compounds* include one or more complexes.

(c) A *coordinate covalent bond* is a shared electron pair of which both electrons originally belonged to the same atom.

(d) The *coordination number* is the number of atoms or groups firmly bound to the central atom.

(e) The *total coordination number* is the number of atoms and sets of unshared electron pairs around a central atom.

Applications

24.107 Oxalic acid removes rust stains because the oxalate ion, $^{-}OOCCOO^{-}$, is a didentate ligand that forms stable chelate complexes with iron(III) ions.

24.111 **(a)**

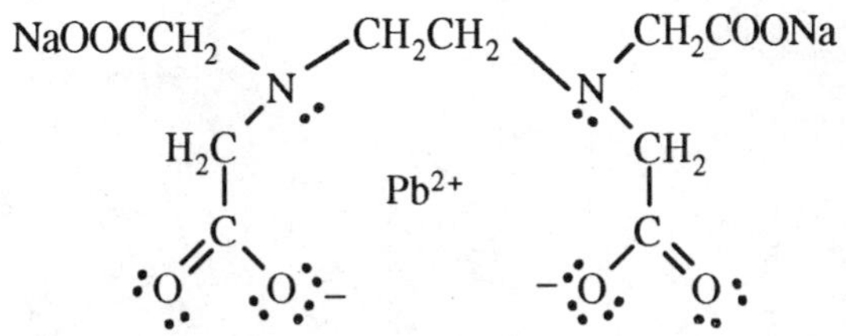

(b) 1.448 g

$$(1.000\ \text{g white lead})\left(\frac{1\ \text{mol white lead}}{775.63\ \text{g lead}}\right)\left(\frac{3\ \text{mol chelate}}{1\ \text{mol white lead}}\right)\left(\frac{374.27\ \text{g chelate}}{1\ \text{mol chelate}}\right) = 1.448\ \text{g}$$

minimum number of grams = 1.448 g

24.115 **(a)** ferromagnetic

(b) Chromium(IV) oxide and Fe_3O_4 are ferromagnetic.

(c) CrO_2

(d) $+\frac{8}{3}$

$$\overset{+8/3}{Fe_3}\ \overset{-2}{O_4}$$
$$+8 \quad -8 = 0$$

24.119 **(a)** metals: Cd, Cr, Pb, Hg

semimetals: As, Sb

(b)

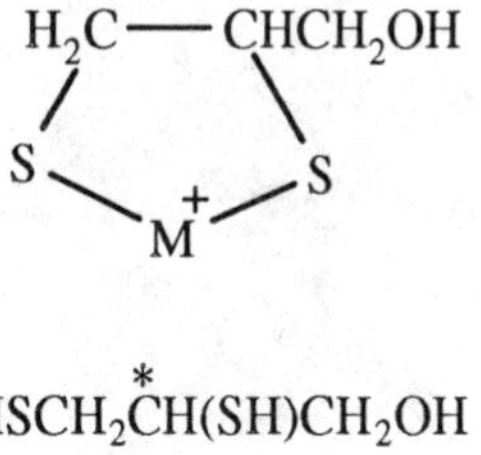

The hydrogens on sulfur are more acidic than those on oxygen, so the negative charges in BAL^{2-} are on sulfur.

(c) $HSCH_2\overset{*}{C}H(SH)CH_2OH$

(d) Hydrogens bonded to sulfur do not form effective hydrogen bonds, so BAL can only form one–third as many hydrogen bonds as glycerol. Since BAL has a larger molecular mass than glycerol, it is not water soluble.

Appendixes

A.1 **(a)** hypobromite ion

Bromine is just below chlorine in Group VIIA of the periodic table. From Table A.3, the formula for hypochlorite ion is ClO^-; therefore, hypobromite ion is the name of the BrO^- ion.

(b) P^{3-}

Phosphorus is just below nitrogen in Group VA in the periodic table. From Table A.3, the formula for the nitride ion is N^{3-}; therefore, the formula for the phosphide ion is P^{3-}.

A.2 **(a)** manganese(II) ion

Manganese is a transition metal and commonly has a variety of oxidation numbers in its compounds. Therefore, a Roman numeral is needed in the name of the Mn^{2+} ion. The old, nonsystematic name for this ion is manganous ion (see Table A.2).

(b) mercury(II) nitrate

Two nitrate ions (NO_3^-) balance the charge of one mercury(II) ion (Hg^{2+}). The name of the metal ion is followed by the name of the anion. The nonsystematic name for this compound is mercuric nitrate.

(c) dinitrogen trioxide

Both nitrogen and oxygen are nonmetals. Several oxides of nitrogen are known. Therefore, Greek prefixes must be used to show which one of them is meant.

(d) diphosphorus pentoxide

Both phosphorus and oxygen are nonmetals. Several oxides of phosphorus are known. Therefore, Greek prefixes must be used to show which one of them is meant. Note that the –a is dropped from penta– in the name, diphosphorus pentoxide.

(e) periodic acid

Both iodine and chlorine are halogens, and their oxo acids are named similarly. $HClO_4$ is perchloric acid (see p. A4 in text); thus, HIO_4 is periodic acid.

(f) potassium iodate

The cation in this compound is the metal ion K^+, and the anion is the iodate ion (IO_3^-). Thus, KIO_3 is named potassium iodate.

(g) lithium permanganate trihydrate

The cation in this compound is the metal ion Li^+, and the anion is the permanganate ion, MnO_4^- (see Table A.3). The water of crystallization in this compound ($3H_2O$) is expressed as trihydrate [or –water (1/3)].

(h) hydrogen bromide

HBr(g) is a molecular compound, and thus its name is hydrogen bromide.

(i) hydrobromic acid

HBr(aq) is an aqueous solution of hydrogen bromide. When dissolved in water, the hydrogen halides ionize, forming hydrogen ions. Thus, HBr(aq) is called hydrobromic acid.

(j) cadmium sulfate–water (3/8)

Although cadmium is a transition metal, its only common oxidation number is +2 (Figure 11.1). Therefore, no Roman numeral is needed. The compound is a hydrate in which eight water molecules are associated with three units of cadmium sulfate, $CdSO_4$. Hence, its name is cadmium sulfate–water (3/8).

(j) magnesium sulfate heptahydrate, magnesium sulfate–water (1/7)

Both of the above names correspond to the hydrate in which seven water molecules are associated with each unit of magnesium sulfate, $MgSO_4$.

B.1 **(a)** 3.652×10^2

To write 365.2 as a number between 1 and 10, the decimal must be moved *two* places to the left:

$$365.2 = 3.652 \times 10 \times 10$$

$$= 3.652 \times 10^2$$

(b) 7×10^{-4}

To write 0.0007 as a number between 1 and 10, the decimal must be moved *four* places to the right:

$$0.0007 = 7 \times 0.1 \times 0.1 \times 0.1 \times 0.1$$

$$= 7 \times 10^{-4}$$

Note that by convention, the decimal point is not shown when there are no significant digits following it (i.e., 7×10^{-4}, *not* $7. \times 10^{-4}$).

(c) 1.42×10^{-2}

$$0.0142 = 1.42 \times 0.1 \times 0.1$$

$$= 1.42 \times 10^{-2}$$

(d) 2.3652×10^{4}

$$23\,652 = 2.3652 \times 10 \times 10 \times 10 \times 10$$

$$= 2.3652 \times 10^{4}$$

(e) 7.5×10^{0}

Since 7.5 is already a number between 1 and 10, the decimal point does not shift to either the right or the left. Equivalently, the decimal point is moved *zero* places:

$$7.5 = 7.5 \times 1$$

$$= 7.5 \times 10^{0} \text{ (recall that } 1 = 10^{0}\text{)}$$

B.2 **(a)** 0.000 000 74

The power of ten, –7, means multiply by 0.1 *seven* times:

$$7.4 \times 10^{-7} = 7.4 \times 0.1 \times 0.1 \times 0.1 \times 0.1 \times 0.1 \times 0.1 \times 0.1$$

$$= 0.000\,000\,74$$

This is equivalent to moving the decimal point seven places to the left.

(b) 39.82

The power of ten, 1, means multiply by 10 *once*:

$$3.982 \times 10^{1} = 3.982 \times 10$$

$$= 39.82$$

This is equivalent to moving the decimal point one place to the right.

(c) 3.27

Since $10^{0} = 1$,

$$3.27 \times 10^{0} = 3.27 \times 1$$

$$= 3.27$$

(d) 0.005 81

$$5.81 \times 10^{-3} = 5.81 \times 0.1 \times 0.1 \times 0.1$$

$$= 0.005\,81 \quad \text{(the decimal point is moved three places to the left)}$$

(e) 745 600 000

$7.456 \times 10^8 = 7.456 \times 10 \times 10 \times 10 \times 10 \times 10 \times 10 \times 10 \times 10$

$= 745\ 600\ 000$ (the decimal point is moved eight places to the right)

B.3 Two, three, or four significant figures are possible. (3.2×10^3, 3.20×10^3, or 3.200×10^3)

The zeros in 3200 may or may not be significant.

In exponential notation,

3.2×10^3 has two significant figures

3.20×10^3 has three significant figures

3.200×10^3 has four significant figures

B.4 five significant figures

Rewriting in exponential notation,

$0.060\ 700 = 6.0700 \times 0.1 \times 0.1$

$= 6.0700 \times 10^{-2}$

The number 6.0700 has five significant figures.

B.5 490

To convert 10^8 to 10^6, multiply by 10^{-2}; to keep the value of the number identical, multiply 4.9 by 10^{+2}:

$4.9 \times 10^{+2} = 490$

Thus, $4.9 \times 10^8 = 490 \times 10^6$

B.6 2.43×10^3

$(8.22 \times 10^7)(3.874 \times 10^{-9})(7.629 \times 10^3)$

$= (8.22 \times 3.874 \times 7.629) \times 10^{[7 + (-9) + 3]}$

$= 243 \times 10^1 = 2.43 \times 10^3$

The product of the numbers, 242.940 0121, is rounded to three significant digits because there are only three significant figures in 8.22.

B.7 **(a)** 3×10^5

$$\frac{(9.6 \times 10^8)}{(3 \times 10^3)} = \frac{9.6}{3} \times 10^{(8-3)}$$

$= 3 \times 10^5$ (one significant figure)

(b) 3.4×10^{13}

$$\frac{(9.42 \times 10^{7})}{(2.8 \times 10^{-6})} = \frac{9.42}{2.8} \times 10^{[7-(-6)]}$$

$$= 3.4 \times 10^{13} \quad \text{(two significant figures)}$$

(c) 3.7×10^{-13}

$$\frac{(8.6 \times 10^{-4})}{(2.3 \times 10^{9})} = \frac{8.6}{2.3} \times 10^{[(-4)-9]}$$

$$= 3.7 \times 10^{-13} \quad \text{(two significant figures)}$$

(d) 6.3×10^{3}

$$\frac{(4.7 \times 10^{-6})}{(7.5 \times 10^{-10})} = \frac{4.7}{7.5} \times 10^{[(-6)-(-10)]}$$

$$= 0.63 \times 10^{4} = 6.3 \times 10^{3} \quad \text{(two significant figures)}$$

(e) 4.3×10^{-11}

$$\frac{(8.76 \times 10^{14})(5.8 \times 10^{9})}{(2.63 \times 10^{22})(4.54 \times 10^{12})} = \frac{(8.76 \times 5.8)}{(2.63 \times 4.54)} \times 10^{[(14+9)-(22+12)]}$$

$$= 4.3 \times 10^{-11} \quad \text{(two significant figures)}$$

B.8 **(a)** 5.27×10^{7}

Convert 3.62×10^{6} to the same power of ten as the larger number to be added, 4.91×10^{7}:

$$\begin{array}{rcl} 3.62 \times 10^{6} & = & 0.362 \times 10^{7} \\ & + & \underline{4.91 \times 10^{7}} \\ & & 5.272 \times 10^{7} = 5.27 \times 10^{7} \quad \text{(three significant figures)} \end{array}$$

(b) 4.31×10^{23}

$$\begin{array}{r} 5.14 \times 10^{23} \\ \underline{-(8.29 \times 10^{22} = 0.829 \times 10^{23})} \\ 4.311 \times 10^{23} = 4.31 \times 10^{23} \quad \text{(three significant figures)} \end{array}$$

(c) 6.2×10^{-13}

$$\begin{array}{r} 6.9 \times 10^{-13} \\ \underline{-(7.2 \times 10^{-14} = 0.72 \times 10^{-13})} \\ 6.18 \times 10^{-13} = 6.2 \times 10^{-13} \quad \text{(two significant figures)} \end{array}$$

B.9 **(a)** 2.8×10^{9}

The fourth power of 2.3×10^{2} is equivalent to

$$(2.3 \times 10^{2})^{4} = (2.3)^{4} \times 10^{2(4)} = 28 \times 10^{8} = 2.8 \times 10^{9} \quad \text{(two significant figures)}$$

(b) 9.00×10^6

To take the square root of 8.10×10^{13}, first rewrite the number so that the power of ten is divisible by *two*:

$$\sqrt{8.10 \times 10^{13}} = \sqrt{81.0 \times 10^{12}}$$

Then take the square root of the number and divide the exponent by two:

$$\sqrt{81.0 \times 10^{12}} = 9.00 \times 10^6 \quad \text{(three significant figures)}$$

(c) 4.0×10^{-4}

To take the cube root of 6.4×10^{-11}, first rewrite the number so that the power of ten is divisible by *three*:

$$\sqrt[3]{6.4 \times 10^{-11}} = \sqrt[3]{64 \times 10^{-12}}$$

Then take the cube root of the number and divide the exponent by three:

$$\sqrt[3]{64 \times 10^{-12}} = 4.0 \times 10^{-4} \quad \text{(two significant figures)}$$

B.10 **(a)** 5

$\log 10^5 = 5$ because $\log (x) = y$, where $10^y = x$. In this case, $x = 10^5$, and thus $y = 5$.

(b) –4

$\log 10^{-4} = -4$. Note that $\log 10^y$ will always be equal to y.

B.11 **(a)** 10^{-8}

antilog $(-8) = 10^{-8}$ because antilog $(x) = 10^x$. That is, the antilog x is the number that has x as its logarithm.

(b) 10^{15}

antilog $(15) = 10^{15}$ because 10^{15} is the number that has 15 as its logarithm.

B.12 **(a)** –11.5

Since $\log (a \times b) = \log a + \log b$,

$$\begin{aligned}\log (3 \times 10^{-12}) &= \log 3 + \log 10^{-12} \\ &= 0.5 + (-12) \\ &= -11.5\end{aligned}$$

Note that the number of significant figures in the original number, 3×10^{-12}, is equal to the number of digits to the right of the decimal point in the logarithm, –11.5.

(b) 16.925

$\log(8.42 \times 10^{16}) = \log 8.42 + \log 10^{16}$

$= 0.925 + 16$

$= 16.925$

(c) 0.0

$\log 1 = \log 10^{0}$ (since $10^{0} = 1$)

$= 0.0$

(d) 3.18×10^{-23}

antilog $(-22.498) = y$, where $10^{-22.498} = y$

$y = 3.18 \times 10^{-23}$

Check: $\log(3.18 \times 10^{-23}) = \log 3.18 + \log 10^{-23}$

$= 0.502 + (-23)$

$= -22.498$

Note that the number of digits to the right of the decimal point in the original number, –22.498, is equal to the number of significant figures in the antilogarithm, 3.18×10^{-23}

(e) 5×10^{5}

antilog $5.7 = y$, where $10^{5.7} = y$

$y = 5 \times 10^{5}$

Note that the number of digits to the right of the decimal point in 5.7 is equal to the number of significant figures in the antilogarithm, 5×10^{5}.

B.13 **(a)** –26.5

$\ln(3 \times 10^{-12}) = -26.5$

Alternatively, the problem could be converted to one involving base ten logarithms:

$\ln(3 \times 10^{-12}) = 2.303 \log(3 \times 10^{-12})$

$= 2.303(\log 3 + \log 10^{-12})$

$= 2.303[0.5 + (-12)]$

$= -26.5$

(b) 38.972

$\ln(8.42 \times 10^{16}) = 38.972$

(c) 0.0

$\ln 1 = \ln(e^0)$ (since $e^0 = 1$)

$= 0.0$

(d) 1.70×10^{-10}

$\text{antiln}(-22.498) = 1.70 \times 10^{-10}$

(e) 3×10^2

$\text{antiln}(5.7) = 3 \times 10^2$

B.14 **(a)** 0.29

To solve $6.9 = 2.0/x$, first multiply both sides of the equation by x,

$$x \cdot 6.9 = x \cdot \frac{2.0}{x}$$

divide each side by 6.9, and simplify:

$$\frac{x \cdot \cancel{6.9}}{\cancel{6.9}} = \frac{\cancel{x} \cdot 2.0}{\cancel{x} \cdot 6.9}$$

$$x = \frac{2.0}{6.9} = 0.29$$

(b) 22.2

To solve $2.84 = x/7.82$, multiply both sides of the equation by 7.82 and simplify:

$$7.82 \times 2.84 = \cancel{7.82} \times \frac{x}{\cancel{7.82}}$$

$$7.82 \times 2.84 = x = 22.2$$

(c) 12.25

To solve $5.82 = x - 6.43$, add 6.43 to both sides of the equation and simplify:

$$(5.82 + 6.43) = (x - 6.43) + 6.43$$

$$5.82 + 6.43 = x = 12.25$$

B.15 **(a)** 2.5

To find the value of x in the equation $2.4^x = 8.6$, take the natural logarithm of both sides and then rearrange:

$x \cdot \ln 2.4 = \ln 8.6$

$$x = \frac{\ln 8.6}{\ln 2.4} = \frac{2.15}{0.88} = 2.4$$

(Best answer, without intermediate round off, is 2.5.)

(b) 2.4

To solve the equation $x = 6^{0.50}$, first take the natural logarithm of both sides,

$\ln x = 0.50 \ln 6 = 0.50 \times 1.8 = 0.90$

and then solve for x by taking the natural antilogarithm of both sides:

$x = \text{antiln } 0.90 = 2.5$

(Best answer, without intermediate round off, is 2.4.)

(c) 1.15

The problem is equivalent to solving the equation $x = (3.00)^{1/8}$. First take the natural logarithm of both sides,

$$\ln x = \frac{1}{8} \times \ln 3.00 = \frac{1}{8} \times 1.10 = 0.138$$

and then solve for x by taking the natural antilogarithm of both sides,

$x = \text{antiln } 0.138 = 1.15$

B.16 $x = 2.9 \times 10^{-2}$ and $x = -6.8 \times 10^{-2}$

To solve $\dfrac{x^2}{(0.050 - x)} = 3.9 \times 10^{-2}$,

first multiply each side by $(0.050 - x)$ to put the equation in standard form:

$x^2 = (0.050 - x)(3.9 \times 10^{-2})$

$x^2 = (2.0 \times 10^{-3}) - (3.9 \times 10^{-2})x$

Rearrangement gives

$x^2 + (3.9 \times 10^{-2})x - (2.0 \times 10^{-3}) = 0$

The coefficients of this equation are than substituted into the quadratic formula:

$$x = \frac{-(3.9 \times 10^{-2}) \pm \sqrt{(3.9 \times 10^{-2})^2 - 4(1)(-2.0 \times 10^{-3})}}{2(1)}$$

Solving, we obtain

$x = 2.9 \times 10^{-2}$ and $x = -6.8 \times 10^{-2}$

B.17 **(a)** 72.46%

$$\text{Percent, \%} = \frac{\text{part}}{\text{whole}} \times 100$$

$$\text{Percent, \%} = \frac{37.97\text{ g}}{52.40\text{ g}} \times 100$$

$$= 72.46\%$$

(b) 6.29 g Fe

$$\text{Percent, \%} = \frac{\text{part}}{\text{whole}} \times 100$$

or, rearranging,

(mass of *whole* sample)(*percent* of sample that is iron)/100 = (*part* of sample that is iron)

$$(67.6\text{ g})\left(\frac{9.31\ \%}{100}\right) = 6.29\text{ g}$$

(c) 29.2 g

$$\text{mass of } \textit{whole} \text{ sample} = \frac{\textit{part} \text{ of sample that is iron}}{\textit{percent} \text{ of sample that is iron}} \times 100$$

$$= \frac{7.18\text{ g}}{24.6\%} \times 100$$

$$= 29.2\text{ g}$$

B.18 **(a)** 250 s

Using the line graph in Figure B.1, the line will cross the y–value 43.6 °C approximately at the x–value 250 s.

(b) 45.0 °C

After 125 s (x–value), the y–value for the line graph in Figure B.1 is approximately 45.0 °C

(c) 2.7 °C

The initial temperature (y–value) of the line graph when time (x–value) is 0 s is 46.3 °C. When the x–value is 250 s, the temperature is 43.6 °C. Thus, the temperature of the water has cooled (46.3 – 43.6) °C = 2.7 °C

B.19 (a) Density of aqueous ethylene glycol at 20 °C:

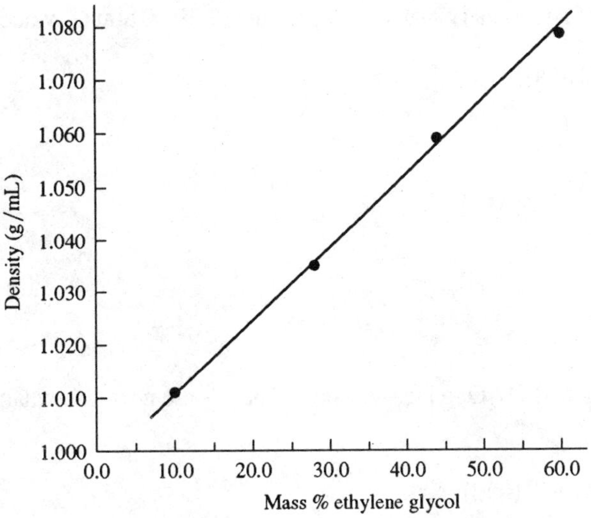

(b) $0.0014 \dfrac{\text{g/mL}}{\text{mass \% ethylene glycol}}$

Note: When the data are graphed on an $8\frac{1}{2} \times 11$ in. sheet of graph paper, the best straight line does *not* go through the data points for 10.0 and 60.0 mass % ethylene glycol. The densities read from this graph are 1.010 and 1.079 g/mL.

$$\lambda = \frac{\Delta y}{\Delta x} = \frac{(1.079 - 1.010)\ \text{g/mL}}{(60.0 - 10.0)\ \%} = 0.0014 \frac{\text{g/mL}}{\text{mass \% ethylene glycol}}$$

(c) 1.030 g/mL

Using the graph in part (a), the density (y–value) at 25.0 mass % ethylene glycol (x–value) is 1.030 g/mL.

B.20 (a) Boiling point of ammonia at different pressures:

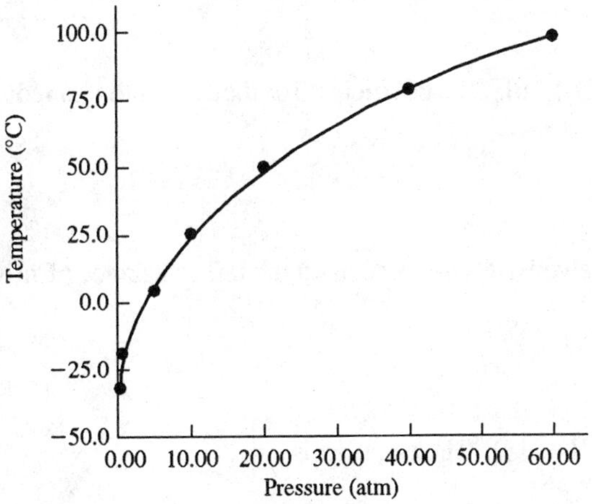

(b) 66.0 °C

Using the graph in part (a), the boiling point of ammonia (*y*–value) at a pressure of 30.00 atm (*x*–value) is 66.0 °C.

F.1 **(a)** $3S(s) + 4HNO_3(aq) \rightarrow 3SO_2(g) + 4NO(g) + 2H_2O(l)$

$\Delta ox = 4,\ 3(4) = 12$

$$\overset{0}{S}(s) + \overset{+1\ +5\ -2}{HNO_3}(aq) \rightarrow \overset{+4\ -2}{SO_2}(g) + \overset{+2\ -2}{NO}(g)$$

$\Delta red = 3,\ 4(3) = 12$

$3S(s) + 4HNO_3(aq) \rightarrow 3SO_2(g) + 4NO(g)$

Only H and O are not balanced. To balance H, add $2H_2O$ to the right side. The O's are now also balanced, and the equation is

$$3S(s) + 4HNO_3(aq) \rightarrow 3SO_2(g) + 4NO(g) + 2H_2O(l)$$

(b) $3Cu_2O(s) + 14HNO_3\ (aq) \rightarrow 6Cu(NO_3)_2(aq) + 2NO(g) + 7H_2O(l)$

$$\overset{+1\ -2}{Cu_2O} + \overset{+1\ +5\ -2}{HNO_3} \rightarrow \overset{+2\ +5\ -2}{Cu(NO_3)_2} + \overset{+2\ -2}{NO}$$

Cu changes oxidation number but is not balanced. Balance it.

$$\overset{+1\ -2}{Cu_2O} + \overset{+1\ +5\ -2}{HNO_3} \rightarrow 2\overset{+2\ +5\ -2}{Cu(NO_3)_2} + \overset{+2\ -2}{NO}$$

The oxidation number of some N remains +5; the oxidation number of other N is reduced to +2. Write HNO_3 twice so that the two types of N can be treated separately.

$\Delta ox = 2(1) = 2,\ 3(2) = 6$

$$\overset{+1\ -2}{Cu_2O} + \overset{+1\ +5\ -2}{HNO_3} + \overset{+1\ +5\ -2}{HNO_3} \rightarrow 2\overset{+2\ +5\ -2}{Cu(NO_3)_2} + \overset{+2\ -2}{NO}$$

$\Delta red = 3,\ 2(3) = 6$

The coefficient for Cu_2O should be 3 [and for $Cu(NO_3)_2$, 6]. The coefficient for the HNO_3 that is reduced should be 2, and the coefficient for NO should also be 2.

$3Cu_2O + 2HNO_3 + HNO_3 \rightarrow 6Cu(NO_3)_2 + 2NO$

Balancing can now be completed by inspection. Twelve HNO_3 are needed on the left to balance $6Cu(NO_3)_2$ on the right.

$3Cu_2O + 2HNO_3 + 12HNO_3 \rightarrow 6Cu(NO_3)_2 + 2NO$

Addition of $7H_2O$ to the right side will balance both H and O. The equation is

$3Cu_2O(s) + 14HNO_3\ (aq) \rightarrow 6Cu(NO_3)_2(aq) + 2NO(g) + 7H_2O(l)$

(c) $5Zn(s) + 12HNO_3(aq) \rightarrow 5Zn(NO_3)_2(aq) + N_2(g) + 6H_2O(l)$

$$\overset{0}{Zn} + \overset{+1\ +5\ -2}{HNO_3} \rightarrow \overset{+2\ \ +5\ -2}{Zn(NO_3)_2} + \overset{0}{N_2}$$

The oxidation number of some N remains +5; the oxidation number of other N is reduced to 0. Write HNO_3 twice so that the two types of N can be treated separately and balance the N that undergo change.

$\Delta ox = 2,\ 5(2) = 10$

$$\overset{0}{Zn} + \overset{+1\ +5\ -2}{HNO_3} + \overset{+1\ +5\ -2}{2HNO_3} \rightarrow \overset{+2\ \ +5\ -2}{Zn(NO_3)_2} + \overset{0}{N_2}$$

$\Delta red = 2(5) = 10$

Change coefficients of Zn and $Zn(NO_3)_2$ to 5.

$$5Zn + HNO_3 + 2HNO_3 \rightarrow 5Zn(NO_3)_2 + N_2$$

Balancing can now be completed by inspection. Ten HNO_3 are needed on the left to balance $5Zn(NO_3)_2$ on the right.

$$5Zn + 10HNO_3 + 2HNO_3 \rightarrow 5Zn(NO_3)_2 + N_2$$

Addition of $6H_2O$ to the right side will balance both H and O. The equation is

$$5Zn(s) + 12HNO_3(aq) \rightarrow 5Zn(NO_3)_2(aq) + N_2(g) + 6H_2O(l)$$

(d) $2Al(s) + 2NaOH(aq) + 6H_2O(l) \rightarrow 2NaAl(OH)_4(aq) + 3H_2(g)$

$\Delta ox = 1(3) = 3,\ 2(3) = 6$

$$\overset{0}{Al} + \overset{+1\ -2\ +1}{NaOH} + \overset{+1\ -2}{H_2O} \rightarrow \overset{+1\ +3\ -2\ +1}{NaAl(OH)_4} + \overset{0}{H_2}$$

$\Delta red = 2(1) = 2,\ 3(2) = 6$

$$2Al + NaOH + 3H_2O \rightarrow 2NaAl(OH)_4 + 3H_2$$

$$2Al + 2NaOH + 6H_2O \rightarrow 2NaAl(OH)_4 + 3H_2$$

(e) $C(s) + 2H_2SO_4(l) \rightarrow CO_2(g) + 2SO_2(g) + 2H_2O(l)$

$\Delta ox = 1(4) = 4,\ 1(4) = 4$

$$\overset{0}{C} + \overset{+1\ +6\ -2}{H_2SO_4} \rightarrow \overset{+4\ -2}{CO_2} + \overset{+4\ -2}{SO_2}$$

$\Delta red = 1(2) = 2,\ 2(2) = 4$

$$C + 2H_2SO_4 \rightarrow CO_2 + 2SO_2$$

$$C + 2H_2SO_4 \rightarrow CO_2 + 2SO_2 + 2H_2O$$

(f) $3C_3H_8O(aq) + 2CrO_3(aq) + 3H_2SO_4(aq) \rightarrow 3C_3H_6O(aq) + Cr_2(SO_4)_3(aq) + 6H_2O(l)$

$\Delta ox = 3(2/3) = 2,\ 3(2) = 6$

$-2\ +1\ -2 \quad +6\ -2 \quad +1\ +6\ -2 \quad -\frac{4}{3}\ +1\ -2 \quad +3\ \ +6\ -2$

$C_3H_8O + 2CrO_3 + H_2SO_4 \rightarrow C_3H_6O + Cr_2(SO_4)_3$

$\Delta red = 2(3) = 6,\ 1(6) = 6$

$3C_3H_8O + 2CrO_3 + H_2SO_4 \rightarrow 3C_3H_6O + Cr_2(SO_4)_3$

$3C_3H_8O + 2CrO_3 + 3H_2SO_4 \rightarrow 3C_3H_6O + Cr_2(SO_4)_3 + 6H_2O$

(g) $H_3AsO_4(aq) + 4Zn(s) + 8HNO_3(aq) \rightarrow AsH_3(g) + 4Zn(NO_3)_2(aq) + 4H_2O(l)$

$\Delta ox = 1(2) = 2,\ 4(2) = 8$

$+1\ +5\ -2 \quad 0 \quad +1\ +5\ -2 \quad -3\ +1 \quad +2\ \ +5\ -2$

$H_3AsO_4 + Zn + HNO_3 \rightarrow AsH_3 + Zn(NO_3)_2$

$\Delta red = 1(8) = 8,\ 1(8) = 8$

$H_3AsO_4 + 4Zn + HNO_3 \rightarrow AsH_3 + 4Zn(NO_3)_2$

$H_3AsO_4 + 4Zn + 8HNO_3 \rightarrow AsH_3 + 4Zn(NO_3)_2 + 4H_2O$

(h) $4H_2O(l) + 6KI(aq) + 2KMnO_4(aq) \rightarrow 3I_2(s) + 2MnO_2(s) + 8KOH(aq)$

$\Delta ox = 2(1) = 2,\ 3(2) = 6$

$+1\ -1 \quad +1\ +7\ -2 \quad 0 \quad +4\ \ -2$

$2KI + KMnO_4 \rightarrow I_2 + MnO_2$

$\Delta red = 1(3) = 3,\ 2(3) = 6$

$6KI + 2KMnO_4 \rightarrow 3I_2 + 2MnO_2$

$4H_2O + 6KI + 2KMnO_4 \rightarrow 3I_2 + 2MnO_2 + 8KOH$

F.2 (a) $Cr_2O_7^{2-} + 3C_2O_4^{2-} + 14H^+ \rightarrow 2Cr^{3+} + 6CO_2(g) + 7H_2O(l)$

$Cr_2O_7^{2-} + C_2O_4^{2-} \rightarrow Cr^{3+} + CO_2(g)$ (acidic solution)

$\Delta ox = 2(1) = 2,\ 6(2) = 12$

$+6\ \ -2 \quad +3\ -2 \quad +3 \quad +4\ -2$

$Cr_2O_7^{2-} + C_2O_4^{2-} \rightarrow 2Cr^{3+} + 2CO_2$

$\Delta red = 2(3) = 6,\ 2(6) = 12$

$2Cr_2O_7^{2-} + 6C_2O_4^{2-} \rightarrow 4Cr^{3+} + 12CO_2$

The elements oxidized and reduced are now balanced. The net charges on each side are
left side: 2(2–) + 6(2–) = 16–
right side: 4(3+) = 12+

The difference is 28 units of charge. Addition of $28H^+$ to the left side will balance charge, and addition of $14H_2O$ to the right will balance H and O.

$2Cr_2O_7^{2-} + 6C_2O_4^{2-} + 28H^+ \rightarrow 4Cr^{3+} + 12CO_2 + 14H_2O$

The coefficients can all be divided by two. The net ionic equation is

$Cr_2O_7^{2-} + 3C_2O_4^{2-} + 14H^+ \rightarrow 2Cr^{3+} + 6CO_2(g) + 7H_2O(l)$

(b) $3H_2O_2(aq) + 2MnO_4^- \rightarrow 3O_2(g) + 2MnO_2(s) + 2OH^- + 2H_2O(l)$

$H_2O_2(aq) + MnO_4^- \rightarrow O_2(g) + MnO_2(s)$ (basic solution)

$\Delta ox = 2,\ 3(2) = 6$

$\overset{+1\ -1}{H_2O_2} + \overset{+7\ -2}{MnO_4^-} \rightarrow \overset{0}{O_2} + \overset{+4\ -2}{MnO_2}$

$\Delta red = 3,\ 2(3) = 6$

$3H_2O_2 + 2MnO_4^- \rightarrow 3O_2 + 2MnO_2$

The elements oxidized and reduced are now balanced. The net charges on each side are
left side: 2(1–) = 2–
right side: 0

The difference is two. Addition of $2OH^-$ to the right side balances charge. Two H_2O must also be added to the right side to balance H and O. The net ionic equation is

$3H_2O_2(aq) + 2MnO_4^- \rightarrow 3O_2(g) + 2MnO_2(s) + 2OH^- + 2H_2O(l)$

(c) $3AsH_3(g) + 4ClO_3^- \rightarrow 3H_3AsO_4(aq) + 4Cl^-$

$\Delta ox = 8,\ 3(8) = 24$

$\overset{-3\ +1}{AsH_3} + \overset{+5\ -2}{ClO_3^-} \rightarrow \overset{+1\ +5\ -2}{H_3AsO_4} + \overset{-1}{Cl^-}$ (acidic solution)

$\Delta red = 6,\ 4(6) = 24$

$3AsH_3 + 4ClO_3^- \rightarrow 3H_3AsO_4 + 4Cl^-$

(d) $Cu(NH_3)_4^{2+} + S_2O_4^{2-} + 4OH^- \rightarrow Cu(s) + 2SO_3^{2-} + 4NH_3(aq) + 2H_2O(l)$

$\Delta ox = 2(1) = 2$

$\overset{+2\ -3\ +1}{Cu(NH_3)_4^{2+}} + \overset{+3\ -2}{S_2O_4^{2-}} \rightarrow \overset{0}{Cu} + \overset{+4\ -2}{SO_3^{2-}} + \overset{-3\ +1}{NH_3}$ (basic solution)

$\Delta red = 2$

$Cu(NH_3)_4^{2+} + S_2O_4^{2-} \rightarrow Cu + SO_3^{2-} + NH_3$ (basic solution)

$Cu(NH_3)_4^{2+} + S_2O_4^{2-} + 4OH^- \rightarrow Cu + 2SO_3^{2-} + 4NH_3 + 2H_2O$

(e) $TlOH(s) + 2NH_2OH(aq) \rightarrow Tl(OH)_3(s) + N_2H_4(aq)$

$\Delta ox = 2$

+1 −2 +1   −1 +1 −2 +1   +3 −2 +1   −2 +1

$TlOH + 2NH_2OH \rightarrow Tl(OH)_3 + N_2H_4$ (basic solution)

$\Delta red = 2(1) = 2$

$TlOH + 2NH_2OH \rightarrow Tl(OH)_3 + N_2H_4$

(f) $4MnO_2(s) + 3O_2(g) + 4OH^- \rightarrow 4MnO_4^- + 2H_2O(l)$

$\Delta ox = 3,\ 4(3) = 12$

+4 −2   0   +7 −2   −2 +1

$MnO_2 + O_2 \rightarrow MnO_4^- + OH$ (basic solution)

$\Delta red = 2(2) = 4,\ 3(4) = 12$

$4\ MnO_2 + 3O_2 \rightarrow 4\ MnO_4^- + 6OH^-$

Balance charge by adding 10 OH^- to the left side, and then balance hydrogen by adding 2 H_2O to the right side:

$4\ MnO_2 + 3O_2 + 10OH^- \rightarrow 4\ MnO_4^- + 6OH^- + 2\ H_2O$

Simplify by subtracting $6OH^-$ from both sides:

$4MnO_2 + 3O_2 + 4OH^- \rightarrow 4MnO_4^- + 2H_2O$

(g) $S_2O_3^{2-} + 4Cl_2(g) + 5H_2O(l) \rightarrow 2HSO_4^- + 8Cl^- + 8H^+$

$\Delta ox = 2(4) = 8$

+2 −2   0   +1 +6 −2   −1

$S_2O_3^{2-} + Cl_2 \rightarrow HSO_4^- + Cl^-$ (acdic solution)

$\Delta red = 2(1) = 2,\ 4(2) = 8$

$S_2O_3^{2-} + 4Cl_2 \rightarrow 2HSO_4^- + 8Cl^-$

$S_2O_3^{2-} + 4Cl_2 + 5H_2O \rightarrow 2HSO_4^- + 8Cl^- + 8H^+$

F.3 (a) $2MnO_4^- + 5H_2C_2O_4(aq) + 6H^+ \rightarrow 2Mn^{2+} + 10CO_2(g) + 8H_2O(l)$

Oxidation numbers can be assigned to all the elements involved in this reaction.

+7 −2   +1 +3 −2   +2   +4 −2

$MnO_4^- + H_2C_2O_4(aq) \rightarrow Mn^{2+} + CO_2(g)$

Let's balance the oxidation half–reaction first:

$H_2C_2O_4 \rightarrow CO_2\ (g)$

Balancing C gives

$H_2C_2O_4 \rightarrow 2CO_2$

O is balanced. Use H^+ to balance H.

$$H_2C_2O_4 \rightarrow 2CO_2 + 2H^+$$

Now use e^- to balance charge.

$$H_2C_2O_4 \rightarrow 2CO_2 + 2H^+ + 2e^-$$

The half–reaction of oxidation is balanced. The reduction half reaction is $MnO_4^- \rightarrow Mn^{2+}$

Mn is balanced. Use H_2O to balance O.

$$MnO_4^- \rightarrow Mn^{2+} + 4H_2O$$

Then use H^+ to balance H.

$$8H^+ + MnO_4^- \rightarrow Mn^{2+} + 4H_2O$$

Add e^- to balance charge.

$$5e^- + 8H^+ + MnO_4^- \rightarrow Mn^{2+} + 4H_2O$$

Now combine the two half–reactions so that electrons cancel. The half–reaction of oxidation must be multiplied by 5, and the half–reaction of reduction must be multiplied by 2.

$$\begin{array}{l} 5(H_2C_2O_4 \rightarrow 2CO_2 + 2H^+ + 2e^-) \\ \underline{+\ 2(5e^- + 8H^+ + MnO_4^- \rightarrow Mn^{2+} + 4H_2O)} \end{array}$$

$$5H_2C_2O_4 + \cancel{10e^-} + \overset{6}{\cancel{16}}H^+ + 2MnO_4^- \rightarrow 10CO_2 + \cancel{10H^+} + \cancel{10e^-} + 2Mn^{2+} + 8H_2O$$

The net ionic equation is

$$5H_2C_2O_4(aq) + 6H^+ + 2MnO_4^- \rightarrow 10CO_2(g) + 2Mn^{2+} + 8H_2O(l)$$

Check: There are 16 H, 10 C, 28 O, 2 Mn, and a 4+ charge on both sides.

The oxidation number of C increases from +3 to +4; carbon is oxidized, and $H_2C_2O_4$ is the reducing agent. The oxidation number of Mn decreases from +7 to +2; manganese is reduced, and MnO_4^- is the oxidizing agent.

(b) $4Zn(s) + 10H^+ + 2NO_3^- \rightarrow 4Zn^{2+} + N_2O(g) + 5H_2O(l)$

Half–reaction of oxidation:	$Zn \rightarrow Zn^{2+} + 2e^-$
Half–reaction of reduction:	$10H^+ + 8e^- + 2NO_3^- \rightarrow N_2O + 5H_2O$
Net ionic equation:	$4Zn(s) + 10H^+ + 2NO_3^- \rightarrow 4Zn^{2+} + N_2O(g) + 5H_2O(l)$

The element oxidized is Zn, and Zn is the reducing agent. The element reduced is N, and NO_3^- is the oxidizing agent.

(c) $As_2O_3(s) + 2NO_3^- + 2H_2O(l) + 2H^+ \rightarrow 2H_3AsO_4(aq) + N_2O_3(aq)$

Half–reaction of oxidation:	$As_2O_3 + 5H_2O \rightarrow 2H_3AsO_4 + 4H^+ + 4e^-$
Half–reaction of reduction:	$2NO_3^- + 6H^+ + 4e^- \rightarrow N_2O_3 + 3H_2O$
Net ionic equation:	$As_2O_3(s) + 2NO_3^- + 2H_2O(l) + 2H^+ \rightarrow 2H_3AsO_4(aq) + N_2O_3(aq)$

The element oxidized is As, and As_2O_3 is the reducing agent. The element reduced is N, and NO_3^- is the oxidizing agent.

(d) $HIO_3(aq) + 8I^- + 5H^+ \rightarrow 3I_3^- + 3H_2O(l)$

Half–reaction of oxidation: $3I^- \rightarrow I_3^- + 2e^-$
Half–reaction of reduction: $15H^+ + 3HIO_3 + 16e^- \rightarrow I_3^- + 9H_2O$
Net ionic equation: $HIO_3(aq) + 8I^- + 5H^+ \rightarrow 3I_3^- + 3H_2O(l)$

The element oxidized is I, and I^- is the reducing agent. The element reduced is I, and HIO_3 is the oxidizing agent.

(e) $3ReCl_5(s) + 8H_2O(l) \rightarrow HReO_4(aq) + 2ReO_2(s) + 15Cl^- + 15H^+$

Half–reaction of oxidation: $4H_2O + ReCl_5 \rightarrow HReO_4 + 5Cl^- + 7H^+ + 2e^-$
Half–reaction of reduction: $2H_2O + ReCl_5 + e^- \rightarrow ReO_2 + 5Cl^- + 4H^+$
Net ionic equation: $3ReCl_5(s) + 8H_2O(l) \rightarrow HReO_4(aq) + 2ReO_2(s) + 15Cl^- + 15H^+$

The element oxidized is Re, and $ReCl_5$ is the reducing agent. The element reduced is Re, and $ReCl_5$ is the oxidizing agent.

(f) $Zn(s) + 2VO^{2+} + 4H^+ \rightarrow Zn^{2+} + 2V^{3+} + 2H_2O(l)$

Half–reaction of oxidation: $Zn \rightarrow Zn^{2+} + 2e^-$
Half–reaction of reduction: $e^- + 2H^+ + VO^{2+} \rightarrow V^{3+} + H_2O$
Net ionic equation: $Zn(s) + 2VO^{2+} + 4H^+ \rightarrow Zn^{2+} + 2V^{3+} + 2H_2O(l)$

The element oxidized is Zn, and Zn is the reducing agent. The element reduced is V, and VO^{2+} is the oxidizing agent.

F.4 **(a)** $C_3H_8O_3 + 20OH^- \rightarrow 3CO_3^{2-} + 14H_2O + 14e^-$

Oxidation is often defined as loss of electrons. The half–reaction

$C_3H_8O_3 + 20OH^- \rightarrow 3CO_3^{2-} + 14H_2O + 14e^-$

is oxidation.

(b) $MnO_4^- + 1e^- \rightarrow MnO_4^{2-}$

Reduction is often defined as gain of electrons. The half–reaction

$MnO_4^- + 1e^- \rightarrow MnO_4^{2-}$

is reduction.

(c) C is oxidized

The oxidation number of C increases from –2/3 to +4. C is oxidized.

(d) Mn is reduced

The oxidation number of Mn decreases from +7 to +6. Mn is reduced.

(e) MnO_4^- is the oxidizing agent

Because Mn is reduced, MnO_4^- is the oxidizing agent (the species that brings about oxidation).

(f) $C_3H_8O_3$ is the reducing agent

Because C is oxidized, $C_3H_8O_3$ is the reducing agent (the species that brings about reduction).

F.5 **(a)** $2Cr(OH)_6^{3-} + 3BrO^- \rightarrow 2CrO_4^{2-} + 3Br^- + 5H_2O(l) + 2OH^-$

Oxidation numbers can be assigned to all the elements involved in this reaction.

$$\overset{+3\ -2\ +1}{Cr(OH)_6^{3-}} + \overset{+1\ -2}{BrO^-} \rightarrow \overset{+6\ -2}{CrO_4^{2-}} + \overset{-1}{Br^-}$$

The half reaction of oxidation is

$$Cr(OH)_6^{3-} \rightarrow CrO_4^{2-}$$

Cr is balanced. Use H_2O to balance O.

$$Cr(OH)_6^{3-} \rightarrow CrO_4^{2-} + 2H_2O$$

Use H^+ to balance H.

$$Cr(OH)_6^{3-} \rightarrow CrO_4^{2-} + 2H_2O + 2H^+$$

Use e^- to balance charge.

$$Cr(OH)_6^{3-} \rightarrow CrO_4^{2-} + 2H_2O + 2H^+ + 3e^-$$

Because reaction takes place in basic solution, OH^- must be added to neutralize H^+. Two OH^- must be added to each side.

$$2OH^- + Cr(OH)_6^{3-} \rightarrow CrO_4^{2-} + 2H_2O + 2H^+ + 2OH^- + 3e^-$$

The $2H^+$ and $2OH^-$ on the right form $2H_2O$, making a total of 4 H_2O on the right.

$$2OH^- + Cr(OH)_6^{3-} \rightarrow CrO_4^{2-} + 4H_2O + 3e^-$$

The half–reaction of oxidation is balanced.

The half–reaction of reduction is

$$BrO^- \rightarrow Br^-$$

Br is balanced. Use H_2O to balance O.

$$BrO^- \rightarrow Br^- + H_2O$$

Use H^+ to balance H.

$$2H^+ + BrO^- \rightarrow Br^- + H_2O$$

Use e^- to balance charge.

$$2e^- + 2H^+ + BrO^- \rightarrow Br^- + H_2O$$

Add $2OH^-$ to each side to neutralize H^+.

$$2e^- + 2OH^- + 2H^+ + BrO^- \rightarrow Br^- + H_2O + 2OH^-$$

The $2H^+$ and $2OH^-$ on the left form $2H_2O$. Simplifying by subtracting 1 H_2O from each side gives

$$2e^- + H_2O + BrO^- \rightarrow Br^- + 2OH^-$$

The half–reaction of reduction is balanced.

The half–reaction of oxidation must be multiplied by 2 and the half–reaction of reduction by 3 for electrons to cancel when the two are added.

$$2(2OH^- + Cr(OH)_6{}^{3-} \rightarrow CrO_4{}^{2-} + 4H_2O + 3e^-)$$
$$\underline{3(2e^- + H_2O + BrO^- \rightarrow Br^- + 2OH^-)}$$
$$\cancel{4OH^-} + 2Cr(OH)_6{}^{3-} + \cancel{6e^-} + \cancel{3H_2O} + 3BrO^- \rightarrow 2CrO_4{}^{2-} + \overset{5}{\cancel{8}}H_2O + \cancel{6e^-} + 3Br^- + \overset{2}{\cancel{6}}OH^-$$

The net ionic equation is

$$2Cr(OH)_6{}^{3-} + 3BrO^- \rightarrow 2CrO_4{}^{2-} + 3Br^- + 5H_2O(l) + 2OH^-$$

Check: There are 2 Cr, 15 O, 12 H, 3 Br, and a 9– charge on both sides.

Chromium is oxidized, and $Cr(OH)_6{}^{3-}$ is the reducing agent. Bromine is reduced, and BrO^- is the oxidizing agent.

(b) $3H_2O_2(aq) + Cl_2O_7(aq) \rightarrow 2ClO_2(aq) + 3O_2(g) + 3H_2O(l)$

Half–reaction of oxidation: $2OH^- + H_2O_2 \rightarrow O_2 + 2H_2O + 2e^-$
Half–reaction of reduction: $6e^- + 3H_2O + Cl_2O_7 \rightarrow 2ClO_2 + 6OH^-$
Net ionic equation: $3H_2O_2(aq) + Cl_2O_7(aq) \rightarrow 2ClO_2(aq) + 3O_2(g) + 3H_2O(l)$

Oxygen is oxidized, and H_2O_2 is the reducing agent. Chlorine is reduced, and Cl_2O_7 is the oxidizing agent.

(c) $3CN^- + 2CrO_4{}^{2-} + 5H_2O(l) \rightarrow 3CNO^- + 2Cr(OH)_4{}^- + 2OH^-$

Half–reaction of oxidation: $2OH^- + CN^- \rightarrow CNO^- + H_2O + 2e^-$
Half–reaction of reduction: $CrO_4{}^{2-} + 4H_2O + 3e^- \rightarrow Cr(OH)_4{}^- + 4OH^-$
Net ionic equation: $3CN^- + 2CrO_4{}^{2-} + 5H_2O(l) \rightarrow 3CNO^- + 2Cr(OH)_4{}^- + 2OH^-$

Carbon and nitrogen are oxidized, and CN^- is the reducing agent. Chromium is reduced, and $CrO_4{}^{2-}$ is the oxidizing agent.

F.6 **(a)** $34K^+ + 14MnO_4{}^- + 20OH^- + C_3H_8O_3(aq) \rightarrow 34K^+ + 14MnO_4{}^{2-} + 3CO_3{}^{2-} + 14H_2O(l)$

$14K^+ + 14MnO_4{}^- + 20K^+ + 20OH^- + C_3H_8O_3(aq) \rightarrow 28K^+ + 14MnO_4{}^{2-} + 6\,K^+ + 3CO_3{}^{2-} + 14H_2O(l)$

(b) $14KMnO_4(aq) + 20KOH(aq) + C_3H_8O_3(aq) \rightarrow 14K_2MnO_4(aq) + 3K_2CO_3(aq) + 14H_2O(l)$

F.7 **(a)** $C_6H_5OH(aq) + 14SO_3(g) \rightarrow 6CO_2(g) + 14SO_2(g) + 3H_2O(l)$

$\Delta ox = 6(14/3) = 28,\ 1(28) = 28$

$-\frac{2}{3}\ +1\ -2 \quad +6\ -2 \quad +4\ -2 \quad +4\ -2$

$C_6H_6O + SO_3 \rightarrow 6CO_2 + SO_2$

$\Delta red = 1(2) = 2,\ 14(2) = 28$

$C_6H_6O + 14SO_3 \rightarrow 6CO_2 + 14SO_2$

$C_6H_6O + 14SO_3 \rightarrow 6CO_2 + 14SO_2 + 3H_2O$

(b) $10FeSO_4(aq) + 2KMnO_4(aq) + 8H_2SO_4(aq) \rightarrow 2MnSO_4(aq) + 5Fe_2(SO_4)_3(aq) + K_2SO_4(aq) + 8H_2O(l)$

$\Delta ox = 2(1) = 2,\ 5(2) = 10$

$+2\ +6\ -2 \quad +1\ +7\ -2 \quad +1\ +6\ -2 \quad +2\ +6\ -2 \quad +3 \quad +6\ -2$

$2FeSO_4 + KMnO_4 + H_2SO_4 \rightarrow MnSO_4 + Fe_2(SO_4)_3$

$\Delta red = 1(5) = 5,\ 2(5) = 10$

$10FeSO_4 + 2KMnO_4 + 8H_2SO_4 \rightarrow 2MnSO_4 + 5Fe_2(SO_4)_3 + K_2SO_4$

$10FeSO_4 + 2KMnO_4 + 8H_2SO_4 \rightarrow 2MnSO_4 + 5Fe_2(SO_4)_3 + K_2SO_4 + 8H_2O$

(c) $8HNO_3(aq) + 3Cu(s) \rightarrow 3Cu(NO_3)_2(aq) + 2NO(g) + 4H_2O(l)$

$\Delta ox = 1(2) = 2,\ 3(2) = 6$

$+1\ +5\ -2 \quad +1\ +5\ -2 \quad 0 \quad +2\ +5\ -2 \quad +2\ -2$

$HNO_3 + HNO_3 + Cu \rightarrow Cu(NO_3)_2 + NO$

$\Delta red = 1(3) = 3,\ 2(3) = 6$

$HNO_3 + 2HNO_3 + 3Cu \rightarrow 3Cu(NO_3)_2 + 2NO$

$6HNO_3 + 2HNO_3 + 3Cu \rightarrow 3Cu(NO_3)_2 + 2NO + 4H_2O$

(d) $2NaCrO_2(aq) + 3NaClO(aq) + 2NaOH(aq) \rightarrow 2Na_2CrO_4(aq) + 3NaCl(aq) + H_2O(l)$

$\Delta ox = 1(3) = 3,\ 2(3) = 6$

$+1\ +3\ -2 \quad +1\ +1\ -2 \quad +1\ -2\ +1 \quad +1\ +6\ -2 \quad +1\ -1$

$NaCrO_2 + NaClO + NaOH \rightarrow Na_2CrO_4 + NaCl$

$\Delta red = 1(2) = 2,\ 3(2) = 6$

$2NaCrO_2 + 3NaClO + 2NaOH \rightarrow 2Na_2CrO_4 + 3NaCl + H_2O$

(e) $H_2O(l) + 3Na_2SO_3(aq) + 2NaMnO_4(aq) \rightarrow 3Na_2SO_4(aq) + 2MnO_2(s) + 2NaOH(aq)$

$\Delta ox = 1(2) = 2,\ 3(2) = 6$

$+1\ +4\ -2 \quad +1\ +7\ -2 \quad +1\ +6\ -2 \quad +4\ -2$

$Na_2SO_3 + NaMnO_4 \rightarrow Na_2SO_4 + MnO_2$

$\Delta red = 1(3) = 3,\ 2(3) = 6$

$$3Na_2SO_3 + 2NaMnO_4 \rightarrow 3Na_2SO_4 + 2MnO_2$$

$$H_2O + 3Na_2SO_3 + 2NaMnO_4 \rightarrow 3Na_2SO_4 + 2MnO_2 + 2NaOH$$

F.8 $3CuS(s) + 8HNO_3(aq) \rightarrow 3S(s) + 3Cu(NO_3)_2(aq) + 2NO(g) + 4H_2O(l)$

copper(II) sulfide	+	nitric acid	→	sulfur	+	nitric oxide
CuS	+	HNO_3	→	S	+	NO

$$3(CuS \rightarrow S + Cu^{2+} + 2e^-)$$
$$2(3e^- + 3H^+ + HNO_3 \rightarrow NO + 2H_2O)$$
$$\overline{3CuS + \cancel{6e^-} + 6H^+ + 2HNO_3 \rightarrow 3S + 3Cu^{2+} + \cancel{6e^-} + 2NO + 4H_2O}$$

$$3CuS + 8H^+ + 8NO_3^- \rightarrow 3S + 3Cu^{2+} + 6NO_3^- + 2NO + 4H_2O$$

$$3CuS(s) + 8HNO_3(aq) \rightarrow 3S(s) + 3Cu(NO_3)_2(aq) + 2NO(g) + 4H_2O(l)$$

G.1 1.0×10^{-5}

We wish to use the relationship between ΔG^0 and the value of the equilibrium constant:

$$\Delta G^0 = -RT \ln K$$

First we calculate ΔG^0 for the equilibrium using the data in Appendix D:

	$Ag_2SO_4(s)$	⇌	$2Ag^+$	+	SO_4^{2-}
ΔG_f^0, kJ/mol	–618.77		77.16		–744.63

$$\Delta G^0 = \left[(2\ \cancel{\text{mol Ag}^+})\left(\frac{77.16\text{ kJ}}{\cancel{\text{mol Ag}^+}}\right) + (1\ \cancel{\text{mol SO}_4^{2-}})\left(\frac{-744.63\text{ kJ}}{\cancel{\text{mol SO}_4^{2-}}}\right)\right] - (1\ \cancel{\text{mol Ag}_2\text{SO}_4})\left(\frac{-618.77\text{ kJ}}{\cancel{\text{mol Ag}_2\text{SO}_4}}\right) = 28.46\text{ kJ}$$

Substituting into $\Delta G^0 = -RT \ln K$ gives

$$\frac{28.46\ \cancel{\text{kJ}}}{-\left(\dfrac{0.008\ 315\ \cancel{\text{kJ}}}{\cancel{\text{K}}}\right)(298.2\ \cancel{\text{K}})} = \ln K_{sp} = -11.48 \text{ and } 1.0 \times 10^{-5} = K_{sp}$$

H.1 No.

If 99.99% of the Si^{2+} is precipitated, (100.00 – 99.99)% = 0.01% must be left. The concentration of strontium ion will be

$$\left(\frac{0.01}{100}\right)(0.10\text{ M}) = 1 \times 10^{-5}\text{ M}$$

From Table 16.6, K_{sp} for $SrSO_4$ is

$$3 \times 10^{-7} = [Sr^{2+}][SO_4^{2-}]$$

For $[Sr^{2+}]$ to be 1×10^{-5},

$$[SO_4^{2-}] = \frac{3 \times 10^{-7}}{1 \times 10^{-5}} = 3 \times 10^{-2}$$

Will $CaSO_4$ precipitate if $[SO_4^{2-}] = 3 \times 10^{-2}$?

From Table 16.6, the value of K_{sp} for $CaSO_4$ is

$$7 \times 10^{-5} = [Ca^{2+}][SO_4^{2-}]$$

and

$$Q = (0.10)(3 \times 10^{-2}) = 3 \times 10^{-3} > K_{sp}$$

Therefore, Ca^{2+} will precipitate. Separation of Ca^{2+} and Sr^{2+} cannot be achieved.

H.2 **(a)**

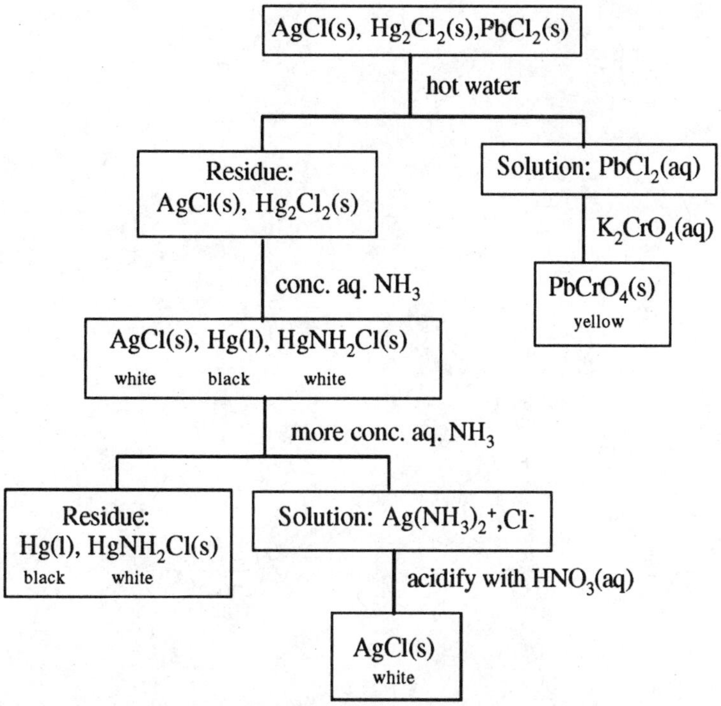

(b) No yellow precipitate will be observed when potassium chromate is added to solution from hot–water extraction. Residue will turn black when treated with concentrated aqueous ammonia. Solution from treatment of black residue with more concentrated aqueous ammonia will give a white precipitate of silver chloride on acidification with nitric acid.

H.3 **(a)** Use flame test. Sodium ion gives yellow color that is invisible when flame is viewed through blue glass; potassium ion gives a purple-violet flame when the flame is viewed through blue glass.

(b) Treat with room temperature water. If solid dissolves, it is KCl. If it does not dissolve, it is $PbCl_2$.

(c) If solution is blue, it is $Cu(NO_3)_2(aq)$; if it is colorless, it is $Ca(NO_3)_2$.

H.4 (a) CoS

Cation of compound is in group 3. Precipitate is CoS.